SERIES

ENTOMOLOGICA

EDITORS

E. SCHIMITSCHEK & K. A. SPENCER

VOLUME 15

DR. W. JUNK B.V. PUBLISHERS THE HAGUE–BOSTON–LONDON 1979

# A SURVEY OF THE LEPIDOPTERA, BIOGEOGRAPHY AND ECOLOGY OF NEW CALEDONIA

Softcover reprint of the hardcover 1st edition 1979

Cover design Max Velthuijs

ISBN-13: 978-94-009-9599-4 e-ISBN-13: 978-94-009-9597-0
DOI: 10.1007/ 978-94-009-9597-0

# A SURVEY OF THE LEPIDOPTERA, BIOGEOGRAPHY AND ECOLOGY OF NEW CALEDONIA

by

J. D. HOLLOWAY

DR. W. JUNK B.V. PUBLISHERS THE HAGUE–BOSTON–LONDON 1979

To my parents

# CONTENTS

# PREFACE

I spent four months in New Caledonia in 1971 with the object of making a quantitative survey of the night-flying macrolepidoptera with light-traps and an assessment of the Rhopalocera and microlepidoptera. This fieldwork was financed by a Government Grant-in-Aid for Scientific Investigations administered by the Royal Society, and by a grant from the Godman Fund. I devoted a further three weeks to sampling on Norfolk I., and, with the help of local naturalists, Mr. and Mrs. F. JOWETT, was able to produce a detailed account of the biogeography and ecology of the moth fauna (HOLLOWAY, 1977).

This book is an account of the results of the New Caledonian work, together with reviews of the geology, phytogeography and general zoogeography presented as background for the Lepidoptera fauna and its geography.

Previous work on the macroheterocera, primarily papers by VIETTE (1948–1971), had recorded not many more than a hundred species, a very low total considering the area of the island relative to that of the Fiji group where the moths were being studied by Dr. G. S. ROBINSON when the New Caledonian expedition was at the planning stage. The Fijian fauna then promised considerably to exceed three hundred species. Evidently many more species awaited discovery in New Caledonia. In addition New Caledonia was geologically both ancient and unusual in the extent of ultrabasic or ultramafic rocks, its flora was renowned for its degree of endemism and apparent antiquity, and mining operations in the ultramafic areas were posing a threat to the very survival of this ancient biota.

All these factors proved a strong inducement for me to visit the island to study the biogeography of the Lepidoptera and to see if techniques of quantitative sampling and numerical analysis could contribute anything towards resolving the conservation problems. I hope this book may in itself make a contribution to the latter by stimulating an interest in the unique biology of New Caledonia, so that ecologists and biochemists may come to appreciate the full value, biological as well as mineralogical, of what the New Caledonian ultramafic rocks have to offer before the opportunity is lost.

The years of work following the expedition would not have been possible without the support and understanding of my parents who provided a roof and sustenance for me during that time. This book is dedicated to them.

In planning the expedition I was extremely grateful for the advice of M. P. COCHEREAU, Mr. P. S. GREEN and Dr. C. N. SMITHERS (who also entertained me in Sydney whilst in transit), and for the approval of the Royal Geographical Society. M. COCHEREAU and his colleagues at the O.R.S.T.O.M. centre at Nouméa, particularly Messrs FABRES, SCHMID and VEILLON, were most hospitable and helpful during my stay, providing free accommodation in Nouméa and much assistance of a technical and administrative nature, and to them the survey owes much of its success. I also received invaluable assistance from the Centre Technique Forestier Tropical and the Département des Eaux et Forêts both in Nouméa and in the field. The International Nickel Co., through their association with the Cofimpac Consortium, were extremely generous with accommodation and supplies in their concession areas of the Plaine des Lacs and also on the Dome de Tiebaghi. Mr. T. E. Hughes of Entech Services kindly equipped me with the portable moth traps at cost. To all these and to many more individuals who showed me kindness during my stay, such as Stan and Joy Clinch of Anse Longue, my deepest thanks.

I am indebted to Mr. R. G. Davies of Imperial College and Professor R. Sibson of the University of Bath who spared time to do the computing of my data for the numerical analyses; Professor Sibson kindly read and criticized a draft of Chapter V. My thanks are due also to Professor M. G. Audley-Charles for his comments on a draft of Chapter I, and to Professors C. P. Alexander and H. H. Ross for vetting the sections on Tipulidae and Trichoptera in Chapter VII.

Work on the New Caledonian material was conducted at the British Museum (Natural History) and I am grateful to the Trustees for enabling this work to take place there. I must pay tribute to the Staff of the Department of Entomology for their patience and help in many matters, particularly to Dr. G. S. Robinson whose parallel work on the Fijian fauna provided a most useful foundation for mine, and to the librarians of the various Departments who were similarly helpful and patient during my searches for literature.

Dr. P. H. Arnaud (California Academy of Sciences), Father J. C. E. Riotte (Bernice P. Bishop Museum, Honolulu) and Dr. P. E. L. Viette (Muséum National d'Histoire Naturelle, Paris) kindly arranged loans of material in their care and my thanks also go to Dr. I. F. B. Common who photographed and dissected several types in the Australian National Insect Collection, Canberra, and provided information on the distribution and status of New Caledonian species in Australia, and to M. A. Delobel who has recently sent me further material from New Caledonia.

Many other individuals helped me with advice and information in person or in correspondence, and I would like to mention in particular Mr. A. Brindle, Professor J. M. Diamond, Dr. J. P. Duffels, Mr. J. S. Dugdale, Mr. P. S. Green, Dr. J. L. Gressitt, Dr. H. Inoue, Dr. R. De Jong, Dr. M. A. Lieftinck and Professor H. E. Moore.

Finally I would like to thank Dr. K. A. Spencer for his help and advice as editor of Series Entomologica.

December 1977

Tillinglea,
Tillingbourne Park,
Wotton,
Dorking,
Surrey RH5 6QL,
England.

# INTRODUCTION

New Caledonia is a French Territory lying within the tropics at approximately 21° S, 166° E. It stands apart from the rest of Melanesia both in geographical position, being isolated from the main chain of archipelagos extending south-east from east of New Guinea, and in its form, a single large land mass rather than an archipelago. In terms of extensive dry land it is very much older than neighbouring island groups and its geological history is also different, relating more to northern New Zealand and south-east New Guinea as will be seen in the next chapter. In particular, much of its area is made up of ultramafic rocks, sea floor overthrust from the east some time in the middle of the Tertiary. This rock is very basic in composition and rich in minerals, giving rise to unusual soil types. Much of the biological history of the island relates to the adaptation of plants and animals to these unusual conditions and their subsequent radiation, resulting in the unique, almost bizarre features of parts of the New Caledonian environment.

The island is long and narrow, its axis running from north-west to south-east. It is approximately 390 km long and, on average, 50 km wide. It is extremely mountainous (Figure 8), the highest point being Mt. Panié (1628 m) in the north. Mé Maoya (1500 m) in the centre of the island and Mt. Humboldt (1618 m) in the south are the principal peaks of the other major massifs.

To the east lies the Loyalty group, a chain of islands parallel to the axis of New Caledonia. They consist of raised coral reefs on a volcanic basement and are probably only of recent emergence. The islands are all low-lying and the most important ones are, from north to south, Ouvéa, Lifu and Maré. Continuing the line and geological structure of New Caledonia to the south-east is the Isle of Pines and to the north-west are a number of smaller islands such as Art and Pott.

New Caledonia lies on the northernmost extremity of the Norfolk Ridge, a submarine structure that extends south via Norfolk I. to the north of New Zealand. To the west is the parallel, but broader Lord Howe Rise that terminates in the Chesterfield Reefs immediately west of New Caledonia. These may have been important in the past as stepping stones for the exchange of plants and animals between New Caledonia and Australia. New Caledonia and the Loyalty group are separated from the New Hebrides to the north-east by a deep oceanic trench where sea floor attached to the Australian tectonic plate is being subducted. This subduction, effectively closing the gap between New Caledonia and the New Hebrides and probably agent of the current uplift of the Loyalty group, will be discussed in detail in the next chapter and has been a factor of great significance in the biological history of New Caledonia.

The land areas from New Guinea to the northern New Hebrides experience heavy rainfall throughout the year with a double maximum (BROOKFIELD & HART, 1971). Elsewhere, to the east and south, there is a single maximum from late December to late March. New Caledonia and the southern New Hebrides are generally drier than the rest of Melanesia except for the western part of Viti Levu in the Fiji group, and have a definite dry season. Fiji is generally intermediate in character.

New Caledonia has far higher rainfall on the eastern coast than on the west, the wettest areas being in the north around Touho and Poindimié and in the extreme south-east. In these areas the average annual rainfall is about 3000 mm. There is a gradual decrease across the width of the island to the west coast where the average annual rainfall ranges from 1000 mm to 1500 mm (BROOKFIELD & HART).

Temperature through the year is similarly uniform over vast areas of Melanesia but seasonal variation increases away from low latitudes. This is illustrated by observations made in the New Hebrides where mean monthly temperatures in the northerly Banks Is. range between 25° and 27° C but in the most southerly island, Aneityum, they range between 21° and 27° C (LEE, 1975). New Caledonia resembles Aneityum, the drier winter months also being significantly cooler.

Cyclones occur in the wetter summer months, arriving mainly from the north-east (COLEMAN, 1972). Otherwise the prevailing winds in those months are the south-east tradewinds. In the winter months New Caledonia can come within the influence of frontal systems moving across the Tasman Sea from Australia. Several were experienced during the survey period, warm fronts bringing heavy rain and possibly even migrant Lepidoptera (pp. 363, 399).

Rainfall is an important determining factor for the pattern of soils in New Caledonia. Most of the central and inland eastern part of the island is characterised by ferruginous tropical soils (latosols) which become increasingly ferrallitic or lateritic on the ultramafic rocks. The east coast and an area to the west of the latosols are characterised by more immature soils, lithosols and regosols. The west coast areas of basalt (Figure 5) have given rise to black tropical soils which support the main areas of savanna. The formation of these black soils and the latosols may have occurred in a period of drier climate than at present (BROOKFIELD & HART, from TERCINIER (1962)). Podsols and rendzinas occur in the Nouméa area, at Anse Longue and at Bourail, areas where a thicket dominated by Leguminosae is frequent (p. 111). Podsols also characterise the dry schist country in the northern central part of the island; their formation may be linked with the extension of savanna vegetation.

There is considerable variation within this broad characterisation of soils, factors such as relief, hydrology, accumulation of organic material and the mineral nature of the substrate all being important. Accumulation of organic material in valleys, depressions and on plateaux is of great importance in the development of rainforest. The soils of hillcrests and summits are usually poor and strongly desaturated, leading to the development of various maquis and 'heath' associations. The most unusual soil types are found on the ultramafic rocks and have been described for the Massif de Boulinda on the west of the island by JAFFRÉ & LATHAM (1974). Overall, such soils have a certain number of features in common. They tend to be very poor in important plant nutrients such as phosphate, potassium and calcium and rich to a variable extent in elements generally considered toxic for plant life, particularly nickel, chromium and cobalt. The soils are predominantly ferralitic and marked by extreme evolution, accumulation of iron in the form of oxides and hydroxides, and elimination of the two other principal constituents of the parent rock, silica and magnesium. The extent of desaturation is extremely variable and is expressed mainly in the concentration of the magnesium cation. These ultramafic soils are thus, for vegetation, the most demanding and their flora is the most highly adapted and unusual in the island.

The diversity of vegetation in New Caledonia (Chapter III) reflects the diversity of physical and edaphic factors. Savannas of *Imperata* grass are extensive on the west coast (Plate 31) and are dominated by niaouli trees (*Melaleuca quinquenervia*) in some areas and leguminous thicket in others. In the north and on the east coast niaouli still dominates the more open habitats but the *Imperata* is largely replaced by a low association of fern, sedge and shrubs. Rainforest develops wherever there is sufficient moisture and accumulation of organic material. The ultramafic areas support a bewildering range of vegetation types from low, sedgy maquis to dense low forest of great floristic diversity. The summit forests of the ultramafic mountains also show immense variety.

New Caledonia is noted for its old, diverse and highly endemic flora. Associated with this is a fauna of equal antiquity and interest. It is therefore extremely unfortunate that the most ancient and most interesting elements of both tend to occur on rocks with the most unusual mineral content. The mountainous areas of ultramafic rock are often referred to as the 'terrains miniers' and have long been prospected and exploited for their mineral wealth. New Caledonia, despite its size, is the third largest producer of nickel in the world after Canada and the Soviet Union. As basic raw materials in the world become scarcer, so exploitation of increasingly lower grade ores becomes commercially attractive. Thus large areas of New Caledonia, particularly those of the greatest biological interest, are threatened by open-cast mining.

Scientists and administrators in the Départment des Eaux et Forêts, ORSTOM (Office de la Recherche Scientifique et Technique Outre Mer) and CTFT (Centre Technique Forestier Tropical) are concerned to safeguard the natural environment against the worst effects of the mining operations but the weight of political and economic power in the island and the emphasis from metropolitan France is behind full exploitation of the mineral resources. The mining companies have generally paid little attention to the environmental problems created by their operations. One exception has been the International Nickel Company which is aware of the problems and has contributed towards the ecological research necessary to mitigate the effects of mining. The Company, as the major participants in a mining consortium (Cofimpac), provided the author with much hospitality and assistance during the survey of the Plaine des Lacs and the Dome de Tiebaghi.

The ecological problems involved were described in a roneo leaflet in circulation in 1971. It was entitled *Problèmes forestiers posés par la degradation des terrains miniers en Nouvelle-Calédonie*, and was produced by CTFT. Until recently the exploitation of nickel was largely restricted to hilltops and ridges (e.g. Plates 31, 34) where the richest ores were located. The environmental problems were those of scrub fires set off to aid prospecting and of erosion by heavy rainfall of soils and rocky debris exposed by the mining. Considerable quantities of this debris are transported in the rivers where, during floods, the silt pollutes low-lying agricultural land through the toxicity of the mineral salts. It finally passes out to sea where the extensive fringing reefs are affected both by the toxicity and by the turbidity of the water. During the moth survey such pollution was observed along the Houailou river valley and in the redness of the seas near the Dome de Tiebaghi.

Most recently technological advances in ore treatment have rendered possible the exploitation of low grade ores occurring over a much wider area, such as in the Plaine des Lacs. The plains contain a great depth of lateritic detritus in the lower levels of which there is a concentration of nickel salts. The CTFT leaflet gave estimates of the needs of a viable commercial operation. This would require an annual production of 50,000 metric tonnes of nickel, and would involve the exploitation of about thirty hectares with soil removed to a depth of thirty metres. Thus both the overburden and the residue of the ore after extraction (concentration of nickel salts is of the order of one or two percent) would represent a vast quantity of unconsolidated material. The problems of erosion and consequent pollution are many times greater than in the traditional methods of mining.

The approach adopted towards solving these problems has focussed both on setting aside areas of particular scientific interest (the Pic du Pin area has recently been designated thus) and on gaining a better understanding of the ecology of the vegetation of the 'terrains miniers', to enable programmes for vegetation regeneration to be put in train to minimise erosion of the disturbed areas.

The general impression gained during the survey was mainly one of 'too little, too late', but this is not to decry the valuable work that is being done, exemplified by the publications of JAFFRÉ (e.g. 1974) and pedologist colleagues. With the political lobby for conservation in New Caledonia so weak, hopes may rest more on economic factors.

Firstly, the unusual flora of the ultramafics may itself be of economic importance. For a long time all the telegraph poles on the island were provided by the 'chêne-gomme', *Arillastrum gummifera* (Plate 24), but prospecting fires have so reduced this resinous tree (Plate 14) that suitable timber now has to be imported. A number of plant species restricted to the ultramafic rocks have yielded substances of medical importance that have not yet been produced artificially. Biochemical studies of these plants and their relationship to the unusual substrate are still in their infancy. Nevertheless it is already known that many of these plants accumulate nickel salts in their tissues.

There are preliminary indications from the moth survey (p. 184) that this major defoliating group of insects has made little headway in exploiting the flora of the more ecologically demanding soils on the ultramafic rocks. There may be biochemical reasons for this and, should it prove generally that phytophagous insects are at very low levels in such vegetation, further research into the phenomenon could be of great economic importance, such as in the development of systemic insecticides. It is therefore of universal importance that significant areas of this vegetation be preserved for posterity.

Nickel is now the mainstay of the New Caledonian economy. Yet in the past the island has been an important producer of both beef and coffee in the Pacific. There are still extensive coffee plantations on the wetter east side of the island and the savannas of the west coast are generally regarded as cattle country. Citrus fruit also grows well. However, manpower is short and in the past labour has been imported from Indochina and Java and is still imported today from the New Hebrides and French Polynesia. Native Melanesians make up only half the total population. Mining is much more highly paid, so the lack of manpower is felt mostly in the coffee and beef producing industries. The current coffee shortage may temporarily reverse this trend. In the future the world price of nickel and the balance between it and the price of coffee and beef may play an important role in determining the pressures on the New Caledonian environment.

In many ways the seventies may prove to have been the ideal period for biological exploration in New Caledonia. The road system is good, enabling most of the important habitats to be reached by car. In 1971 a series of bridges was being built across the estuaries of the east coast, promising to eliminate the frustrations of numerous ferry crossings. Mining companies were expanding their exploration programmes into many previously inaccessible areas, giving biologists also an opportunity to explore them before any mineral exploitation should occur. Few do, so the opportunity may be missed to make an adequate record of the unique flora and fauna before much of it is irretrievably lost through threatened mining operations.

The purposes of the moth survey described in this book were twofold. Firstly, the fauna appeared to have been little studied. The number of species listed by VIETTE (1949, 1951) was very low compared with that for Fiji revealed by the work of ROBINSON (1975) and the unique flora held out the possibility of a correspondingly unusual fauna. Studies of literature on other faunal groups tended to reinforce this possibility. The moth faunas of neighbouring areas were relatively well known or were about to be investigated (the New Hebrides by ROBINSON), so it would be possible to obtain a fairly complete picture of the zoogeography of New Caledonian species. This could in turn be related to the floristics and overall biological history.

The second aim was to continue investigating the possibility that quantitative sampling of moths with light-traps might provide a valuable method of environmental surveying and monitoring. Moths are readily sampled in large numbers, enabling a quantitative analysis to be made of distributional and seasonal patterns and of ecological diversity. Identification is relatively easy, so there is potential for involving non-specialist helpers in a survey. The basic taxonomy of the macrolepidoptera at least is comparatively advanced. The phytophagous nature of the majority of moths, often highly specific, leads to the prediction that aspects of the ecology of the group, such as the localisation and diversity of associations of species, will reflect the nature of the vegetation of a given area. An earlier study in Borneo (HOLLOWAY, 1970, 1973a, 1976) and a current one on Norfolk I. (HOLLOWAY, 1977) have both lent some support to this possibility. The contribution of the New Caledonian survey will be discussed in Chapter VIII but the weak characterisation of the floristically richer ultramafic vegetation by the moth species suggested that the correlation between vegetation and moth fauna is not universal.

# I. THE GEOLOGICAL HISTORY OF NEW CALEDONIA IN ITS AUSTRALASIAN CONTEXT

New Caledonia is one of the older land masses in Australasia. Its stratigraphic column ranges from the Permian to the Miocene with relatively few breaks in the sequence (LILLIE & BROTHERS, 1970). The first emergence of land in the area was probably in the Upper Jurassic, an event possibly associated with the Rangitata Orogeny of New Zealand. The sedimentary sequence in some areas is broken then, with some evidence of metamorphism, and does not resume until the Upper Cretaceous (GUILLON, 1974). Where Cretaceous deposits do occur they contain coal (Formation à Charbon) in some strata (LILLIE & BROTHERS, 1970: figure 2), suggesting alternation of land and marine facies until further episodes of uplift occurred in the Eocene. From then onwards land in the area was probably significant to substantial, as will be seen when the geology is discussed in more detail.

The possibility of persistence of land in New Caledonia since the middle of the Mesozoic means that it would have been in existence contemporaneously with the events of the dispersal of Gondwanaland from the beginning of the Cretaceous onwards. This might be reflected in the composition of the flora and fauna.

## Plate tectonics and island arcs in Australasia

There are many general treatises available on the theory of plate tectonics and its role in the movement of continental masses (e.g. SMITH, BRIDEN & DREWRY, 1973; TARLING & RUNCORN, 1973) and a further account here would be superfluous, though it must be stated that a minority of geologists still question the main conclusions of plate tectonic theory. There is general agreement about the order and dating of the various events in the fragmentation and dispersal of Gondwanaland (KEAST, 1973; SMITH, BRIDEN & DREWRY) and the events within the Indian Ocean, including the separation of Australia from Antarctica, have been particularly closely studied (MCKENZIE & SCLATER, 1971; JOHNSON, POWELL & VEEVERS, 1976).

From about −130 my (million years) the southern Indian Ocean and the South Atlantic opened up, effectively separating Australasia/Antarctica from Africa and India. The −75 my reconstruction of MCKENZIE & SCLATER (1971) is shown in Figure 1 with the New Zealand area added following GRIFFITHS & VARNE (1972) and HAYES & RINGIS (1973). It represents the situation just after the termination of this initial episode of spreading (−80 my) which resulted in the formation of the South Wharton Basin west of Australia and other areas of ocean floor south-east of India and north of the Prydz Bay area of Antarctica (JOHNSON, POWELL & VEEVERS). The reconstruction suggests that, at that time, there may still have been archipelagic connections between Australia and Africa via Kerguelen and the Ninetyeast Ridge, India, the Mascarene Plateau and Madagascar that might have served as a route for the dispersal and exchange of land organisms. Kerguelen and Ninetyeast supported a flora later in the Cretaceous and early in the Tertiary (p. 34) and might well have done so earlier. Connection between South America and Antarctica was unbroken, though there are no indications as to whether they were united by continuous land or an archipelago.

The pole position indicated in Figure 1 is the Australian palaeomagnetic pole. If this represents the true pole then it suggests that the Indian Ocean coast of Antarctica and the western and northern part of Australia experienced a climate much more favourable for life than did the Pacific coast of Antarctica, New Zealand and south-east Australia.

The Atlantic is thought to have opened more rapidly in the south than in the north so that contact or proximity of Brazil to West Africa continued much later than that between southern Africa and Patagonia. Biogeographical relationships are consistent with this interpretation of events (p. 32).

From −80 my to −53 my spreading in the Indian Ocean occurred along an east-west axis resulting in a north-south separation of India from Australia/Antarctica. Spreading was rapid (as much as 17.5 cm/year) and

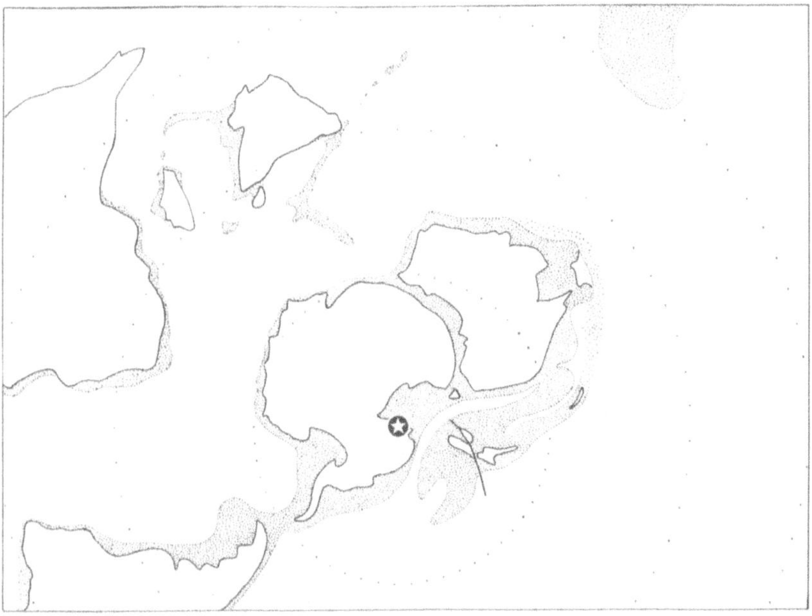

*Fig. 1.* Reconstruction of the southern hemisphere at −75 my after MCKENZIE & SCLATER (1971). Continental shelf area is stippled and the enclosed star indicates the Australian palaeomagnetic pole on which 30° circles of latitude marked are based.

resulted in the formation of much of the present Indian Ocean and the northern part of the Wharton Basin. The previous spreading episode had occurred roughly along a south-west to north-east axis between India and Antarctica. Roughly simultaneously (−80 to −60 my) the North Tasman Sea was formed and spreading occurred between New Zealand and Antarctica (HAYES & RINGIS, 1973; SUGGATE, 1971), effectively separating the New Zealand continental shelf area, including the Norfolk Ridge (with New Caledonia) and the Lord Howe Rise, from the still joined Antarctica and Australia. This separation is shown in its early stages in Figure 1. Once achieved, this separation probably persisted more or less unchanged up to the present day.

After −53 my until −32 my the north-south spreading rate in the Indian Ocean slowed considerably and during this period the separation of Australia from Antarctica began, albeit slowly, along a spreading axis distinct from that in the Indian Ocean. The spreading axis in the Indian Ocean probably became disjunct along a transform fault running north-south along the Ninetyeast Ridge (SCLATER & FISHER, 1974). To the east of this ridge it became displaced to the north of its western counterpart and has now largely been subducted at the Java Trench following more recent spreading events. The result is that the magnetic lineations of the sea floor to the west of the Ninetyeast Ridge, derived from the more southerly portion of the spreading axis, get older from south to north, but this is reversed on the eastern side. This latter period of slower spreading east of Ninetyeast Ridge may be relevant to spreading events in Melanesia, discussed on p. 18.

The reconstruction of events presented by JOHNSON, POWELL & VEEVERS is pertinent to the controversy as to whether S.E. Asia and the Sunda Shelf were ever attached to Gondwanaland and moved north with the Indian land mass. Indeed their reconstruction appears to discount such a hypothesis. The hypothesis was put forward by RIDD (1971, 1972) and supported by TARLING (1972). The evidence presented is partly palaeontological; *Glossopteris* flora fragments have been found in Thailand yet are primarily characteristic of Gondwanaland. The main evidence is geological in that there was a cratonic source of sediments at the present day west coast of S.E. Asia in the Palaeozoic that possibly persisted until the Triassic. The proponents of the hypothesis indicated a fit of this west coast with eastern India and western Australia in Gondwanaland.

The weight of evidence is against this hypothesis. Palaeomagnetic evidence indicates that in the late Palaeozoic Malaya was at 15° N, a result incompatible with association with Gondwanaland (MCELHINNY *et al.*, 1974). STAUFFER & GOBBETT (1972) stated that the Malaysian flora in the Permian and Carboniferous was predominantly Cathaysian, and tropical conditions were evident in S.E. Asia during the Palaeozoic at a time when Gondwanaland was extensively glaciated. STAUFFER (1974) expanded on this, suggesting that the blocks from Malaya to S. China have behaved as one since the Triassic. To the west and south the blocks are truncated and show no continuation with adjacent oceanic crust of Cretaceous and Tertiary age, in contrast with the situation in southern India. There is evidence of convergent plate boundaries in the area in the Cretaceous, with subduction from Sumatra to Borneo. Mesozoic floras indicate a continuation of the Palaeozoic warm climate and STAUFFER cast doubt on the *Glossopteris* records from S.E. Asia.

Thus throughout the period of geological time that concerns the biological history of New Caledonia it is probable that much of S.E. Asia was in approximately the same position as it is today, though probably lacking initially the major fringing archipelagos of Indonesia. It is indicated in this position in Figures 1 and 2. If S.E. Asia had drifted north prior to, or in conjunction with,

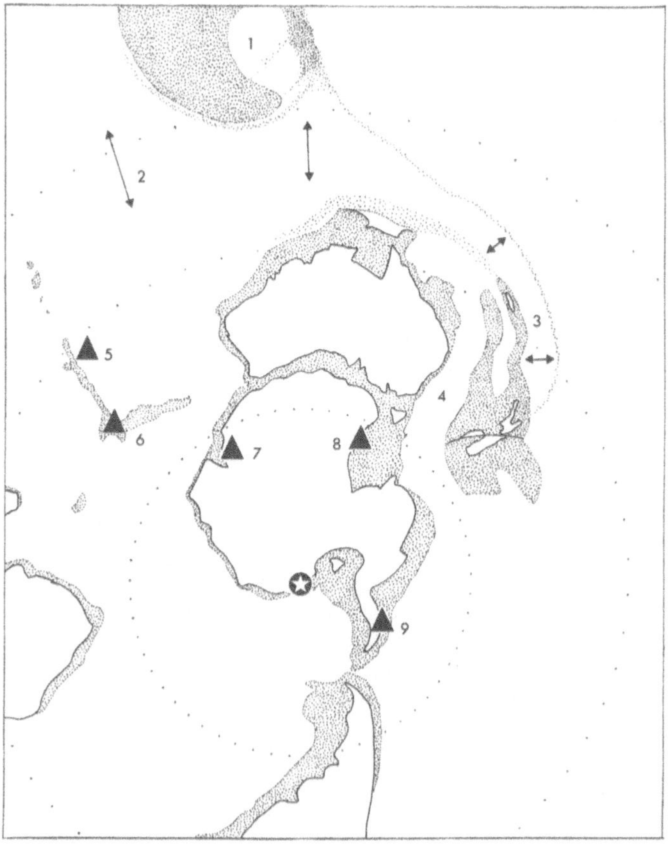


*Fig. 2.* Reconstruction of the southern hemisphere at −45 my, mainly after McKENZIE & SCLATER (1971). S.E. Asia (1) is shown in its present day position. The arrows indicate a hypothetical spreading axis extending from the Wharton Basin east of Ninetyeast Ridge (2) to the South Fiji Basin (3). The Tasman Sea (4) has opened to its full extent. Triangles indicate the sites of late Cretaceous and early Tertiary austral pollen floras, on Ninetyeast Ridge (5), Kerguelen (6), Prydz Bay (7), McMurdo Sound (8), and Seymour I. (9). other conventions are as in Figure 1.

southern India then archipelagic connections from Australia via India to Africa would have been very much more extensive in the early stages of such movement. But there are no indications in Malaysian floras of that time of association with the austral elements that may have been dispersing in Gondwanaland.

JONES (1971) and MCKENZIE & SCLATER (1973) dated the separation of Australia from Antarctica at −45 my, though creation of ocean floor between the two began as early as −53 my according to JOHNSON, POWELL & VEEVERS (1976). A −45 my reconstruction based on that of MCKENZIE & SCLATER (1971) is shown in Figure 2, with the New Zealand continental shelf and associated ridges and rises to the north in their present position relative to Australia. At about −32 my the Indian and Australian plates became united with movement along Ninetyeast Ridge and spreading north of the Wharton Basin ceasing. A more rapid rate of spreading in the southern Indian Ocean ensued, separating Australia and Antarctica at a greater rate. Most of New Zealand and New Caledonia moved with this combined plate and this movement continues at the present. KENNETT *et al.* (1972) suggested that free movement of ocean currents between Australia and Antarctica was not possible until towards the end of the Oligocene (−25 my approximately).

A change in direction of spreading of the Pacific plate, on which almost all Polynesia and Micronesia are situated, may have been connected with the changes in the Indian Ocean. Spreading changed from a north-easterly to an easterly direction in the tropical Pacific. Dates given for this change vary from −25 my (JACKSON *et al.* (1972), from the bend in the Hawaiian-Emperor chain of volcanoes) to −40 my (UYEDA & BEN-AVRAHAM (1972), from a consideration of bends in Pacific island chains generally). BURKE & WILSON (1976) supported the latter date. Though these dates do not tie up precisely it is possible to speculate that all these events were part of a general readjustment of spreading geometry in the world at about −32 my. This period may also have seen the cessation of spreading in the South Fiji Basin, the emplacement of ultrabasic rocks in New Guinea, the Solomons and New Caledonia, and an episode of vulcanism in Fiji.

For a biogeographer it is not enough to know the history of the association and separation of the major land masses. Such facts may determine some of the distribution patterns of the least dispersive elements of the flora and fauna, patterns that may persist to the present. But the majority of taxonomic groups or members of them show at least some capability for dispersal along island chains or even over very great distances. The endemic biota of Hawaii is illustration enough of this point. The spread and interaction of floras and faunas along island chains will depend on factors such as the relative richness and competitiveness of elements of the source biotas, as well as dispersal and the configuration of the chains themselves. The chains may also change in geological time, stepping stones for organisms becoming more or less frequent along them, or they may become ecologically more or less acceptable to the dispersing organisms.

Thus information on past archipelagic conditions in any area is crucial to understanding the biogeography. In this respect the history of the area between the Australian and Asian land masses and to the east of Australia is extremely complex. Over the past few years much more information has become available and it is possible to produce a speculative synthesis, though there are still areas of obscurity such as the history of Sulawesi and the relationship of south-east New Guinea to the Coral Sea, Rennell I., and New Caledonia. The following account is an attempt to obtain a consensus from the literature so as to provide a working hypothesis with which biogeographical patterns can be compared. Biogeographical considerations have been excluded except where they suggest solutions in areas of obscurity. Such an account obviously runs the risk of oversimplification, especially as written by a biologist with only a limited geological training, and should therefore be regarded primarily as an introduction to the extensive literature referred to.

Much of the history of the area is concerned with recent ideas on marginal seas and remnant arcs, particularly those marginal to the subduction zone of the Pacific plate. Marginal seas are thought to open when subducting plate boundaries create friction and consequent mantle convection, leading to ocean crust formation behind or within island arcs between the subduction zone and the interior of the other plate involved; the latter is often a continental area (UYEDA & BEN-AVRAHAM, 1972). Such island arcs are curved, the convexity facing the plate being subducted (KARIG, 1972), as exemplified by the Philippines, the Marianas and Japan. Between the Philippines and the Marianas is an extinct spreading ridge that created the intervening sea (UYEDA & BEN-AVRAHAM). The formation of the various seas and basins between Australia and the Kermadec Trench is thought to have involved this process.

SOLOMON & GRIFFITH (1972), discussing subduction activity at the eastern margin of Australia, suggested that very complex cycles may have occurred with blocks becoming detached and then recolliding, rather than that there had been a simple progression of the subduction zone eastwards. But at least since the beginning of the Cretaceous the pattern has been, according to CAREY (1970), 'a simple spreading with the progressive separation of the marginal orogen from the continental block with the development of disjunctive seas behind.' KARIG (1972) suggested a more complex pattern of events where New Caledonia formed through the collision of separate arcs in the early Tertiary. The three longitudinal geological zones in the island described by GUILLON (1974), discussed further below, could lend support to this hypothesis. It does not seem likely that such constituent arcs could ever have been more widely separated than, say, the arcs of the New Hebrides or the Solomons (where an analogous longitudinal zonation occurs; p. 16), as the main episodes of ocean floor formation at that time were to either side of New Caledonia.

VAN DER LINDEN (1969) first suggested that the N. Tasman Sea and S. Fiji Basin had been created by sea floor spreading processes and identified ridge structures in each that could be extinct spreading ridges. Not enough is known about the magnetic lineations of the S. Fiji Basin to conclude whether the Three Kings Rise is an extinct spreading ridge, though recent literature has tended to support VAN DER LINDEN's general thesis. The N. Tasman Sea has been shown to have been formed by sea floor spreading by HAYES & RINGIS (1973). The spreading did not occur about the Dampier Ridge as suggested by VAN DER LINDEN but was identified further west. The Dampier Ridge may be an associated marginal structure equivalent to an island arc, though probably never emergent. The actual spreading ridge may be more closely referable to the Tasmantid Guyots identified as 'hot spots' by VOGT & CONNOLLY (1971).

The magnetic lineations described by HAYES & RINGIS have only been traced as far north as Brisbane and at that point the spreading ridge is much closer both to the Lord Howe Rise and to the Australian coast. The oldest lineations there are −70 my rather than −80 my, suggesting either that spreading was later or slower there or that there was subduction to the west or vulcanism to the east. The oldest volcanics identified on Lord Howe I. are −30 my (SUTHERLAND & RITCHIE, 1974) but the Late Cretaceous and Eocene orogenesis in eastern Australia mentioned by DAVID (1950) might have been associated with subduction. Further orogeny in eastern Australia occurred in the late Miocene and Late Pliocene (DAVID), that of the Late Miocene being the most intense (BROWN, CAMPBELL & CROOK, 1968). It might have been contemporaneous with the later volcanics on Lord Howe I. dated at −8 my by SUTHERLAND & RITCHIE. The vulcanism at this time could conceivably have been more extensive along the Lord Howe Rise western margin if there were general compression in the area. Emergent volcanic islands, since eroded, could have been created on the Middleton and Elizabeth Reefs and perhaps on other sea mounts near the surface extending up to the plateau of the Chesterfield Is. The biogeographic implications of this are discussed on p. 30. The Late Pliocene orogeny in E. Australia may correspond similarly with the Norfolk I. volcanics of that period dated by JONES & MCDOUGALL (1973).

It is debatable whether the Coral Sea to the north of the Tasman Sea was formed by a similar spreading process at the same time. PACKHAM (1973), in a footnote, stated that the Coral Sea basement was at least pre-Eocene, and MILSOM (1974) referred to fan-shaped magnetic lineations there that were possibly of Eocene date. Resolution of this problem will be important in determining the relationship of New Caledonia with S.E. New Guinea (p. 15), and in testing the validity of palaeogeographic reconstructions (e.g. AUDLEY-CHARLES, CARTER & MILSOM, 1972) involving the rotation and migration of the central part of New Guinea from a more south-easterly position and axial trend as indicated in Figure 3.

In another footnote PACKHAM referred to drilling in the New Caledonian Basin between the Norfolk Ridge and the Lord Howe Rise. There the basement had not been reached in sediments dated at Middle Eocene. These two ridges may well have separated prior to this date with the intervening basin forming in a manner comparable to the Lau-Tonga-Kermadec system discussed on p. 17. In Figure 2 (Middle to Upper Eocene) they have been shown as already separated, but in Figure 1 (Late Cretaceous) they have been shown as adjacent. The possibility that the opening of this basin and the formation of the Coral Sea were part of a single episode of crustal extension early in the Tertiary requires investigation.

New Caledonia and New Zealand have been associated geologically over a long period. Their Mesozoic sedimentary successions are very similar (LILLIE & BROTHERS, 1970) and in both there was an orogenic period, the Rangitata Orogeny, with metamorphism in the Late Jurassic (AVIAS, 1973). Prior to that there was a long period of geosynclinal deposition, probably adjacent to the Lord Howe Rise abutting on to the east coast of Australia. The south-east and central part of New Guinea may also have been involved in this and the geosyncline may have extended along the north of the Campbell Plateau to the Chatham Is. in association with the Antarctic coast (SUGGATE, 1971).

The separation of New Zealand from Antarctica has been discussed in a series of papers in HAYES (1972). The separation started at −80 my. The area between the Campbell Plateau and the Antarctic Ridge has magnetic lineations running back to an age of at least −79 my (SUGGATE, 1971). There are some indications of a change in direction of this spreading from NNW to ENW at −62 my (CHRISTOFFEL & HAYES in HAYES, 1972). These dates correspond with those established for the opening of the N. Tasman Sea and hence it is reasonable to infer that the New Caledonia/Lord Howe/New Zealand/Chatham block separated completely from Australia/Antarctica at that time as a unit.

Tectonic movement in New Zealand was at a minimum from the Palaeocene to the Oligocene with erosion to near peneplain conditions and some marine transgression (SUGGATE). Vulcanism was sporadic until the Miocene, when marine regression began. The major period of uplift and faulting started in the Pliocene and continues at the present, though there were precursor movements in the Miocene. This change from quiescence to vulcanism and uplift follows the Oligocene change in plate movements discussed on p. 11. Today the Australian and Pacific plates abut at the north-trending Macquarie Ridge where the relative motion of the Australian plate is northwards against the Pacific plate (WEISSEL & HAYES, 1972). This relative motion continues along the Great Alpine Fault of Southland which runs approximately along the south-west to north-east diagonal. It is a transcurrent fault and has experienced a displacement of between 300 km to 660 km approximately since the Miocene, but perhaps mostly over the last 10 my (HAYES & TALWANI, 1972; CHRISTOFFEL & VAN DER LINDEN, 1972; WELLMAN, 1973). This shear is gradually translated into the situation at the Tonga Trench where the easterly moving Pacific plate is subducted beneath the eastern margin of the northerly moving Australian plate.

The New Caledonian area would thus appear to have been stable in position relative to Australia and New Zealand throughout most of the Tertiary (Figures 2-4). Its relationship to New Guinea and areas to the north is more complex. SPENCER (1974) edited a compendium of information on Mesozoic and Cainozoic orogenic areas which includes data on most of Indonesia and Melanesia. The geographical history of the area in terms of emergent land is somewhat inscrutable. Geologists are often understandably reluctant to commit themselves on the position and extent of dry land at any time.

New Guinea is thought to have evolved through the collision of two separate structures in the latter part of the Tertiary (BAIN, 1973). The blocks and ranges of mountains to the north of the Mamberamo, Sepik and Markham river systems, such as the Finisterre Range, together with the islands of Biak, Japen, and possibly part of the Vogelkop are probably a continuation of the Bismarck (particularly New Britain) volcanic island arc system (HAMILTON, 1971). This may be represented to the west of New Guinea by Waigeo I. and part of the North Moluccas (Halmahera). The central range of mountains to the south probably only began to emerge from the sea in the Lower Miocene (HERMES, 1974; MILSOM, 1974) with further intensive uplift in the Pliocene and Pleistocene. These mountains are formed of thick sediments deposited at the northern margin of the Australian continental shelf and thus there is a Palaeozoic basement. On the outer margin of this basement a mobile belt of intrusive rocks, ultramafics and metamorphics formed in the Tertiary. This central part of New Guinea was probably significantly separated from Australia by shallow seas until the Pleistocene.

The south-eastern extension of this central range has had a somewhat different history. The tectonic regime changed abruptly at the end of the Mesozoic, a change possibly related to the opening of the Tasman and Coral Seas. Folding, thrusting and metamorphism occurred in the early Tertiary with overthrusting of ultramafics from the east within the period from the end of the Cretaceous to the Middle Oligocene (MILSOM, 1974). Pillow basalts occur to the Australian side of these ultramafics. This early Tertiary history is very similar to that of New Caledonia (AVIAS 1973). The previous sedimentary history also shows some parallels with that of New Caledonia and New Zealand, though not as close as the resemblance between the latter (LILLIE & BROTHERS, 1970).

One might therefore expect a degree of continuity from the Norfolk Ridge to the south-east of New Guinea. In fact there is a large gap of probable ocean floor between the two with detached fragments such as the Rennell group. A hypothesis that crustal extension had occurred in this region, increasing the separation between S.E. New Guinea and New Caledonia, might find support from the following observations published on the area.

THOMPSON (1972) referred to a hypothesis of KARIG that some extension of ocean floor might have occurred within the immediate vicinity of S.E. New Guinea, associated with the Woodlark and Pocklington (from the Louisiades to Rennell) island arcs. CAREY (1970) and HAMILTON (1971) both discussed extensive left-lateral faulting within New Guinea that displaced fragments of the older basement rocks as far west as the Sula Is., east of Sulawesi. This faulting was due to a shear movement that developed as a result of the westward movement of the Pacific plate against the northward moving margin of the Australian plate. Sea floor extension in the Fiji Plateau (p. 19) might also be related to this movement. 1300 km of sinistral movement is apparent within New Guinea itself (CAREY). AUDLEY-CHARLES, CARTER & MILSOM (1972) produced a Middle Miocene reconstruction showing the central part of New Guinea displaced to a more easterly position and at a greater angle to latitude much as indicated in Figure 3. The old, now easterly displaced fragments of the New Guinea basement such as the Sula Is. are shown arranged around the north of the Australian continental shelf in conjunction with Tenimber and Timor further west, a continuity referred to elsewhere (AUDLEY-CHARLES, 1975) as the Westralian Geosyncline. The migration of New Guinea to its current position as a result of the shear movement would have necessitated the creation of some sea floor to the south and east in adjustment. There are also some biogeographical pointers to a closer association of New Caledonia and New Guinea (excluding the northern part) as an Inner Melanesian Arc in the middle of the Tertiary which will be discussed in later chapters.

There are more concrete geological indications of association of island groups along the Outer Melanesian Arc from the Bismarcks to Tonga and further west in the northern part of New Guinea, as mentioned above. In the Bismarcks the New Britain basement is Eocene, consisting of part terrestrial, part marine island arc volcanics. The volcanics are dated mainly from Upper Eocene to Lower Miocene and from Upper Miocene to Pliocene (MILSOM, 1974).

Accounts of the geology of the Solomons may be found in COLEMAN (1966), THOMPSON & HACKMAN (1969), HACKMAN (1973) and COLEMAN & HACKMAN (1974). The islands are considered to have evolved from a series of Late Mesozoic oceanic welts which started to shoal at the end of the Oligocene or in the Lower Miocene and sustained rapid uplift until the present day. The ultramafics in the Solomons were probably thrust-emplaced at the end of the Oligocene, as post-Oligocene sediments contain fragments from them and erosion surfaces have been recognised on some islands. The Solomons have been divided into three longitudinal geological provinces. The southern part, nearest to the current subduction zone, is recent volcanic. The central part has a possibly Late Cretaceous igneous and metamorphic basal complex, though the metamorphism may be Eocene. This is covered by Eocene to Oligocene basaltic lavas up to 1000 m thick. These are in turn overlain with Miocene to Recent reef sediments and products of the erosion of the volcanics. The ultramafic belt is between the central and Pacific provinces. The latter consists of submarine basalts overlain by organic calcareous sediments. The threefold structure of the island group is reminiscent of the situation in New Caledonia.

The continuity of the Solomons with the New Hebrides, Fiji and Tonga as regards general history has been stressed by DICKINSON (1967), KARIG & MAMMERICKX (1972), GILL & GORTON (1973) and WARDEN & MITCHELL (1974). The New Hebrides consist essentially of eastern and western island chains with Miocene to Recent and possibly pre-Miocene sedimentary and volcanic rocks, the latter with parallels in Fiji. As in the Solomons the oldest rocks are on the north-east or Pacific side. There are ultramafics mainly in the north-east, separated from the south-west by a chain of Pliocene to Quaternary volcanoes (WARDEN & MITCHELL).

The land area of the New Hebrides was suggested by MALLICK (1975) to have been at most a tenth of its present extent until the Pleistocene when uplift was rapid and considerable, presumably due to interaction with New

Caledonia. There was some uplift in the Late Pliocene and some, followed by erosion and submergence, in the Middle Miocene (−15 my), the latter being referred by MALLICK to the Burdigalian tectonism. This earlier event could mark the inception of the current subduction zone to the south-west of the archipelago.

The Fiji group has mainly been built up by vulcanism through the latter half of the Tertiary (GREEN & CULLEN, 1973) but an early period of igneous activity in Viti Levu lasted from the Upper Eocene to the Middle Miocene (GILL, 1970). The petrology of this earlier period is consistent with that of an island arc. CHASE (1971) gave a date of −50 my (Lower Eocene) for these Viti Levu volcanics. RODDA (1974) supported this date and dated subsequent deformations in the Oligocene, Miocene and possibly Pliocene times. These deformed rocks are overlain by unfolded strata that extend back perhaps to the Upper Miocene. In the Middle Miocene there was much folding as in the New Hebrides, followed by prolonged erosion and deposition of sediments in shallow and deep basins (RODDA). Basaltic vulcanism occurred from −10 my to −4 my. Pliocene and Pleistocene rocks are unfolded and untilted but uplifted 1000 m in places. ROBINSON (1975) suggested that significant land in the Fiji area only occurred from about −4 my but there are indications from the accounts consulted that this period could have been much longer.

The Tonga, Lau and Kermadec island groups probably also date in their basement structures from the end of the Cretaceous or earliest Tertiary (KARIG, 1970; HOCHSTEIN, SCHOFIELD & SHOR, 1974). They are also typical volcanic island arcs.

Thus there is a certain continuity running from northern New Guinea and the Bismarcks to the Kermadecs, suggesting a common history commencing at the end of the Cretaceous or early Tertiary. Subduction is occurring actively today to the south of this chain from the Bismarcks to the southern New Hebrides, and along the Tonga and Kermadec Trenches. The history of the S. Fiji Basin and of the Fiji Plateau is integral to an understanding of the situation. The two ocean basins are by no means contemporaneous in formation and the partial subduction of the former has probably led to the formation of the latter.

The Lau-Havre Basin is also of relatively recent derivation. It separates the Tonga-Kermadec island chain from the Lau-Colville chain. The basin contains oceanic crust with high heat flow and only thin deposits of sediment and is probably no older than latest Tertiary (KARIG, 1970). It was suggested by HAWKINS (1976) to have formed since the Middle to Late Miocene by crustal dilation disrupting a single arc or separating two closely spaced parallel ones. GILL & GORTON (1973) dated the separation as commencing at −5.5 my. A borehole south-east of the Lau group in the Lau Basin reached crust of this age (KENNETT *et al.*, 1972).

A further borehole somewhat to the east of the centre of the S. Fiji Basin encountered a Middle Oligocene basement (−32 my approx.) (KENNETT *et al.*), suggesting that the period of formation of the basin must include that date, probably towards the conclusion of its creation. Thus the Tasman Sea, the New

Caledonia Basin (possibly) and the S. Fiji Basin formed as marginal seas progressively later. Perhaps this process is being continued today in the Lau Basin. If the formation of these basins was successive and not overlapping, then the S. Fiji Basin may have commenced to form from about −50 my, the date of the New Caledonia Basin, and would have continued to dilate until just after −32 my. This correlates approximately with the slower phase of spreading in the Indian Ocean and between Australia and Antarctica.

The continuity of the Outer Melanesian Arc suggests that this might have been the northerly margin of a much more extended S. Fiji Basin between the Inner and Outer Melanesian Arcs, a continuation that has now largely been subducted to the south of the Outer Arc. It is tempting to link the spreading ridge involved with that active at the same time to the north of the Wharton Basin and east of the Ninetyeast Ridge, and to suggest that the spreading axis was continuous right across the north of Australia, the Outer Arc connecting to the north through the North Moluccas and the Talaud Archipelago (or the most northerly part of it) to the Philippines. This hypothesis is illustrated in Figure 2.

Some authors (GILL & GORTON, 1973; KARIG & MAMMERICKX, 1972) have suggested that the Outer Melanesian Arc faced north-east with subduction of the Pacific plate from that direction during the early Tertiary but others (e.g. MALLICK, 1975) have disagreed. The predominantly northerly motion of the Pacific plate at that time and the suggested northerly motion of the Outer Arc might have been such that relative motion between the two was slight.

At about −32 my the motion of the Pacific plate changed and the major north to south extension of ocean crust in the Australasian area became exclusive to the Antarctic Rise south of Australia. Spread in the S. Fiji Basin probably ceased. The overthrust of ultramafics in New Caledonia, New Guinea and, on the other side of the basin, in the Solomons and perhaps the New Hebrides could have marked the end of this spread, resulting from an overall compression of the basin between the now more convergent major plates. This compression may have subsequently been translated into subduction south of the Outer Arc as seen today. This speculative synthesis of events requires further hard evidence from the field, particularly with regard to precise dating, before it can be accepted but it provides a working hypothesis for subsequent biogeographical discussion.

Today the Outer Arc appears fragmented and distorted relative to the hypothesis of continuity presented above. In places there are indications of overlap such as at Feni I., an extension of the Solomon group off New Ireland. This might be attributed to the shearing effect of the westerly moving Pacific plate along such structures as the Vityaz Trench north of the Fiji Plateau.

The major break in the Arc, however, is that separating Fiji from the southern New Hebrides. Between Hunter I. in the south New Hebrides and Viti Levu in the Fiji group there is a transform fault. To the north-west of this fault subduction of the S. Fiji Basin has occurred but not to the south-east. Thus the S. Fiji Basin south of the transform has not been diminished by subduction since it was formed but to the north it has been consumed in the

New Hebrides Trench and its counterparts to the north-west. This implies that New Caledonia and Fiji have remained stable relative to each other but the New Hebrides have moved relatively from continuity with Fiji to their present proximity to New Caledonia (KARIG & MAMMERICKX, 1972; MALLICK, 1975). The New Hebrides/Fiji continuity was maintained at least until the Lower Miocene (−23 my) (GILL & GORTON 1973) and the separation may have commenced in the Middle Miocene (CHASE, 1971).

CHASE stated that the geometry of the thrusting of Fiji north-eastwards relative to the New Hebrides required that extension should have occurred in the Fiji Plateau area between the two island groups. He identified three active

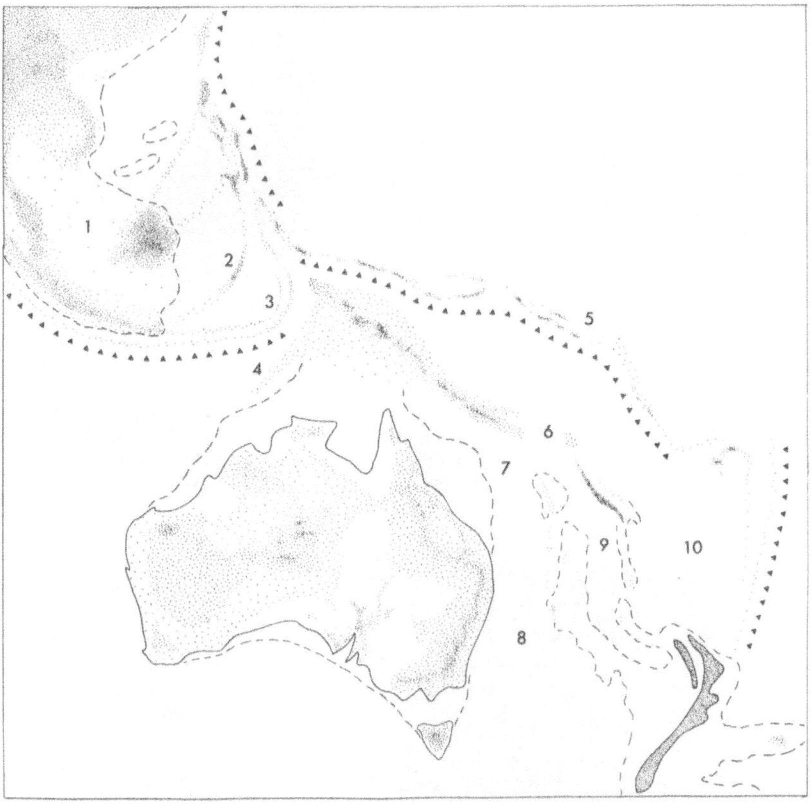


*Fig. 3.* Reconstruction of Australasia at −15 my as hypothesised in the text. Present day continental shelf areas are indicated by dashed lines and the intensity of stipple is indicative of the probability of land except within Australia where present day relief is indicated. Rows of triangles illustrate the major subduction zones. The numbers indicate the position suggested for the Sunda Shelf (1), the western part of Sulawesi (2), the Banda arcs (3), the Westralian portion of the Outer Banda Arc (4), the Solomons (5), Rennell I. (6), the Coral Sea (7), the Tasman Sea (8), the New Caledonia Basin (9), and the South Fiji Basin (10). The land suggested for New Zealand is after FLEMING (1975).

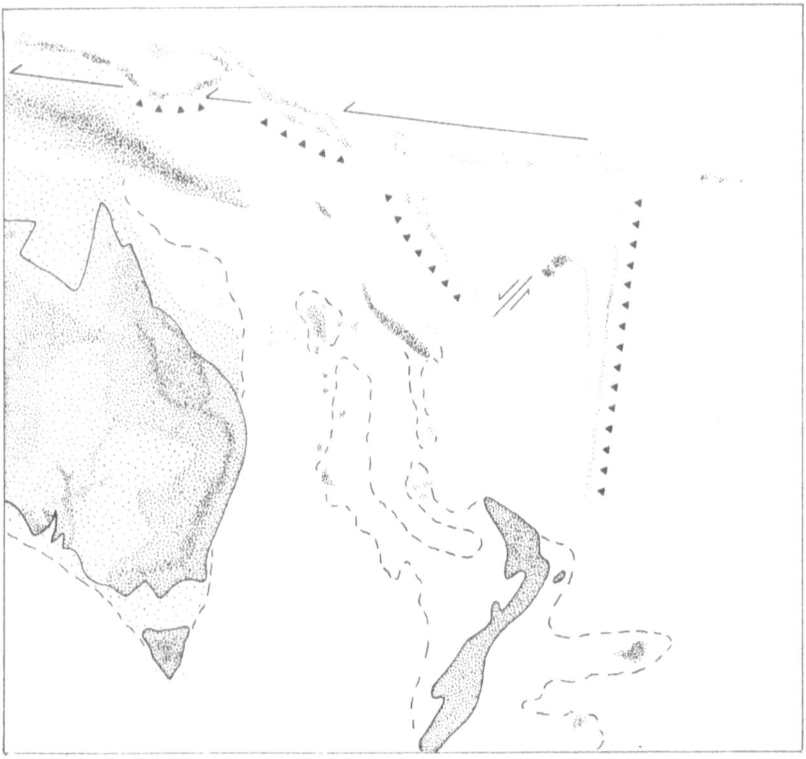

*Fig. 4.* Hypothetical reconstruction of eastern Australasia at −5 my with conventions as in Figure 3. The megashear zone across the north of New Guinea and the shear zone of the Hunter fracture between Fiji and the New Hebrides are illustrated.

sea floor spreading centres on the Plateau. The Plateau is bounded on the north by the Vityaz Trench, to the south of which is a chain of small islands and seamounts that include Rotuma. CHASE suggested this chain of islands was an intruded, older, western continuation of the Samoa-Wallis chain. A date of −4.3 my for the Rotuma volcanics (ROBINSON, 1975) is suggestive of a more recent history but does not preclude the hypothesis of CHASE.

In Figures 3 and 4 speculative reconstructions for −15 my and −5 my are shown, based largely on those of PACKHAM (1973) for the Outer Melanesian Arc. In the former a tentative palaeogeography of Indonesia is given, based on the literature mentioned in the following discussion. Land present in the New Zealand area follows FLEMING (1975). Within the Australian continent density of stipple indicates present day relief but elsewhere it is used to suggest the probability of dry land in the island arcs.

The tectonic evolution of the Indonesian Archipelago from the late Palaeozoic to the Pliocene was said by KATILI (1975) to have proceeded with subduction and accompanying vulcanism spreading systematically in ever-widening arcs away from the Asian continent, in effect much as seems to have occurred to the east of Australia.

The Late Cretaceous subduction zones were generally close to the present day Sunda Shelf though much of north-east Borneo was excluded (HAMILTON, 1971; KATILI). At the end of the Cretaceous tectonism brought about uplift in the Philippines which have probably been present as a significant archipelago since that time (GERVASIO, 1973; RUTLAND & WALTER, 1974). Presumably the Sulu and Palawan arcs and their associated marginal basins are of a similar date.

KATILI has suggested that the subduction zone migrated at the Oligocene/Miocene boundary to a zone extending from the S. Philippines through the Sangihe Is., the northern and southern arms of Sulawesi and the Balabalangan Is. to the vicinity of Java. In the Middle and Upper Miocene he suggested it migrated again to its present position associated with the Banda Arcs. Such migrations could have been associated with the opening of the associated marginal seas, the Sulawesi and Banda Seas respectively. In the later Tertiary the central part of New Guinea began to interact with the Outer Melanesian Arc with uplift in both areas and intensification of the left-lateral shear along this northern margin of the Australian plate. This shearing has complicated the geology of the area between the North Moluccas and Sulawesi. Continental fragments, probably torn from the New Guinea basement, occur in the Sula Is. (HAMILTON, 1971). The westward thrust involved may have contributed much to the current uplift and complex shape of Sulawesi.

The history of Sulawesi, of the Lesser Sunda Is. and the S. Moluccas has been the subject of a certain degree of controversy. The four arms of Sulawesi have somewhat different histories though the north and south arms and the western central part have fundamental features in common distinguishing them from the south-east and east arms, and the two parts are separated by gravity anomalies (AUDLEY-CHARLES, 1974). The more easterly half has been suggested by AUDLEY-CHARLES to represent a continuation of the Outer Banda Arc via Ceram and Buru that extends north from Sulawesi via the islands of Tifore and Maju and the westerly part of the Talaud group to the southern Philippines. The eastern part of the Talaud group, the Nenusa Is., may be referable to Halmahera and the westerly part of the Outer Melanesian Arc.

The Lesser Sunda Is. also consist of two approximately parallel arcs, the Banda Arcs. There is a consensus that the Inner Banda Arc (Bali, Flores, Wetar, Wulur, Manuk and the Banda Is. marking its course) has always been part of the Asian continental margin as regards its geological history. The Outer Banda Arc (Sumba, Timor, Tenimber and Kei to the South Moluccas) was suggested by KATILI to be of a similar nature, continuing the line of the Nias and Mentawi groups to the west as does the Inner Arc that of Sumatra and Java; in this hypothesis subduction was always to the south of both arcs.

An alternative hypothesis proposed that the Outer Banda Arc from Sumba eastwards was part of the Australian continental shelf, with oceanic accretions from the north (AUDLEY-CHARLES, CARTER & MILSOM, 1972; AUDLEY-CHARLES, 1975). The lithology of these islands shows closer association with that of Australia than with the marginal structures of the Sunda Shelf and there are discontinuities in the vicinity of Sumba in several linear features to the west such as gravity anomalies, the earthquake pattern and aspects of submarine morphology. It was suggested that these islands formed a Westralian geosynclinal system that has become deformed and uplifted by interaction with a subduction zone associated with the Inner Banda Arc.

A more recent publication (CARTER, AUDLEY-CHARLES & BARBER, 1976) presented a hypothesis combining elements of these two alternatives. It explained the presence of old Australian rocks in the Outer Banda Arc overlain by more recent Asian facies in terms of a collision between an Outer Banda Arc associated with the Sunda Shelf and the north-west margin of the Australian continent.

CARTER *et al.* presented palaeogeographic reconstructions for the Middle Miocene and Late Pliocene showing the Banda Sea opening as a marginal basin over these periods. The western half of Sulawesi is shown as adjacent to the coast of Borneo throughout this period. The reconstruction in Figure 3 presents a more generalised picture but for biogeographical reasons discussed later the western part of Sulawesi is shown as an evenly curved arc convex to presumed subduction from the east; it is suggested here that it then reached its present shape and position as a result of deformation by the left-lateral shear across New Guinea. This would serve to decrease the water gap between Sulawesi and Borneo rather than increase it as suggested by the reconstruction by CARTER *et al.* Any proximity to Borneo would have preceded the opening of the Sulawesi Sea as a marginal basin as suggested on p. 21.

The major uplift of the Banda Arcs was probably during the late Pliocene and Pleistocene. Thus it is only recently that the Banda Arcs have presented significant archipelagic connections between Sundaland and Australasia, permitting the interchange of organisms along them. Even the more dispersive groups such as birds and butterflies show predominantly recent patterns of exchange along the arcs and there is little indication of endemic development in either the floras or the faunas of the islands (HOLLOWAY & JARDINE, 1968; HOLLOWAY, 1973b).

For the same dispersive groups Sulawesi shows little of the strong association with Borneo that would be expected from its present position. A closer association would be expected had the western part persisted as significant land over a long time in the position suggested by CARTER *et al.* If their hypothesis is correct then this part of Sulawesi is likely to have remained submerged until very recent interaction with the more easterly half. Yet the degree of endemism within the Sulawesi biota is suggestive of a relatively extensive subaerial history for at least part of the island.

Its closest biogeographical associations are with the Philippines and, via the Sula Is. which have a predominantly Sulawesian fauna (HOLLOWAY & JARDINE), with the South Moluccas. The level of endemism and the degree of faunal dissimilarity with these two areas suggests that interaction even with them has been from a position of moderate isolation. The major period of uplift in Sulawesi was from the Miocene onwards, culminating in the late Pliocene and Pleistocene.

HOLLOWAY & JARDINE identified an old route of faunal exchange between Asia and Australia via the Philippines and the Outer Melanesian Arc. Sulawesi was only weakly associated with this; groups epitomising distribution along this old spread route have low (pachyrrhynchine curculionids, p. 237) or no (*Platymantis* frogs, p. 243) recorded representation in Sulawesi. Connection with such dispersals has probably been mainly as a side-track from the Philippines from where most of the Asiatic elements in the Sulawesi fauna were derived, rather than from Borneo (HOLLOWAY & JARDINE). Exchange with Borneo and with the South Moluccas appears to be predominantly recent (HOLLOWAY, 1973b), though there are intriguing indications of an old exchange between Sundaland and New Guinea in a few plant groups that appear not to have involved the Outer Melanesian Arc. They are discussed on p. 75. The route may have been more southerly but there are no related groups of species in Sulawesi today at least in the palm groups mentioned; these would be expected if the exchange had been subsequent to the uplift of extensive land in Sulawesi as may have occurred in the cicada genera discussed on p. 235 where the association is with the Outer Melanesian Arc. Possibly these plant groups illustrate an early exchange via minor archipelagoes that did not persist sufficiently to facilitate endemic radiation.

The Middle Miocene reconstruction of Figure 3 is consistent with the biogeographic patterns though it obviously requires geological confirmation, probably from study of the ocean floor of the Sulawesi and Banda Seas as suggested by CARTER *et al.* The northern arm of Sulawesi shows obvious structural association with Mindanao via the Sangihe Is. but the role of the easterly part in the biogeography of the area is obscure, especially if it is related geologically to Ceram and Buru. The biogeography of Sulawesi requires much more detailed investigation through new fieldwork; palynological studies particularly may help to unravel its biological history. Study of the Sula Is. and the Sangihe and Talaud groups would also be very valuable.

The relationship of New Caledonia to other land masses and the potential access routes for its flora and fauna can be summarised as follows. In the late Cretaceous the previously close association between the New Caledonia/New Zealand land mass and Australia/Antarctica was brought to a close by the opening of the Tasman Sea and the Antarctic south of the Campbell Plateau. Any relationship between New Caledonia and New Zealand would not have been affected to any great degree by this event though there are no indications as to whether it was any more intimate in the past. Relationships with S.E. New Guinea and the condition of the Lord Howe Rise at that time are obscure. The Outer Melanesian Arc, if it existed prior to the formation of the S. Fiji Basin, may have formed a minor fringing archipelago on the Pacific side of the Inner Arc. The Coral Sea and New Caledonian Basin may have been formed in the Eocene, further isolating New Caledonia from Australia but not from any incipient land in S.E. New Guinea. Thereafter until the Middle Oligocene the Inner and Outer Melanesian Arcs probably became separated as illustrated in Figure 2, and the combination of the two could have provided an early opportunity for a trickle of Indo-Malayan floral and faunal elements to reach Australia, New Caledonia and New Zealand. Access to Australasia may have been possible prior to this for the most mobile taxa, given even a sparse array of Polynesian-sized atolls between the two areas. The end of this period saw uplift in S.E. New Guinea, New Caledonia and New Zealand. The first volcanic episodes occurred on Lord Howe I. and there could have been uplift in other localities on the Norfolk Ridge though there is no evidence for this. Thus an augmentation of archipelagic links along the Inner Melanesian Arc may have occurred and persisted into the Miocene.

The course of the Miocene saw the northern margin of Australia approach and interact with the Outer Melanesian Arc to the north through central New Guinea and, eventually, with the Banda Arcs to the west. The Outer Melanesian Arc could have been a fairly continuous archipelago from the end of the Oligocene, though certainly more attenuated. It was probably less emergent towards the east, especially in the area of the New Hebrides. Exchange between the two arcs would have been greatest in the New Guinea area. Any interaction between New Caledonia and the arcs is liable to have been through Inner Arc New Guinea at that time.

Towards the end of the Miocene and in the Pliocene the two parts of New Guinea would have approached, ultimately to fuse, breaking down the degree of biogeographical independence of the two arcs. At the same time the association between New Caledonia and S.E. New Guinea may have declined if the latter rotated and migrated westwards from the position indicated in Figure 3. Figures 3 and 4 illustrate these changes in the association of lands within and between the arcs.

Similarly over about the past 10 my the New Hebrides, acting as a channel perhaps even when of limited extent for organisms dispersing into the Pacific from the west, have become more isolated from Fiji and closer to New Caledonia and the Loyalty Is.

Finally, the relationship between New Caledonia and Australia through the Tertiary must be considered. Their relative positions have probably remained unchanged, so facility of biological exchange at any period will depend on the occurrence of stepping stones between the two. Figure 6, showing land exposed by a sea-level drop of about 120 m during the last glaciation, reveals many areas only just covered by the sea that, given an episode of general uplift or vulcanism or of falling sea-level, might break the surface and provide a route of exchange for organisms. The areas exposed include parts of the Great Barrier Reef and the Queensland Plateau, the Chesterfield Reefs at the northern end of the Lord Howe Rise, and a series of sea mounts extending south from the Chesterfields on the eastern edge of the Rise to culminate in Lord Howe I. itself.

Apart from during periods of low sea-level in the Pleistocene there are two periods when increased biological exchange may have been facilitated by these stepping stones. One is in the Late Miocene when there was uplift in E. Australia, vulcanism on Lord Howe I. and uplift (Cycle II) in New Caledonia (LILLIE & BROTHERS, 1970; AVIAS, 1973), though the last event is given a later date (Plio-Quaternary) by GUILLON (1974). Other volcanics may have occurred on the sea mounts to the north of Lord Howe I. and this is suggested, together with uplift in the Chesterfield Reef area, in Figure 4. These areas would then have become submerged following erosion. Exposure of land in the Norfolk I. area at this time is unlikely (HOLLOWAY, 1977).

An earlier period of general uplift could have been coincident with the earliest volcanics (−30 my) known on Lord Howe I. and the overthrusting of ultramafics in New Caledonia and S.E. New Guinea, perhaps all resulting from compression of the area as the pattern of sea floor spreading in the Pacific and Indian Oceans changed (p. 11). But assumption of the contemporaneity of these events is probably stretching available geological evidence too far. The date for the emplacement of the New Caledonian ultramafics has not been fixed precisely and AVIAS (1973) and DUBOIS *et al.* (1973) have suggested it occurred in the Upper Eocene, about 10 my earlier. On the other hand DUBOIS *et al.* have dated the Loyalty Is. volcanics at −29 my within a range of error of 8 my, indicating the possibility of tectonism at the time in question.

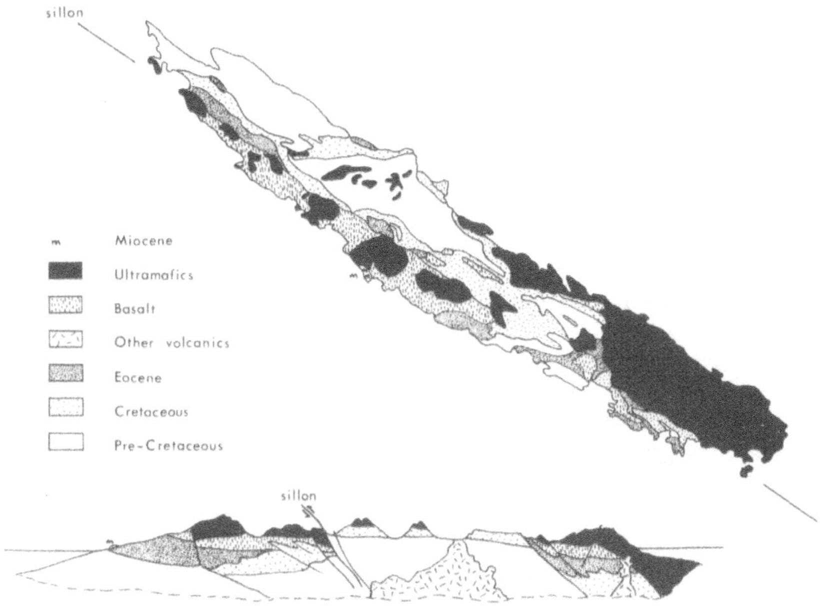


*Fig. 5.* Geological map and section of New Caledonia after LILLIE & BROTHERS (1970) and GUILLON (1974).

## The geological structure of New Caledonia

The patterns of sedimentation, orogenesis and igneous activity in New Caledonia have been described by LILLIE & BROTHERS (1970), AVIAS (1973), BROTHERS (1973) and GUILLON (1974). The geological map and generalised section in Figure 5 are based on those by LILLIE & BROTHERS and GUILLON.

Sedimentation started in the Permian and was continuous from the Triassic to the Oxfordian (early Upper Jurassic) in a trough bordering a stable sialic block to the west, resulting in a thick series of greywackes. There was orogenesis, uplift and metamorphism in the Upper Jurassic (corresponding to the Rangitata Orogeny in New Zealand), sedimentation only resuming in the Upper Cretaceous. These Cretaceous sediments were mainly conglomerates and clays but there were rhyolitic tuffs and coarse sandstones indicative of continental-type sedimentation in the Nouméa Basin. Coal was also formed. Overlying these formations are cherts and limestones. Basalts were discharged in the Palaeocene (Montian) to cover large parts of the western slope of the island, at a time when the opening of the Tasman Sea was ceasing. These basalts are discordantly overlain by greywackes, products of the erosion of complex Eocene fold structures which involved nappe formation on the western side of the island. There was metamorphism in the Late Eocene and the Eocene/Oligocene boundary saw a paroxysm of orogenesis, sediments being folded parallel to the axis of the island.

The ultramafics were emplaced late in this orogenic phase, being ocean floor overthrust from the east some time during the period from the end of the Eocene to the end of the Oligocene. AVIAS (1973) and BROTHERS (1973) dated the climax of this orogenesis from the Middle Oligocene to the Lower Miocene. The sheet of ultramafics is thought to have been much more extensive when first emplaced, having been eroded and dissected subsequently. Radioactive dating of these ultramafics may not give any indication of when they were emplaced but could show when the ocean floor at the western margin of the S. Fiji Basin was formed, giving a pointer to the time sea floor spreading commenced there.

High pressure metamorphism in the north has been dated by BROTHERS (1973) at between −38 and −21 my, a period which covers most of the Oligocene and centres on the date of the Loyalties volcanics and of the change of sea floor spreading patterns. The intensity of this metamorphism was greatest in the north-east where phyllites and schist in the centre of the island grade to the gneiss which forms the Mt. Panié range (COLEMAN, 1967).

Fragments of the ultramafics in derived sediments appear no earlier than the Lower Miocene (at Népoui). There are only localized outcrops of Miocene age. Subsequently there are only continental lagoonal and fluviolacustrine deposits of the Plio-Quaternary. During the latter half of the Tertiary some erosion and peneplanation occurred but there was further uplift at the end of the Miocene (AVIAS, 1973) or the Plio-Quaternary (GUILLON, 1974) forming the mountains that now run along the spine of the island.

The detailed structure of New Caledonia is as yet little understood. There is a line (sillon) along the axis (Figure 5) that separates the western basalts from the older central zone. The line is an old structure, coinciding with a large, north-east dipping reversed fault that formed before the deposition of the Cenomanian, moved again in the Late Cretaceous and once more in the Eocene. In the south it coincides with the south-west edge of the major ultramafic massif.

Thus from the end of the Eocene New Caledonia persisted as a large land mass. It was probably extensively subaerial prior to that, at least from the Late Cretaceous and possibly from the Upper Jurassic. The ultramafics were extensive when first emplaced and may have made up most of the land surface. Subsequent erosion has led to a more diverse rock basement until today the ultramafics occupy only a third of the land surface. Their unusual mineral content and that of their derived soils contrasts greatly with soils derived from the sedimentary and metamorphic rocks and this difference is reflected in the associated vegetation types. This past predominance of ultramafics and their later decline may provide an important dating pointer for colonisation by some elements of the flora and fauna (p. 68).

The Loyalty Is. (DUBOIS *et al.*, 1973) have a volcanic basalt basement which has been dated at $-29 \pm 4$ my. These basalts are associated with tuffs containing limestone pebbles of probable Oligocene (Aquitanian) age. Submergence then led to the formation of extensive coral reefs on top of the volcanics. Current uplift commenced on Maré in the Lower Pleistocene and was greater in the south where the chain is closest to the New Hebrides Trench. DUBOIS, LAUNAY & RECY (1975) showed that the differential uplift along the Loyalties chain and on the mainland coast near Yaté fitted theoretical predictions concerning bulging of the lithosphere immediately prior to subduction zones. The rate of uplift has not yet been determined.

## Tertiary climates and sea-level changes

The passage of Australia, New Caledonia and New Zealand northwards during the Tertiary brought them ever closer to the equator which, given constant climatic conditions through the Tertiary, should have led to progressively ameliorating climates in each area. In fact this was far from the case and global climatic changes during the Tertiary and more local effects produced by changes in ocean currents have been significant. In turn these climatic changes affected the biological development of the areas concerned. Geographical changes brought about by glacio-eustatic fluctuations in sea-level have probably had a marked effect on biogeographic patterns by exposing or submerging areas of low relief that could act as routes of faunal exchange between more permanent land areas.

GILL (1961) considered that, in Australia, temperature increased over the early part of the Tertiary to a mid-Tertiary maximum and then declined to the Pleistocene minima. FLEMING (1975) showed temperature in New Zealand to have risen until the end of the Eocene, a rise that was followed by a marked drop to a low level that extended over the first half of the Oligocene (−38 to −30 my). In the latter half of the Oligocene and the early Lower Miocene it rose again, though not to Eocene levels, but may have fallen for another period of a few million years at the end of the Lower Miocene. The beginning of the Middle Miocene was very warm and thence there was an overall decline in temperature culminating in the Pleistocene glaciations. Palaeontological evidence has indicated a period of low temperature in New Zealand in the middle of the Pliocene.

The Eocene rise in temperature may reflect the separation of New Zealand from Antarctica, moving from a warm temperate regime into a subtropical one. The general Middle Oligocene to Miocene rise may reflect the more recent northward motion of Australasia as a whole, though this effect has been dominated since the Middle Miocene by an overall global cooling.

In New Zealand there was gradual marine transgression until the Middle Oligocene and Oligocene seas there were cool (STEVENS, 1973). Further light has been cast on this situation by the work of KENNETT *et al.* (1972) in their work investigating a general uncomformity in the Lower Oligocene both in cores of deep sea sediments from the south-west Pacific and in the marine succession of sediments now uplifted in New Guinea. From the end of the Eocene well into the Oligocene occurred one of the principal episodes of Cainozoic cooling, associated with intense Antarctic glaciation similar to that of today. Other, less intense episodes of glaciation are thought to have occurred in the Lower and Middle Eocene. The Oligocene glaciation was suggested by KENNETT and colleagues to have intensified the Antarctic ocean currents, resulting in greater erosion of bottom sediments. They suggested that at that time there was a primitive circum-polar current that passed round the north of Australia and down through the Coral and Tasman Seas or, perhaps less likely, currents sweeping northwards through the Tasman Sea. These currents would have been cold. The cessation of this erosion of bottom sediments may have occurred as a result of the breakdown of either of these proposed current systems following the Late Oligocene rifting of Australia from Antarctica. The circum-polar current would then have adopted its present course south of Australia.

Truly marine conditions did not develop over all of south Australia until the Late Oligocene. In Victoria two major transgressions are evident, one in the Palaeocene (perhaps reflecting the initial −55 my rift) and the second commencing in the Eocene and culminating in the Miocene. This latter transgression may reflect the final separation of Australia from Antarctica when current plate motions became prevalent.

One may therefore speculate that when the ultramafics were being emplaced in New Caledonia and S.E. New Guinea, and the first episode of vulcanism was occurring in the Lord Howe area, the New Caledonian climate was relatively cool and sea-level may have been at a minimum. This, then, is perhaps a period when land areas and archipelagic connections were more extensive, favouring faunal and floral dispersal and exchange.

From the Lower Miocene onwards a more detailed picture emerges. HAYES *et al.* (1973) regarded the period from −20 to −4 my as one of much glacial activity in Antarctica, with a peak at −5 my. From −15 to −10 or −9 my (Middle Miocene) there was a glacio-eustatic drop in sea-level with corresponding subaerial erosion on Laysan I. of the Hawaiian group and elsewhere in the Pacific (SCHLANGER & GILLETT, 1976). This was followed by a period of submergence until −6 my in the Hawaiian group, corresponding presumably to a warmer period. From −6 to −5 my the Messinian Regression occurred in the Mediterranean (SCHLANGER & GILLETT) and cooling (Upper Miocene) in New Zealand (STEVENS, 1973). This would correspond with the peak in Antarctic glaciation at -5 my.

GLENIE, SCHOFIELD & WARD (1968) referred an apparent fall of almost 100 m in sea-level in New Zealand to the 'early Pliocene' and a rise of about 150 m to the latter part of the Pliocene, with a maximum in the early Pleistocene. The Pliocene temperature changes in Antarctic seas suggested to have occurred by CIESIELSKI & WEAVER (1974) indicate a different pattern, if it is legitimate to related cool periods to falls in sea-level. Over most of the first half of the Pliocene (−4.30 to −3.95 my) conditions were probably much warmer than today, too warm for the existence of the West Antarctic Ice Sheet. There followed a major cooling trend to −3.70 my following an Antarctic glacial advance, the coolness persisting until −2.7 my (when further erosion

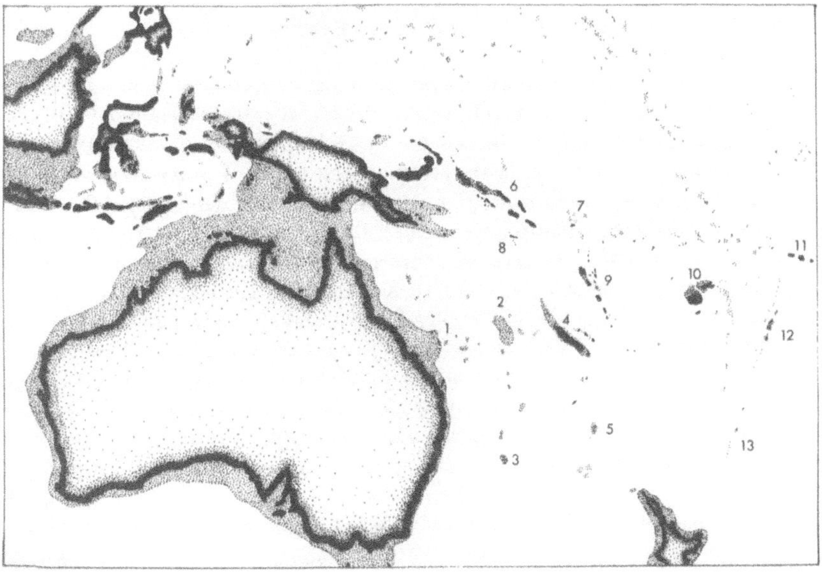


*Fig. 6.* A map of the Indo-Australian area indicating the land that would probably be exposed with a 100 m fall in sea-level as probably occurred during the Pleistocene glaciations, after ROBINSON (1975) and HOLLOWAY (1977). The numbers indicate the Great Barrier Reef and Queensland Plateau area (1), the Chesterfield Reefs (2), Lord Howe I. (3), New Caledonia (4), Norfolk I. (5), the Solomons (6), the St. Cruz Is. (7), Rennell I. (8), the New Hebrides (9), Fiji (10), Samoa (11), Tonga (12), the Kermadecs (13).

on Laysan occurred (SCHLANGER & GILLETT)). JONES & MCDOUGALL (1973), from examination of the quench zone of the Norfolk I. volcanics, produced mainly during this cool period, suggested that sea-level was not very different from that of today. The period of low sea-level followed by a rise above the present level described by GLENIE, SCHOFIELD & WARD could therefore either refer to the Messinian Regression and early Pliocene warming or to the later Pliocene cooling and a possible warmer period after that. During the Messinian Regression the second series of Lord Howe volcanics would have suffered little erosion and in Figure 4 the possibility of their extensive emergence and that of other areas to the north is indicated.

It is not clear from the literature whether a warmer period did occur after the later Pliocene cooling (−2.7 my) but BROTHERS (1954) and ZEUNER (1959) both regarded the Pleistocene as a period in which there was an overall net fall in sea-level from a Late Pliocene high level to that of the present day. On this net fall were imposed the eustatic fluctuations caused by the Pleistocene glaciations. Sea-level minima of perhaps 100 m (QUINN, 1972; CHAPPELL, 1974), at most 283 m (STEVENS, 1973), occurred at approximately −20,000 to −16,000 years and at −165,000 years. In Figure 6, based on present day bathymetry, the land exposed by a drop of 100 m is indicated. Archipelagic connections become marked between Australia and New Caledonia, moderate between New Caledonia, Lord Howe I. and New Zealand, and a chain of atolls emerges, extending from the St. Cruz Is. via Rotuma to Samoa, noted already by ROBINSON (1975).

The climate over the tropical Pacific during the glaciations may have been much drier than it is today (QUINN, 1972). The intensification of oceanic circulation and the land exposed by the fall in sea-level may have led to meteorological conditions favouring the persistence of a dry zone across the Pacific. The extensive arid zone of Australia probably experienced its greatest increase during the Pleistocene (GILL, 1961), though BEARD (1977) considered it to have developed gradually through the Tertiary. Thus the organisms exchanged between Australia and New Caledonia via the Chesterfield Reefs during the glaciations are likely to have been those characteristic of more arid vegetation types. The exchange would have occurred at a time when rainforest in N.E. Australia was probably much more restricted than it is today and not extending so far south. The black tropical soils supporting the dry west coast vegetation associations in New Caledonia may have evolved largely during a drier period in the past (p. 2) and these west coast habitats may have developed and been more extensive during this period. One might also speculate that they may reflect to some extent the sort of vegetation likely to have developed on the Chesterfield Reefs and on the extensive tongue of reef-clad land that was exposed to the north-west of New Caledonia (Figure 6). These west coast habitats will be seen to support moth species that include a high proportion of Australian elements of probable recent derivation (p. 186).

# II. THE PHYTOGEOGRAPHY OF NEW CALEDONIA

There are dangers in interpreting biogeographic patterns solely in terms of broad events in the geological history of an area though such patterns often show good correlation with geography (HOLLOWAY & JARDINE, 1968). HOLLOWAY (1969, 1974a) related patterns in the distribution of butterflies in the Indian area to the drift of India northwards in the Tertiary. He suggested that these patterns were consistent with land contact between peninsular India and Asia having been made in the late Tertiary, resulting in the breakdown of a barrier to floral and faunal exchange between the two. Geological literature then available suggested that this barrier might have been a steadily decreasing water gap.

More recent literature suggests that this barrier may have been ecological rather than geographical. Land contact between the two continents may have been as early as Eocene (SAHNI & KUMAR, 1974; MOLNAR & TAPPONIER, 1977). Floristic evidence tends to support this (LAKHANPAL, 1970). The probably early Eocene• Intertrappean Flora of southern India contained podocarps, araucarians, Proteaceae, *Casuarina*, many palms (31 spp.) including *Nypa*, aquatic plants and also elements of the modern tropical American flora. Generally there was a mixture of austral elements and the early Tertiary London Clay Flora, the latter suggested by LAKHANPAL to have been a dispersive west Tethyan maritime or estuarine flora. There were no dipterocarps or Pandanaceae. The Neogene floras were more typically Asian with dipterocarps occurring in southern India in the Miocene. Present day podocarps in India are probably of Malesian derivation.

Biological exchange with Asia was therefore possible from the Late Oligocene at least. However, the Siwalik flora (S. Himalaya, Miocene to Pliocene) is indicative of a dry interior to the continent at that time (AXELROD, 1974) which may have precluded significant radiation of Asian rainforest and its associated fauna in India. Even today there is considerable difference between the floral and faunal diversity of the wetter south-west part of India and Sri Lanka and the drier central and eastern parts.

Some Australasian biogeographic patterns may have resulted from ecological factors rather than through differential contact of the various land areas involved.

## Past southern hemisphere floras: the fossil record

The pre-contact austral flora of India is involved in the broad relationship of southern hemisphere biotas to events in the dispersal of Gondwanaland, including New Caledonia. General treatments of this theme have been written by KEAST (1971, 1973), RAVEN & AXELROD (1972, 1974), GASKIN (1970, 1975) and SMITH (1974). Both RAVEN & AXELROD and GASKIN modified their later accounts in the face of new geological evidence and GASKIN's first publication was based on the assumption that continental drift had not occurred as, largely, was that of DARLINGTON (1965).

The general conclusion is that certain archaic southern disjuncts may reflect in their current distribution patterns the separation of the southern continents. Many of the faunal groups showing this will be mentioned in Chapter VII. Links between South America, Australia and New Zealand generally appear to postdate those between these lands and Africa when the morphology of the organisms is considered.

RAVEN & AXELROD (1974) suggested that contacts between Africa and Australia persisted until −100 my and thence may have diminished rather than ceasing abruptly, for there could have been subsequent archipelagic connections via Madagascar, India and the Kerguelen Plateau as suggested in Figure 1, showing a −75 my reconstruction of the geography of the southern hemisphere. The work of JOHNSON, POWELL & VEEVERS (1976) has indicated that this separation may have commenced much earlier at −130 my but the hypothesis of archipelagic exchange continuing to a much later date is still valid.

RAVEN & AXELROD also proposed that western Gondwanaland (Africa, S. America) was the most likely site for the initial radiation of the angiosperms, rather than the Oriental Region as suggested by THORNE (1963).

Angiosperm pollen first appears in Australian deposits dated as late Middle Albian (about −115 my), somewhat later than in Europe and the Americas (DETTMANN, 1973). The assemblages are less diverse than those of the Americas and West Africa, suggesting migrations into the Australian mid-Cretaceous floras rather than evolution within them.

Pollen of dicotyledonous angiosperms first appears in the Albian of New Zealand also, entering a primarily podocarpaceous forest vegetation. *Nothofagus* and Proteaceae first appear in the Cenomanian (−110 to −100 my) with, somewhat later, Cruciferae, Liliaceae, Casuarinaceae and Myrtaceae (COUPER, 1960a).

No fossil floras for New Caledonia have been discovered as yet (VIROT, 1956), so one can only surmise its floral history through reference to the present flora. Its early history may have paralleled that of New Zealand, as it has a remarkably diverse flora of podocarps and araucarians and great diversity also in the dicotyledonous angiosperm families recorded earliest in New Zealand. If it had only acquired significant land area in the later Cretaceous or early Eocene when probably largely isolated from Australia and New Zealand through the opening of the Tasman Sea, then greater disharmony amongst these families might have been expected.

The fossil floras of Africa and South America become increasingly dissimilar through the Cretaceous and are largely distinct from the Palaeocene onwards (MULLER, 1970; KEAST, 1971; RAVEN & AXELROD, 1974), a result of the opening of the Atlantic. The gradual decline of shared taxa, noted primarily amongst those of tropical latitudes, may have resulted from the geometry of the two continents in separation, the north coast of South America moving laterally relative to the south coast of the western bulge of Africa. Volcanic islands in the Atlantic may have offered stepping stones for continued exchange (RAVEN & AXELROD). Late Cretaceous and Early Tertiary exchange of temperate elements between Australasia and South America via Antarctica would have excluded Africa.

By the Upper Cretaceous the angiosperms had come to dominate the fossil floras of the world at the expense of the Mesozoic gymnosperm and fern floras (BARNARD, 1973). HUGHES (1973) suggested that in pre-Cretaceous times the conifers were probably a varied but universal group spreading through the tropics and into low latitudes across Pangaea (the continents of the world as unified until the Cretaceous). Separation of conifers into northern and southern groups possibly began from the mid-Cretaceous when the angiosperms began to fill the equatorial lowlands and present a substantial biological barrier. The present situation may then have arisen through the differential extinction of conifer groups in each hemisphere, the differentiation possibly being due to some ecological factor such as the difference in the proportion of maritime to continental habitats in the two hemispheres. Elements of the northern conifer flora continued to be lost in the Tertiary but the Podocarpaceae and Araucariaceae persisted in isolation in the south. A similar pattern of replacement by radiation from the tropics of more competitive taxa could have occurred with the early angiosperms. This process would have been fragmentary in the southern hemisphere through the break-up and isolation of the various land masses. The replacement of older groups by younger ones would have been more disrupted, leading to conservation of the former.

HUGHES suggested that the present day podocarps in S.E. Asia are relict rather than redistributive. The reconstructions of Pangaea before −100 my of SMITH, BRIDEN & DREWRY (1973) show S.E. Asia and northern Australia in proximity, almost enclosing the Tethys Ocean. Archipelagic exchange between the two areas may well have been feasible if the reconstruction is correct, but the dating of events by JOHNSON, POWELL & VEEVERS suggests that at least some adjustments may be necessary. There are indications from Borneo (the palynological work of MULLER, reviewed briefly by HOLLOWAY, 1970) that northern conifers predominated before the Pliocene and only then do southern elements appear in the fossil record.

There are, however, many references to southern floral elements in the fossil record of the northern hemisphere. ZAKLINSKAYA (1964) mentioned *Agathis, Araucaria, Dacrydium, Phyllocladus, Casuarina, Eucalyptus* and Proteaceae from the Upper Cretaceous and Lower Tertiary in S. Eurasia, Kazakhstan and W. Siberia. MULLER (1970) referred to proteaceous pollen from the Late Cretaceous of Canada. But CRANWELL (e.g. 1963) has cast doubt on many such records from the northern hemisphere. Wind-pollinated groups such as the conifers and *Nothofagus* produce large quantities of pollen that may be dispersed over huge distances. Chilean *Nothofagus* pollen has been recorded from Tristan da Cunha and Gough Island (CRANWELL, 1964b) and KORONEVA

(1964) found samples of sediment from the sea bottom 100–150 km east of North Island, New Zealand, to be rich in pollen of genera such as *Dacrydium*, *Phyllocladus*, *Podocarpus* and the introduced *Pinus*. Long distance transport of such pollen in the past cannot be ruled out. Indeed, it might be possible to explain the observations of ZAKLINSKAYA in terms of pollen brought into central Asia by monsoonal wind systems passing over a still drifting, austral, Late Cretaceous India.

At the end of the Cretaceous the Australian-Antarctic flora is distinct from the Malesian (Bornean sediments; no proteaceous pollen recorded) which again appears to have had only limited exchange with the European (MULLER, 1970). Records from Antarctica, notably Seymour I. (CRANWELL, 1974b), the McMurdo Sound (MCINTYRE & WILSON, 1966) and Prydz Bay (KEMP, 1972) reveal a flora for that time including such taxa as *Nothofagus*, the conifers *Acmopyle*, *Podocarpus*, *Dacrydium* and *Phyllocladus*, Proteaceae such as *Knightia*, *Drimys* (Winteraceae), Lauraceae and *Laurelia* (Monimiaceae). *Acmopyle* is now only found in New Caledonia and Fiji.

Palaeocene sediments from a borehole now at 11° S on the Ninetyeast Ridge contained carbonaceous and lignitic clays and sands interbedded with volcanoclastics. The pollen flora, though poor, had strong austral connections including *Dacrydium*, *Podocarpus*, *Phyllocladus* and *Microcachrys* (HARRIS, 1974). *Dacrydium* and *Microcachrys* have been recorded from the mid-Tertiary of Kerguelen (COUPER, 1960b). These and the Antarctic localities are indicated by triangles on the −45 my reconstruction in Figure 2.

Neither podocarps nor *Nothofagus* have yet been recorded from the Tertiary sediments of New Guinea (COUPER) but one would have expected at least the more mobile 'insular' elements of this Australian-Antarctic flora to have reached an emergent New Caledonia prior to the opening of the Tasman Sea and New Caledonian Basin.

RAVEN & AXELROD (1974) listed a number of angiosperm families that may have been amongst the earliest arrivals in Australia-Antarctica, perhaps derived from Africa through subtropical south Tethyan dispersals. To the Proteaceae, Winteraceae and Monimiaceae above they added the Santalaceae, Icacinaceae, Thymeleaceae, Epacridaceae, Dilleniaceae, Casuarinales and the ancestors of the New Caledonian endemic Strasburgeriaceae and Oncothecaceae.

The Early Tertiary flora of Australia included Proteaceae with *Beauprea*-like pollen, Sapindaceae and Casuarinaceae (COOKSON, 1964). *Agathis* grew in Victoria, *Araucaria* in Tasmania (GILL, 1961). The south coast of Western Australia supported a tropical or subtropical rainforest with *Nothofagus* dominant and Proteaceae and *Casuarina* subdominant (BEARD, 1977). Probably all Australia enjoyed a more humid, subtropical to tropical climate (GILL; SMITH, 1974).

The Miocene flora of southern New South Wales, discussed by MARTIN (1973), included *Nothofagus* of the *brassii* pollen group, *Phyllocladus*, *Dacrydium*, *Podocarpus* sect. *Dacrycarpus*, *Microcachrys*, *Drimys* and *Quintinia* (Saxifragaceae), rainforest taxa with counterparts today in New Caledonia, northern New Zealand and elsewhere in Melanesia. This rainforest flora disappeared from New South Wales later in the Miocene, due probably to decreasing precipitation that may have resulted from the glaciations in Antarctica discussed on p. 29. It gave way to an association dominated by Myrtaceae and *Casuarina*. In the Pliocene the rainforest reappeared (probably following climatic amelioration; p. 30), depleted, lacking *Nothofagus brassii*, but did not totally replace the myrtaceous association. With the development of modern arid conditions in the Pleistocene the Myrtaceae and *Casuarina* again became predominant with the addition of Gramineae and Compositae giving a savanna flora. KERSHAW (1974) described a similar interchange of forest and sclerophyll vegetation in response to wetter and drier periods during the later interglacials and glacials of the Pleistocene in Queensland (the Atherton Tableland).

It would appear that climatic changes in Australia have tended to reduce the similarity of its vegetation to that of New Caledonia and northern New Zealand. The older elements of the New Caledonian vegetation are dominated by many of the genera of the old Australian rainforest association. *Casuarina* and Myrtaceae are also characteristic of more sclerophyllous vegetation in New Caledonia. The relationship of these vegetation types in New Caledonia to hydrology, relief and the accumulation of organic material in the soils suggests a similar situation to that inferred by MARTIN for Miocene New South Wales, in that the rainforest remnants there today require high precipitation, shelter and fertile soils.

The New Zealand flora has also lost a number of taxa during the Tertiary that are still found in New Caledonia, such as the *Nothofagus brassii* group, and some further taxa are restricted to the extreme north such as *Xeronema* (Liliaceae) with one species on Three Kings I. and another (*moorei*, Plate 1) montane in New Caledonia.

The Cretaceous vegetation of New Zealand has already been referred to. Palynological evidence (COUPER, 1960a; FLEMING, 1975) has indicated that Malvaceae (*Plagianthus*, *Hoheria*; part of a complex shared with Australia), Chloranthaceae (*Ascarina*), Araliaceae, Epacridaceae (*Dracophyllum*) and Sapindaceae (as in Australia) were present in the Palaeocene or Eocene. Other generally austral genera such as *Laurelia* do not appear till much later in the fossil record though this does not exclude their occurrence much earlier.

Several Indo-Malayan elements appear in the New Zealand fossil record as early as the Eocene. This would appear to be at odds with the geographical reconstructions presented in the previous chapter unless these taxa were characteristically mobile, capable of long distance dispersal along the Melanesian Arcs. Such powers would be indicated if the taxa concerned were present on islands that have always been extremely isolated such as Norfolk I., the Kermadecs (HOLLOWAY, 1977) or those of the eastern Pacific, particularly Hawaii. The distributions for most of the genera given by BALGOOY (1971) suggest that this is the case.

The earliest Indo-Malayan genus recorded is *Beilschmiedia* (Lauraceae), a questionable record from the end of the Cretaceous. The genus is now centred

*Plate 1.* Top left, *Dracophyllum involucratum* at Grand Lac; top right, *Spathoglottis*, a common orchid of *Baeckea*-sedge maquis; bottom left, *Astelia* in the *Podocarpus* forests of Mt. Panié; bottom right, *Xeronema moorei* on Mt. des Sources.

in Malesia with a subcentre in New Guinea. It also occurs in the Solomons, E. Australia and S. America, as well as New Zealand. There are no strong indications of mobility.

In the Eocene the Loranthaceae appeared in the form of *Elytranthe* (*Iliostylus* in BALGOOY). Four genera of this family are endemic to New Zealand and *Elytranthe* has extended to Norfolk I. where another genus in the family also occurs, an indication of mobility. Three of the endemic genera are stated to be of palaeotropical affinity by BALGOOY. No endemic genera or generic subcentres occur in New Caledonia in the Loranthaceae but the presence of *Delias ellipsis*, a taxonomically isolated butterfly species of a genus that is virtually restricted to Loranthaceae for its larval foodplants (D'ABRERA, 1971) except for some species feeding on Santalaceae in Australia (COMMON & WATERHOUSE, 1972), might indicate that the family has been present there for a considerable time.

The genus *Dysoxylum* of the Meliaceae also appeared in New Zealand in the Eocene. It is widespread, centred in Malesia with subcentres in E. Australia and New Caledonia. It is cauliflorous, the strongly scented flowers being attractive to Lepidoptera in New Caledonia, such as the butterflies *Delias ellipsis* and *Hasora khoda* (HOLLOWAY & PETERS, 1976) and the sphingid *Macroglossum corythus* (p. 356). A species occurs on Norfolk I., indicative of good dispersal powers.

In the Oligocene *Myrsine* (Myrsinaceae), *Typha* (Typhaceae) and *Epilobium* (Onagraceae) appeared. *Myrsine* is pandemic, extending to Rapa I. and Hawaii, and has a subcentre in New Caledonia. A species usually referred to the closely related genus *Rapanea* occurs on Norfolk I., and a derived genus *Elingamita*, related to the New Caledonian *Tapeinosperma* (GODLEY, 1975), is restricted to Three Kings I. *Typha* is widespread and a species occurs on Norfolk I. *Epilobium* is similarly widespread and now has many species in New Zealand thought by RAVEN (1972) to have radiated since the Pliocene following invasions from Australia; the seeds are light and wind-dispersed.

The interesting lily genus *Astelia* also appeared in the New Zealand Oligocene. In BALGOOY it is divided into three subgenera: *Astelia s. s.* is centred in New Zealand but species occur in the Mascarenes, New Guinea, eastern Australia, the Subantarctic Is. of New Zealand and the Society Is.; *Asteliopsis* has a species in New Caledonia (Plate 1), one in New Zealand and one each on Rapa, the Marquesas, in South America and the Falklands, and six on Hawaii; *Tricella* occurs in S.E. Australia, New Zealand and the Chatham Is. The genus is evidently mobile and its disjunct distribution may be attributable to poor competitive abilities in the face of modern Malesian elements. The New Caledonian species is most frequently encountered in montane cloud forest dominated by *Podocarpus* species.

Oligocene macrofossils have been attributed to *Pittosporum*, *Carpodetus* and *Geniostoma*. *Pittosporum* (Pittosporaceae) is a widely distributed Palaeotropical genus with subcentres in Malesia, Australia, New Caledonia, New Zealand and Hawaii. *Carpodetus* (Escalloniaceae) is today recorded from New Guinea, the Solomons and New Zealand and therefore is not truly Indo-Malayan, perhaps originally austral in character. *Geniostoma* (Loganiaceae), recorded as a microfossil later in the Miocene, is widespread with a derived genus in Hawaii and a centre in New Caledonia.

Many more Malesian and more widely distributed genera appear in the Miocene fossil record. *Potamogeton* (Potamogetonaceae), *Elaeocarpus* (Elaeocarpaceae) and *Alectryon* (Sapindaceae) all extend to Hawaii today. *Rhipogonum* and *Cordyline* (Liliaceae) both have species on Norfolk I. *Weinmannia* (Cunoniaceae), centred in S. America, widespread in Malesia with a subcentre in New Caledonia, extends to the Marquesas but does not occur in Australia. It is a prominent feature of some New Guinea forests (ROBBINS, 1961) as well as those of New Caledonia and New Zealand and is distributed generally along the Melanesian arcs. Similarly the pandemic *Coriaria* (the only genus of the Coriariaceae) is recorded in the Indo-Australian area only from E. Asia, Luzon, New Guinea, the New Hebrides, Fiji, the Society Is., Samoa, the Kermadecs, Chathams and New Zealand. This is primarily an Outer Melanesian Arc distribution. It is montane in the tropics.

Another Miocene arrival was *Macropiper* (Piperaceae), recently revised by SMITH (1975). It is related to the widespread genus *Piper*. There are six species on Fiji of which three are endemic, and others in Micronesia and Polynesia. The New Zealand species is distinctive and extends to Norfolk I., Lord Howe I. and the Kermadecs. There are no species in New Caledonia. Like *Astelia* it may be more restricted today through competition from more vigorous Malesian elements, possibly *Piper* itself.

In the Upper Miocene there was climatic cooling in New Zealand (FLEMING, 1975), new tropical Indo-Malayan plants cease to appear in the fossil record, many Tertiary taxa such as *Nothofagus brassii* died out and others became restricted to the extreme north. Relative to New Caledonia New Zealand today is very poor in genera (344 compared with 655 in New Caledonia), especially when its greater area is considered. The extinctions in the later Tertiary, augmented during the Pleistocene glaciations, may have contributed much to this generic poverty. Perhaps also New Zealand has always been much more isolated, lacking the stepping stone facilities that may have been afforded to New Caledonia in the past by the Chesterfield Reefs and New Guinea. The New Caledonian flora may also have been augmented recently through its approach to the Outer Melanesian Arc through the New Hebrides. Another factor that may be responsible for the richness of the New Caledonian flora is the dissimilarity between the ecological regimes of the ultramafic and sedimentary rocks which may virtually permit the existence of two floras in close proximity but with little competitive interaction between the two.

The late Tertiary cooling in New Zealand was not associated with increasing aridity as in Australia and thus there is no equivalent to the dry sclerophyll forests dominated by Myrtaceae and *Acacia*, though the latter did appear in the fossil record briefly in the Pleistocene (FLEMING). The forest vegetation of New Zealand today is dominated by *Nothofagus*, conifers and genera such as *Weinmannia* and *Metrosideros* (Myrtaceae) in varying associations according to latitude and relief (GODLEY, 1975). Over half the genera recorded from New

Zealand are endemic or otherwise restricted to Australasia and over half the 30% of worldwide genera are primarily temperate in nature (e.g. *Epilobium*) (BALGOOY, 1971). The flora lacks typical and important Australian genera in the families Myrtaceae, Rutaceae and Proteaceae, as well as lacking or having lost the families Casuarinaceae and Mimosaceae.

The history of the New Zealand flora has been summarised by FLEMING (1975). Palaeoaustral conifers were prominent in the vegetation throughout the Jurassic and Cretaceous. These were joined by typically southern angiosperms in the latter period. These elements remained strong in the early Tertiary but the Malayo-Pacific (mainly Indo-Malayan) element became increasingly important to dominate in the Miocene. This latter element died away after a climax in the Early Miocene and many of its immigrants became extinct, a decline likely to have been due to climatic deterioration mitigating against colonisation by essentially tropical organisms.

Interpretation of the relationships of the older Australasian floras must take into account the possibility that differences and resemblances between vegetation elements or whole floras are as much due to historical effects of climate as to past geography. On the one hand latitudinal correlations between Australia and a hypothetically more integral Inner Melanesian Arc from New Guinea to New Zealand might arise through polarisation of temperate and tropical elements during a general climatic deterioration or one brought about by a change in overall latitude following movement of tectonic plates. There are indications from the New Zealand fossil record discussed above that such an event may have served to decrease the similarity between the floras of New Zealand and New Caledonia.

On the other hand there are indications that the development of greater aridity on the Australian continent has led to the decline of more humid vegetation types similar to those of New Caledonia and New Zealand today. Genera common to all three areas might have been lost in Australia which, unless recorded as fossils, would lead to apparently exclusive relationships between New Zealand and New Caledonia.

SMITH (1974), with reference to other literature, wrote:

'It seems likely that parts of South America and New Guinea, suffering considerable orogeny, and the smaller island fragments of Gondwanaland origin (including Madagascar, Ceylon, New Zealand and New Caledonia), maintained more moist isothermal conditions throughout their histories. There is therefore a contrast between regions which have experienced increasing or fluctuating aridity in late Tertiary and Pleistocene times, and those which have never experienced serious aridity. This seems to have been the main factor involved in the selection of different taxa in different areas from the originally common, though not necessarily uniform, flora of late Cretaceous Gondwanaland. Thus while taxa like the Philesiaceae, Proteaceae and Restionaceae are typically common in the drier areas of southern land-masses, Winteraceae, *Nothofagus* and *Podocarpus* typify the wetter more isothermal regions. The Indian element in northern Australia is the result of selection, on both continents, of the same elements from the same basic Gondwanaland flora, under the same (monsoonal) conditions. Similarly the formerly mysterious floristic relationship between Madagascar and New Caledonia becomes clear when both islands are seen as fragments of the same original landmass, having experienced similar climatic conditions since their separation. Most southern plant disjunctions can be well explained in this fashion, by considering the effects of climate upon the evolution of an originally whole but subsequently sundered flora.'

Present day distributions of plants in Australasia should be examined with this in mind as well as in relation to the working hypothesis of geological history presented in the previous chapter.

## Distribution patterns in the present day Pacific flora

Before examining the flora of New Caledonia in detail it is useful to consider the distribution patterns of Pacific plants generally.

The Pacific fern floras were discussed by BROWNLIE (1965). The overall pattern is of a Malaysian flora becoming progressively depleted eastwards and southwards. Prolific development of ferns occurs when cloud forest is present, which excludes much of Polynesia and Micronesia. There is minor diversification in Fiji with 20% specific endemism, though the New Hebrides support a young flora derived mainly from Fiji and the Solomons (BRAITHWAITE, 1975) with specific endemism of 4.4%. There is some extension of 'Antarctic' New Zealand and Australian groups into the Pacific, such as that of *Polystichum* to the Society Is., but this was suggested to be a recent trend by BROWNLIE, spores from the source areas possibly being carried over long distances by the strong westerly upper winds (discussed by HOLLOWAY (1977)).

The Loyalty Is. have some thirty species, a typical weed flora with none of the New Caledonian endemic groups. Of New Caledonian fern species 42% are endemic and one quarter of these are truly distinct, the rest being mostly of Malesian derivation with minor diversification as in Fiji. In the lindsaeoid ferns there is a cytologically distinct old group that is associated with the ultramafics and other poor soils and a more modern Malesian group associated with richer soils on sedimentary rock, a trend that is found generally throughout the flora. A small third element of the fern flora is related to eastern Australian taxa and is absent from island groups to the north; this element has parallels in many other plant and animal groups.

In New Zealand there is also elaboration of ancient groups and a similar cytological distinction in *Lindsaea*. One third of the species are endemic but the relationships of those not belonging to ancient groups are with S.E. Australia and Tasmania. Half the species overall are shared with those areas.

Lord Howe I. also has a predominantly Australian fern flora. It has some endemic development in the Hymenophyllaceae, a phenomenon associated only with the older areas of the Pacific (E. Australia, New Zealand, New Caledonia and, to a much lesser extent, Fiji).

BROWNLIE divided the Pacific fern floras into four main elements: an ancient group of primitive families, relics of an old worldwide flora, that persists in New Caledonia and New Zealand; an intermediate flora, possibly derived at an early stage from Malesia, including the lindsaeoids and the Hymenophyllaceae mentioned, now based in E. Australia, New Caledonia and New Zealand; a recent Malesian group forming the bulk of Pacific floras, extending to E. Australia and thence probably to New Zealand; finally, a recent southern element evolved and based in E. Australia and New Zealand but extending into the Pacific. These broad categories are reflected in the geography of the rest of the flora and in most faunal groups.

The most complete assessment of Pacific phanerogam distribution is that of BALGOOY (1971). The distributions of all genera are tabulated with indications of centres of speciation, the various island groups are classified into botanical regions and provinces, and each group is discussed in terms of the representation of genera of various distribution types. BALGOOY produced his classification of island groups on the basis of genera shared, using a number of coefficients of resemblance, notably that of KROEBER. Figure 7 illustrates a single-link cluster analysis dendrogram and a minimum spanning tree (p. 153) derived from an array of PRESTON (1962) coefficients for the island groups using BALGOOY's data, but omitting the island groups with the smallest floras (consisting almost entirely of widespread genera of strand species) as discussed on p. 152. The results are very similar to those gained from the KROEBER coefficient but the PRESTON measure tends to give a more satisfactory comparison of floras where one is much larger than the other (p. 150 *et seq.*).

BALGOOY did not include comparisons with New Guinea or areas of Australia or South America in his figures so it is not possible to 'anchor' the classification to the principal source areas for the Pacific floras. However, from figures in THORNE (1965) and GOOD (1974), it is possible to derive coefficients between New Guinea and New Zealand (0.75, and, of the 165 genera shared, 120 are also in Tasmania), New Guinea and New Caledonia (0.65) and Queensland and New Caledonia (0.53) which can be compared with those for the BALGOOY data in Table 1. The second is lower than the first, implying greater similarity of the New Guinea flora to that of New Caledonia. This similarity is also greater than that between the New Caledonian and New Zealand floras (0.80), but less than that between the New Caledonian flora and that of most of the rest of Melanesia and Polynesia. The link between New Caledonia and Queensland is lower, indicating greater similarity, than any other New Caledonian link except those with the Loyalties and the New Hebrides. This suggests that the distinction of New Caledonia from the rest of Melanesia and Polynesia is at least partially due to significant and largely exclusive interaction with the flora of Australia.

Figure 7 shows up several groupings of Pacific archipelagoes, groupings that are also obtained from a single-link analysis of the table of KROEBER coefficients in BALGOOY. Notable groups are: the Bismarcks and Solomons; Fiji, Tonga, Samoa and the New Hebrides (more recent collecting and research in the New Hebrides (CHEW, 1975) has increased the number of genera recorded and reduced the coefficients with Fiji, Tonga and Samoa relative to that with New Caledonia and those with the Bismarcks and Solomons, thereby enhancing this grouping, i.e. 0.34, 0.36, 0.35 relative to 0.44, 0.46 and 0.40 respectively as compared with Table 1); most of Polynesia except Rapa I.; the Carolines and Marianas in Micronesia; the New Zealand continental shelf

*Table 1.* Matrix of PRESTON coefficients for Pacific island groups in terms of their floras.

| Sols | StCr | NHeb | NCal | Loys | LdHw | Norf | Kerm | NZld | SAnt | Chat | Boni | Mari | WCar | ECar | Fiji |
|---|---|---|---|---|---|---|---|---|---|---|---|---|---|---|---|
| 0.29 | 0.53 | 0.47 | 0.60 | 0.58 | 0.74 | 0.78 | 0.88 | 0.90 | 0.98 | 0.97 | 0.72 | 0.53 | 0.47 | 0.49 | 0.48 |
| | 0.40 | 0.40 | 0.59 | 0.58 | 0.78 | 0.79 | 0.87 | 0.90 | 0.97 | 0.94 | 0.72 | 0.52 | 0.45 | 0.47 | 0.40 |
| | | 0.59 | 0.55 | 0.75 | 0.91 | 0.96 | 0.98 | 0.95 | 0.98 | 0.99 | 0.88 | 0.75 | 0.67 | 0.68 | 0.56 |
| | | | 0.45 | 0.47 | 0.79 | 0.79 | 0.81 | 0.87 | 0.98 | 0.92 | 0.69 | 0.53 | 0.52 | 0.50 | 0.36 |
| | | | | 0.10 | 0.63 | 0.64 | 0.71 | 0.80 | 0.89 | 0.86 | 0.70 | 0.52 | 0.61 | 0.58 | 0.50 |
| | | | | | 0.68 | 0.67 | 0.83 | 0.88 | 0.96 | 0.91 | 0.71 | 0.55 | 0.62 | 0.63 | 0.54 |
| | | | | | | 0.53 | 0.64 | 0.70 | 0.84 | 0.76 | 0.72 | 0.79 | 0.82 | 0.83 | 0.73 |
| | | | | | | | 0.57 | 0.69 | 0.92 | 0.78 | 0.79 | 0.78 | 0.81 | 0.82 | 0.79 |
| | | | | | | | | 0.53 | 0.76 | 0.63 | 0.84 | 0.88 | 0.90 | 0.90 | 0.84 |
| | | | | | | | | | 0.20 | 0.10 | 0.87 | 0.89 | 0.93 | 0.91 | 0.88 |
| | | | | | | | | | | 0.57 | 0.97 | 0.98 | 0.97 | 0.98 | 0.98 |
| | | | | | | | | | | | 0.92 | 0.99 | 0.94 | 0.97 | 0.94 |
| | | | | | | | | | | | | 0.63 | 0.71 | 0.73 | 0.66 |
| | | | | | | | | | | | | | 0.43 | 0.49 | 0.48 |
| | | | | | | | | | | | | | | 0.36 | 0.51 |
| | | | | | | | | | | | | | | | 0.47 |

areas, together with the Kermadecs, Lord Howe I. and Norfolk I. The Juan Fernandez Is. link in at a high level with the Chatham Is. and generally have their lowest coefficients with the New Zealand group (though it has not been possible to make a similar numerical comparison with temperate South America), perhaps a reflection of the general 'Antarctic' nature of the flora. The New Zealand group is distinguished at a higher level in the analysis from a broadly tropical Pacific cluster including all areas from the Bismarcks to New Caledonia, Rapa I. and the Marquesas.

Though the New Caledonia-Loyalties pair falls within this tropical Pacific cluster in the single-link analysis it is somewhat distinct from the main west to east trend from Melanesia to Polynesia. It is distinguished by the large number of endemic genera and much stronger ties with Australia and, to a lesser extent, New Zealand. It unites the New Zealand cluster with the Pacific cluster in the analysis. This distinctiveness becomes more apparent at values of $k$ greater than 1 in the $B_k$ (overlapping) classification system described on p. 152. Melanesian and Polynesian areas continue to cluster together prior to the association of New Caledonia and the Loyalties with them. For example, at dissimilarity level 0.45, when New Caledonia is linked with the New Hebrides, the latter is already a member of a $B_3$ cluster with the Solomons, Fiji, Tonga and Samoa. At level 0.47 the Loyalties and New Caledonia form a completely linked triplet with the New Hebrides that overlaps through the last with a $B_3$

Number of genera in each is indicated by figures in brackets after the name of each area.

| Samo | Tnga | Cook | Scty | Tbai | Rapa | NTua | STua | Marq | Hwii | ReGi | Ccos | Glpg | JFrn | |
|---|---|---|---|---|---|---|---|---|---|---|---|---|---|---|
| 0.49 | 0.50 | 0.58 | 0.62 | 0.68 | 0.76 | 0.65 | 0.70 | 0.71 | 0.81 | 0.87 | 0.88 | 0.86 | 0.98 | **Bismarcks (632)** |
| 0.41 | 0.48 | 0.58 | 0.58 | 0.67 | 0.73 | 0.66 | 0.68 | 0.65 | 0.81 | 0.88 | 0.88 | 0.88 | 0.97 | **Solomons (654)** |
| 0.63 | 0.67 | 0.82 | 0.77 | 0.80 | 0.89 | 0.83 | 0.83 | 0.83 | 0.91 | 0.98 | 0.96 | 0.95 | 1.00 | **St. Cruz (126)** |
| 0.38 | 0.42 | 0.49 | 0.51 | 0.56 | 0.68 | 0.59 | 0.64 | 0.61 | 0.76 | 0.88 | 0.89 | 0.87 | 0.98 | **New Hebrides (396)** |
| 0.54 | 0.48 | 0.48 | 0.60 | 0.51 | 0.67 | 0.56 | 0.61 | 0.62 | 0.74 | 0.84 | 0.92 | 0.83 | 0.93 | **New Caledonia (655)** |
| 0.56 | 0.49 | 0.61 | 0.62 | 0.60 | 0.71 | 0.58 | 0.63 | 0.70 | 0.78 | 0.86 | 0.88 | 0.83 | 0.95 | **Loyalties (262)** |
| 0.77 | 0.75 | 0.77 | 0.77 | 0.76 | 0.71 | 0.83 | 0.62 | 0.76 | 0.77 | 0.90 | 0.92 | 0.88 | 0.86 | **Lord Howe (138)** |
| 0.73 | 0.73 | 0.72 | 0.75 | 0.79 | 0.73 | 0.80 | 0.78 | 0.80 | 0.78 | 0.88 | 0.89 | 0.85 | 0.85 | **Norfolk (104)** |
| 0.81 | 0.85 | 0.80 | 0.80 | 0.82 | 0.75 | 0.87 | 0.83 | 0.83 | 0.80 | 0.90 | 0.93 | 0.88 | 0.78 | **Kermadecs (61)** |
| 0.89 | 0.90 | 0.86 | 0.86 | 0.88 | 0.81 | 0.91 | 0.88 | 0.89 | 0.82 | 0.94 | 0.95 | 0.89 | 0.76 | **New Zealand (344)** |
| 0.96 | 0.99 | 0.97 | 0.91 | 0.97 | 0.88 | 0.98 | 0.96 | 0.95 | 0.85 | 0.98 | 1.00 | 0.98 | 0.79 | **Subantarctic Is. (78)** |
| 0.94 | 0.97 | 0.93 | 0.91 | 0.94 | 0.86 | 0.95 | 0.95 | 0.93 | 0.85 | 0.96 | 1.00 | 0.94 | 0.75 | **Chatham Is. (106)** |
| 0.69 | 0.68 | 0.68 | 0.68 | 0.72 | 0.70 | 0.76 | 0.73 | 0.70 | 0.74 | 0.88 | 0.89 | 0.83 | 0.94 | **Bonins (164)** |
| 0.51 | 0.46 | 0.52 | 0.54 | 0.54 | 0.67 | 0.53 | 0.58 | 0.62 | 0.75 | 0.83 | 0.86 | 0.79 | 0.95 | **Marianas (215)** |
| 0.50 | 0.49 | 0.58 | 0.62 | 0.61 | 0.74 | 0.63 | 0.67 | 0.69 | 0.83 | 0.89 | 0.88 | 0.86 | 0.97 | **W. Carolines (336)** |
| 0.47 | 0.53 | 0.60 | 0.61 | 0.65 | 0.74 | 0.70 | 0.72 | 0.71 | 0.82 | 0.87 | 0.87 | 0.87 | 0.97 | **E. Carolines (228)** |
| 0.26 | 0.22 | 0.44 | 0.47 | 0.58 | 0.70 | 0.53 | 0.60 | 0.57 | 0.75 | 0.85 | 0.90 | 0.84 | 0.95 | **Fiji (476)** |
| | 0.33 | 0.43 | 0.44 | 0.54 | 0.66 | 0.54 | 0.58 | 0.56 | 0.76 | 0.89 | 0.87 | 0.84 | 0.97 | **Samoa (302)** |
| | | 0.44 | 0.48 | 0.52 | 0.70 | 0.47 | 0.52 | 0.61 | 0.77 | 0.86 | 0.87 | 0.83 | 0.96 | **Tonga (263)** |
| | | | 0.34 | 0.42 | 0.58 | 0.47 | 0.46 | 0.50 | 0.71 | 0.84 | 0.86 | 0.84 | 0.96 | **Cook Is. (126)** |
| | | | | 0.40 | 0.55 | 0.43 | 0.37 | 0.39 | 0.68 | 0.85 | 0.89 | 0.84 | 0.92 | **Society Is. (201)** |
| | | | | | 0.59 | 0.47 | 0.43 | 0.49 | 0.73 | 0.85 | 0.90 | 0.88 | 0.95 | **Tubuai Is. (88)** |
| | | | | | | 0.76 | 0.63 | 0.57 | 0.69 | 0.83 | 0.89 | 0.85 | 0.87 | **Rapa I. (93)** |
| | | | | | | | 0.29 | 0.56 | 0.74 | 0.82 | 0.88 | 0.88 | 0.98 | **N. Tuamotus (70)** |
| | | | | | | | | 0.52 | 0.69 | 0.82 | 0.89 | 0.83 | 0.96 | **S. Tuamotus (81)** |
| | | | | | | | | | 0.66 | 0.83 | 0.87 | 0.83 | 0.94 | **Marquesas (113)** |
| | | | | | | | | | | 0.82 | 0.90 | 0.78 | 0.86 | **Hawaii (226)** |
| | | | | | | | | | | | 0.86 | 0.66 | 0.94 | **Rev. Gigedos (86)** |
| | | | | | | | | | | | | 0.82 | 0.98 | **Cocos (60)** |
| | | | | | | | | | | | | | 0.90 | **Galapagos (190)** |
| | | | | | | | | | | | | | | **Juan Fernandez (89)** |

cluster that contains all Melanesia, most of Polynesia and the W. and E. Carolines. At level 0.50 the pair is included with the tropical Pacific cluster when $k = 3$ but excluded when $k = 4$; the six Melanesian archipelagoes show complete linkage at that level. A similar result is obtained using the KROEBER coefficient.

A classification of this nature is not very informative considering that most island floras are a disharmonic blend of elements derived from a variety of source areas. Thus the Bonins are separated from the rest of Micronesia by virtue of a Palaearctic element. It is more fruitful to adopt BALGOOY's other approach and classify the plant genera into distribution types. Each area can then be considered in terms of these.

BALGOOY's distribution types were derived subjectively but there are numerical techniques available whereby genera can be classified in terms of both their overall distribution and their centres of species richness (HOLLOWAY, 1969, 1974). The number of plant genera involved in the Pacific (1,666) renders a complete analysis impracticable but there might be scope for performing analyses on more restricted sets of data such as random samples of a more manageable size. Alternatively some of BALGOOY's subjectively derived groups could be analysed to test their distinctiveness from each other or their homogeneity. Such an analysis is beyond the scope of this book and BALGOOY's categories will be taken at face value though the segregation of

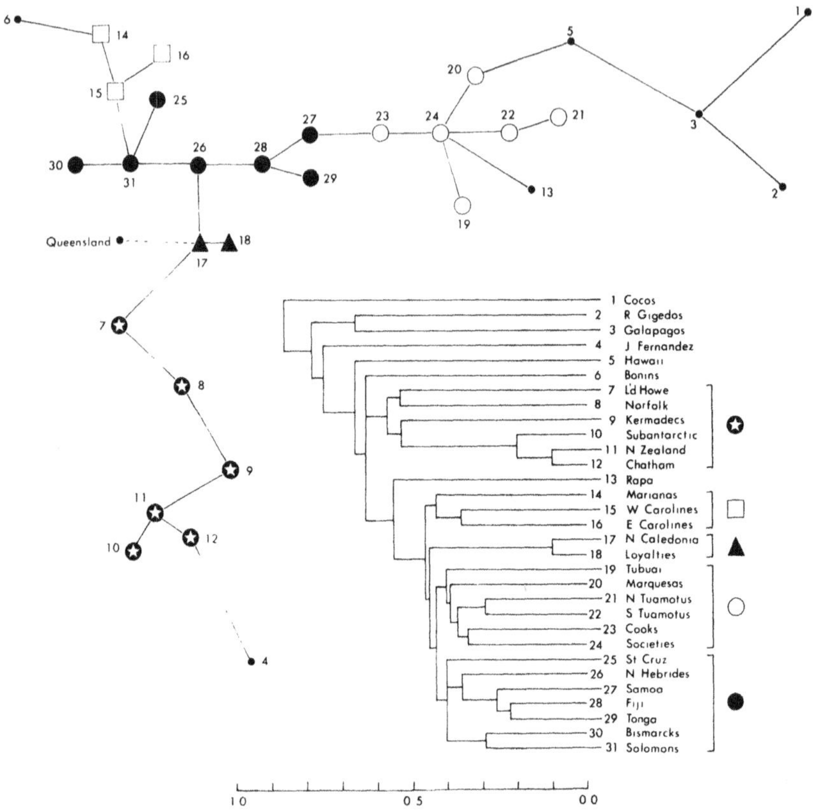


*Fig. 7.* A single-link dendrogram and minimum spanning tree derived from an array of PRESTON coefficients (Table 1) indicating the dissimilarities of Pacific island groups in terms of plant genera. The various clusters on the dendrogram are indicated by symbols in the minimum spanning tree where the lengths of the lines connecting the points indicate the dissimilarity values. The dissimilarity value between New Caledonia and Queensland is included for comparison. The scale at the bottom indicates the dissimilarity levels for the dendrogram but the minimum spanning tree is on a smaller scale.

some Melanesian genera (categories 5b and 6a) will be examined more critically (p. 72); numerical confirmation of the validity of these would be valuable.

For the purposes of this chapter BALGOOY's groups 1–5a can be taken together, being categories of more or less widespread tropical genera but including also a few temperate and tropical montane taxa. These all extend through the Indo-Australian tropics and are, from the New Caledonian viewpoint, probably mostly of Malesian derivation. These account for 397 New Caledonian genera (60.6%) out of 655. A further 51 (7.8%) are distinctly Australian genera. Thirty-six (5.5%) are Australian/Melanesian genera but not specifically centred in either area and a further 14 (2.1%) are Melanesian genera with a distribution extending broadly from the Moluccas to Fiji. These two categories are poorly distinguished from each other and several ambiguous cases occur. The Subantarctic genera (categories 7, 7a and 7b) represent the more widely distributed temperate austral elements but only 21 (3.2%) such genera occur in New Caledonia. There are two genera characterised by BALGOOY as Neotropical (*Epistephium* of the Orchidaceae and *Lindenia* of the Rubiaceae).

Finally there are 31 genera (4.7%) that are found in two or more Pacific island groups and 104 (15.8%) endemic genera. Many of the former are shared only with the Loyalties or extend weakly to the New Hebrides and have probably expanded from a centre of radiation in New Caledonia. Fifteen do not occur in Australia or Malesia, five are represented in both, six only in Malesia and five only in Australia. Thirty-two of the endemics are allied to genera of the Malesian and more widespread categories, nine are of Australian affinity and 15 are allied to Pacific or Subantarctic genera. Three endemics are of Neotropical affinity and 46 are of obscure alliance.

The widespread genera are probably mostly late Tertiary colonists of New Caledonia if one can draw parallels with the history of New Zealand, and it is perhaps within the genera of more restricted distribution and amongst the families known to have an ancient history in Australasia that patterns representing more ancient dispersals and geographical associations can be detected.

## Ancient austral families in the New Caledonian flora

In the absence of a fossil record for the New Caledonian flora the constitution of the flora at the end of the Cretaceous and beginning of the Tertiary can only be guessed at by examining the present day representation of families that are known to have radiated or been present in Australasia at that time or earlier. Notable in this respect are the southern conifers, *Nothofagus* and the Proteaceae. Also present, mentioned at the beginning of this chapter, were the Cruciferae, Liliaceae, Lauraceae (in Antarctica and possibly *Beilschmiedia* in the Late Cretaceous of New Zealand), Casuarinaceae and Myrtaceae. Sapindaceae were present in Australia at the beginning of the Tertiary and the Malvaceae, Chloranthaceae, Araliaceae, Epacridaceae, Palmae, Meliaceae and Loranthaceae were present in Eocene New Zealand. RAVEN & AXELROD (1974) suggested further families had been involved in Cretaceous austral radiations, namely the Santalaceae, Icacinaceae, Epacridaceae, Thymeleaceae, Winteraceae, Monimiaceae, Dilleniaceae and the ancestors of the New Caledonian endemic Strasburgeriaceae and Oncothecaceae.

The conifers, *Nothofagus* and the Proteaceae will be discussed in detail towards the end of this section and the Loranthaceae were mentioned on p. 37.

The Cruciferae have four genera endemic to New Zealand, one shared between New Zealand and Tasmania but none in New Caledonia that do not belong to widespread genera. The New Zealand genera are restricted to South

Island today and, if descendants of the ancient radiation, suggest that the Cretaceous austral Cruciferae favoured temperate climates and thus would have been unlikely to have been successful in a tropical New Caledonia.

The Liliaceae include *Astelia*, mentioned on p. 37, and *Xeronema* (Plate 1), discussed on p. 35. The New Caledonian species of *Xeronema* is restricted to low montane maquis on the ultramafic rocks. There are also three Australian genera that extend to New Caledonia (*Eustrephus*, *Stypandra* and *Lomandra*) but these might be later colonists.

The New Caledonian Lauraceae have recently been revised by KOSTERMANS (1974). Almost all species are endemic, but the majority belong to genera that are widespread in the Malesian tropics today. They are all trees or shrubs. Two species of *Beilschmiedia* occur in New Caledonia (out of about 250 in the genus), both growing on, and one restricted to, the ultramafics. *Cryptocarya*, a worldwide tropical genus (including Hawaii), has 17 species out of a total of 350 in New Caledonia, mostly characteristic of forest both on sedimentary and ultramafic rocks. *Litsea* is similarly widespread, though absent from Hawaii, and is centred in Malesia with subcentres in New Guinea and New Caledonia. The genus contains 500 species with 14 endemics in New Caledonia that are almost entirely restricted to the ultramafic rocks and many are very localised, for example restricted to a few of the western massifs. The genus *Adenodaphne* is endemic but related to *Litsea*. It contains four species largely restricted to the ultramafics. *Endiandra* is restricted to Malesia and Australasia with subcentres in New Guinea and E. Australia. It contains 90 species of which six are endemic to New Caledonia, growing both on the ultramafics and sedimentary rock. There are indications both of mobility and of a long Australasian history (perhaps as part of the earliest Indo-Malayan penetration) in the current distribution of the family, though there is nothing to point to exclusively austral radiation at this very general level of examination. Such radiation as has occurred is perhaps best referred to the 'new' adaptive radiation discussed in the next section.

The Casuarinaceae include *Casuarina s. s.* and *Gymnostoma*. The former, if one excludes the extremely widespread strand species *C. equisetifolia* which extends from E. Africa to the Marquesas, is centred in Australia and extends to New Guinea and Malesia in the west and to New Caledonia in the east (e.g. *Casuarina collina* in Plate 33). *Gymnostoma* has centres in New Guinea and New Caledonia and also extends from Malesia to Fiji, but has only one species in Queensland. In New Caledonia its species dominate several of the ultramafic vegetation associations (Plates 16, 18, 19, 35) and it is an important component of vegetation of 'Australasian' character elsewhere in its range (p. 83). The situation in the Casuarinaceae is open to interpretation as arising from a relatively recent expansion of genera that had radiated for some time in isolation in Australia (*Casuarina*) and in the tropical part of the Inner Melanesian Arc (*Gymnostoma*). However, the fossil record indicates that Casuarinaceae were more widely distributed in southern floras in the past and the present situation may represent only a relict of a much more complex distribution pattern.

The Myrtaceae of New Caledonia are predominantly, as in New Zealand, of the more widespread, and therefore possibly more dispersive, Myrtoideae, but the Leptospermoideae, such as *Callistemon* and *Xanthostemon*, are also prominent. *Callistemon*, with four endemic species (figures for endemic species are from GUILLAUMIN (1964) unless otherwise stated), is shared only with S. and E. Australia. *Xanthostemon* (Plate 2) is centred in New Caledonia with 31 endemic species, many associated with the ultramafics. It only occurs in the north of Australia but extends through much of the Papuan Region and is recorded from the Philippines. It could be regarded as an expanded Inner Melanesian Arc genus similar to *Gymnostoma*. Other Melanesian genera are *Cloezia* (*Mooria*), centred in New Caledonia (12 endemics) but occurring in the Philippines and New Guinea (one species each) and Fiji, and *Mearnsia* (one endemic; Plate 2) with a more even Melanesian distribution.

On the other hand *Austromyrtus* is equally centred in New Caledonia and E. Australia with weak extension to the New Hebrides and Fiji. *Austromyrtus* and the endemic *Myrtastrum* and *Uromyrtus* are segregates of *Myrtus* (BALGOOY) and together total 29 endemic species in New Caledonia. The genera *Baeckea* (six endemics) and *Melaleuca* (one endemic) show a similar relationship with Australia though are more definitely centred there. The former (Plate 2) has species dominant in habitats of maquis vegetation throughout New Caledonia and is a component of 'Australasian' vegetation associations in Malesia. The latter includes *M. quinquenervis* (niaouli, Plate 3), a dominant of many different vegetation associations in New Caledonia, including the savannas of the west.

There are three other endemic myrtaceous genera. *Arillastrum* (Plate 24) and *Purpureostemon* are monotypic but *Cupheanthus*, probably a derivative of *Eugenia*, has six species. *Eugenia* is worldwide in distribution with subcentres in the Oriental Region and New Guinea and has 87 species endemic to New Caledonia (including the genera *Jambosa*, *Syzygium* and *Caryophyllus* as in BALGOOY). Though New Caledonia parallels New Zealand in the proportions of subfamilies represented and in speciation in the genus *Metrosideros*, it has generally experienced more interaction with Australia as far as Myrtaceae are concerned.

The Sapindaceae contain a similar mixture of genera to the Myrtaceae though it is more predominantly Malesian in nature. There is a subcentre of the Malesian genus *Guoia* (nine endemics). The endemic *Podonephelium* and *Gongrodiscus* (two species each) are related to the Malesian *Alectryon* (one species endemic to New Caledonia) and *Mischocarpus* respectively. *Cupaniopsis* is Australasian with centres in Australia, New Guinea and New Caledonia (21 endemics). The endemic monotypic *Loxodiscus* is of Australian affinity and *Storthocalyx* (four species) is related to *Euphorianthus* (Philippines, Moluccas, New Guinea, Bismarcks, New Hebrides), a Melanesian Arc association. The family is primarily tropical (HUTCHINSON, 1973).

The Malvaceae have no marked representation in New Caledonia but there is a group of related genera in Australia and New Zealand that is probably part of an old radiation. New Caledonia is similarly excluded from apparently ancient radiations (from current distribution patterns) in the Stylidiaceae,

*Plate 2.* Top left, ***Strasburgeria*** on Mt. Mou; top right, ***Baeckea ericoides;*** bottom left, ***Mearnsia*** in flower on Mt. Panié; bottom right, ***Xanthostemon aurantiacum*** beside Grand Lac. The plate should be viewed sideways.

Centrolepidaceae, Restionaceae and Philesiaceae. There may be ecological reasons for such exclusion (SMITH, 1974 and p. 39).

The Chloranthaceae are a tropical and south-temperate family (HUTCHINSON, 1973) but BALGOOY recorded only the Malesian *Chloranthus* and *Ascarina* for the Pacific. Only *Ascarina* occurs in New Caledonia. It has a Melanesian Arc and Pacific distribution that excludes Australia but extends from the Philippines to New Zealand and the Marquesas.

The Araliaceae have a number of endemic genera in New Caledonia and several others are centred there. Three endemics are monotypic (*Botryomeryta*, *Pseudosciadium* and *Enochoria*), *Apiopetalum* (Plate 3) has four mainly montane ultramafic species, *Eremopanax* has ten species and *Myodocarpus* thirteen. *Meryta* is centred in New Caledonia but has single species in New Guinea, Queensland, the north of New Zealand, Norfolk I., eastern Melanesia and Polynesia to Rapa I. It shows some parallels with *Ascarina*, *Metrosideros* and *Astelia* in its overall distribution and these genera may represent an oceanic element, possibly fairly old, that is today unable to compete in larger areas with richer floras nearer the Malesian source or centre except perhaps in montane localities. *Pseudopanax* (*Nothopanax*, *Neopanax*) would have all the appearance of an old austral genus, centred in New Zealand with species in New Caledonia and South America, were it not for the occurrence of a species in China. Its presence also on the Kermadecs suggests mobility and the present distribution may therefore be relict from a wider, possibly Melanesian Arc distribution.

There are other genera with interesting distributions that do not occur in New Caledonia. *Osmoxylon* and *Plerandra* have Melanesian Arc distributions that exclude Australia. Hawaii has three endemic genera of Malesian affinity (*Tetraplasandra*, *Dipanax*, *Munroidendron*), a genus primarily centred there but shared with the Marquesas (*Cheirodendron*) and one (*Reynoldsia*) shared only with the Marquesas, Samoa and the Society Is.

The Araliaceae suggest from their general distribution that they have good dispersal abilities and, if not represented already in the ancient austral floras, may have provided some of the earliest Malesian colonists of Australasia, particularly of the outlying archipelagoes and, indeed, the Hawaiian group. The genus *Pittosporum* (Pittosporaceae) may also have been an early arrival, having subcentres in Africa, the Orient, Australia, New Caledonia (42 endemic species), New Zealand and Hawaii. The Meliaceae and the Loranthaceae may be added to this category. The former includes *Dysoxylum*, mentioned on p. 37, with a subcentre in New Caledonia (25 endemics), and the endemic *Anthocarapa* which has palaeotropical affinities.

Similar early radiations may have occurred in the Palmae. In the Eocene of India (before contact with Asia) there were possibly 31 species (LAKHANPAL, 1970). Pollen referable to *Rhopalostylis* has been found in the Eocene of New Zealand (FLEMING, 1975), suggesting that modern genera could have been in the process of segregation early in the Tertiary.

The world palm flora has been reviewed by MOORE (1973a, b) who has also collected in New Caledonia and published on the flora there (MOORE, 1976,

*Plate 3.* Top left, *Apiopetalum* (left) and *Greslania montana* on Mt. des Sources; top right, niaouli (*Melaleuca quinquenervia*); bottom left, *Araucaria bernieri* and *A. muelleri* marking a water seepage amid *Baeckea*-sedge maquis near Grand Lac; bottom right, *Araucaria montana* at an intermediate altitude on Mt. Panié.

1978). The following discussion is based on this literature and on comments on the taxonomy and ecology of New Caledonian palms kindly sent to the author in correspondence by Professor MOORE.

Palms are mostly poorly adapted to truly arid situations but are often the principal components of vegetation types associated with wetter situations. Apart from the strand genus *Cocos*, the New Caledonian palms are either coryphoids or arecoids, mostly the latter. Coryphoids are found from low to high elevations, often in areas of seasonal contrast or where arid conditions prevail. They are also found as swamp or rainforest palms, particularly in the Indo-Australian tropics. The arecoids are rainforest palms usually found at low or moderate elevations but occasionally in cloud forest at higher elevations.

The only coryphoid palm in New Caledonia is the monotypic *Pritchardiopsis*, part of the pantropical *Livistona* alliance which is widespread in the Indo-Australian tropics. *Pritchardiopsis* was found in the vicinity of the Bay of Prony in the south of the island and is now presumed extinct.

MOORE (1973b), in his distribution maps of the various arecoid alliances, indicated that three were recorded from New Caledonia but subsequently (*in litt.*) he excluded the *Ptychosperma* alliance. BALGOOY recorded one species of *Veitchia*, a genus of this alliance, from New Caledonia, but it is clearly an introduction. *Veitchia* is centred on Fiji (6–8 species) with six in the New Hebrides and one in each of Tonga and the Philippines (Palawan). The genera of the alliance together have a distribution that correlates well with the presumed past geography of the Outer Melanesian Arc (p. 18). *Ptychosperma* itself is centred in New Guinea but extends from the Philippines to the Solomons and into Micronesia, *Ptychococcus* extends from E. Malesia to the Solomons, *Balaka* is centred on Fiji and Samoa and *Drymophloeus* extends from E. Malesia to Samoa. The distribution of the alliance as mapped by MOORE excludes E. Malesia.

The two main alliances found in New Caledonia are the *Clinostigma* alliance and the *Archontophoenix* alliance. The *Clinostigma* alliance includes a monotypic genus in the Mascarenes, one broadly Malesian, two in the Oriental Region and the rest, perhaps including the major development of the alliance, in the Melanesian Arcs (including as a northern outlier the monotypic genus *Satakentia* in the Ryukyu Is.). In areas of the Sunda Shelf it occurs with the related *Oncosperma* alliance which could have restricted its diversification there. There is a monotypic genus in the Marquesas (*Pelagodoxa*), two in Fiji (*Neoveitchia*, *Goniocladus*) and one on Lord Howe I. (*Lepidorrhachis*). The largest genus, *Heterospathe*, is broadly Melanesian but centred in New Guinea, and *Clinostigma* and *Physokentia* are eastern Melanesian in distribution, polytypic, the latter related to the New Caledonian genera. *Cyphosperma* and *Clinosperma* are shared between New Caledonia and Fiji and there are a few other genera centred in New Guinea. The alliance has adaptations for long distance dispersal.

In New Caledonia the *Clinostigma* alliance is represented by eleven named genera. *Alloschmidia*, *Campecarpus*, *Clinosperma*, *Cyphosperma*, *Lavoixia* and *Veillonia* are represented by single species, *Cyphokentia*, *Cyphophoenix* (shared with the Loyalties), *Brongniartikentia* and *Burretiokentia* with two, and *Basselinia* with several. A probable undescribed genus in the north has about four species. *Campecarpus, Clinosperma, Cyphokentia* and *Basselinia* are restricted to, or centred on the southern ultramafics, the last genus also occurring on the western massifs (JAFFRÉ, 1974). *Brongniartikentia* and *Burretiokentia* occur in the forest of both south and north while *Alloschmidia, Lavoixia* and *Veillonia* are found in the Panié and Ouaième ranges in the north.

The *Archontophoenix* alliance includes the genus *Rhopalostylis* and thus has been present in Australasia since the Eocene, probably representing the oldest radiation of palms in the area that persists to the present day. *Rhopalostylis* is found in New Zealand, the Chatham Is., the Kermadecs and Norfolk I. The monotypic *Hedyscepe* is restricted to Mt. Gower on Lord Howe I. *Archontophoenix* has two species and extends from northern New South Wales to northern Queensland. Four endemic genera occur in New Caledonia. The monotypic *Actinokentia* is largely southern and appears to be restricted to the ultramafics. *Chambeyronia* has two, possibly three, species and occurs throughout the island. *Kentiopsis* is also monotypic, limited to three populations in south-west central New Caledonia while *Mackeea* is northern.

The distribution of the alliance shows parallels with that of the genus *Dracophyllum* (Epacridaceae) discussed below, of the skink genus *Leiolopisma* (p. 244) and of the gekko tribe Carphodactylini (p. 245). Such distributions may be common to rainforest taxa that radiated throughout Australasia in the Late Cretaceous and Early Tertiary prior to the development of extensive archipelagic links with Malesia along the Melanesian Arcs in the Miocene. Present day distributions of such taxa in Australia are probably much more restricted than in the past due to the expansion of the arid zone and diminution of the tropical rainforest. The distributions of the *Clinostigma* and *Ptychosperma* alliances suggest radiation in Australasia after the links with Malesia had been established. Absence of the older austral radiations from New Guinea may be due to extinction through competition or they may have been precluded by the geography of New Guinea when the radiation occurred; its major emergence may not have commenced.

Lord Howe I. has endemic genera referable to both the New Caledonian palm alliances and a third, *Howea*, with two species, that falls into the arecoid *Linospadix* alliance with three other genera containing over fifty species distributed over N.E. Australia and New Guinea. Lord Howe has had a long geological and possibly subaerial history, the latter dating at least from the Upper Miocene (p. 24). Perhaps detailed examination of the taxonomic affinities of its palms, skink, *Dracophyllum* and other older endemics such as *Negria* (Gesneriaceae) will reveal whether the divergence of these taxa is consistent with participation at the onset of the old radiations to which they belong or whether subsequent colonisation after the Miocene volcanics is more likely.

The Santalaceae in New Caledonia include the monotypic *Amphorogyne* and *Daenikera*, endemic genera of Palaeotropical affinity. Both *Exocarpos* and

*Santalum* are broadly Australasian and Pacific with subgenera in Hawaii. *Exocarpos* has one section endemic to New Caledonia (*Terminales*) and two more of Melanesian distribution that occur there (*Sarcocalyx* and *Xyllophyllos*). The family may have been one of the earlier to radiate in the Melanesian tropics.

The tropical family Icacinaceae is widespread in Malesia and the two New Caledonian endemics, *Gastrolepis* and *Anisomallon*, are indicated by BALGOOY to be of Palaeotropical affinity.

The Thymeleaceae, a worldwide family, are perhaps more strongly Australasian with monotypic endemic genera *Deltaria* and *Solmsia* in New Caledonia. The genus *Lethedon* (*Microsemma*) is centred in New Caledonia with twelve endemic species but occurs also in the Loyalties and New Hebrides. There is one species in Queensland. This is one of four instances where there may have been spread, perhaps recent, from a New Caledonian centre across the Coral Sea to Queensland. The exchange has been mostly in the opposite direction as far as it is possible to tell. The other genera are *Balanops* (of the monobasic family Balanopaceae), with seven species in New Caledonia and one in each of Fiji, the New Hebrides and Queensland, *Coronanthera* (Gesneriaceae), with a centre in New Caledonia and species in the Solomons and Queensland, and *Baloghia* (Euphorbiaceae), weakly centred in New Caledonia with species in E. Australia, Lord Howe I. and Norfolk I. In the last case the spread could have been in the opposite direction but the Norfolk species is shared with New Caledonia, testimony of mobility amongst the New Caledonian members of the genus. However, one cannot exclude the possibility that these patterns could have resulted from the extinction of a comparable radiation in Australia, especially where rainforest taxa are involved. An example of this may be the genus *Argophyllum* (Escalloniaceae) with three species in the highlands of Queensland and seven in New Caledonia (*Pacific Plant Areas*, 2: 84). The taxonomic affinities of the species concerned could throw light on this but the author has been unable to locate literature on the genera concerned.

The Monimiaceae *s. l.* as set out by BALGOOY include *Laurelia*, known as a fossil from Cretaceous Australasia and now found only in New Zealand and South America, the New Caledonian monotypic endemics *Carnegiodoxa* and *Canaca*, and *Nemuaron* (two species), of Australian affinity. The monotypic endemic genus *Amborella* is often placed in a distinct family, Amborellaceae (e.g. by GUILLAUMIN, 1964). BALGOOY included the Trimeniaceae in his Monimiaceae but HUTCHINSON (1973) regarded them as separate. They include the African genus *Xymolos*, the Australian genus *Piptocalyx* and two Melanesian genera, *Trimenia* and *Sphenostemon*. *Sphenostemon* is perhaps referable to the Inner Melanesian Arc, occurring in New Guinea, N. Queensland, the Bismarcks and New Caledonia. *Trimenia*, extending through Sulawesi, New Guinea, the Bismarcks, Solomons, New Caledonia, Fiji, Samoa and the Marquesas, is more extensive in the Outer Arc. Finally the Monimiaceae include the genus *Hedycarya*, centred in New Caledonia with sixteen species but also occurring in E. Australia, New Guinea and Melanesia as far east as Samoa and Tonga, and New Zealand. The family, or group of families, predominates in the tropics and subtropics of the southern hemisphere (HUTCHINSON, 1973) and is probably a good example of an austral group strongly represented in New Caledonia.

The Dilleniaceae are a widespread tropical family that includes the genus *Hibbertia*, centred in Australia (100 species) and extending to the Moluccas and New Guinea (one species). There is a subcentre of 25 species in New Caledonia, one of which extends to Fiji, and there is a single species in Madagascar (*Pacific Plant Areas*, 2: 240).

Links between New Caledonia and New Zealand have been limited so far in the discussion to *Xeronema* and the affinities between *Elingamita* and *Tapeinosperma* in the Myrsinaceae. Even *Tapeinosperma*, though centred in New Caledonia with 40 species, has subcentres in the Solomons and Fiji and extends also to the New Hebrides, New Guinea, E. Australia and Borneo. Other examples are few, even within the really ancient austral families. Outside these families the following are of interest.

The tribe Coronantherinae of the Gesneriaceae (BURTT, 1962) includes *Coronanthera*, centred in, and *Depanthus*, endemic to New Caledonia, the monotypic *Negria* on Lord Howe I., and *Rhabdothamnus* in northern New Zealand. The occurrence of *Coronanthera* in the Solomons and Queensland suggests that this again may be a group that has been restricted from a more extensive austral distribution by increasing aridity in Australia. The affinities of this tribe are with South American taxa.

In the small parasitic family Balanophoraceae the endemic monotypic *Hachettea* of New Caledonia is most closely related to the endemic monotypic *Dactylanthus* of New Zealand. But the Malesian genus *Balanophora* extends into the Pacific as far as Rapa I. and has a subcentre in Australia, so the endemic genera could well be divergent relicts of the earliest radiation of the family into Australasia.

Finally there is the family Alseuosmiaceae, included with the Caprifoliaceae by BALGOOY but regarded as separate by GOOD (1974). This is comprised of two genera, *Memecyclanthus* and *Periomphale* (*Pachydiscus*), endemic to New Caledonia with a total of eleven species, and one genus of eight species, *Alseuosmia*, endemic to New Zealand.

Coming now to the oldest austral angiosperm families and the conifers, the Epacridaceae are almost entirely Australasian apart from the monotypic genus *Lebetanthus* in South America (VIROT, 1975). Of the 21 genera only five occur in New Zealand and two in New Caledonia. They all occur in Australia and include 400 species there. The New Caledonian genera are the two most widespread, *Dracophyllum* and *Styphelia*, and the species have recently been treated in detail by VIROT.

The New Caledonian Epacridaceae all require acid soils and are absent from limestone. They are primarily found on the ultramafics and their derivatives, secondarily on the metamorphic rocks. Almost all are species of the maquis associations.

The main centre of development in *Dracophyllum* today is in New Zealand where there are 35 species. The single Lord Howe I. endemic has its closest relative there (GOOD, 1974). Two species occur in Australia and Tasmania and seven in New Caledonia. One of these New Caledonian species, *D. involucratum* (Plate 1), has been placed in an endemic subgenus, *Cordophyllum,* by VIROT and is restricted to the southern ultramafics. The remainder are placed in the subgenus *Dracophyllum* which includes one of the Australian species and eight of the New Zealand ones. The majority are restricted to the ultramafics but *verticillatum* (Plates 43, 45) also occurs on metamorphic and sedimentary rocks and *ouaiemense* is found in maquis only on the summit of the Roches d'Ouaïème. The genus could be cited as an instance of a general Australasian distribution subsequently much restricted in Australia by the decline of moister habitats.

The genus *Styphelia* is divided into three sections, two of which occur in New Caledonia. One is monotypic and endemic (*Cyathopsis*), restricted to the western and southern ultramafics. The ten species of section *Leucopogon* are all restricted to ultramafic areas except *pancheri* and *cymbulae* which occur also on metamorphic rocks and are widespread throughout the island. The latter species occurs also in the southern New Hebrides and Fiji and is related to the widespread Malesian *malayana.* The section *Leucopogon* is centred in Australia where there are about 100 species but it occurs also in New Guinea and New Zealand and extends weakly into Malesia. The third section, *Cyathodes,* is centred in New Zealand but occurs in Australia, New Guinea and scattered localities in Polynesia as far east as Rapa, the Marquesas and Hawaii.

The Winteraceae (SMITH, 1945; BALGOOY) have their centre of diversification and specialisation in Australasia, their probable centre of origin. Six known genera occur there and only one in the New World. *Exospermum* and *Zygogynum* (seven species) are endemic to New Caledonia and *Belliolum,* related to *Bubbia,* is shared with the Solomons. *Bubbia* is centred in New Caledonia (twelve endemics) and occurs also on Lord Howe I., in Queensland, New Guinea and the Bismarcks. There is an outlying species in Madagascar suggested by RAVEN & AXELROD (1974) to have resulted from a recent long distance dispersal. The plausibility of this hypothesis will rest on the taxonomic relation of the Madagascan species to the Australasian ones. Both *Bubbia* and *Drimys,* recorded from Cretaceous Antarctic floras, are well represented in the highlands of New Guinea. *Drimys* now occurs in South America, Australia, the Moluccas and the Philippines as well as New Guinea. *Pseudowintera,* endemic to New Zealand, is the sixth genus though GOOD (1974) related the Fijian monobasic family Degeneriaceae to the Winteraceae.

The Proteaceae have recently been treated in detail by JOHNSON & BRIGGS (1963, 1975) and the New Caledonian representatives have been described by VIROT (1968). The early members of most tribes were probably trees of mesothermic, closed (Gondwanan) forests, xeromorphic characters in the present day flora being secondary and of multiple origin in the various southern hemisphere centres of the family. The tribal and subtribal differentiation at least occurred during the Cretaceous and thus predated most of the events in the break-up of Australasia-Antarctica. The New Caledonian flora contains representatives, usually endemic genera, of a high proportion of the tribes recognised and could therefore be said to be harmonic and consistent with the hypothesis of it having been derived from Australia prior to the opening of the Tasman Sea.

Again there are no exclusive links between New Caledonia and New Zealand. In the Knightiinae the New Caledonian *Eucarpha* (previously placed in *Knightia s. l.*) has two species widespread in the island, mainly on the ultramafics and often in forest associations. It is related to the New Zealand genus *Knightia* (which featured in the Antarctic flora) but also included in the subtribe is the Australian genus *Darlingia*, now restricted to the north-eastern tropical rainforest. The distribution of the subtribe parallels that of the gekko tribe Carphodactylini and other plant and animal taxa mentioned on p. 52.

Association with Australian tropical rainforest genera is found also in the New Caledonian Hicksbeachiinae. The genus *Virotia* (species previously referred to *Macadamia*) contains six species of diverse ecology but several belong to forest associations or are restricted to ultramafic localities. One species (*angustifolia*) is restricted to the small area of low, dense forest on the Dome de Tiebaghi (p. 118, Plate 38). There is one species from the Australian tropical rainforest tentatively referred to *Virotia* by JOHNSON & BRIGGS. Other genera of the subtribe, *Hicksbeachia* and *Athertonia*, also occur in this rainforest. *Malagasia* from Madagascar and *Heliciopsis* from Malesia complete the subtribe. The latter may have been derived from the austral flora of pre-contact southern India (JOHNSON & BRIGGS).

*Sleumerodendron* is a monotypic endemic genus, *S. austrocaledonicum* being recorded from scattered localities on the ultramafic and metamorphic rocks. It is related to *Gevuina*, a genus found in Australia and South America. In Australia *Gevuina* is restricted to the tropical rainforest. It occurs in New Guinea with distinctive outliers in Fiji, the New Hebrides and Samoa, possibly a relatively old Outer Arc dispersal. There is confusion in the literature as these species are referred to *Kermadecia* by BALGOOY.

*Kermadecia* is endemic to New Caledonia with four species found in a variety of ecological situations. The genus is placed by JOHNSON & BRIGGS in the Roupalinae with the tropical American genus *Roupala*.

*Garnieria* is another monotypic endemic restricted to the ultramafics. It is a member of the Persooniinae which have a monotypic genus *Toronia* in New Zealand and three in Australia that include a majority of non-rainforest species.

*Beauprea* (13 species) and *Beaupreopsis* (monotypic, a species of open forest and maquis of the southern ultramafics) belong to the Cenarrheninae in a subgroup completed by *Cenarrhenes*, a Tasmanian monotypic endemic associated with cool *Nothofagus* rainforest. Most *Beauprea* species are restricted to the ultramafics and several are mainly montane. There is palynological evidence of *Beauprea*-like taxa in New Zealand in the Cretaceous and Tertiary (COUPER, 1960a).

The two remaining New Caledonian genera are shared with Australia. *Grevillea* is centred in Australia but has an interrelated group of three species associated with the ultramafics in New Caledonia. The relation of these to Australian taxa suggested a Tertiary dispersal to JOHNSON & BRIGGS. *Grevillea* has extended to savanna habitats in New Guinea and Sulawesi, probably in the Pleistocene. *Stenocarpus* has twelve species in New Caledonia, predominantly associated with the ultramafics. The genus also occurs in New Guinea, north and eastern Australia and has affinities with the western Australian genus *Strangea*.

The distribution of the southern beech genus, *Nothofagus*, has long attracted the attention of phytogeographers (e.g. STEENIS, 1971; SMITH, 1974) and some work has been done on the associated fauna (SCHLINGER, 1974; DUGDALE, 1975). Modern species appear unable to disperse over more than very short distances as the seeds are not readily windborne and sink in sea water (CRANWELL, 1963; PREEST, 1963). The group has therefore been considered to provide strong evidence of past land connections, either through landbridges (STEENIS) or continental drift (SCHLINGER). The genus in the past and today has an exclusively austral distribution in Antarctica, South America, New Zealand, Australia, New Caledonia, New Guinea and New Britain. Records from the northern hemisphere of both *Nothofagus* and Podocarpaceae have been regarded by STEENIS (1971) and CRANWELL (1963) as erroneous or resulting from long distance wind dispersal of pollen.

In New Caledonia only the *brassii* group is represented of the three pollen types recognised. The absence of a fossil record means that the history of the group in New Caledonia can only be guessed at. The five or more species are restricted to the ultramafics where they tend to dominate various rainforest associations. They are generally montane and some are extremely local, such as *N. baumannii* (Plates 28, 29) on the summit of Mt. Mou. During the moth survey *N. codonandra* was noted in forests on the Mt. des Sources and beside the R. Bleue (Plates 26, 30), and *N. aequilateralis* was found in extensive stands in the valley bottoms of Pic du Pin (Plates 21, 22). JAFFRÉ (1974) recorded both *N. codonandra* and *N. balansae* as forest dominants in similar valley situations on the western Massif de Koniambo.

The *brassii* group is represented by its pollen or extant species throughout the range of the genus but today only occurs in New Guinea and New Britain as well as in New Caledonia. The pollen record in New Guinea (south-east) dates from the Upper Miocene. The *brassii* group became absent from the pollen record in South America in the Oligocene and in Australia and New Zealand in the Pliocene. The indications are that climatic deterioration and increasing aridity led to this extinction (p. 35), and the *brassii* group therefore probably requires humid tropical or subtropical conditions. Its largely montane habit suggests that it is tolerant of coolness but perhaps not of an extended cold season as has occurred throughout New Zealand in the recent past.

The origins of the New Caledonian and New Guinea *Nothofagus*, all of section *Bipartitae* and characterised by *brassii*-type pollen, can only be guessed at, as little is known of their means of dispersal which may have been very different in the early history of the genus. The *brassii* group was dominant over the species of the more temperate (today) *menziesi* and *fusca* pollen groups throughout Australasia until fairly late in the Tertiary, and today its sixteen or so New Guinea species dominate the montane forests there. The New Caledonian species are also forest dominants. STEENIS, from observations of regeneration of *Nothofagus* in New Guinea, has suggested that the *brassii* group also has poor means of dispersal. However, the New Britain species, *starkenborghii*, is shared with, and obviously derived from, New Guinea and must have crossed moderate water gaps to extend its range thus.

All these factors suggest two possible origins for New Caledonian *Nothofagus*. Firstly, they could be relicts from the general Australasian-Antarctic flora derived before the opening of the Tasman Sea. This would indicate the persistence of land in New Caledonia from prior to −80 my. The present isolation from Australia must be presumed to exceed by far the dispersal range of the genus. Secondly, they could have been derived from New Guinea if, recently emergent, it derived its *Nothofagus* from Australia over a comparatively small water gap in the Late Oligocene or Miocene before the *brassii* group became extinct there. From New Guinea the genus could have spread to New Caledonia along the Inner Melanesian Arc as suggested by GASKIN (1970), if it occurred in the form suggested in Figure 3 with only relatively small water gaps breaking the chain of land.

In the latter case one might expect a stronger floristic relationship between New Caledonia and New Guinea today than in fact occurs. Many exclusively Inner Arc relationships would no doubt have developed if a relatively sedentary genus such as *Nothofagus* could have spread from one to the other. Indeed the crossing would have occurred twice at least if the resemblance between two New Caledonian species and two New Guinea ones (including *starkenborghii*) noted by STEENIS (1971) is indicative of phylogeny. Against this apparent similarity STEENIS has noted that the New Caledonian species can be separated from the New Guinea ones by spiral versus distichous phyllotaxy.

On balance the first hypothesis is the more likely. The New Guinea species may well have been derived from Australia as suggested, unless it could be shown that land in south-east New Guinea had persisted from the Late Cretaceous. Similarities between the two *brassii* floras could be attributed to conservatism in evolution, especially if several *brassii* lines were involved in the initial colonisation of New Caledonia and persisted in Australia to spread ultimately to New Guinea. Pollen data from bottom sediments off New Caledonia might be very valuable in resolving this question and indeed might throw light on the history of other elements of the flora such as the conifers.

The New Caledonian conifers have recently been revised by LAUBENFELS (1972) who also (1969) produced a wider treatment of part of the family Podocarpaceae. The conifers have been cited as providing evidence of old adaptive radiation in New Caledonia by CARLQUIST (1974) who discussed their ecology in some detail.

Four families are represented. The Araucariaceae and Podocarpaceae are the most numerous but the Cupressaceae have three genera and the Taxaceae have a single endemic monotypic genus, *Austrotaxus*. *Austrotaxus spicata* is found in forests on sedimentary and metamorphic rocks and thus differs from the other conifer genera which tend to be centred on the ultramafics. Its closest relative is the S.E. Asian genus *Amentotaxus*. *Taxus* occurs today as far east as Sulawesi and the New Caledonian genus may therefore be descended from an old Malesian colonist.

The Cupressaceae include three species of *Libocedrus s. s.*, all forest species of the southern ultramafics. The genus also occurs in New Zealand, one of the few exclusive links between the two areas. Two species of the Australian-centred genus *Callitris* are again both restricted to the south, one associated with crests and plateaux, the other with fluviatile conditions. Probably derived from *Callitris* or the Australian genus *Actinostrobus* is the monotypic endemic *Neocallitropsis pancheri* which is both fluviatile and a feature of montane maquis on the southern slopes of the Mt. des Sources (Plate 4).

*Plate 4.* *Neocallitropsis* amid montane maquis on the Mt. des Sources.

The Araucariaceae are represented by both *Araucaria* and *Agathis*. *Agathis* occurs from Sundaland to New Zealand. From Sundaland to the Bismarck Is., excluding the Lesser Sundas, there is probably only one variable species and in the St. Cruz Is., the New Hebrides and Fiji there are three allopatric ones (WHITMORE, 1977). Three species grow in Queensland and one in the north of

*Plate 5.* *Agathis lanceolata* emergent from a skein of low gully forest in the valley of the R. Blanche.

New Zealand. No less than five occur in New Caledonia. *A. montana* is restricted to high altitudes on Mt. Panié (Plates 6, 49) and *A. ovata* (Plates 15, 21) and *A. lanceolata* (Plate 5) to the southern ultramafics, the former in open habitats usually near springs or ravines and the latter mainly in rainforest. Two other species, *A. moorei* and *A. corbassonii*, occur in central and northern forests on sedimentary and metamorphic rocks. Though found elsewhere in Australasia and having extended to the south of Australia earlier in the Tertiary, *Agathis* today is predominantly a genus of the Melanesian Arcs and its extension westwards into Malesia with other austral taxa suggests that it is moderately dispersive.

In part of its range the seeds are bored into by the larvae of moths of the monobasic family Agathiphagidae (WHITMORE, 1977). The family has morphological characteristics that are generally considered primitive (COMMON, 1970) and the pupae have active mandibles to assist the emergence of the adult from the seed. The genus *Agathiphaga* has two species, one in Queensland and the other extending from the St. Cruz Is. through the New Hebrides to Fiji. None have been recorded from New Caledonia to date.

The presence of *Araucaria* on Norfolk I. and on the recently emergent Loyalty Is. also suggests moderate to good ability to cross water gaps, at least for some species. The Loyalty species is *columnaris* (*cookii*) which is a

*Plate 6.* *Araucaria columnaris* on the coast at Port Boisé.

widespread coastal species on the main island and on the Isle of Pines. The Norfolk species has affinities with it.

The genus today is less extensive in Australasia than *Agathis*, having failed to penetrate the Outer Melanesian Arc and having died out in New Zealand and most of Australia. It is limited in the latter to two maritime species in Queensland (*Pacific Plant Areas*, 2: 188). There are two species in South America and four in New Guinea. No less than thirteen are recognised in New Caledonia, a number due largely to the recent revision by LAUBENFELS. The previous treatment by GUILLAUMIN & VIROT (1962) recognised only eight, as did CARLQUIST (1974) in his discussion of the ecology and radiation of the New Caledonian species.

These eight species are as follows. *A. columnaris* (Plates 6, 10) is essentially a coastal species. Apart from the localities already mentioned it also occurs on low coral islands of small size such as Améré, located between the mainland and the Isle of Pines. *A. subulata* (formerly *A. balansae*) grows at medium altitudes in the south, including areas off the serpentine (Plate 47). The very similar *A. bernieri* (Plates 3, 7 (probably), 17) is a forest emergent on southern

*Plate 7.* *Araucaria bernieri* emergent from valley rainforest on Pic Busé; a single *Agathis lanceolata* can be seen in the right foreground.

ultramafic hills above about 500 m, especially on the Mt. des Sources and a broad area to the south-east to Grand Lac. *A. biramulata* is local in the south, *A. humboldtensis* occurs at high altitudes on Mt. Humboldt and the Mt. des Sources, and *A. montana* grows on the more northerly western massifs and on the north-eastern mountains (Plate 3). *A. rulei* (Plate 29) is the counterpart of *montana* in the south, growing on ultramafic hills between 600 and 1200 m. *A. muelleri* (Plates 3, 17, 21, 23, 30) occurs at all altitudes on the southern ultramafics, can be associated with springs (as in Plate 3) and forms particularly dense and extensive stands on the Mt. des Sources where it is often codominant with *Nothofagus codonandra* (p. 105, Plate 30).

The five new species recognised by LAUBENFELS are either extremely local, such as *schmidii* from the summit platform of Mt. Panié, *nemorosa* from Port Boisé and *laubenfelsii* near Yaté, or are found in scattered coastal localities (*luxurians, scopulorum*).

The diversity of species within New Caledonian *Araucaria* led CARLQUIST (1974) to suggest that they must have evolved and diversified on a larger land mass than occurs at present. Some of this radiation could have occurred before the opening of the Tasman Sea, New Caledonia being colonised by several lines that have since become extinct in Australia. But New Caledonia could have been more extensive in the past. Shallow seas extend along a tongue of the Norfolk Ridge well to the north-west of New Caledonia. This tongue could have been exposed at times in the past during orogenesis or by a relatively small drop in sea-level as suggested in Figures 4 and 6. At other times higher sea-levels or a more 'dispersed' geological structure could have resulted in a more archipelagic New Caledonia. Both conditions could have led to diversification or speciation.

The Podocarpaceae are represented by three genera, *Acmopyle, Dacrydium* and *Podocarpus,* though LAUBENFELS has divided up the last two into a number more. These are referred to as subgenera in the treatment below; they do not necessarily tally with those in BALGOOY's distribution table or those referred to by CARLQUIST (1974).

HAIR (1963) has suggested an Antarctic origin for many podocarp groups, borne out by the fossil record (p. 34). The distributions of some genera and subgenera today may be largely relictual. *Acmopyle* only occurs today in New Caledonia (*pancheri,* a dry forest species restricted to the ultramafics and the Panié range) and Fiji (*sahniana,* a montane species) but is known as a fossil from South America, Antarctica and possibly southern India.

The podocarps are perhaps potentially the most dispersive of the southern conifers, their fleshy fruit being suited to dispersal by birds (PREEST, 1963). CARLQUIST (1974) noted the great ecological diversity of the New Caledonian representatives but the presence of several subgenera and sections that are generally more widespread indicates that this diversity has not arisen through radiation within the island but is possibly more a reflection of the harmonic nature of an old, colonising Cretaceous conifer flora.

In *Dacrydium* two subgenera are represented. *Falcatifolium* contains four largely allopatric species, one widespread in Sundaland, E. Malesia and the Philippines and the others endemic to Borneo, New Guinea and New Caledonia. The New Caledonian species is *taxoides,* a forest species particularly common at higher altitudes. The *Dacrydium* species *sensu stricto* fall into the tropical section which is centred on the Melanesian Arcs and extends into S.E. Asia. It is absent from Australia. Four species occur in New Caledonia, largely restricted to the ultramafics. One, *lycopodioides,* occurs also in Fiji and another, *araucarioides,* is an important component of certain low forest associations on the southern ultramafics (p. 99, Plates 8, 21).

Five subgenera of *Podocarpus* are present according to the subdivisions of LAUBENFELS (1972). The subgenus *Parasitaxus* is endemic and monotypic. *P. ustus* (Plate 8) is the only parasitic conifer in the world. It grows to a height of one or two metres, is purplish blue in tint and probably parasitic on other conifers. It is mainly found in the region of the Mt. des Sources and R. Blanche but also grows on the Panié range where other podocarps are common.

*Plate 8.* Top left, ***Podocarpus ustus*** in the forest on Mt. Pouédihi; top right, ***Podocarpus minor*** on the shore of Grand Lac; bottom left, ***Agathis montana*** on Mt. Panié; bottom right, ***Dacrydium araucarioides*** on Pic du Pin.

Subgenus *Dacrycarpus* is centred in New Guinea (five species) but extends to the Oriental Region. One species extends from S.E. Asia to Fiji and another is endemic to New Zealand. None occur in Australia and one, *vieillardii*, related to a New Guinea species, is endemic to New Caledonia where it is widespread on the ultramafics. *Prumnopitys*, distributed through Malesia and also in New Zealand and South America, includes one New Caledonian endemic of New Zealand affinity, *ferruginoides*, that is found mainly in forests on the ultramafics. *Decussocarpus* includes an African section of three species, an intriguing Oriental Region section of five species (perhaps akin to *Heliciopsis* in the Proteaceae; p. 56) and section *Decussocarpus* itself with three species. One, *vitiensis*, extends from the Moluccas through New Guinea to Fiji and the closely related *comptonii* is found in a variety of habitats in New Caledonia. The third species, *minor*, is also endemic to New Caledonia but morphologically distinct from the other two. It is restricted to lacustrine and riverine habitats in the extreme south (Plate 8, on the strand of Grand Lac).

Three sections of subgenus *Podocarpus* itself are represented in New Caledonia according to CARLQUIST (1974). LAUBENFELS described an interrelated group of three (the montane *polyspermus* and *longefoliatus* and the localised southern *decumbens*) that have affinities with a group widespread from Malesia to Fiji, *sylvestris*, a widespread forest species related to *neriifolius* (Nepal to Fiji), *lucienii* and *novaecaledoniae*, largely southern forest species but the former occurring on Mt. Panié, and *gnidioides*, localised in the south on rocky crests and related to Australian and New Zealand species.

Podocarpaceae were a principal component of Cretaceous and Early Tertiary Australo-Antarctic floras and their present restriction must be explained in terms of climatic changes and/or competition from more vigorous modern elements in the floras of the various areas. A combination of these factors was suggested by CARLQUIST in interpreting the history of the Australian flora. Apart from the increasing restriction of old rainforest taxa brought about by climatic changes (p. 35) there persist in the temperate forests of Tasmania several conifer genera that are absent from similar habitats on the mainland. CARLQUIST suggested that mainland Australia has acted as a greater target area for dispersing angiosperm groups, an area that has also functioned to increase diversification, especially of those groups adapting to arid habitats. The climate and somewhat greater isolation of Tasmania have perhaps shielded the older taxa from some of the pressures that resulted in their demise on the mainland.

The same argument can be applied to New Caledonia. Though several characteristically austral families are absent from the island (p. 47) it is evident, both from the range of taxa present and the degree of radiation and diversification in some, that it shared in the early development and radiation of the Australasian flora and that its isolation subsequently has led to the conservation of old Australasian warm rainforest elements that have become extinct or restricted in Australia through increasing aridity and likewise in New Zealand through climatic cooling. A further factor that may have contributed to this conservation is the adaptation of such elements to the unusual mineral regimes of the ultramafic soils. This adaptation would have already progressed considerably by the time potentially more competitive Malesian rainforest taxa began to colonise in numbers in the later Tertiary, affording the older taxa a degree of ecological isolation. New Guinea has probably served a similar function of conservation though to a lesser extent, the effect of its larger area and therefore holding capacity being outweighed by its geological youth and greater accessibility to Malesian taxa; areas of ultramafic rock are also relatively less extensive. Nevertheless a number of what are probably old Australasian rainforest taxa have enjoyed some success and have radiated, especially in montane habitats. In certain ecological situations they have even spread westwards into the Oriental Region, for example *Agathis* and some of the podocarp groups and Myrtaceae.

It is possible that a great number of other taxa dating from the Cretaceous have become extinct in New Caledonia. Of the genera endemic to New Caledonia (including the Loyalties) approximately half are monotypic and two thirds have three species or less. The most polytypic genera in the flora are those with widespread geographical distributions as illustrated in Table 2. The really distinctive elements of the New Caledonian flora where the relationships are obscure are almost entirely monotypic. THORNE (1965) listed fifteen such genera, namely *Amborella, Austrotaxus, Canacomyrica* (Myricaceae), *Dugezia* (Primulaceae), *Maxwellia* (Bombacaceae), *Neocallitropsis, Memecyclanthus* and *Periomphale* (Caprifoliaceae or Alseuosmiaceae), *Oceanopapaver* (Capparaceae), *Oncotheca* (HUTCHINSON (1973) consigned this genus to oblivion as far as family was concerned by means of a circular cross-reference), *Paracryphia* (Eucryphiaceae), *Phelline* (Aquifoliaceae), *Serresia* (Violaceae), *Solmsia* (Thymeleaceae) and *Strasburgeria* (Strasburgeriaceae) (Plate 2). *Periomphale* has two species and *Phelline* eleven but the rest are monotypic. Some authors (e.g. GOOD, 1974) place many of these in distinct families. The Strasburgeriaceae, Amborellaceae and Alseuosmiaceae have already been mentioned but GOOD referred also to the Oceanopapaveraceae, Oncothecaceae and Paracryphiaceae. The older elements of the rainforest flora may have been ousted progressively by subsequent more vigorous colonists though not to the same extent as may have occurred in Australia and New Guinea.

## The 'new' adaptive radiation in New Caledonia

The sixteen genera with 25 or more species endemic to New Caledonia are listed in Table 2, together with their distribution category from BALGOOY (1971) and other notes on distribution. The majority (eleven) are extremely successful widespread genera, six of which extend to Hawaii. Two, *Xanthostemon* and *Tapeinosperma*, are 'Melanesian Arc' genera and only one, *Hibbertia*, is Australian. The remaining two, *Oxera* and *Pancheria*, are endemic. *Oxera* belongs to the Verbenaceae, a worldwide family containing many widespread genera. *Pancheria* is in the family Cunoniaceae, primarily Aus-

*Table 2.* List of genera with 25 or more species endemic to New Caledonia. The BALGOOY distribution category is given, together with other comments on distribution. The categories are: 1. Worldwide; 4. Widespread in Old World; 4a. Indo-Australian widespread; 6. Australian genus hardly in Malesia; 6a. Australo-Papuan; 8. Pacific widespread. Species figures are from GUILLAUMIN (1964). *Eugenia* is treated in *sensu lato* as in BALGOOY.

| *Genus and family* | *Species* | *Distribution* |
|---|---|---|
| *Eugenia* (Myrtaceae) | 87 | 1 |
| *Psychotria* (Rubiaceae) | 82 | 1 New Caledonian subcentre |
| *Phyllanthus* (Euphorbiaceae) | 64 | 1 New Caledonian subcentre |
| *Dendrobium* (Orchidaceae) | 45 | 4a |
| *Pittosporum* (Pittosporaceae) | 42 | 4 Subcentres New Caledonia, New Zealand, Hawaii |
| *Tapeinosperma* (Myrsinaceae) | 40 | 8 Melanesian Arc, centred New Caledonia |
| *Alyxia* (Apocynaceae) | 34 | 4 |
| *Planchonella* (Sapotaceae) | 32 | 1 |
| *Parsonsia* (Apocynaceae) | 32 | 4a Major centre in New Caledonia |
| *Xanthostemon* (Myrtaceae) | 31 | 6a Melanesian Arc, centred New Caledonia |
| *Pancheria* (Cunoniaceae) | 30 | Endemic |
| *Ficus* (Moraceae) | 29 | 1 Subgroup centre |
| *Elaeocarpus* (Elaeocarpaceae) | 27 | 4 |
| *Hibbertia* (Dilleniaceae) | 26 | 6 Subcentre in New Caledonia |
| *Oxera* (Verbenaceae) | 26 | Endemic |
| *Dysoxylum* (Meliaceae) | 25 | 4a Subcentre in New Caledonia |

tralasian but with a few Malesian, S. African and S. American taxa. The family may thus be one of the old austral groups. It is particularly well represented in the Melanesian Arcs, several genera being localised in tropical Australasia and one, *Weinmannia*, being much more widespread and an important component of Melanesian vegetation. *Pancheria* has been cited by CARLQUIST (1974) as a good example of adaptive radiation in New Caledonia. Its species are most often encountered in the ultramafic maquis.

These and others are the principal genera that have participated in what CARLQUIST referred to as the 'new' adaptive radiation of the New Caledonian flora, arising from more recent episodes of immigration and colonization (the Cunoniaceae may be exceptions and much older). Most appear to be, or be derived from, genera that are widespread in the tropics. Their arrival in New Caledonia was probably part of the influx of Indo-Malayan elements into Australasia in the early and mid-Tertiary discussed in relation to New Zealand on p. 35. They are generally of the same dispersive character and three (*Dysoxylum*, *Pittosporum* and *Elaeocarpus*) appear in the New Zealand fossil record in the period in question.

CARLQUIST noted the great edaphic diversity of New Caledonia arising from the combination of geological diversity with the range of humidity, rainfall and insolation encountered, as well as relief. As mentioned above, the peculiar, mineral-rich soils arising from the ultramafic rocks may have prevented the establishment of many groups of plants, leading to the conservation of older elements. Today these ultramafics and their derivative lateritic soils occupy only a third of the total area of the island but in the past they were probably much more extensive. This situation would have arisen after their emplacement, probably towards the end of the Oligocene. Subsequently there would have been a gradual change to the present day situation as erosion reduced the extent of the ultramafic sheet. The Miocene saw the trickle of Malesian immigrants to Australasia become a flow as land in New Guinea and other intervening areas became uplifted (RAVEN & AXELROD, 1974). The colonists prior to this flow and in its early stages could well have been confronted with the alternatives of adaptation to, and subsequent radiation on, the extensive ultramafic terrain or restriction to very limited areas of other rock types.

The old austral taxa described earlier, such as the conifers, Proteaceae and Epacridaceae, are restricted to, or commonest in, vegetation associations on the ultramafic rocks. When they occur away from these rocks it is usually in association with poor, lateritic soils, usually at high altitudes. It might therefore be expected that the older and more successful of the later colonists might show a similar restriction and that more modern elements would predominate in the richer forests on sedimentary and metamorphic rocks. On the whole this would appear to be so both for the flora and the associated fauna (p. 184) but there are significant exceptions such as in *Ficus*. There is scope for further rigorous testing of this hypothesis in the field.

JAFFRÉ (1974), studying the western Massif de Koniambo, noted the marked contrast between the flora of this ultramafic massif and that of the surrounding terrain, sedimentary rocks and basalt. He related this difference to the ecological barrier posed by the unusual soils of the massif. The flora of the massif was composed almost entirely of endemic species most of which are not found off the ultramafics. Compared with the flora as a whole he noted the relatively low representation of Leguminosae and Gramineae and relatively high representation of Epacridaceae, Dilleniaceae and Cyperaceae, families which are abundant on the ultramafics generally. Other families better represented on Koniambo than in the flora as a whole were the austral Myrtaceae, Proteaceae and Cunoniaceae and also the Euphorbiaceae, Apocynaceae, Rutaceae and Sapotaceae. All these last are widespread tropical families and all have several genera endemic to New Caledonia. They also include four of the genera listed in Table 2 (*Phyllanthus, Alyxia, Planchonella, Parsonsia*) and JAFFRÉ made particular note of the richness of *Phyllanthus* and *Alyxia* on Koniambo. In contrast, as well as the Leguminosae, the Rubiaceae, Orchidaceae and Araliaceae are more poorly represented than in the flora as a whole. The largely epiphytic Orchidaceae can be excluded from consideration by virtue of their ecology, but the Rubiaceae and Araliaceae, like the other families, include several genera endemic to New Caledonia and the former includes the genus *Psychotria*, the second most successful in terms of species in New Caledonia. However, the representation of the Rubiaceae on Koniambo is drawn from these genera rather than the more widespread ones on the whole. The New Caledonian Araliaceae consist almost entirely of genera endemic to or centred on New Caledonia, so other factors may be involved. The Leguminosae have very few genera endemic to New Caledonia, the family being represented by a majority of widespread genera.

Some indication of the validity of the hypothesis can also be gained from examination of genera and families where the New Caledonian flora has been studied in detail such as in parts of the new Flora already published (e.g. AUBRÉVILLE (1967) on Sapotaceae and KOSTERMANS (1974) on Lauraceae).

AUBRÉVILLE recognised seven genera in the Sapotaceae endemic to New Caledonia and the monotypic *Pichonia*, a strand plant, endemic to the Loyalty Is. (the only such genus). BAEHNI (1965, reviewed by BALGOOY in a distribution table) produced a markedly different classification of the New Caledonian flora, also including a number of endemic genera. It is only possible to study the ecology of the various taxa in the treatment of AUBRÉVILLE so that will be adopted here.

The widespread genera *Manilkara* and *Mimusops* have one strand species each in New Caledonia. Four of the endemic genera are restricted to the ultramafic maquis, namely *Leptistylis* (eight species), *Trouettea* (two species), *Sebertia* (two species, one a nickel accumulator) and *Corbassona* (two species). *Pyriluma*, a monotypic endemic, is a lowland forest taxon and *Pycnandra* (thirteen species) and *Ochrothallus* (seven species) have species evenly distributed between the maquis, lowland forest and montane forest. *Planchonella*, widespread in the Indo-Australian area and extending to Hawaii, has sixteen species two of which are strand species, and the rest are equally divided between the maquis and lowland forest. *Iteiluma* is shared with Queensland and has two maquis species but *Niemeyera*, shared with E. Australia, has one forest species. Two Melanesian Arc genera, *Bureavella* (two species) and *Rhamnoluma* (three species), are restricted to lowland forest and maquis respectively. AUBRÉVILLE did not specify the rock substrate in his lowland and montane forest categories so the degree of segregation is not fully clear but, on balance, it could be said that the older elements of the Sapotaceae are concentrated on the ultramafics. Similarly the genera *Litsea*, with a New Caledonian subcentre, and *Adenodaphne* (endemic) in the Lauraceae (KOSTERMANS) are largely restricted to the ultramafics, whereas species of the widespread and less strongly represented genera *Cryptocarya* and *Endiandra* are not, tending to characterise rainforest throughout the island.

The palms on the other hand, with indications of a long history and endemic radiation in New Caledonia, are associated with the development of rainforest and are not limited or segregated to any great degree by rock type; the humus-rich soils associated with rainforest are less demanding chemically on the ultramafics than those supporting the maquis and low *Gymnostoma* forest associations. Similarly the interrelated group of eighteen species in the subgenus *Pharmacosycea* of *Ficus* (CORNER, 1970) are all found generally in lowland rainforest apart from *dzumacensis* from Mt. Mou and *maialis* from the southern ultramafics. These species all belong to the section *Austrocaledonicae* which is restricted to New Caledonia apart from one species in the New Hebrides (*granatum*). Obviously further information on the ecology of other genera is needed before the hypothesis can be confirmed or discarded.

Adaptation to the ultramafics and their unusual mineral regimes has proceeded to a marked degree in a number of taxa in that they accumulate nickel salts in their sap. New Caledonia has eleven out of the thirteen species known to have this property (JAFFRÉ *et al.*, 1976), the other two being *Alyssum bertolonii* in Italy and an Australian species of *Hybanthus* (Violaceae). *Hybanthus* is a worldwide genus with a subcentre in New Caledonia, one species extending to the Loyalties and New Hebrides. Otherwise it is absent from the Melanesian Arcs except for New Guinea. Two of the four New Caledonian *Hybanthus* accumulate nickel. The other species are *Sebertia acuminata, Psychotria douarrei,* and seven out of twenty-two endemic species of *Homalium* (Flacourtiaceae), a widespread tropical genus that extends to Polynesia, though it is poorly represented in Australia (Queensland only).

## The Melanesian Arcs

The major question remaining concerns the relationship of New Caledonia to New Guinea and the Melanesian Arcs and the possibility that plant distribution patterns may provide evidence for the geological hypothesis of Inner and Outer Arcs that have recently converged, making contact in the New Guinea area in the Late Miocene or Pliocene and approaching more recently in the New Caledonia and New Hebrides area.

The flora of the Bismarck and Solomon groups is markedly Papuan in character and the association of Fiji, the New Hebrides, Tonga and Samoa has a relatively rich and diverse flora that is also of Papuan affinity, though Fiji has several unusual endemic genera and has probably acted as a minor secondary centre of radiation for the Pacific (THORNE, 1963, 1969). Fiji has 69% specific endemism (1266 species, roughly half the New Caledonian total) in its flora and 2.1% generic endemism (SMITH, 1951; BALGOOY, 1971), considerably less than New Caledonia. The generic endemics are mainly related to widespread taxa, five out of ten being Rubiaceae and three more being palms. *Degeneria,* often regarded as the sole member of the Degeneriaceae, is related to the Winteraceae and may be one of the oldest elements in the Fijian flora. Three genera are shared exclusively with New Caledonia, one being the conifer *Acmopyle.* Some Australian-New Caledonian taxa extend to Fiji but overall the floras are dissimilar and indicate a high degree of isolation from each other and, in their overall affinities with other areas, different routes of derivation. Several genera have their centres or subcentres in Fiji. The other related areas, the New Hebrides, Tonga and Samoa, have virtually no endemic genera (one of the Rubiaceae in Samoa) and no centres or subcentres. SMITH (1955) noted that a large number of genera had their eastern limits in Fiji, 26.1% of those represented according to BALGOOY, these being mainly the more sedentary taxa of Malesian rainforest.

RAVEN & AXELROD (1972) and WHITMORE (1973) were the first authors to attempt to relate Pacific phytogeography to the plate tectonic aspects of the geological history of the area. The former authors worked to a general geological hypothesis similar to that presented in Chapter I, though they invoked an Eocene connection between Australia and Fiji before the subsequent formation of the S. Fiji Basin, followed by permanence of land areas to the present day, in order to explain the presence on Fiji of the conifers, *Degeneria, Gymnostoma* and Proteaceae. Proximity to Australia after the Cretaceous is ruled out by geological evidence. The presence of as much land as exists today in the Fiji group throughout the Tertiary is unlikely but even the persistence of land the size of Lord Howe I. (138 plant genera, excluding lower plants) could have ensured the survival of some ancient elements. Some of these older elements, such as *Degeneria* and *Acmopyle*, could have been derived from association with New Caledonia before the S. Fiji Basin opened, and then conserved in this manner. Long distance dispersal, particularly for the bird-dispersed podocarps, cannot be ruled out and some of the older taxa could have been derived from the New Guinea area by dispersal along the Outer Arc.

WHITMORE suggested that present day distribution patterns in the southwest Pacific arose through migration across a continuous land surface or on islands lying closely offshore from such land in the Cretaceous or earliest Tertiary. The present situation then arose through the opening of ocean basins and fragmentation of the island chains. The problem with this theory and the more extreme one of a now sundered 'Melanesian Foreland' adjacent to the north of the present Outer Arc (CORNER, 1975) is that they run counter to geological evidence, particularly as they envisage land decreasing from a relatively continuous state to the present archipelagic condition through the Tertiary, rather than vice versa as appears to be the case from purely geological evidence. However, many botanists have maintained (e.g. CORNER, 1967, 1975) that the phytogeography of Melanesia and Polynesia implies more land in the past than geology is able to provide from subsurface investigation.

The crux of this disagreement is the difference of opinion amongst botanists over the dispersal powers of plants generally or of specific genera, or, relatively speaking, the water gaps necessary to restrict or prevent floral exchange between two ecologically similar areas. Resolution of this dispute is hampered by the lack of hard evidence on the potential modes of dispersal in the form of observation of seeds in sea bird excreta or preened from feathers, seeds washed up on a shore or extracted from air currents, or seed-bearing branches of species with heavy, submersible seeds borne aloft at sea by hurricanes or floating in rafts and so on. Summaries of what evidence there is may be found in CARLQUIST (1967, 1974); BALGOOY (1971) categorised Pacific plant genera according to potential dispersal method. Plants do arrive on extremely isolated islands where land connections are precluded on strong geological grounds such as Norfolk I. (HOLLOWAY, 1977), Hawaii and a number more in high latitudes that have regained vegetation following total glaciation. THORNE (1963) suggested that it is establishment, not dissemination, that is the critical phase of dispersal.

An alternative approach is to compare plant distributions with what is known or suspected about past geography in the absence of any presupposition about dispersal powers other than the general one that these will vary, perhaps considerably, from genus to genus and thus may reflect different aspects of the geographical history of an area. In a similar manner different powers of dispersal in the elements of the flora and fauna have led to differences in the representation of elements from the three source areas (Australia, New Zealand and New Caledonia) from group to group in the biota of Norfolk I. (HOLLOWAY, 1977).

The low generic endemism or precinctiveness in the Outer Melanesian Arc east of New Guinea is consistent with the geological suggestion of land area limited in extent until relatively recently. Fiji is somewhat exceptional to this and the Solomons slightly so, and this again is consistent with indications of a slightly longer subaerial history in each area relative to the New Hebrides, Tonga or the Kermadecs.

A survey of the generic distributions collated by BALGOOY (1971) lends some support to the hypothesis of two separate Melanesian Arcs, later convergent in New Guinea. The generally Melanesian genera included in distribution types 5b, 6a and 8 can partly be divided into two groups referable to one or other arc and a third distributed over the whole area. The suggested fusion of the two arcs in New Guinea would have caused any clearer segregation in existence beforehand to have become confused by Inner Arc elements spreading eastwards along the Outer Arc from New Guinea, and exchange more recently between New Caledonia and the New Hebrides may also have blurred the patterns. Relationships with more westerly areas cannot be predicted with any confidence as the geological picture is unclear, particularly with regard to Sulawesi. But the South Moluccas might have been associated with the Inner Arc, and the North Moluccas and the Philippines with the Outer. Outer Arc genera might be expected to have spread more to Micronesia and Polynesia and Inner Arc genera would have had closer association with Australia.

The segregation can perhaps be epitomised by the related genera *Acsmithia* and *Spiraeanthemum* in the Cunoniaceae. The former has subcentres in New Guinea and New Caledonia and extends to E. Malesia (the Moluccas and Sulawesi) and Queensland. The latter is centred in Fiji and ranges from the Bismarcks to Samoa.

Apart from *Spiraeanthemum* there are sixteen other genera, absent from New Caledonia, that could owe their current distribution to past restriction to the Outer Arc though eight (asterisked) are centred in New Guinea. They are *Finschia** (Proteaceae), *Rejoua* (Apocynaceae), *Plerandra* (Araliaceae), *Haplolobus** (Burseraceae), *Merremia* subgenus *Wavula* (Convolvulaceae), *Pullea* (Cunoniaceae), the orchid genera *Glomera**, *Glossorhyncha*, *Mediocalcar** and *Pseuderia**, the Rubiaceae *Amaracarpus**, *Badusa*, *Dolicholonium* and *Mastixiodendron*, *Wenzelia** (Rutaceae) and *Burckella* (Sapotaceae). All extend to Fiji except *Finschia* (New Hebrides). Seven are recorded from the Caroline Is., the Micronesian archipelago to the north of the centre of the Outer Arc, but only one (*Pullea*, one species) has species in Queensland. Eight are in E.

Malesia and five in the Philippines. This distribution pattern is also shown by a number of faunal groups, discussed in Chapter VII.

Exclusively Inner Arc distributions are shown by *Deplanchea* (Bignoniaceae), *Sloanea* subgenus *Antholoma* and *Dubouzetia* (Elaeocarpaceae). Other possible Inner Arc genera, now expanded into other areas but still centred on the Inner Arc like *Acsmithia*, are *Gymnostoma* (Casuarinaceae), *Alphandia* (Euphorbiaceae), *Sphenostemon* (Monimiaceae), *Xanthomyrtus* (Myrtaceae), *Rhamnoluma* (Sapotaceae) and, centred in New Caledonia, *Delarbrea* (Araliaceae), *Durandea* (Linaceae) and the Myrtaceae *Cloezia*, *Mearnsia* (Plate 2) and *Xanthostemon* (Plate 2). Of these thirteen genera seven occur in Queensland but only two in Fiji and none in Micronesia. Nine occur in E. Malesia and five in the Philippines. The Inner Arc was probably never as continuous as the Outer and hence there would have been more independent radiation in New Caledonia and central New Guinea, and consequently more generic endemism. Inner Arc New Guinea would have interacted to some degree with the rainforest areas of Queensland, more extensive until recently, and indeed the latter may have provided the major route of exchange between New Guinea and New Caledonia, when and if the Chesterfield Reefs were emergent in the mid-Tertiary and if the Inner Arc had always been as fragmented as today rather than as in Figure 3. The role of tropical Queensland in Inner Arc relationships is more apparent in the Lepidoptera groups discussed on p. 220.

If the reconstruction of Miocene geography in Figure 3 and the earlier Tertiary geography in Figure 2 is passably accurate then the Philippines and East Malesia, particularly the North Moluccas, would have been the main route for the dispersal of Asian taxa into the Outer Arc and possibly also into the Inner Arc. These might therefore have been derived preferentially from China through Taiwan rather than from the Malaysian area. This northern route of exchange between the two areas has been discussed elsewhere by HOLLOWAY & JARDINE (1968) with regard to Lepidoptera and birds, and to some extent by THORNE (1963). The former authors also discussed faunistic evidence for the fusion of two separate arcs in New Guinea.

In an inevitably cursory survey of plant generic distributions of this nature it is not possible to examine the relationships of the species within each genus; indeed in a large number of cases these will not have been studied to any depth. But it is perhaps at that level that most indications of segregated evolution in the two arcs will be found if it occurs. There are about twenty other genera distributed over both arcs, such as *Trimenia*, *Tapeinosperma* and *Guillainia* (Zingiberaceae), where specific affinities might prove interesting. Many of the more widespread genera that have radiated in Melanesia could also be studied.

The genus *Ficus* has been studied in great detail by CORNER (1965, 1967, 1970, 1975) and it is thus possible to examine the various subgenera, sections, series and subseries and their distributions relative to the hypothesis of two arcs. This involves the perhaps unwarrantable assumption that these taxonomic units represent natural, monophyletic groupings, indeed the whole survey of plant distribution presented in this chapter rests on the assumption that, *in general*, current taxonomic groupings reflect phylogeny. But if significantly similar patterns of distribution are detected in a variety of taxonomic groups there is justification both for regarding such groups as probable natural entities and seeking explanations for their geographical limitation.

In *Ficus*, series *Malvanthereae* of subgenus *Urostigma* is centred strongly in New Guinea and Australia with only a few apparently mobile species such as *F. glandifera* and *F. obliqua* extending into Outer Arc areas; none occur in New Caledonia.

In subgenus *Pharmacosycea* section *Oreosycea*, series *Vasculosae* is mainly Oriental and series *Austrocaledonicae* is virtually restricted to New Caledonia but extends to the New Hebrides. The third series, *Nervosae*, contains sixteen species that have a predominantly Outer Arc distribution though three are Oriental. There are four in each of the Philippines and Sulawesi, three in the Moluccas, Bismarcks and Solomons, eight in New Guinea and one, shared with the Solomons, in Fiji and the New Hebrides.

Subgenus *Ficus* has several subgroups of interest. Section *Rhizocladus* has one series (*Distichoideae*) centred in New Guinea, only one of four species extending to adjacent areas, and another series (*Ramentaceae*) where three subseries include thirteen species virtually restricted to New Guinea and three others are predominantly Oriental.

Series *Scabrae* of section *Sycidium* has an extensive, largely Melanesian Arc distribution, containing a large number of species. Three occur on the Sunda Shelf and another is restricted in that area to northern Borneo. The series occurs in other areas as follows: Taiwan (3), Ryukyu (2), Philippines (9), Sulawesi (11), N. Moluccas (2), S. Moluccas (4), New Guinea (7), N. Australia (5), Bismarcks (5), Solomons (8), New Hebrides (4), Rotuma (1), Fiji (7), Samoa (5), Tonga (1), Niue (2), New Caledonia (2), Kei Is. (2). Thus there is no marked centre in New Guinea despite significant penetration of Australia and the species are distributed evenly over the Outer Arc which here includes the Philippines and Taiwan, analogous to the distributions of faunal groups discussed on p. 23. The two New Caledonian species are amongst the more widespread members of the group. One (*scabra*) is found also in the Loyalties, New Hebrides, Fiji, Samoa, Tonga and Rotuma, a distribution typical of a number of moth species (p. 210). The other (*fraseri*) occurs through Queensland to N. New South Wales and also in the New Hebrides, a typical lepidopteran Coral Sea distribution (p. 204).

Section *Adenosperma* of subgenus *Ficus* appears intermediate in distribution between the Inner and Outer Arc type but it is strongly centred in New Guinea (sixteen species), with no more than five species occurring in any other locality: Sulawesi (3), N. and S. Moluccas and Bismarcks (5), Solomons (4), New Hebrides, Queensland and Kei (2), Rotuma and Tenimber (1). Thus numbers decrease rapidly away from New Guinea and endemism is restricted mainly to New Guinea. The two New Hebrides species, *austrina* and *adenosperma*, are widespread in Melanesia, the latter being the single Rotuman species. The section may therefore be a group that radiated initially in Inner Arc New

Guinea and has subsequently expanded along the Outer Arc after the approach and fusion of the two.

Section *Sycocarpus* has three subsections. *Auriculisperma* has five species in the Solomons and one in Fiji. *Dammaropsis* has one is New Guinea. *Papuasyce* has two in New Guinea and one in the Bismarcks. It is tempting to interpret this as the result of localised distribution and minor radiation in the Outer Arc (*Papuasyce* and *Auriculisperma*) with a representative in Inner Arc New Guinea (*Dammaropsis*) akin to the situation in the more polytypic cicada genera discussed on p. 235, but this would be hard to confirm.

In *Ficus* there appears to be a clear dichotomy, with the possible exception of *Adenosperma*, between groups with an extensive Melanesian (putative Outer Arc) distribution and those with speciation virtually restricted to New Guinea or, in one instance, New Caledonia. One group restricted to New Guinea is associated with three Australian ones in series *Malvanthereae*. In the palms discussed earlier the dichotomy is less clear but one might contrast the *Ptychosperma* and *Clinostigma* alliances in the arecoids, predominantly Outer Arc, with the genera *Orania* and *Cyrtostachys* and the *Linospadix* alliance, also arecoids. *Orania* and *Cyrtostachys* have their greatest development in New Guinea (ten and seven species respectively) but the former has six species in Sundaland and the latter one. The *Linospadix* alliance has great diversity in New Guinea but is also strongly represented in the Australian tropics.

This dichotomy does not often appear to give rise to related sister groups like *Acsmithia* and *Spiraeanthemum*, though examples for the butterflies are given on p. 219 and for the cicadas on p. 235. Evidence of such sister groups will be crucial for supporting the hypothesis that the apparent dichotomy of distribution patterns in Melanesia is due to past geography rather than a result of different ecological characteristics favouring the radiation of some taxa in the more archipelagic parts of Melanesia and not others (dispersal for example).

*Orania*, *Cyrtostachys* and series *Ramentaceae* in *Ficus* show radiation restricted to New Guinea that has its closest relationships with lands on the Sunda Shelf (Sundaland). The possibility of a southern route whereby Sunda Shelf (Malesian) taxa colonised Inner Arc New Guinea without any association with the Outer Arc is intriguing. Figure 3 suggests a rather disjunct route might have been possible through parts of emerging Sulawesi and the S. Moluccas but the geological history of this area is obscure and it is beyond the compass of this work to examine the topic in detail.

## Comparison with New Guinea

New Guinea, or its previously separate parts, has provided the hub for the intermingling of floras and faunas, their evolution and their dispersal in Melanesia and the south-west Pacific. It is worthwhile to examine its phytogeography in detail, especially in comparison with New Caledonia where a potentially similar intermingling, evolution and dispersal through the Tertiary has been affected by much greater geographical isolation. This long isolation is reflected in the generic (15.9%) and specific (92% of about 2660 species) endemism. THORNE (1963) wrote of New Caledonia, 'Probably no richer nor more peculiar, archaic and endemic relict seed-plant flora can be found elsehwere compressed into such a small area'.

Oriental elements reaching Australia in the Tertiary must have done so through New Guinea, especially the Inner Arc portion, though the land area involved as a stepping stone was much smaller in the early stages of this dispersal. THORNE (1963) and GOOD (1958, 1963) both suggested that exchange between Australia and New Guinea has been predominantly from the Pliocene onwards though there are indications that some minor exchange has continued over an appreciable time. Evidence for such early exchange may be provided primarily by palynological studies in northern Australia, for many taxa once shared with New Guinea may have become extinct there with the decline of areas of rainforest. The *brassii* group of *Nothofagus* is a good example.

Today coastal Queensland, northern New South Wales and Arnhemland are in the area of overlap between the Australian and Papuan biotas. The recent exchanges have been mainly from New Guinea to Australia (GOOD, 1958). Apart from Pleistocene exchanges across the Torres Straits when exposed by Pleistocene low sea levels, mostly indicated by the presence of the ecosystems of one area in the other (Papuan rainforest in N. Australia and Australian savanna in S. New Guinea), the floras show many more differences than would be expected from their modern proximity, leading GOOD (1963) to suggest that the effects of continental drift may have increased this proximity relatively recently. The uplift of the southern lowlands of New Guinea around the Fly and Digul river systems would effectively have halved the water gap between Australia and the central part for instance. Elements would have dispersed earlier or later between the two according to their dispersal powers. But against this must be considered the indications that a sedentary genus such as *Nothofagus* reached New Guinea from Australia at an earlier date. Previous periods of low sea-level may have occurred, such as at −5 my (p. 29), and one such is indicated in Figure 4 with the suggestion that part of the Torres Straits may have been exposed. It cannot be discounted that the physical barrier between Australia and New Guinea may be more or less coincident with an ecological barrier, marking a change from arid, seasonal habitats to more uniform, humid, rainforest habitats. Rainforest taxa shared between the two areas prior to the Pleistocene glaciations and the consequent increase in aridity in Australia would not only have been prone to extinction because of that aridity but, faced by an influx of Malesian rainforest taxa facilitated by exposure of the Torres Straits, might also have been ousted by competition. The taxa of austral character in New Guinea had access to considerable areas of montane habitat wherein they could hold their own in the face of such competition. Their hypothetical Australian counterparts had not.

New Caledonia has 50 genera shared with Australia that do not occur in New Guinea (BALGOOY, 1971) and the closest affinity at the generic level is with Queensland rather than with New Guinea or eastern Melanesia (excluding the relationship with the New Hebrides where the majority of exchange was probably recent). Several more New Caledonian genera that occur in New Guinea are centred in Australia and are more likely to have been derived directly from the latter. Also many Inner Arc genera shared between New Guinea and New Caledonia today may have attained this distribution following exchange through northern Australia followed by disjunction through extinction in the latter.

New Guinea has 1349 flowering plant genera of which three quarters are widespread and were probably derived at some stage from the Oriental Region. There are 62 Australian genera, 125 localised Pacific (Melanesian Arc etc.) genera and 140 endemics. These categories are listed in Table 3, the figures being taken from GOOD, 1974.

*Table 3.* Generic distribution types and endemism in New Guinea, after GOOD (1974).

| *Distribution type* | *Number of genera* | *Percentage with endemic species* | *Mean number of species per genus in genera with endemic species* |
|---|---|---|---|
| Widespread tropical | 500 | 63 | 11.9 |
| Indo-Malaysian | 495 | 51 | 14.3 |
| Asiatic American | 27 | 77 | 8.7 |
| Local Pacific | 125 | 70 | 6.1 |
| Endemic | 140 | 100 | 2.5 |
| Australian | 62 | 39 | 2.6 |

New Guinea has been a major centre of plant evolution and geological evidence would indicate that most of this evolution has taken place since the Oligocene, though it may have had a longer history in the Outer Arc section and in the south-east. A high percentage of the widespread genera contain endemic species and the mean number of endemic species per genus of this percentage is also high (Table 3). This also applies to a lesser extent to the localised Pacific genera where a higher percentage include endemics but these are fewer per genus. By contrast both these figures for the Australian genera are very low. The number of species per genus for the endemic genera is also very low. The endemic genera contain only 4% of all endemic species. These figures could be interpreted as indicating that an old endemic radiation, and to some extent that of localised Melanesian genera, has given way through extinction to a more recent radiation of vigorous Malesian rainforest taxa that, through poorer powers of dispersal, only reached New Guinea after their spread was facilitated by the uplift of intervening land in the later Tertiary.

Half the 64 species endemic to New Guinea belonging to Australian genera are in *Olearia*, *Trachymene*, *Haloragis*, *Cladium* and *Helichrysum* (GOOD, 1960). *Cladium* (Cyperaceae) is better regarded as a widespread genus (BALGOOY, 1971). *Olearia* (Compositae) and *Haloragis* (Haloragaceae) have both penetrated Polynesia as far as Rapa I., the former also growing in the Marquesas. *Helichrysum* (Compositae) also occurs in New Caledonia and New Zealand. *Trachymene* (Umbelliferae) occurs in Malesia, and New Caledonia. Many of the other genera range further to Malesia, New Caledonia and New Zealand and could be classed amongst the more mobile of Australian-centred genera. Of the sixteen genera exclusive to New Guinea and Australia there are seven endemic species in New Guinea compared with about 300 in Australia. There are also eighteen small genera balanced in distribution between Australia and New Guinea. Eight are monotypic, six have a species in each area and three have two in each. Only in *Toechima* (Sapindaceae) are there four in each. All figures are from GOOD. Perhaps many of these genera represent a recent extension of arid habitat taxa from centres in Australia into surrounding archipelagos as climate has deteriorated.

This is in contrast to the marked development of Australian genera in New Caledonia. The relationship is strong in the orchids, but many of these are also in New Zealand and west wind dispersal of the light diaspores could have excluded New Guinea. The relationship is also exemplified in the Rutaceae, Proteaceae and such genera as *Lomandra* (*Xerotes*, Xanthorrhoeaceae but included under Liliaceae in BALGOOY), *Campynema* (Amaryllidaceae), *Hibbertia* and *Castanospermum* (Leguminosae) (GOOD, 1974).

This greater representation is surprising in view of the greater isolation of New Caledonia. Some of the associations may date from a closer association between the two areas in the Cretaceous (the Proteaceae for example). Others may be referable to subsequent exchanges along stepping stone links involving the Chesterfield Reefs.

A further factor, perhaps more important than the others, is the greater access Malesian taxa have had to New Guinea since the Miocene orogenies created significant links through the Indonesian Archipelago. In New Guinea as in New Caledonia these genera have been the most successful in radiations of endemic species. But in New Caledonia the characteristically Australian genera have far more endemic species and the endemic genera have a mean number of species of approximately 4 (GUILLAUMIN, 1964) as compared with 2.5, and contribute 16% of the specific endemism as compared with 4% in New Guinea. There are no families endemic to New Guinea and the percentage of endemic genera (10.3) is less than that of New Caledonia (15.9%). Widespread genera make up 60.6% of the New Caledonian total whereas in New Guinea they make up 76%. The 'new' endemic radiation in New Guinea appears to have proceeded far further at the expense of the old than it has in New Caledonia, though the old radiation in the geologically younger area may not have been so considerable to start with.

Thus New Guinea, or the Inner Arc part of it, may have had much stronger relationships with Australia in the past, and greater development of endemic taxa in parallel with New Caledonia, but these would have been eclipsed and partially extinguished by ever increasing invasion by highly successful and competitive Malesian rainforest taxa. Some parallels with Australia may also have been lost through the extinction of rainforest taxa as a result of increasing aridity on that continent and the greater effects of competition from Pleistocene Malesian invaders in the absence of montane refugia.

GOOD (1960) noted that the families especially characteristic of New Guinea were the Araliaceae, Cunoniaceae, Elaeocarpaceae, Icacinaceae and Winteraceae. It is perhaps significant that these families, especially the first two, contain a high proportion of Melanesian Arc genera. The Araliaceae include such genera as *Boerlagiodendron, Mackinlaya, Meryta, Osmoxylon* and *Plerandra* as well as those discussed on p. 49 and a few genera precinctive to Hawaii and Polynesia. These are probably some of the older elements of the Melanesian and Pacific floras. Genera such as *Ficus, Elaeocarpus, Eugenia* and *Psychotria* are important components of the floras of both New Caledonia and New Guinea in terms of species number.

## Conclusions

The phytogeography of New Caledonia may be summarised as follows. There are many old elements, presumed relicts from a previously more extensive and general Cretaceous Australasian-Antarctic flora that probably reached New Caledonia before the opening of the Tasman Sea. The subsequent isolation of the island has resulted in it being partially shielded from the flow into Australasia and the Pacific of competitive rainforest taxa from the Oriental tropics. The ecologically demanding soil of the extensive ultramafic rocks may also have had a conservative effect, favouring the older elements with a longer history of adaptation to them. This conservation through isolation may explain the apparently ancient and faint relationships between New Caledonia and Madagascar noted by GOOD (1950), the latter having had a parallel history relative to Cretaceous floras in Gondwanan Africa, close contact followed by isolation. Exclusive parallels with the New Zealand flora may have arisen in a similar manner if the Australian and New Guinea counterparts of such shared groups became extinct through climatic changes and competition from new invaders.

More modern elements did reach New Caledonia in the Tertiary however, though not on the scale of the colonisation of New Guinea. Some spread directly from Australia, possibly assisted by the periodic exposure of the Chesterfield Reefs, and some may have come through association with the Melanesian Arcs, particularly Inner Arc New Guinea.

Indo-Malayan elements began to penetrate Australasia in the early Tertiary but not to any great degree until the closing stages of the Oligocene. Up to, and during, the Miocene it is possible that access to Australasia was primarily from China via Taiwan to the Philippines and the Outer Melanesian Arc, thence to the elements of the Inner Arc and Australia. Australian rainforest elements such as Araucariaceae, Podocarpaceae, Casuarinaceae, some Myrtaceae (especially *Metrosideros*) and Proteaceae, Cunoniaceae and Winteraceae would also have extended into the Melanesian arcs, perhaps more predominantly into the Inner Arc, where they would have mingled with the Oriental

Araliaceae, Rubiaceae, Rutaceae, Sapindaceae, Santalaceae, Sapotaceae, Apocynaceae and Palmae. This Melanesian Arc flora would have been the principal source for the colonisation of Polynesia until late in the Tertiary. Thus taxa radiating in Hawaii include a high proportion of Araliaceae, *Pritchardia* in the Palmae, Rubiaceae and Santalaceae.

There are indications that there may have been some access to the Inner Arc in New Guinea for taxa centred on the Sunda Shelf at a date early enough to permit considerable radiation there. Sulawesi or its component parts could have facilitated such dispersal.

The great increase of land area in Indonesia in the Pliocene and Pleistocene led to the spread of numerous more competitive but less dispersive Oriental plant taxa to Melanesia, probably ousting many elements of the older Melanesian flora, particularly in New Guinea, or restricting them to certain habitats such as high altitudes or poor soils. In such habitats the flora has held its own and even advanced westwards into the Oriental Region.

A reconstruction of past events of this nature from present day distribution patterns must remain conjectural until hard evidence is available from palynological investigations. Where there is no long terrestrial sequence of pollen-bearing strata it might prove fruitful to examine offshore deposits where there may be significant quantities of windblown pollen at least.

Other information of value may be gained from more restricted deposits with pollen, such as the deposit of coal on Rapa I. (CRANWELL, 1964a). This coal probably dates from the later Tertiary. It was probably laid down in marshy conditions as it overlies a clay, but above and below are lava flows that should permit more precise dating in due course. Microfossils positively identified are from tree ferns, Cyperaceae, Myrtaceae (*Metrosideros* is present today), Sapindaceae and Rubiaceae (*Coprosma* is present today), an assemblage of families characteristic of the Melanesian Arc flora.

# III. THE VEGETATION OF NEW CALEDONIA

Similarities between the vegetation of subtropical New Zealand, montane New Caledonia and highland New Guinea have been noted in the past (ROBBINS, 1961; DAWSON, 1963; THORNE, 1965). Elements of this vegetation include a majority of the Melanesian Arc and austral genera mentioned in the previous chapter as well as a number of more widely distributed genera such as *Pittosporum* that have subcentres in Melanesia. These taxa may represent the genral flora prevalent in the Melanesian Arcs and the tropical parts of Australia and New Zealand in the Miocene and early Pliocene prior to the invasion of more competitive Oriental elements.

The older Melanesian flora has to some extent remained distinct from the more recent Oriental lowland rainforest elements, being restricted largely to montane habitats, the poorer soils or, in the case of the New Caledonian ultramafics, soils of stringent mineral regimes to which it was preadapted. Indeed the flora appears to have been successful in spreading westwards into the Oriental Region where such habitats occur.

THORNE (1965) mentioned a large number of genera that are prominent in such elements of the vegetation of at least two of New Zealand, New Caledonia and New Guinea, namely *Agathis*, *Podocarpus*, *Dacrydium*, *Libocedrus*, *Araucaria*, *Oncinia*, *Astelia*, *Cordyline* (Liliaceae), *Ascarina*, *Eucarpha/Knightia*, *Muehlenbeckia* (Polygonaceae; Australasian), *Hedycarya*, *Quintinia* (Saxifragaceae; Philippines, E. Malesia, Australasia, including Miocene New South Wales (p. 34)), *Pittosporum*, *Weinmannia*, *Nothofagus*, *Corynocarpus* (Corynocarpaceae; Australasian), *Metrosideros*, *Styphelia* subgenus *Cyathodes*, *Meryta* (Araliaceae; Australasia and Pacific), *Schefflera* (Araliaceae; widespread with Melanesian radiation), *Dracophyllum* and *Geniostoma* (Loganiaceae; Malesian with a New Caledonian centre). Almost all of these genera fall into the 'Old Melanesian' category and a large number are known to have had a long history in the area.

ROBBINS (1961) wrote of the New Guinea montane forests:

> 'The fascination of these montane forests lies in their affording a key to past southern floras. Thus the mixed lower montane forests of the highlands may be regarded as tropical outliers, now relict, of New Zealand's podocarp broadleaf forests .....'

He mentioned that they contain many genera and species lost in New Zealand since the Miocene such as the *Nothofagus brassii* group.

The lower montane forests of New Guinea have a canopy composed primarily of Podocarpaceae, Elaeocarpaceae, Cunoniaceae, Lauraceae and Myrtaceae, all families that dominate New Zealand forests but are generally much richer in genera in New Guinea. *Olearia, Ascarina, Melicope* (Rutaceae; Malesian to New Zealand) and *Weinmannia* are also present, with smaller trees such as *Rapanea* (*Myrsine*), members of the Winteraceae, and shrubs such as *Piper, Schefflera, Pittosporum* and *Geniostoma.*

The beech forests have an upper stratum of *Nothofagus*, Cunoniaceae, *Quintinia, Podocarpus, Cryptocarya* (Lauraceae; pantropical), *Zanthoxylum* (Rutaceae; pantropical, including Norfolk I.), and *Alphitonia* (Rhamnaceae; Australasia and Pacific). The second stratum contains *Weinmannia, Ficus*, Myrsinaceae and *Evodia* (Rutaceae; Malesian and Pacific to Rapa I. and Norfolk I.). The high cloud forest in New Guinea is dominated by Podocarpaceae and Myrtaceae.

A close comparison with New Guinea has not been made for the New Caledonian vegetation. DAWSON (1963) compared it with the New Zealand lowland forests. These resemble New Caledonian montane forests, perhaps particularly those of the wetter north-east such as are found on the Mt. Panié range. In both cases *Weinmannia, Metrosideros* and *Elaeocarpus* are canopy dominants but conifers are less prevalent in New Caledonia and, apart from *Araucaria*, are not canopy emergents as they are in New Zealand.

The proportion of monotypic genera in the two floras is similar but the number of species per genus for shared forest genera averages at 11.4 for New Caledonia compared with 3.0 in New Zealand. This difference is likely to have been less extreme before warm climate taxa became extinct in New Zealand at the end of the Tertiary and during the Pleistocene glaciations. DAWSON listed these genera and the following examples represent the extremes, New Caledonian species being the first number in each case: *Dysoxylum* 26: 1, *Elaeocarpus* 29: 2, *Eugenia s.s.* 30: 1, *Geniostoma* 17: 1, *Hedycarya* 17: 1, *Litsea* 12: 1, *Myrsine* 28: 9, *Parsonsia* 36: 2, *Planchonella* 32: 1, *Schefflera* 21: 1, *Freycinetia* 15: 1, *Weinmannia* 5: 2, but in contrast there are *Astelia* 1: 10, *Pittosporum* 42: 26, and *Metrosideros* 10: 11. The proportion favours New Zealand for shared genera of open habitats, e.g. *Dracophyllum* 7: 35, *Cotula* 1: 24, *Wahlenbergia* 1: 10, *Carex* 7: 55, *Juncus* 2: 17, *Ranunculus* 1: 43, and *Senecio* 1: 16, but an exception is the more characteristically tropical *Solanum* 21: 3.

JAFFRÉ & LATHAM (1974) described what might almost be termed typical Melanesian montane forest on top of Mt. Boulinda, a central western ultramafic massif, at an altitude of over 1000 m. On the summit is a low cloud forest dominated by *Metrosideros, Quintinia, Weinmannia* and *Cunonia*. The herb layer is absent but there are numerous epiphytic mosses and ferns together with *Astelia*. The subsummit slopes immediately surrounding support a taller dense forest of Myrtaceae, Podocarpaceae and Lauraceae, much richer in species. The canopy includes trees of the genera *Decussocarpus, Dacrydium, Cryptocarya, Metrosideros*, and *Schefflera*. The second stratum features the genera *Hedycarya, Rapanea* (*Myrsine*), *Podocarpus, Psychotria, Dacrydium, Phelline, Tapeinosperma, Pandanus* and *Basselinia*.

There are descriptions of montane vegetation in other Melanesian islands. SMITH (1951) and ROBINSON (1975) mentioned *Agathis, Podocarpus, Dac-*

*rydium*, *Metrosideros* and *Elaeocarpus* as important components of the higher montane forests of Fiji and there is cloud forest with Myrtaceae predominant, especially *Metrosideros* and *Eugenia* (*Syzygium*). Prominent among the forests of lower altitudes are members of the Rubiaceae, Lauraceae, Meliaceae, Sapotaceae and Myrtaceae, and genera such as *Ficus*, *Vitiphoenix* (Palmae), *Piper* (incl. *Macropiper*), *Myristica* (Myristicaceae; Malesian with a New Guinea centre), *Pittosporum*, *Elaeocarpus*, *Evodia*, *Eugenia*, *Cyrtandra* and *Psychotria* (SMITH).

In the New Hebrides the *Agathis* forests have two Podocarpaceae associated with them and other associates are *Calophyllum* (Guttiferae; pantropical), *Hernandia* (Hernandaceae; pantropical) *Dysoxylum*, *Syzygium*, *Spiraeanthemum*, *Geissois* (Cunoniaceae; E. Australian and S.W. Pacific), *Elaeocarpus*, *Planchonella* and *Ficus*, and the understorey is largely composed of Araliaceae, Rubiaceae, Sapindaceae, Sapotaceae and *Myristica* (BEVERIDGE, 1975). This forest was referred to as 'forêt dense climacique à *Calophyllum* et *Agathis*' by SCHMID (1975). GILLISON (1975) noted similarities between this association and *Araucaria hunsteinii* forests on acid volcanoes in New Guinea.

The New Hebridean summits and high ridges support an association dominated by *Metrosideros* and *Weinmannia* with *Quintinia*, *Ascarina*, *Eugenia* and *Dysoxylum* as important components and also including *Coriaria*, *Astelia*, *Balanops* and *Freycinetia* (GILLISON & BEVERIDGE, 1974; BEVERIDGE, 1975; GILLISON, 1975).

The association between *Podocarpus*, *Dacrydium* and *Agathis* is found further west in Malaya and Borneo (BEVERIDGE, 1975). such associations are either montane or restricted to poor sandstone soils in the lowlands. On Mt. Kinabalu they are joined by many more Australasian genera such as *Phyllocladus* and *Leptospermum*. ASHTON (1971) described the scrub vegetation of poor sandstone soils in Sarawak and a photograph (his Plate 57a) of *Dacrydium*-dominated low forest shows vegetation similar in structure to that dominated by *Dacrydium araucarioides* and *Gymnostoma deplancheana* in the Plaine des Lacs, New Caledonia (Plates 8, 18). *Gymnostoma* is present in these Bornean lowland associations and also *Baeckea* (Myrtaceae; Australian centred) and *Tristania* (Myrtaceae; Malesian), both New Caledonian maquis genera. The genera *Podocarpus*, *Styphelia* subgenus *Leucopogon* and *Austrobuxus* (*Longetia*, Euphorbiaceae; Australasian but with its main centre in New Caledonia) also feature and species of *Eugenia* and *Calophyllum* are also associated to a certain extent.

The importance of studying vegetation associations as well as floristics in phytogeography has been stressed by GILLISON (1975). There is an apparent ecological segregation of Melanesian and Malesian forest types throughout Melanesia and the impression gained from literature on the pollen record of changes in the Australian flora (p. 35) is of vegetation associations advancing and retreating more or less as units in response to climatic changes. A detailed study of vegetation throughout the Indo-Australian tropics with this in mind could do much to increase understanding of the history of the Melanesian flora.

Though the forest vegetation of sedimentary and metamorphic rocks in New Caledonia has its counterparts in the rest of Melanesia, that of the ultramafics and their derivative laterites has many features that are unique, such as the high proportion of Australian elements, strong dominance by genera such as *Gymnostoma* and a general sclerophyllous structure determined by the water relationships of the laterite or parent rock. The rainforests on the ultramafics are perhaps more typically Melanesian, such as that on Mt. Boulinda referred to above, and that in the valley of the R. Bleue where Cunoniaceae, such as *Geissois*, Sapotaceae and *Nothofagus* are prominent. The niaouli savanna also lacks counterparts in Melanesia, being more referable perhaps to *Eucalyptus* associations in Australia. The niaouli (*Melaleuca quinquenervia*) also occurs in New Guinea but there it is more characteristically a swamp species (BROOKFIELD & HART, 1971). It was therefore interesting to note during the moth survey that it was virtually absent from the ultramafic localities visited in the central southern massif apart from a narrow fringe of trees at the edge of the herbaceous marsh association in the valley of the R. Blanche near Mt. Pouédihi.

The rest of this chapter will be devoted to an account of the New Caledonian vegetation as a background to the moth sampling programme and the analysis of its results. It will be seen in later chapters that the distribution of Lepidoptera within the island shows marked correlation with vegetation type. It was not possible to collect floristic data for the various associations sampled apart from noting any particularly dominant species and obtaining identifications at a later date from specimens or photographs taken. M. SCHMID of ORSTOM, Nouméa helped with much of this identification and further identifications were provided by Mr P. S. GREEN of the Royal Botanic Gardens, Kew.

No really comprehensive account of the vegetation has been published since the work of GUILLAUMIN (1952) and VIROT (1956), though detailed studies of two of the western ultramafic massifs have been published by JAFFRÉ (1974) and JAFFRÉ & LATHAM (1974). The following account is based largely on these works.

The classification system of FOSBERG (1967) is applied where possible to communicate something of the structure of each vegetation type to complement the accompanying photographs. The FOSBERG system has been criticised by GILLISON & BEVERIDGE (1974) and GILLISON (1975) for failing to give a sensitive classification of rainforest types. These authors used an unpublished system devised by L. J. WEBB in their work in the New Hebrides. But the greater structural diversity of the New Caledonian vegetation means that the FOSBERG system can be very useful and variations in forest structure can be conveyed by description and photographs where possible.

GUILLAUMIN recognised three broad categories of vegetation, namely maquis, forest and niaouli savanna. VIROT added to these the halophile associations of coasts and estuaries and the herbaceous associations that occur mainly in marshy situations.

VIROT divided the halophile associations into mangrove, suffrutescent or herbaceous strand vegetation and a marginal woody association. The first and last are associated in the New Hebrides, being successional stages correlated with distance and elevation from the sea. A *Rhizophora-Avicennia* zone gives way to a *Ceriops* zone with some *Acacia spirorbis*, which in turn gives way to more species-rich strand woodland (MARSHALL & MEDWAY, 1974). The mangrove areas in New Caledonia are much less extensive, occurring thinly along most coasts but being particularly associated with estuaries on the east coast. The zones are much compressed compared with those in the New Hebrides, consisting primarily of *Rhizophora-Avicennia*, the *Ceriops* zone only developing to any great extent in the Diahot estuary in the north (M. SCHMID, *per comm.*). VIROT also recognised a zone with *Bruguiera*, mainly along the fringes of sheltered lagoons. An extensive area of *Rhizophora-Avicennia* mangrove can be seen in Plate 9.

The strand woodland also is not highly developed, but again occurs along the sides of the estuaries. It is dominated by *Barringtonia* but also includes *Calophyllum*, *Guettarda* and *Acanthus*. VIROT recognised differences in the strand woodland depending on whether it occurred in association with a sandy or rocky shore and, if the latter, whether the rocky substrate was ultramafic or sedimentary rock. On the sandy shore the herbaceous strand vegetation is

*Plate 9.* Mangrove, *Leucaena* (slope in foreground) and grassland on the west coast to the north of Col Nassiboué.

*Plate 10.* Strand vegetation at Port Boisé; *Cocos*, *Pandanus* and *Barringtonia* in the foreground.

usually well developed and at Anse Longue (Plate 32) a scattering of *Barringtonia* trees grows along the strand with a herbaceous undergrowth (*Ipomoea*, *Canavallia*, *Atriplex*, *Euphorbia*).

The richest strand woodland encountered during the survey was at Port Boisé (Plates 6, 10) where a zone of *Barringtonia*, *Pandanus*, *Cocos*, *Scaevola* and other shrubs grows at the back of the beach and grades up a slope of ultramafic rock into low forest dominated by *Araucaria columnaris*. Associated with it were *Cycas* and *Crinum*.

No moth samples were taken specifically from mangrove though those from Col Nassiboué (63), Anse Longue (32), Cap Colnett (49) and R. Amoa (51) were in the neighbourhood of elements of the mangrove succession and the Port Boisé site 73 overlooked the strand woodland.

The strand and mangrove floras and their component genera are widespread in the Indo-Australian and Pacific tropics. The moth survey did not reveal any species strongly associated with them.

VIROT subdivided his forest category into six sub-categories.

**Forêts littorales climaciques.** These forests generally include all lowland coastal forests whether on the slopes of hills or in plains where degradation by erosion is not intense and where, in the coolest months, the temperature does not drop below 15°C. Such forest varies from low xerophyllous to mesophyllous and usually grades on hillsides into drier niaouli or *Acacia spirorbis* associations. There is little development of humus or ground cover and the association is generally dry. This forest type was not identified with confidence during the moth survey. VIROT indicated it had its greatest development in the Loyalty Is., the Isle of Pines and along the stretch of coast south from Yaté to Goro. The forest at Port Boisé and by the R. Fausse Yaté (sites 74 (Plate 20) and 67) sampled during the survey might belong to this category, as might that at Grottes de Koum (43; Plate 40), though the last is not strictly coastal.

**Forêts photophiles paludicoles.** Such forests are characteristic of marshy hollows inundated during the wet season and of the alluvial plains or rivers. They are dominated exclusively by niaouli, in this case paralleling its behaviour in New Guinea, with an undergrowth mainly of ferns and Cyperaceae. These forests are not abundant and tend to be associated with the more mature reaches of the larger rivers. Extensive areas of such forest were not encountered during the survey but niaouli was generally found to be taller, more luxuriant in association with rivers, and the herbaceous marshy association of the R. Blanche, seen in Plate 15, was fringed in places by a narrow strand of niaouli, the only locality in the central part of the southern ultamafics where this tree was noted during the survey.

**Forêts vallicoles mesophiles.** These forests develop on hillsides and slopes where valleys or ravines enable humus-rich, moist soils to accumulate. They are characteristic of such conditions over most of the drier part of the island on all rock substrates and most of the forest in New Caledonia is of this nature or of the next category. There is undoubtedly a degree of floristic variation depending on rock substrate and altitude, as will be seen when individual areas are discussed in more detail. The restriction to valleys and ravines is well illustrated in Plate 11, a view of the hills near Sarraméa from the Plateau de Dogny, though the effect has probably been accentuated by the degradation of the vegetation of the upper slopes and ridges by human activity, mainly fires.

**Forêts ombrophiles.** Similar rich multistratal evergreen forests (1A1.1 in the FOSBERG classification) develop away from the valleys where precipitation is

*Plate 11.* Hill near Sarraméa with valley forest interspersed with niaouli savanna on the ridges. The pale clear area in the centre at the bottom is site 29.

high enough, especially over the northern half of the east coast. The major areas of forest indicated in Figure 8 are mainly of this nature and were encountered during the survey on the slopes and hills near the Col d'Amieu (Plate 51) but more particularly on Mt. Aoupinié (Plate 12).

**Forêts oro-néphélephiles.** These are high altitude forests found above 800 m, to some extent cloud forests as they are frequently enveloped by clouds and bear the brunt of strong winds. They are floristically very diverse and are perhaps most characteristic of the higher mountains of the ultramafic massifs. The higher altitudes and cloud forest conditions permit more luxuriant growth of the shrubby vegetation of the associated crests and ridges, this ultimately being augmented by canopy dominants such as *Nothofagus* and *Podocarpus* species and canopy emergents such as *Agathis* and *Araucaria* species. Examples surveyed by the author were on the Mt. des Sources (Plate 30), Mt. Mou (Plates 28, 29) and on the summit of Mt. Panié (Plate 49). JAFFRÉ & LATHAM (1974) described two associations referable to this category on the summit of Mt. Boulinda.

**Forêts photophiles photo-xerophiles.** This category includes the more open forests of drier regimes, on cliffs and ravines, encompassing associations dominated by niaouli as well as more floristically diverse ones. Indeed they only have in common their more open canopy that permits an understorey of

*Plate 12.* Rich rainforest on sedimentary rock, Mt. Aoupinié.

shrubby species to occur. They will be discussed in more detail later but the various dominants include *Arillastrum gummiferum*, *Agathis ovata*, *Gymnostoma deplancheana* or *G. chamaecyparis*, and *Dacrydium araucarioides* as well as niaouli. Where the last is dominant there can be great variety in the type of understorey encountered.

VIROT divided his category of shrubby maquis associations into two. The first, *maquis sclerophylles-xerophiles edaphiques*, includes all the different types of non-forest vegetation on the ultramafics except those of his second category, *maquis orophiles*, characteristic of higher altitudes. VIROT recognised three zones in the island in these associations on floristic grounds. The first includes the extensive southern massif and those areas of ultramafic rock extending up the east coast. He listed three species of *Grevillea* and three of *Hibbertia* that characterise it. A number of species of both maquis and forest are restricted to this zone, some, such as *Callistemon pancheri* being virtually restricted to the Plaine des Lacs. The species include *Grevillea gillivrayi*, *Dacrydium araucarioides* (Plate 8), *Gymnostoma deplancheana* (Plate 18), *Agathis ovata* (Plate 15) and *Araucaria muelleri* (Plate 23). However, recently the *Grevillea* has been recorded from the western Massif de Boulinda (JAFFRÉ & LATHAM, 1974) and the *Gymnostoma* from the more northerly Massif de Koniambo (JAFFRÉ, 1974). These massifs are part of the second zone of VIROT, representing an area of overlap between northern and southern floristic zones.

The Massif de Koniambo is perhaps the most southerly of the massifs characteristic of the northern floristic zone which extends from Koné northwards. This zone also has characteristic *Grevillea* and *Hibbertia* species, together with species of *Gardenia, Myrtopsis* and *Soulamea.* The ecology of the ultramafic rocks is so different from that of the surrounding sedimentary rocks and basalts that their flora could be regarded as being distributed amongst islands of an archipelago. Varying mineral regimes in each massif such as the richness in chrome of the Dome de Tiebaghi could enhance this effect. The floristic zonation observed could have arisen in response to this as the ultramafic sheet was fragmented by erosion.

Broadly the non-rainforest ultramafic vegetation can be divided structurally into types of low, almost closed forest or woodland, often dominated by species of *Gymnostoma* or, especially on the western massifs, by *Acacia spirorbis*, and more open associations of low shrubs interspersed with Cyperaceae such as *Costularia* and *Baumea* and ferns, mainly *Pteridium* and *Gleichenia.*

The type of vegetation that develops in any locality is determined by the combined effects of relief, precipitation, hydrology and edaphic factors. JAFFRÉ & LATHAM recognised three principal phyto-edaphic groupings in the vegetation, namely (1) serpentiniphilous magnesicole vegetation on magnesium-rich soils with high capacity for base exchange, (2) oligotrophic vegetation on soils with low capacity for base exchange, and (3) hygrophilous vegetation on soils with an accumulation or organic matter. The first has high floristic specificity and includes the *Gymnostoma*-dominated associations, the second embraces most of the more open associations and the third includes the *forêts vallicoles* and *oro-néphéliphiles* of VIROT.

JAFFRÉ & LATHAM stated that the specificity of the floras of the different biotopes they examined on the Massif de Boulinda decreased from the magnesium-rich soils with high capacity for base exchange to those of low capacity for exchange, highly desaturated and poor in bases, and to those with accumulations of organic material. One passes in effect from a flora rich in species, exclusive to the substrates closely associated with the ultramafic serpentine rocks, to a flora approaching those characteristic of podsols on phthanites, or to a flora rich in common rainforest species; this gradation parallels an increasingly distant association between the character of the soil and that of the parent rock.

Within the three broad categories secondary groupings can be recognised that are related to relief, mineral regimes, climate, hydrology and altitude.

It is of great interest that the group of moth species closely associated with the vegetation of the ultramafics is generally much richer where the second and third categories of vegetation prevail. It is most characteristic of the second, the open associations, as species typical of rainforest generally tend to predominate in the third.

Maquis approaching that of the second category occurs on ridges and crests of the sedimentary and metamorphic rocks. It is generally much poorer floristically than its counterparts on the ultramafics, but this may in part be due to the activities of man, burning it periodically in hunting. It is probably also associated with the development of base-poor soils, podsols and laterites. It is particularly well developed in the northern part of the Panié range such as in the vicinity of the Col d'Amoss.

VIROT's *maquis orophiles* are best regarded as high altitude variants of the second category of JAFFRÉ & LATHAM. On the higher mountains in the southern massif, such as the Mt. des Sources, Mt. Humboldt and small areas of Mt. Mou, a dwarf association grows, adapted to windier conditions. A number of species are restricted to it such as *Xeronema moorei* (Plate 1) and *Greslania montana* (Plate 3). The *Baeckea ericoides* dominant in these associations is a prostrate form that is also found in the maquis of the adjacent lowlands, such as by the R. Bleue near site 10 (Plate 19). A species of moth, *Caledasthena montana*, belonging to an endemic monotypic genus, was found to be especially abundant in samples from this association on the Mt. des Sources.

JAFFRÉ (1974) described somewhat different *orophile* associations on the Massif de Koniambo at about 800 to 900 m on plateaux. It was dominated by *Tristania guillainii* but unlike similar associations lower down this was joined by characteristically orophilous species in the genera *Guioa* (Sapindaceae), *Styphelia* and *Metrosideros*. Elsewhere the orophilous species were dominated by an open stand of *Araucaria montana*.

The herbaceous associations described by VIROT included a variety of savanna types on the west coast that are usually more or less dominated by trees or shrubs such as niaouli, *Leucaena* and species of *Acacia*. Also there are a few more peculiar types such as those of the Plateau de Dogny and the Plaine des Lacs that will be described in detail later.

The scheme of classification applied by VIROT is somewhat too broad and loose to give a detailed impression of the variety of vegetation types to be found and of their interrelation. VIROT discussed several transects in various parts of the island that illustrated this to some degree and there are also the detailed accounts by JAFFRÉ and JAFFRÉ & LATHAM of two of the western

*Plate 13.* View from hills south of Grand Lac of Lac-en-Huit and extensive plains of *Baeckea*-sedge maquis extending to Pic du Pin.

ultramafic massifs. This diversity of vegetation will be illustrated by reference to these more detailed studies and to areas investigated during the moth survey.

The first and last months of the survey were spent in the Plaine des Lacs and its environs. Around Grand Lac and Lac-en-Huit, and between these lakes and Pic du Pin are extensive plains of deep lateritic and haematitic sediment derived from the ultramafics. The plains are divided up by ridges and hills of ultramafic rock as shown in Plate 13. Towards Nouméa in the west, and south towards Port Boisé the terrain is more undulating and dissected by valleys (Plate 14). As on the western massifs, the type of vegetation that develops depends on the water regime and edaphic factors.

Where water is permanently near the surface in the plains such as at Grand Lac (Plate 16) or in the flood plain of the R. Blanche (Plate 15) or R. Yaté, the herbaceous marshy associations (lMl.4, evergreen shortgrass and sedge bog in the FOSBERG system) mentioned by VIROT occur. There is a dense cover of Cyperaceae such as *Baumea* and *Costularia* (primarily Australasian) dominated by *Xyris* species (Xyridaceae; pantropical) and small shrubs, primarily Myrtaceae. The strand of Grand Lac and associated rivers supports a number of characteristic plants such as *Boronella pancheri* (Rutaceae; endemic with Australian affinities), *Podocarpus minor* (Plate 8), *Tristania* species and *Xanthostemon aurantiacum* (Plate 2). The R. Bleue is also bordered by a narrow

*Plate 14.* Skeletons of burnt *Arillastrum gummiferum* at Les Dalmates amid the more bushy maquis characteristic of this area.

strand of shrubby vegetation (seen in Plate 26) which includes another species of *Xanthostemon.*

With better drainage the laterite deposits of the plains support a shrubby maquis that also occurs on the steeper hillsides nearby and elsewhere on the southern ultramafics. At high altitudes it grades into the more dwarfed mountain maquis described by VIROT and seen on the Mt. des Sources in Plate 4, amid *Neocallitropsis. Baeckea ericoides* (Plate 2) is usually dominant and there is extensive ground cover of Cyperaceae, particularly an abundant *Costularia,* and ferns. This association, seen particularly in Plates 16, 17 and 18, could be described as a mesophyllous open evergreen broad sclerophyll scrub (1G1.2b) and its montane counterpart as open evergreen orthophyll dwarf scrub (1H1.1). On the western massifs such associations tend to be found on moderately to strongly desaturated base-poor soils (JAFFRÉ & LATHAM). For brevity they will be referred to henceforth as *Baeckea*-sedge maquis. *Dracophyllum* is also an important component and the terrestrial orchid, *Spathoglottis* (Plate 1), is often characteristic.

In the undulating terrain towards Nouméa (Plate 14) around Les Dalmates and the Col Mouirange and at Mine Anna-Madeleine, in the Yaté valley and at Houailou the maquis is taller, more diverse, yet still very open with perhaps

*Plate 15.* View of the headwaters of the R. Blanche, the valley bottom pale with the herbaceous marshy association (p. 92). A tree of *Agathis ovata* is in the foreground and in the background is the Mt. des Sources.

only 50% ground cover in places. *Baeckea* is less important and at Houailou the association includes a *Gymnostoma* (Plate 19), possibly *chamaecyparis* which is a dominant in some associations on the western massifs. Such more open associations might be termed mesophyllous evergreen sclerophyll steppe forest (2A1.2b). The maquis of intermediate slopes visited on Mt. Kaala and the Dome de Tiebaghi is similar and comparison with the published accounts of other western massifs suggests that these vegetation types may be generally characteristic of soils intermediate in desaturation and base content. They may also characterise rockier soils as compared with the finer textured soils favoured by the *Baeckea*-sedge maquis in its typical form. Some variations of this more open maquis might be termed evergreen sclerophyll shrub steppe savanna (2E1.2).

Associated with ravines, presumably wet seasonal water courses, and springs on slopes where closed forest does not develop, the open maquis becomes dominated by *Agathis ovata* as at Anna-Madeleine, on slopes east of Grand Lac and by springs on Pic du Pin (Plate 21). In the first two cases the ravines are cut deep into rocky lateritic deposits accumulated on shallower slopes with a very open maquis, perhaps as little as 25% ground cover. In the last case the springs emerge from steeper slopes amid typical *Baeckea*-sedge maquis.

*Plate 16.* Top left, small depression amid *G. deplancheana* association mentioned on p. 99; top right, *Baeckea*-sedge maquis on hills with *Gymnostoma* association in upper part of gully forest; bottom left, *Gymnostoma* and herbaceous associations on haematite; bottom right, herbaceous marsh association at Grand Lac.

*Plate 17.* Grand Lac. *Araucaria bernieri* and *A. muelleri* emergent from middle zone of gully forest. *Baeckea*-sedge maquis on near side of lake and the paler area beyond the lake is an extent of the herbaceous marsh association.

*Plate 18.* *Gymnostoma deplancheana* with lichen beneath at Grand Lac.

*Plate 19.* Top left, open maquis by the R. Bleue; top right, maquis at Houailou with *Gymnostoma* on the left and *Costularia* in the foreground; bottom left, *G. chamaecyparis* association in the Massif de Kopeto; bottom right, maquis degraded to *Pteridium*, Dome de Tiebaghi.

Most of the area between Grand Lac and Port Boisé, surrounding the valley of the R. Koué, is clad in thick, low forest dominated by *Gymnostoma deplancheana. Dacrydium araucarioides* is often a prominent component and the understorey contains a rich diversity of shrubs. A small fragment of this association near Grand Lac is shown in Plate 18 and some of the understorey shrubs can be seen in Plate 16, top left. In the FOSBERG terminology it is a non-resinous (largely) evergreen narrow (partially) sclerophyll forest (1A1.7b). The substrate generally appeared to be haematite rather than the more gravelly laterite of the plains near Grand Lac; it occurred there on elevated bands of haematite near the lakeside marsh association (Plate 16, bottom right). Where this vegetation occurred at Grand Lac the ground beneath was largely covered by lichens of the genera *Cladonia* and *Cladina*, and these occurred also in the more extensive stands of the association to the south-east. Some of this lichen can be seen in Plate 18. Beneath it there was usually an accumulation of humus that remained moist even after several days of intense sunshine, and it is possible that this lichen cover has a crucial role to play as regards the wellbeing of the association.

JAFFRÉ & LATHAM described a similar association dominated by *G. deplancheana* in the Massif de Boulinda (their station B5). It is probably the most northerly occurrence of the association and was found localised at the foot of a slope near permanent running water on base-rich soil (especially magnesium). There seemed to be a marked correlation between this vegetation type and the availability of water in the Plaine des Lacs also.

Plate 16 (bottom left) shows a fairly level area of haematite south-east of Grand Lac. In the foreground the haematite is low enough to be inundated periodically and its pockets contain elements of the Grand Lac marsh community together with *Drosera neocaledonica.* On slightly elevated areas in the background grows the dense *Gymnostoma* association. A further illustration of this relationship with the water table was found at site 5 where there was a large, steep-sided depression about 10 m deep amid an expanse of the *Gymnostoma* association. It was probably created by past mining operations and, when visited during the dry season (May), had a small pond in its bottom. But there were indications that it flooded almost to the brim at times as its sides were bare of vegetation apart from small seedlings of *Gymnostoma* and *Costularia.* Presumably these seedlings are killed each year when the water table rises in the wet season. A similar but very much smaller depression nearby, only one metre deep, was also almost bare of vegetation (Plate 16, upper left).

The vegetation of hillside ravines and water courses also illustrates the importance of water. Small, deeply cut ravines, dry in winter, are often marked only by the occurrence of *Agathis ovata* as mentioned earlier. More extensive and permanent water courses permit a denser forest association to develop. Typically this consists of rich rainforest at the bottom where the water course attains a greater gradient at the edge of the plain and humic material includes *Agathis lanceolata*; examples are found on the hillsides to the south of Grand Lac. The upper reaches of the rich forest contain emergent *Araucaria bernieri* and *A. muelleri* (Plate 17) and grade into the *Gymnostoma* association in the steeper, upper part of the water course where perhaps the parent rock is nearer the surface and has more influence on the vegetation (Plate 16, top right). The spring line is often marked by *Agathis ovata* or *Araucaria muelleri* (Plate 3, here seen unusually with *A. bernieri*).

In the R. Koué and Port Boisé area the *Gymnostoma* association again grades into richer forest where humic material accumuates. The forest at Port Boisé is a good example (Plate 20). It is virtually unistratal with little ground cover. The foreground of the plate shows regrowth of *Trema* (Ulmaceae; pantropical) and *Alphitonia austrocaledonica* (Rhamnaceae; Australasian) in a mining track cut through the forest. These small trees, particularly the latter, are important in the early stages of forest succession in New Caledonia and the *Alphitonia* features in other vegetation types as well. This forest grades into the coastal fringe vegetation with *Araucaria columnaris* discussed earlier.

Gully forest associations are widespread in the Plaine des Lacs, in the valley of the R. Blanche to the north, at Pic du Pin and westwards towards Nouméa. Sometimes, as on the relatively dry western side of the Col Mouirange, only the *Gymnostoma* association develops. In other areas, where hollows or the gentle relief of a valley permit humus to collect, the *Gymnostoma* may be absent and the richer forest develops directly. Examples were noted in a hollow near the summit of Mt. Pouédihi (site 66; *Podocarpus ustus* was discovered there) where the rich forest contained little *Gymnostoma* and was separated from the surrounding maquis by a fringe of *Alphitonia* (Plate 21, top right), and near the head of the R. Blanche (Plate 5) where majestic *Agathis lanceolata* rose from a skein of richer vegetation marking the course of a stream through the maquis.

The most complex development of hillside gully forest was observed at Pic du Pin (Plates 22, 23) and is worth describing in detail. In the valley bottom, where it flattens out into the plain of *Baeckea*-sedge maquis, there is a large area of almost pure *Nothofagus*, probably *N. aequilateralis* (Plates 21, top left, and 22). This may be a regenerative stand following some disaster such as burning, as the trunks are all very narrow and densely spaced, the only large trunks being fallen and decayed. Where the slopes begin to rise behind, a more mixed forest occurs with *Agathis lanceolata* as at Grand Lac. Yet higher, this forest forms a zone with emergent *Araucaria bernieri* which grades upwards into a *Gymnostoma* zone with emergent *Dacrydium araucarioides*. The undergrowth of the latter was dense, virtually impenetrable, and rich in palms and *Freycinetia*. The upper limits grading into the *Baeckea* maquis that characterises most of the hillside are marked by an association of gnarled shrubs with *Agathis ovata*, *Dacrydium araucarioides* and *Araucaria muelleri* (Plate 21, lower right). A stand of pure *Dacrydium* was noted near the head of a more minor, unforested valley (Plate 8, bottom right). The fringes of these forests and other unforested declivities slightly moister than the surrounding slopes are characterised by the pale foliage of *Greslania rivularis* (seen over the ravine in the foreground of Plate 22), the bamboo suggested to be the

*Plate 20.* Rainforest at Port Boisé, with *Trema* and *Alphitonia* regenerating in foreground.

foodplant of the butterfly *Austroypthima* by HOLLOWAY (1974b). Pic du Pin is the most important locality for this butterfly so far discovered. In the plain beyond the *Nothofagus* the waters run in a single deep ravine that is marked in its passage by a skein of darker, richer vegetation in the maquis (Plate 23).

*Plate 21.* Top left, area of *Nothofagus* in valley at Pic du Pin; top right, forest on Mt. Pouédihi, fringed by *Alphitonia* and with *Araucaria bernieri* emergent; bottom left, *A. muelleri* forest on the Mt. des Sources; bottom right, heads of valleys on Pic du Pin with *Agathis ovata* (especially foreground) and *Greslania rivularis* (paler growth).

*Plate 22.* Pic du Pin. *Baeckea*-sedge maquis in foreground and on far hills; *Nothofagus* forest in middle distance with more mixed forest with emergent *Araucaria bernieri* behind. *Greslania rivularis* hangs over the ravine in the foreground.

Steeper, rockier sided valleys in the south may permit the development of a more open forest with a maquis undergrowth dominated by *Arillastrum gummiferum* (Myrtaceae; endemic). This tree is readily killed by fire, having a resinous sap, and it has been seriously depleted by bush firing during prospecting. Skeletal forests can be seen in the valley of the R. Koué and at Les Dalmates (Plate 14). At R. Koué the understorey is basically the *Gymnostoma* association and at Les Dalmates the tree is regenerating amid a fragment of valley forest with *Agathis lanceolata* surrounded by tall, shrubby maquis. The *Arillastrum* association is at its most spectacular on the steep slopes of the Yaté Gorge (Plate 24) where it grows as part of an open evergreen orthophyll forest (1D1.1) co-dominant with *Cocconerion minor* (Euphorbiaceae; endemic). The bright green leaves of the *Arillastrum* and the grey green rosettes of leaves of the *Cocconerion* make a pleasing combination. Beneath these trees grows typical *Baeckea*-sedge maquis.

The flat, open valley of the R. Blanche is wet enough to exclude forest

*Plate 23.* Pic du Pin. *Araucaria muelleri* in the foreground with the main area of mixed forest below with emergent *Araucaria*. The *Nothofagus* grows in the narrow tongue in the middle distance just before the forest dwindles to the skein that follows the stream through the maquis.

except where tributaries enter it and support the herbaceous community already described for Grand Lac and seen in Plate 15. The neighbouring R. Bleue runs in a narrower valley, the slopes meeting the water's edge; multistratal evergreen rainforest (1A1.1) seen in Plate 25 has developed in the valley bottom. It grades into a sparse, open maquis on higher, drier ground (Plate 19, top left). *Nothofagus codonandra* is a frequent canopy component of this forest and the families Cunoniaceae (with *Geissois*) and Sapotaceae are also well represented. Occasional specimens of *Araucaria bernieri* are emergent. The skein of shrubby species fringing the more open parts of the river has already been mentioned and can be seen at the base of the forest in Plate 26.

A similar rich multistratal evergreen rainforest occurs in the valleys on the south and west slopes of Mt. Koghis. It was not feasible to explore this up to the summit but it appears (Plate 27) to grade to a more dwarfed habit on the higher ridges and towards the summit. On the lower ridges overlooking Nouméa the forest is often abruptly replaced by a poor (fire-impoverished) open *Baeckea*-sedge maquis dominated, unusually, by niaouli.

The localised and singular forests of the high ultramafic peaks and plateaux

*Plate 24.* The Yaté Gorge with slopes clad with the *Arillastrum* and *Cocconerion* association. In the left foreground is a tree of *Arillastrum gummiferum.*

are not easy to relate to the FOSBERG system. That on the summit of Mt. Mou grades from *Baeckea*-sedge maquis of increasing density, with *Pancheria hirsuta, Myodocarpus crassifolius* (Araliaceae; endemic, a genus used by CARLQUIST (1974) to exemplify insular radiation) and *Dracophyllum verticillatum* prominent, to a gnarled avergreen forest (1A1.3) dominated by *Nothofagus baumannii*, with *Araucaria rulei* emergent (Plates 28, 29). Also prominent in this forest are *Metrosideros oreomyrtus, Hibbertia baudouini, Beauprea pancheri* and a species of *Apiopetalum. Strasburgeria* also occurs (Plate 2). There is an undergrowth of pteridophytes, Cyperaceae and *Astelia.* The summit is usually cloud-covered and is marked by much taller forest grading towards 1A1.1 but with the canopy rather open, dominated by *Podocarpus* species such as *P. longefoliatus.*

The forested summit ridges of the Mt. des Sources are much steeper than those of Mt. Mou and the flatter parts between the summit and Pic Busé are clad in the dwarf montane maquis described earlier, punctuated by stands of *Neocallitropsis.* The forest is dominated by a fairly open canopy of *Nothofagus*

*Plate 25.* Rainforest in the valley of the R. Bleue.

*Plate 26.* Rainforest by the R. Bleue, showing the shrubby association of the riverbank.

*Plate 27.* Rainforest on Mt. Koghis.

*Plate 28.* Rainforest on the summit of Mt. Mou with *Nothofagus baumannii* (background), *Hibbertia baudouini* (right) and *Gleichenia* (bottom left).

*Plate 29.* View over the canopy of the *Nothofagus* forest of Mt. Mou, with emergent *Araucaria rulei.*

*codonandra* and *Araucaria muelleri,* the former predominating on the western slopes (Plate 30) and the latter, densely, on the eastern (Plate 21, lower left). Thus, respectively, these forests might be said to grade from gnarled evergreen sclerophyll forest (1A1.3b) almost to resinous 'narrow' (i.e. conifer) sclerophyll forest (1A1.7a), though the latter category is more strictly applied to northern hemisphere forests dominated by needle-bearing conifers. Where the *Nothofagus* association extends onto gentler slopes the tree is not so tall and *an undergrowth of Greslania montana* occurs, this bamboo being common also in the montane maquis (Plate 3, upper left). These forests are less diverse floristically than that of Mt. Mou. Lower down, for example on the slopes of Pic Busé, the valley forest is more microphyllous with myrtaceous trees such as *Metrosideros* prominent. The *Nothofagus* is still an important component but the emergent *Araucaria* is (probably) *bernieri* (Plate 7).

The western massifs arise from extensive undulating plains of grassland on the black tropical soils of the basalts. Plate 31 shows the Massif de Kopéto arising out of *Imperata* grassland with a shallow valley that is filled in the foreground with *Lantana* and in the background with a dense stand of niaouli. These extensive areas of grassland are much modified by man through cattle

*Plate 30.* *Araucaria muelleri* and *Nothofagus codonandra* forest near the summit of the Mt. des Sources.

*Plate 31.* West coast grasslands on basalt with the ultramafic Massif de Kopeto behind.

ranching. They are often dominated by niaouli to give the characteristic niaouli savanna (1J1.1, evergreen orthophyll low savanna) but areas are taken over by a variety of shrubs such as the introduced lantana or yellow guava (*Psidium guajava*), or the widespread Pacific shrub of disturbed habitats, *Leucaena leucocephala* (Leguminosae). This is often dominant and dense (Plate 9), giving microphyllous evergreen scrub (1B1.8a) or microphyllous evergreen forest (1A1.9) depending on the age of the stand. Areas visted around Ouitchambo and on the limestone crags of the Grottes de Koum approach the latter state.

Other areas of this grassland become dominated by a more diverse assemblage of Leguminosae, principally *Acacia farnesiana*. The herbaceous stratum also tends to be more diverse. This association attains its greatest development at Anse Longue (Plate 32) and areas encountered during the survey are illustrated by triangles in Figure 8. It is more open and lower than that dominated by *Leucaena* and is a microphyllous evergreen thorn scrub (1B1.8b). In wetter areas it becomes taller and more diverse floristically (darker areas in the photograph), almost an evergreen broad sclerophyll scrub

*Plate 32.* Leguminous thicket at Anse Longue.

(1B1.4). These west coast grassland associations are perhaps paralleled by the dry talasinga grassland of Fiji described by ROBINSON (1975).

A more extensive evergreen broad sclerophyll scrub is that dominated by *Acacia spirorbis* (gaiac). It characterises what are possibly secondary associations on the western ultramafics from La Foa northwards and occurs in small areas on the northern slopes of the gabbro rock of the Col de Petchécara, the territory of the Tribu de Poupa. One of these stands is shown in Plate 33. Most of the associations examined were almost pure *A. spirorbis*, perhaps with a few *Alphitonia* in the canopy and *Baeckea virgata* growing below (at Petit Gouli and Poupa only). There is only poor ground cover, usually of Cyperaceae (Plate 42, top left). The association is found extensively at low altitudes around the bases of the western massifs and has been described from two of them by JAFFRÉ (1974) and JAFFRÉ & LATHAM (1974). There *Codia montana* (Cunoniaceae; Australian) can be an important component and *Baeckea* is scarce; the *Acacia* is a component of some of the other associations of the western massifs and occurs infrequently elsewhere. Where it is dominant on Koniambo it occurs on deep ferralitic soils and on Boulinda it favours hydromorphic soil on old alluvium. The association was observed during the survey in the Massif de Kopeto and at the foot of Mt. Kaala near Kaala-Gomèn. Where alluvial fans emerge from the western massifs such as in the

*Plate 33.* The *Acacia spirorbis* association at Poupa. The tree on the left is *Casuarina collina.*

Plaine des Gaiacs in the Népoui area it occurs in a particularly luxuriant association together with a *Pandanus* species. On Ouen Toro near Nouméa the *Acacia spirorbis* and *Acacia farnesiana* grassland associations are intermingled. The flat summit plateau of the Dome de Tiebaghi is covered by almost pure *A. spirorbis.* Areas visited where the species is abundant are indicated by squares in Figure 8.

*A. spirorbis* occurs in the New Hebrides, mainly in the west of Erromango. There it is associated with brown soils on calcareous rocks or with ferralitic soils on basalts (SCHMID, 1975).

The lower slopes of the western massifs (observed during the survey at Mt. Krapé and in the Népoui valley in the Massif de Kopéto) are clad with an association of similar stature and, to some extent, structure to the *A. spirorbis* association but floristically very much more diverse. It covers by far the largest area of the Massif de Koniambo and is extensive on the Massif de Boulinda. It was termed 'groupement paraforestier à *Gymnostoma chamaecyparis*' by JAFFRÉ (1974). The *Acacia* is still a minor component and there are intergrades with the *A. spirorbis* association. The *Acacia* is also important in valley

*Plate 34.* Ouaté valley in the Massif de Kopéto. To the right is the association with *Pandanus* characteristic of valleys and the slopes on the left are clad with the *Gymnostoma chamaecyparis* association.

bottoms amid this association where *Casuarina collina*, niaouli and *Pandanus* are frequent (Plate 34). The typical *Gymnostoma* association is characteristic of base-rich (especially magnesium) soils with a high exchange capacity. The *Gymnostoma* generally exceeds the other shrubs in stature and its brighter green balls of needles are prominent in the canopy. This can be seen in Plate 19 (lower left) where the association is relatively tall (3-4 m), bushy, with an open ground layer. Plate 35 illustrates a more dwarfed, denser version about 1.5 m tall from the higher slopes of Mt. Krapé.

On Mt. Krapé in the area of site 55 there is a development of rainforest in a col and the head of the valley of the R. Kraoui (Plate 36). It is tallest and most developed near the valley bottom (Plate 37) but grades gently up the slopes into the *Gymnostoma* association as seen in Plate 35. Scattered *Araucaria* (probably *montana*) grow out of the maquis just above the forest.

JAFFRÉ (1974) described valley forests on the Koniambo massif to the north. They show great variety, occurring mainly at higher altitudes and having as dominants either the more gregarious *Nothofagus* (*codonandra* and *balansae*) or a species of *Cocconerion* near *minor.* Another forest type has *Araucaria bernieri* as a canopy emergent, *A. montana* being characteristic of the dwarf, shrubby orophilous association mentioned on p. 91.

The vegetation of Mt. Kaala is generally similar to that of the Massif de Koniambo (VIROT, 1956).

*Plate 35.* Cross section of low, dense *Gymnostoma chamaecyparis* association on Mt. Krapé, the *Gymnostoma* being emergent (the crowns top left and right).

*Plate 36.* Gully rainforest on Mt. Krapé, grading into the *Gymnostoma* association up the slope.

Of the more northerly massifs of the west coast only the Dome de Tiebaghi was studied in detail during the survey. The ultramafic substrate is particularly rich in chrome. The summit plateau is clad entirely with *Acacia spirorbis* with the occasional *Alphitonia neocaledonica*; this could be a secondary association following burning during prospecting. The slopes of the Dome are mainly covered by an open, bushy maquis as seen in the foreground of Plate 39. The floristic components of this were not ascertained but it probably parallels the more open, bushy associations of the slopes of Boulinda and Koniambo

*Plate 37.* Gully rainforest at its most developed state on Mt. Krapé.

described by JAFFRÉ and JAFFRÉ & LATHAM. *Tristania guillainii* is present and important in places. This maquis has been impoverished by burning in some areas, marked by extensive *Pteridium* (Plate 19, lower right). *Gymnostoma* appeared virtually absent, a few plants being observed in the western ravine descending to Paagoumène (probably *chamaecyparis*).

On the upper slopes this ravine flattens and broadens to contain a low forest association unique to Tiebaghi. It is exceedingly dense with a thick even canopy where *Tristania guillainii* is frequent (Plate 38). *Acacia spirorbis* is also present. It is probably best classed as an evergreen hardwood orthophyll forest (1A1.4). At its fringes it is probably 3-4 m tall and appeared to be in an area with numerous springs and seepages which coalesce to produce the stream running through the ravine below.

The top of the Dome de Tiebaghi affords a view eastwards across the north of the island that conveys an impression of the terrain of those parts (Plate 39). In the middle distance is an extensive area of dry, deeply dissected schist country and in the far distance can be seen the long ridge of gneiss that culminates further south in Mt. Ignambi and Mt. Panié.

Emergent from the schist country just inland from Koumac are pinnacles and crags of limestone. They are somewhat more richly forested than the surrounding terrain though this forest has been degraded to the *Leucaena* association by the activities of man in places. Plates 40 and 42 (bottom left) show one of these crags at Grottes de Koum beside the Koumac River. A tall,

*Plate 38.* View of the forest association of the Dome de Tiebaghi. The bushes in the foreground are *Tristania guillainii.*

*Plate 39.* View east across the north of the island from the Dome de Tiebaghi. The dry schist hills are in the middle distance and the gneiss ridge of the northern part of the Panié range in the background.

though much disturbed forest with little ground layer grows at the base of the crags, becoming more stunted as the gradient increases up the crag (Plate 42). Streams, dry when visited, issue from the base of the crags to join the river. The forest contains amongst other species a large *Ficus* and an important component is a large, pale grey barked deciduous tree seen in both plates but most particularly in the foreground of Plate 40. It was tentatively identified from this photograph as a species of *Gyrocarpus* (Hernandiaceae; widespread) by M M. SCHMID. Similar limestone outcrops occur at Linderalique and Hienghène on the east coast but it was not possible to examine these closely.

A typical expanse of the schist country is shown in Plate 41, by the valley of the R. Bambou. It represents probably the driest and poorest (on poor podsols) extreme of a range of vegetation types dominated by niaouli, a single specimen of which is illustrated in Plate 3, upper right. These niaouli associations are probably mostly secondary, arising after fire damage (GUILLAUMIN, 1952). Near habitation in the north and in most areas on the west coast it has

*Plate 40.* Rainforest by limestone crags, Grottes de Koum.

an undergrowth of *Imperata* grass, occasionally with a plant or two of *Baeckea virgata*. *Imperata* has also penetrated areas of the east coast but only in more level areas suitable for cattle and in the immediate vicinity of villages. Elsewhere the ground layer is composed of pteridophytes, particularly *Pteridium* and *Gleichenia*, Cyperaceae and *Baeckea virgata* or, on the northern schists and gneiss, *B. parvula*. Other elements characteristic of the *Baeckea*-sedge maquis such as *Dracophyllum*, *Scaevola* and the orchid *Spathoglottis* (Malesian; widespread) may also occur. In the schist hills this association is impoverished and open as can be seen in Plate 41. The niaouli is stunted on the slopes and ridges but grows taller and more densely in the valley bottoms where it mingles with *Casuarina collina* and trees more typical of the forests. This gradation in stature of the niaouli can also be seen in Plate 42, bottom right, showing a valley running down the eastern slopes of the gneiss range near the Col d'Amoss. These niaouli associations could be referred to FOSBERG categories such as evergreen broad sclerophyll low savanna (1J1.2) where ferns predominate or evergreen sclerophyll shrub savanna (1K1.2) where shrubs are abundant in the undergrowth.

On the crest of the Col d'Amoss occurs a variety of vegetation types that could broadly be termed evergreen broad sclerophyll dwarf steppe scrub (2C1.2). In places there are outcrops of the parent rock, gneiss with quartz.

*Plate 41.* Vegetation of the dry schist country. Low open scrub on the hills give way to a taller forest in the valley.

Niaouli is still dominant but of gnarled and stunted habit. The scrub is of variable richness depending on how recently it has been burnt. At its poorest it consists of niaouli with *Gleichenia* and Cyperaceae but where relatively undisturbed it is enriched with *Baeckea parvula, Dracophyllum verticillatum* and other shrubs; several species of terrestrial orchid also occur. An expanse of this scrub is shown in Plate 43. Included is a skein of forest vegetation (with a species of *Gymnostoma* frequent) that follows the course of a stream. The niaouli grows taller beside the stream. Further upstream is a deep ravine that is much more thickly forested, the forest including *Ficus* and palms.

A similar scrub occurs on the Plateau de Dogny at about 1000 m in the centre of the island. Both areas of such scrub are indicated by stars in Figure 8. Both were notable in that they supported the endemic satyrid butterfly, *Paratisiphone lyrnessa*, a species most characteristic of the ultramafic *Baeckea*-sedge maquis. The Dogny scrub is also dominated by dwarfed niaouli and a *Baeckea* species, either *virgata* or *ericoides* but of a prostrate form. There is evidence of burning in places and consequent impoverishment. There are areas

*Plate 42.* Top left, structure of the *Acacia spirorbis* association at Poupa; top right, the dense, shrubby association of the ultramafic outlier on the Col de Boa; bottom left, the vegetation of the limestone crags by the Grottes de Koum; bottom right, valley east of the Col d'Amoss showing niaouli increasing in stature with descent into the valley, with some forest elements at the bottom.

*Plate 43.* Dwarf scrub on the plateau north of the Col d'Amoss, with forest vegetation following a stream. On the left foreground is *Dracophyllum verticillatum*.

with almost pure *Gleichenia* or *Gleichenia* with *Lycopodium cernuum* and others rich in Cyperaceae such as *Costularia*. At its richest the scrub includes *Dracophyllum* and the *Spathoglottis* species characteristic of the southern maquis. An area of the richer scrub can be seen in the foreground of Plate 45 and it covers the hill in the background of Plate 44. Flatter, marshy areas are vegetated with a pure sward of a reed, possibly a *Juncus* or a plant of similar habit, and such an area can be seen in the middle distance of Plate 44. As on the Col d'Amoss numerous small streams on the Plateau are marked by skeins of forest, also seen in Plate 44. This forest and that on the upper slopes of the sides of the Plateau (Plate 45) is lower and more microphyllous than that of the lower slopes and is rich in Myrtaceae, particularly *Metrosideros*.

All these niaouli associations with ferns, *Baeckea* and other shrubs that characterise the hills of the east coast and crests and ridges in the centre of the island would appear to be secondary, probably following burning. Near habitation they become altered by the invasion of *Imperata* sward and the encroachment of introduced shrubs such as lantana and the yellow guava. Evidence of

*Plate 44.* Associations of the Plateau de Dogny. Dwarf scrub in foreground and on hill in background, marshy association in the middle distance and skeins of rainforest following the streams.

recent burning is frequently encountered. It was interesting to note that *Imperata* had only penetrated the dwarf scrub for perhaps 1 km to the north of the road over the Col d'Amoss before being replaced by Cyperaceae. The rich forests of the gullies and larger valleys evidently escape this degradation by burning, as illustrated by the skeins of forest amid the scrub mentioned and the

*Plate 45.* The edge of the Plateau de Dogny, showing the dwarf scrub with *Dracophyllum verticillatum* (right foreground) and the low forest of the upper slopes.

gully forests shown in Plate 11. Plate 46 shows the almost vertical north face of the roches d'Ouaième. In the foreground is a hillock with the niaouli, fern and scrub association but the cliffs are clad with pure fern apart from the forested gullies. The niaouli, fern and scrub association is pre-European and probably very ancient; it was noted by Captain COOK on parts of the east coast in 1774 (BEAGLEHOLE, 1955–1969).

This prompts the question of whether there was a natural association that characterised the ridges and crests and their associated poor soils on the mountains of sedimentary and metamorphic rock, an association that has now largely been lost by degradation following the activities of man. Elsewhere in Melanesia these habitats are often vegetated with low *Metrosideros* forest and there are indications on Mt. Dogny and other high mountains that this may have been more widespread in New Caledonia as well; perhaps the generally drier climate rendered it more vulnerable to fire damage. This drier climate may also have led to the development of a more sclerophyllous and fire-prone association in more extreme conditions on the steeper slopes and narrower ridges, especially on the drier western side of the island. Such sclerophyllous associations probably contained elements in common with the *Baeckea*-sedge maquis of the ultramafics; this appears to be indicated by the present remnants and the associated moth species.

*Plate 46.* The Roches d'Ouaième. The niaouli dominated fern and shrub association typical of much of the east coast is seen on the slope in the foreground.

The only relatively undisturbed non-forest vegetation on sedimentary rock on the drier side of the island observed during the survey was that on the steeper slopes and crags east of the Col d'Amieu by the Bréhoa and Ouambéo rivers. Plate 47 shows such a steep slope clad in what might be termed evergreen sclerophyll scrub with trees (1E1.2), the trees being *Araucaria subulata* (probably) and the principal component of the scrub being *Baeckea virgata*.

The question of the original vegetation of the peaks and ridges now occupied by niaouli scrub must await more detailed investigations by plant ecologists. It is possible that some of the more swampy parts of the Plateau de Dogny might yield cores with pollen profiles that would provide some of the answers. A detailed investigation of the relationship of floristic richness to frequency or recency of burning would also be valuable.

VIROT (1956) also discussed this problem and suggested that some of the more inaccessible areas of the metamorphic rocks in the north-east might still

*Plate 47.* Shrubby association with emergent *Araucaria subulata* on a steep slope near the Ouambéo R.

support a rich maquis. He noted an association of the margins of the valley forests of the southern slopes of the Roches d'Ouaième that included *Beauprea, Podocarpus, Calophyllum, Austrotaxus, Agathis,* Palmae and tree ferns that graded into a more dwarfed maquis with niaouli, *Xanthostemon, Baeckea, Nepenthes* and various pteridophytes. The former association resembles that found on the higher slopes of the Panié range and the dwarf maquis was stated by VIROT to grow on a reddish lateritic clay, a similar substrate to that of the Plateau de Dogny as far as could be told from a superficial examination.

The lower forests of the Panié range are typical of valley forests on sedimentary rocks though they include few really large trees. The upper forests, gnarled evergreen forests (1A1.3), are dominated by *Podocarpus* and have a rich undergrowth of palms, *Pandanus* and *Freycinetia.* On the older palm fronds on Mt. Ignambi near the summit there are encrustations of moss and other small epiphytes, testimony of the frequent cloud cover. These

*Plate 48.* Part of the lower, more shrubby type of vegetation of Mt. Panié, with abundant *Freycinetia.*

Ignambi forests are relatively tall but the generally steeper slopes of Mt. Panié are often clad in a more open vegetation, a sort of maquis with more microphyllous shrubs such as *Metrosideros*, an abundant tangle of *Freycinetia* (Plate 48) or emergent *Araucaria montana* (Plate 3, lower right). Some areas near the summit are almost entirely forested with *Pandanus* and the narrow summit plateau is covered with a magnificent forest of *Agathis montana* with an understorey of large *Pandanus* (Plate 49). A characteristic understorey shrub of the more open spaces of this summit zone is a pink-flowered *Mearnsia* (Plate 2). The *Agathis* is also a frequent canopy emergent in the subsummit *Podocarpus* forests (Plate 8, lower left).

These high altitude forests of the wetter north-east of the island are probably most closely analogous to those of high altitudes elsewhere in Melanesia.

Two further vegetation types sampled during the moth survey require description. They resembled each other in having a thick undergrowth of diverse shrubs of which *Baeckea virgata* was the principal component. Both appeared to be on fairly level outcrops of ultramafic rock with only shallow soils, that on the Col de Boa (inland from Poya) mossy and relatively rich in humus. On the Col de Boa *Acacia spirorbis* is dominant with some *Casuarina collina* and niaouli (Plate 42, upper right). The other, near the mouth of the R.

*Plate 49.* The summit plateau forest of Mt. Panié; *Agathis montana* with *Pandanus*.

Amoa, is dominated by relatively tall niaouli and is adjacent to very open maquis with tall bushes.

A rather cursory account such as this by a zoologist cannot hope to do justice to the complexity and variety of the New Caledonian vegetation and the ecological factors that lie behind it. The fact remains that few plant ecologists have worked in New Caledonia until quite recently. This is all the more surprising when one considers the biochemical discoveries that might be made through study of the adaptation of a unique assemblage of plants to a variety of unusual mineral regimes. Also the current activity in the mining field calls out for corresponding activity in the plant ecological field in order that a programme for conservation or regeneration of the vegetation can be developed. The work of botanists at ORSTOM in Nouméa, particularly that of T. JAFFRÉ in conjunction with soil experts, promises to fill this gap if augmented and adequately financed. It is a pity that much of this work (references in JAFFRÉ (1974)) is not published in a form more accessible than mimeographed reports.

# IV. THE SAMPLING PROGRAMME

The author visited New Caledonia for four months from early May to early August, 1971. Most of this time was spent in the field sampling moths from as many different types of vegetation as possible.

Samples were made with light-traps (Plate 50) incorporating lamps with a high proportion of emission in the ultra-violet. A Robinson-pattern mercury vapour moth trap was used in sites accessible by road, a restriction imposed by the weight of the Honda E300 generator used to power it. As the generator would only run for about six hours it had to be refuelled at midnight, a process occasionally jeopardised by oversleeping.

Elsewhere a small, collapsible, portable trap was used, incorporating a 6 watt actinic tube run from a 12 volt rechargeable motorcycle battery. This trap, based on a design by HEATH (1965), is made and marketed by Entech Services of Liverpool who kindly supplied the two traps used at cost. An additional feature of these traps, discovered during the survey, is that, with muslin clipped across the top of the central funnel with clothes-pegs and a paper extension taped to the bottom, they make serviceable Berlese funnels. Extractions were made in New Caledonia by placing the litter in the muslin, covering the top of the funnel and the litter with a glass plate (culinary), air being able to circulate beneath this because it was lifted up by the clothes pegs, and leaving the apparatus in the sun for a day. Several such extractions were made from litter from a variety of vegetation types and further litter samples were extracted at the ORSTOM laboratories in Nouméa. All these extractions are now with Professor J. BALOGH of the University of Budapest (who joined the survey for a few days at Sarraméa and Kopéto).

The actinic light-traps produced catches about half the size or less of those from the Robinson-pattern trap but nevertheless proved most useful in a number of areas not accessible by road and, when the Robinson trap was run near the vehicle, could be taken off for half an hour or so on foot to another site to be collected the next morning. The survey of Norfolk I., described by HOLLOWAY (1977), is being continued and conducted entirely with such a trap.

Though the survey concentrated on night-flying macroheterocera, a qualitative collection of microlepidoptera was made and butterflies and day-flying moths were also collected. Accounts of the butterflies have already been published by HOLLOWAY (1974b) and HOLLOWAY & PETERS (1976). Samples were also made of the large numbers of Trichoptera and Tipulidae that came to the light-traps. The former are being worked on by Professor H. H. ROSS and representatives of the latter have been sent to Professor C. P. ALEXANDER after preliminary determination at the BM(NH) by Mr P. CRANSTON. Psocoptera were collected by sweeping and beating and are being worked on by Dr. C.

*Plate 50.* The light traps used. Left, the portable, battery-run Entech trap on Mt. Mou in the summit forest; right, the Robinson-pattern mercury vapour trap amid the *Gymnostoma* association of the Plaine des Lacs.

N. SMITHERS and Professor I. W. B. THORNTON in Australia in conjunction with their own extensive collections (e.g. THORNTON & SMITHERS, 1974).

The survey was conducted from a Volkswagen Combi van (seen in Plates 33 and 40) that served as both caravan and laboratory. It coped with most New Caledonian road conditions when tolerably dry. It had to be engineered out of a muddy pothole by the R. Bleue, survived a hair-raising slippery retreat off the Mt. des Sources in torrential rain but foundered in a muddy lay-by near Le Crouen before regaining the road virtually carried by six large Melanesians. It gave a much smoother ride than a Land Rover, important when boxes of pinned specimens are involved.

The sampling of macroheterocera was quantitative and many common species were counted in the field. This was on the whole satisfactory, though duplexes in the genera *Anomis*, *Tiracola*, *Lacera* and *Poecilasthena* were not noted till later, nor was the nature of the genus *Fabresema* suspected. In most of these cases sufficient material was retained to give a good subjective assessment of the broad ecological distribution of the species and to get a rough quantification for inclusion in the species/abundance curves of Figure 27.

Sampling in the cool, dry season has the advantage that the roads are in good condition, permitting access to most of the localities they serve. A disadvantage is that there are numerous cool, clear, often windy, moonlit nights, conditions that are not conducive to obtaining large samples.

Maximising the coverage of vegetation types during the survey resulted in many small samples from numerous localities and a number of moderately large ones when conditions were favourable (usually occurring every few weeks when frontal systems from the west brought warm, cloudy weather, usually with torrential rain). The optimum periods for sampling would probably be those months preceding and following the worst of the wet season (late December to March).

The first month was spent sampling localities in the southern ultramafic massif, particularly the Plaine des Lacs area. The second saw sampling in the west coast habitats at Anse Longue and various localities with rich rainforest or the *Acacia spirorbis* association on the road system from Thio to La Foa via Canala and the Col d'Amieu. At the end of the month a transect on Mt. Mou was sampled. The third month was devoted to the north of the island, covering ultramafic areas on both east and west coasts, various vegetation associations in the extreme north of the island and localities at high altitudes on the Mt. Panié range. Towards the end of the month samples were made in the area where the most luxuriant rainforest on sedimentary rock grows (Mt. Aoupinié), followed by further samples from west coast localities. The final month was spent sampling further localities in the southern ultramafic area, resampling old localities there to assess seasonal changes, and making a few final samples from Ouen Toro near Nouméa.

In the list of localities following, the sites are each given a number and are indicated by this number in Figure 8. Several samples were rejected for the numerical analyses described in the next chapter as being too small and others were combined, having been made close to each other overlooking the same type of vegetation. They therefore had to be renumbered for the numerical analysis and such numbering is given in *italic*. Each locality is identified by a UTM grid reference. The nature of the vegetation is indicated briefly, with cross-reference to the previous chapter. The number of individuals and species taken, the dates of sampling and type of trap used are noted, as are weather conditions. Use of the Robinson pattern trap is indicated by 'MV' and of the portable trap by 'acl'. The proportions of species referable to the three major ecological elements identified in the next chapter are given as percentages thus: Forest & Secondary; Ultramafic; West coast/*A. spirorbis*. Finally species taken only at a particular locality are noted together with others of interest such as those with abnormally high representation.

## List of localities

**Site 1.** *1.* 06980/75358. 250 m. Grand Lac. In area backed by hillside with *Baeckea*-sedge maquis including some *Gymnostoma* association and an open maquis with much bare laterite exposed (e.g. Plate 18). 11.5.71, acl, cloudy, warm, rainy; 12.5.71, acl, cool, clear, full moon; 20.5.71, MV, clear, coolish. 68 individuals, 26 species. 46; 50; 4.

*Table 4.* Gazetteer of place names mentioned in text, with UTM grid references. Sampling localities are numbered in bold type.

Amieu, Col d' (05246/76488) **26**
Amoa R., valley (05298/76860) **51**
Amoss, Col d' (04390/77548) **45–47**
Anna-Madeleine (06903/75404) **76**
Anse Longue (06218/75543) **15, 16, 32**
Anse Vata, *see* Nouméa
Aoupinié, Mt. (05298/76594) **62**
Bâ, near (05531/76586) **59**
Bambou R. (04304/77369) **44**
Blanche, R. (06720/75500)
Bleue, R. (06704/75544) **8–10**
Boa, Col de (05250/76490) **57**
Boguen, Col (05656/76089) **37**
Boisé, Port (07039/75276) **73, 74**
Boulinda, Mt., Massif (05156/76530)
Bourail (05509/76130)
Bréhoa R. (05853/76139) **24**
Busé, Pic (06644/75493) **13, 14**
Canala (05994/76196)
Ciu, Cascade (06016/76153) **21**
Colnett, Cap (04742/77333) **49**
Coulée, La (06620/75402)
Crouen, Le (05917/76179)
Dalmates, Les (06714/75424) **77, 78**
Diahot, R. (04396/77490)
Dogny, Mt. (05902/76088) **31**
Dothio (06156/06132) **18, 19**
Fausse Yaté, R. (06984/75478) **67**
Foa, La (05853/75985)
Fonwhary (05827/76024)
Goro (07069/75344)
Grand Lac (06975/75356) **1–4, 71, 72**
Hienghène (04934/77117)
Houailou (05644/76458) **58**
Humboldt, Mt. (06470/75790)
Humboldt, R. (06419/75725)
Ignambi, Mt. (04574/77374) **48**
Kaala-Gomén (04351/77158) **38**
Kaala, Mt. (04351/77163)
Kavatch (04857/77068) **50**
Koghi(s), Mt. (06550/75456) **79, 80**
Koh (05866/76177) **22**
Koné (04852/76606)
Koniambo, Massif (04800/76700)
Kopéto (05102/76542) **30**
Koué, R. (07030/75307) **75**
Koum, Grottes de (04314/77286) **43**
Koumac (04258/77260)
Krapé, Mt. (05276/76375) **55, 56**
Kuenthio, R. (06153/75982) **17**
Laguerre, Port (06365/75495)
Linderalique (04960/77119)
Louis, St. (06590/75409)
Me Maoya (05346/76370)
Mou, Mt. (06386/75576) **33–36**
Mouirange, Col (06701/75407) **65**
Nassiboué, Col (06334/75497) **63**
Népoui (04971/76389)
Nouméa (06487/75324)
Ouaïème, Roches de (04850/77173)
Ouambéo R. (05854/76132) **25**
Ouatou R. (05863/76151) **23**
Oubatche (04619/77398)
Ouen Toro (06497/75321) **64, 81, 82**
Ouitchambo (06033/75854)
Paagoumène (04140/77342)
Païta (06467/75517)
Panié, Mt. (04770/77260) **60, 61**
Petchécara, Col de (06128/76135)
Petit Gouli (05852/76047) **28**
Pic du Pin (06879/75386) **7, 69, 70**
Plaine des Lacs (07020/75363) **5, 6**
Plum (06706/75329)
Poindimié (05340/76843)
Ponérihouen (05415/76690)
Pouébo (04557/77447)
Pouédihi, Mt. (06728/75474) **66**
Poupa (06120/76150) **20**
Poya (05156/76402)
Probokonoro (05447/76357) **53**
Prony (06860/75306)
Roussettes, Col des (05467/76304) **54**
Sarraméa (05888/76065) **29**
Sources, Mt. des (06649/75520) **11, 12**
Tamoa, La (06347/75601)
Tao (04790/77265)
Therèse, Ste. (05238/76815) **52**
Thi, Forêt, valley (06582/75446)
Thio (06255/76092)
Tiebaghi (04168/77382) **39–42**
Tiébo, Col (04524/77426)
Tinchialit = ? Col Tiébo
Tipindjé (05000/77065)
Toma, Col (05813/76123) **27**
Tontouta, R. (06231/75685)
Touho (05242/77018)
Voh (04688/76822)
Yahoué (06538/75437)
Yambé (04640/77385)
Yaté Gorge (06955/75499) **68**

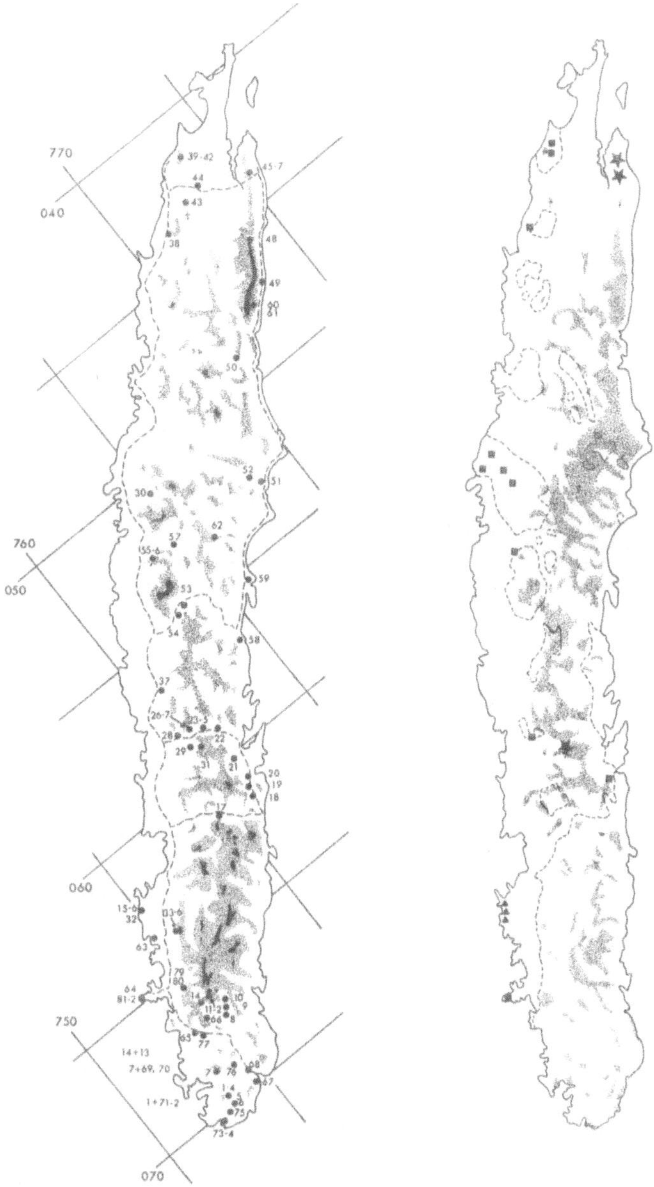


*Fig. 8.* Maps of New Caledonia showing (left) the UTM grid, major roads (dashed lines), land over 500 m (light stipple), land over 1000 m (heavy stipple) and the moth trapping localities (numbered), and (right) the areas of rainforest (stipple), ultramafic rock (broken lines), dominant *Acacia spirorbis* encountered during the survey (squares), west coast leguminous thicket encountered (triangles), and dwarf maquis scrub on sedimentary or metamorphic rock (stars).

**Site 2.** *2.* 06975/75356. 250 m. Grand Lac. Overlooking extensive *Baeckea*-sedge maquis seen surrounding lake in Plate 17. 13.5.71, acl, clear, moonlit; 14.5.71, MV, overcast then clear; 15.5.71, MV, overcast, warm, rain later; 16.5.71, MV, very windy with rain; 9.8.71, MV, overcast but light from moon, drizzle; 10.8.71, MV, cool, cloudy; 11.8.71, MV, cool, clear, getting cold and foggy. 830 individuals, 95 species. 55; 40; 5. The single *Chloroclystis macroaedeagus* was taken here and one of two *Thallarcha multifasciata.*

**Site 3.** *3* (+72). 06965/75355. 260 m. Grand Lac. On hillside of *Baeckea*-sedge maquis between two area of gully rainforest as described on p. 99–100. 17.5.71, MV, cool, clear; 18.5.75, MV, overcast, warm. 320 individuals, 68 species. 52; 47; 1. Singletons of *Stictoptera describens* and *Theretra nessus* were taken here and at site 2. They are likely to be more widespread and are perhaps rare during the dry season.

**Site 4.** *4* (+71). 06980/75367. 250 m. Grand Lac. At edge of herbaceous marsh association beside *Gymnostoma* association, i.e. in the middle distance of Plate 16, lower right. 19.5.71, MV, clearish and warm at first, sporadic rain later. 132 individuals, 44 species. 49; 47; 4. This locality and site 71 produced abundant *Adeixis inostentata,* which was not recorded elsewhere. *Egone bipunctalis* was also common.

**Site 5.** *5.* 07020/75362. 240 m. Plaine des Lacs. In an extensive area of *Gymnostoma* association; the MV trap in Plate 50 is located in site 5. 20.5.71, MV, overcast, warm, light to moderate rain. 67 individuals, 25 species. 56; 44; 0. One of two *Caledophia pauli* was taken here but the catch was very low considering the conditions and range of the outlook; it is roughly half the site 4 catch made in similar conditions. Possible reasons for this are discussed on p. 184.

**Site 6.** 07020/75364. 240 m. Plaine des Lacs. Sited in similar vegetation to site 5 but overlooking the water-filled excavation described on p. 99. Plate 16, top left, shows some of the vegetation nearby. 20.5.71, acl, weather as for 5. 4 individuals, 2 species. The species were *Teara circumfumata* and *Leucocosmia nonagrica.* This was another surprisingly small catch.

**Site 7.** *6.* 06879/75383. 240 m. Pic du Pin. In *Baeckea*-sedge maquis by extensive *Nothofagus* forest as in Plate 22. The trap was sited just where the forest narrows to a skein in the middle distance of Plate 23. 23.5.71, MV, part overcast early, then clear, cool; 24.5.71, MV, weather as above; 7.8.71, MV, clear, cool, bright moon; 8.8.71, MV, overcast, warm, but bright from moon. 699 individuals, 89 species. 58; 42; 0. The single female *Micronia cynthiae* was taken here.

**Site 8.** 7 (+9). 06755/75530. 150 m. R. Bleue. Ultramafic rainforest as described on p. 104. 25.5.71, MV, Clear, cold. 48 individuals, 30 species. 72; 27; 0.

**Site 9.** 7 (+8). 160 m. 06734/75544. R. Bleue. Up the valley from site 8, the rainforest somewhat richer (Plate 25), taller. 26.5.71, MV, Clear, cold. 57 individuals, 25 species. 67; 29; 4.

**Site 10.** *8.* 06704/75544. 180 m. R. Bleue. Further up the valley still with rainforest present as in Plate 26 but areas of maquis (Plate 19, upper left) were closer. 27 and 28.5.71, MV, cold, clear. 111 individuals, 43 species. 58; 42; 0.

**Site 11.** *9.* 06650/75512. 760 m. Mt. des Sources. Overlooking broad valley of montane maquis (p. 91) as in Plate 4. 29.5.71, MV, clear with setting moon, then high cloud; 30.5.71, acl, warm, clearish at first, cloudy later. 283 individuals, 54 species. 55; 41; 4. The single *Adeixis montana* was taken here and both this and the next site yielded large numbers of *Caledasthena montana* and *Chloroclystis lunifera*. A single *Agrotis munda* taken here on the first night (and conceivably also the satyrid butterfly, *Melanitis leda*, found in the trap at the same time), together with the *A. ipsilon* taken at the next site on the second night may have been Australian vagrants associated with the frontal system that arrived on the morning of 31.5.71.

**Site 12.** *10.* 06653/75513. 760 m. Mt. des Sources. Not far from the previous site but overlooking the area of *Araucaria muelleri* forest described on p. 109 and shown in Plate 21, lower left. Montane maquis was extensive on slopes all around. 29.5.71, acl, and 30.5.71, MV, with weather as noted for site 11. 172 individuals, 36 species. 68; 32; 0. One of two *Eupithecia rigida* was taken here.

**Site 13.** 06635/75513. 500 m. Mt. des Sources. Valley forest with emergent *Araucaria bernieri* and *Nothofagus codonandra* on the slopes of Pic Busé as described on p. 109 and illustrated in Plate 7. *Baeckea*-sedge maquis was also present. 31.5.71, acl, wet, warm, windy. The top of the trap was blown off during the night and extinguished. 7 individuals, 5 species.

**Site 14.** *11.* 06636/75488. 500 m. Mt. des Sources. Slopes of Pic Busé as for 13 but a broader outlook and more *Baeckea*-sedge maquis. 31.5.71, MV, weather as for 13. Trapping from 1800 to 2130 only as generator failed. 771 individuals, 66 species. 76; 19; 4.

**Site 15.** *12* (+16). 06230/75551. 30 m. Anse Longue. In col in low range of hills seen on right of Plate 32, overlooking extensive leguminous thicket and savanna as described on p. 111. 4.6.71, acl until 0100 hrs when battery ran out, warm, heavy rain. 30 individuals, 12 species. 40; 10; 50.

**Site 16.** *12* (+15). 06239/75536. 20 m. Anse Longue. The trap was sited on the grassy knoll in the foreground of Plate 32 with a broad outlook of the leguminous thicket vegetation. 5.6.71, MV, clear, cool, moonlit; 28.6.71, acl, part overcast, a little rain, windy. 148 individuals, 38 species. 57; 2; 41. The only two *Eublemma anachoresis* and the singleton *Helicoverpa armigera* were taken here.

**Site 17.** *13.* 06153/75982. 120 m. R. Kuenthio. Overlooking the valley of the R. Kuenthio virtually on the northern margin of the southern ultramafic massif. To the south the hillside sloped up into typical *Baeckea*-sedge maquis but to the north the valley was vegetated with niaouli-*Imperata* savanna, small patches of gully rainforest and secondary scrub such as *Leucaena* and lantana. 6.6.71, MV, clear, moon almost full. 140 individuals, 46 species. 77; 14; 9. The single *Eublemma glaucochroa* and one of two *E. baccalix* were taken here.

**Site 18.** *14.* 06176/76128. 90 m. Dothio. The trap was sited in a small col between the Dothio and Nekouitouru rivers. Poor *Baeckea*-sedge maquis with much *Pteridium* alternated with small patches of gully forest; on ultramafic rock. 7.6.71, MV, warm, full moon with thinnish cloud cover. 72 individuals, 38 species. 69; 25; 6. The single *Neogyne elongata* was taken here.

**Site 19.** *15.* 06133/76135. 400 m. Dothio. South-facing slopes above a deep valley full of rainforest on gabbro rock. It was a tribal area with cleared patches of *Imperata* sward, niaouli savanna and secondary scrub (lantana, guava) fringing the forest just on the south side of the Col de Petchécara. 8.6.71, MV, clear, cool, full moon; 9.6.71, MV, moderately warm, moon lightly overcast, 188 individuals, 71 species. 93; 6; 1. One of two *Gymnoscelis sara* was taken here.

**Site 20.** *16.* 06120/76150. 200 m. Poupa. On gabbro-diorite on the north (drier) side of the Col de Petchécara. The area sampled is shown in Plate 33. Niaouli-*Imperata* savanna in the foreground is divided from dense *Acacia spirorbis* association in the background (p. 112) by a fragmentary skein of rainforest marking the course of a small stream. 10.6.71, MV, warm, moon obscured by ridge and then by cloud. 280 individuals, 74 species. 74; 11; 15. This was one of two sites where *Westermannia poupa* was taken. *Cleora acacicola* was abundant and six *Prasinocyma albicosta* were taken, both species showing marked association with the *Acacia.*

**Site 21.** 06016/76153. 250 m. Cascade de Ciu. Sedimentary rock. The trap was sited in the clearing at the top of the cascade amid an amphitheatre of rainforest. 11.7.71, acl, warm and rainy. 18 individuals, 13 species. 100; 0; 0.

**Site 22.** *17.* 05866/76177. 140 m. Koh. The trap had an extensive outlook over the sides of a valley with much niaouli-*Imperata* savanna, secondary scrub (*Leucaena*, lantana etc.), with a few small patches of gully rainforest. 12.6.71, MV, warm and hazily overcast at first, then cool, clear, moonlit. 1000 individuals, 93 species. 88; 7; 5. Large catches of *Callopistria maillardi, Oxyodes ochreata, Mocis trifasciata, Rhodogastria albivitrea,* the two commoner *Leucania* species and *Gnathothlibus erotus,* all species widespread in the island and possibly associated with open, disturbed habitats on sedimentary rock. The single *Anisodes kohensis* was taken here.

**Site 23.** *18.* 05863/76151. 320 m. Ouatou R. Similar to the previous site but with rainforest extensive in the valley overlooked by the trap. 13.6.71, MV, clear, moonlit later. 239 individuals, 69 species. 89; 5; 6. *Ctenoplusia placida* was commonest here.

**Site 24.** *19.* 05853/76139. 320 m. Bréhoa R. As 23 but rainforest much more extensive. 14.6.71, MV, clear, cool; 15.6.71, cloudy at first, clear, cool later. 675 individuals, 98 species. 85; 10; 5. The single *Chloroclystis pyrsodonta* was taken here.

**Site 25.** *20.* 05854/76132. 320 m. Ouambéo R. Similar to site 24 but close to the cliffs with *Araucaria* and maquis described on p. 126 and illustrated in Plate 47. 16.6.71, acl, clear at first but overcast with rain later. 85 individuals, 51 species. 82; 16; 2. A few of the ultramafic maquis species were recorded here.

**Site 26.** *21.* 05828/76117. 320 m. Col d'Amieu. Trap set in front of the forestry station overlooking cultivated ground, stands of forestry species, mainly conifers, some niaouli amid a general area of rainforest on sedimentary rock. 17.6.71, MV till midnight, 9 cm of rain, trap flooded with both moths and water; 18.6.71, MV, overcast but little rain. 1245 individuals, 102 species. 87; 9; 4. Five female *Xanthorhoe sodaliata* taken on the first night could have been Australian vagrants travelling with the frontal system that brought the rain. A single *Anoratha paritalis* was taken here.

**Site 27.** *22.* 05813/76123. 460 m. Col Toma. Hillcrest vegetation with niaouli and fern and shrub undergrowth, though sparse and often planted with *Pinus elliotti.* This graded into taller shrubby vegetation with *Alphitonia, Trema* and one or two maquis shrubs, fringing the extensive rainforest. 18.6.71, acl, overcast but little rain; 19.6.71, acl, cloudy, a little rain. 391 individuals, 80 species. 77; 19; 4. This site was notable for the occurrence of several ultramafic maquis species, particularly eight *Adeixis major,* nine *Horisme grisearia* and seven *Aeolochroma hypochromaria.* The only *Simplicia toma* were taken here and five out of seven *Ziridava xylinaria.*

**Site 28.** *23.* 05852/76047. 30 m. Petit Gouli. The trap was sited in a narrow, pastured valley between two hills of ultramafic rock clad with pure *Acacia spirorbis* association; *Baeckea virgata* and *Alphitonia* were frequent. 20.6.71, MV, clear, cool. 193 individuals, 58 species. 70; 15; 15. Two out of four *Collix ghosha* were taken here and six *Prasinocyma albicosta.*

**Site 29.** *24.* 05888/76065. 160 m. Sarraméa. Valley of the Fo Ouia near the farmhouse of M. Julien Bonnard. The trap was set in the area of pasture seen in the centre, bottom of Plate 11. The area was surrounded by extensive rainforest and there was a clump of large *Albizzia* trees nearby (Plate 51). 21.6.71, MV, clear, cool; 22.6.71, MV, clear, cool; 23.6.71, MV, rain at first then clear; 27.6.71, acl, overcast, warm. 588 individuals, 102 species. 86; 5; 9. 21 *Rhesalides asphalta* taken here had probably been feeding on the *Albizzia* trees as larvae. One out of two *Eupithecia rigida,* the single *Eoasthena sarramea* and *Eublemma rivula* were taken here as well as two out of four *Cleora dothionis.*

**Site 30.** *25.* 05102/76542. 160 m. Kopéto. Ouaté valley, Massif de Kopéto. The trap was set in an area similar to that seen in Plate 34, covering part of the more bushy, open form of the *Gymnostoma chamaecyparis* association and the valley vegetation with *Acacia spirorbis* described on p. 113–4. 24.6.71, MV, clear, cool. 182 individuals, 58 species. 67; 14; 19. One out of two *Digama marmorea* was taken here.

**Site 31.** *26.* 05902/76088. 950 m. Plateau de Dogny. The trap was sited on the edge of the plateau covering both the dwarf scrub and the low forest of the upper slopes described on p. 121 and seen in Plate 45, 26.6.71, acl, cloudy with occasional moderate rain. 116 individuals, 28 species. 75; 21; 4. Several maquis species were taken here, most notably 19 *Horisme grisearia* but the *Baeckea*-feeding *Syneora strixoides* was also common, as was *Tiracola circularis.*

**Site 32.** *27.* 06218/75543. 5 m. Anse Longue. The trap was set in an area of leguminous thicket vegetation by moister, almost forest growth in a small valley. The location can be seen in Plate 32 where the darker skein of

*Plate 51.* The pasture slopes at Sarraméa where the trap was set for site 29. The extensively forested slopes of Pic Vincent behind the Col d'Amieu can be seen in the background and the picture is framed by *Albizzia* trees.

vegetation meets the centre of the left margin of the picture just above the curve of the beach. 29.6.71, acl, partly overcast, warm at first. 214 individuals, 37 species. 65; 5; 30. One of two *Digama marmorea* was taken here and *Traminda mundissima* was exceptionally abundant, together with *Scopula undilinea*.

**Site 33.** *28*. 06387/75574. 400 m. Mt. Mou. Lower slopes with *Baeckea*-sedge maquis just above, two areas of valley forest on either side (the lower, richer zones) and disturbed cultivated and secondary habitats below. Ultramafic rock. 1.7.71, MV, overcast, misty, with long spells of light rain. 2728 individuals, 104 species. 80; 16; 4. The single *Hippotion boerhaviae* was taken here.

**Site 34.** *29* (+36). 06386/75576. 500 m. Mt. Mou. Uphill from site 33 in extensive *Baeckea*-sedge maquis. 1.7.71, acl, weather as for 33. 302 individuals, 37 species. 73; 22; 5.

**Site 35.** 06385/75594. 1110 m. Mt. Mou. The trap was set on the summit ridge amid the *Nothofagus* forest but near the *Podocarpus* forest (p. 105, Plates 28, 29, 50). 3.7.71, acl, mist and drizzle probably. 9 individuals, 7 species. 25; 75; 0. Maquis species predominated, namely *Euphyia perseverata*, *Epiplema lateritica* and *Poecilasthena dimorpha*. The single *Macaduma cristata* was taken here and a number of microlepidoptera not noticed elsewhere. The most interesting capture was two out of four *Chlorocoma octoplagiata*. The other two were taken at the highest Mt. Panié site and the feature common to the two areas is *Podocarpus* cloud forest.

**Site 36.** *29* (+34). 06383/75584. 800 m. Mt. Mou. Ridge of *Baeckea*-sedge maquis overlooking upper part of gully forest with *Gymnostoma* association (p. 100). 4.7.71, acl, clear, cool, moonlit. 37 individuals, 18 species. 50; 44; 6. *Caledasthena montana* was more abundant here than at lower altitudes.

**Site 37.** 05656/76089. 250 m. Col de Boguen. Overlooking valley dominated by niaouli, but also with *A. spirorbis*, *Baeckea virgata*, lantana and grasses present. There was evidence of extensive recent burning. 7.7.71, acl, cool, variably overcast, fullish moon. 14 individuals, 7 species. 100; 0; 0. The species were almost entirely drawn from the more widespread, perhaps secondary growth associated extreme of the forest and secondary element.

**Site 38.** 04351/77158. 80 m. Kaala-Gomén. Near town rubbish dump at the foot of Mt. Kaala. Weedy secondary vegetation in the immediate vicinity but much *Acacia spirorbis* and some *Gymnostoma chamaecyparis* association on the slopes further off. 8.7.71, MV, partly overcast, warm, but bright with full moon. 32 individuals, 14 species. 36; 7; 57. Species of the west coast/*Acacia spirorbis* association predominated.

**Site 39.** 04168/77382. 440 m. Dome de Tiebaghi. Almost pure *Acacia spirorbis* on the summit plateau; *Alphitonia* and *Baeckea parvula* present. 9.7.71, acl, windy, cold, flurries of rain with moonlit periods. The bulb was blown off the base of the trap during the night. 5 individuals, 5 species, namely *Prasinocyma albicosta*, *Anisodes niveopuncta*, *Eois tiebaghi*, *Cleora acacicola* and *Parosteodes grisea*. Except for the second and last, these species show correlation in their local distribution with the *Acacia*.

**Site 40.** 04174/77379. 420 m. Dome de Tiebaghi. Overlooking the low, open bushy maquis described on p. 116 and seen in the foreground of Plate 39 which illustrates the outlook of the trap. The plateau *A, spirorbis* was also near. 10.7.71, MV, weather as for 39. 23 individuals, 13 species. 77; 15; 8.

**Sites 41 and 42.** *30.* 04156/77366 and 04153/77365. 330 m. Dome de Tiebaghi. Overlooking the dense valley forest described on p. 118 and shown in Plate 38. 10.7.71, MV and acl, warm, overcast but sky brightened by moon. 112 individuals, 40 species and 20 individuals, 9 species. 47; 38; 15, and 33; 43; 23. Three *Eois tiebaghi* were taken and the single *Hemithea acaudata.*

**Site 43.** *31.* 04314/77286. 20 m. Grottes de Koum. Rainforest at foot of limestone crags with secondary vegetation including many Leguminosae nearby at riverside (p. 118, Plate 40). 11.7.71, MV and acl, clear, warmish, moonlit later. 149 individuals, 37 species. 78; 5; 17. *Idaea scintillans* was frequent.

**Site 44.** *32.* 04304/77369. 50 m. Bambou R. A locality in the dry northern schist hills. The outlook is shown in Plate 41, the open niaouli association grading to a taller forest in the valley being described on p. 120. 12.7.71, MV, clear, getting cold and moonlit later. 350 individuals, 53 species. 76; 8; 16. The site was notable for the only specimen of *Ophisma gravata* and a very large catch of *Platysenta illecta.*

**Site 45.** *33* (+46). 04390/77548. 370 m. Col d'Amoss. The trap was sited near the road amid the low niaouli scrub described on p. 120. There was taller niaouli-*Imperata* savanna further down the slopes. 13.7.71, MV, cool, windy, overcast at first. 71 individuals, 22 species. 86; 14; 0.

**Site 46.** *33* (+45). 04378/77545. 440 m. Col d'Amoss. The trap was sited on the plateau to the north of the col on the slope in the right foreground of Plate 43 amid the low niaouli scrub. 13.7.71, acl, weather as for 45. 45 individuals, 11 species. 91; 9; 0.

**Site 47.** *34.* 04368/77540. 500 m. Col d'Amoss. A little further north than site 46, again in the niaouli scrub but overlooking the deep, forested ravine described on p. 121. 14.7.71, acl, overcast with a little rain. 106 individuals, 25 species. 76; 24; 0. 21 of the ultramafic maquis species *Euphyia perseverata* were taken.

**Site 48.** 04574/77374. 180 m. Mt. Ignambi. The trap was sited in a small grassy clearing amid montane forest; *Podocarpus* probably dominant. 16.7.71, acl, strong wind, overcast and misty, rain at times. 18 individuals, 10 species. 100; 0; 0.

**Site 49.** *35.* 04742/77333. 5 m. Cap Colnett. Trap set at foot of cascade amid the niaouli, fern and shrub association (p. 120). Some rainforest trees occurred near the water and there was secondary scrub and grasses by the road and towards the seashore. 17.7.71, MV, warm, overcast. 127 individuals, 43 species. 79; 12; 9. *Argina cribraria* was common.

**Site 50.** *36.* 04857/77068. 40 m. Kavatch. On hillock behind the Castex forestry station, Hienghène valley. Pasture in valley, niaouli-*Imperata* savanna on hillock and fragments of rainforest nearby. 18.7.71, acl, warm, overcast with a little rain. 151 indivduals, 49 species. 89; 7; 4.

**Site 51.** *37*. 05300/76859. 10 m. Amoa. 2 km inland from main road. Niaouli with thick *Baeckea virgata* and maquis undergrowth adjacent to more open, tall bushy maquis on ultramafic rock as described on p. 128. Mangrove not far distant. 19.7.71, MV, warm, humid, overcast with some rain. 490 individuals, 89 species. 85; 10; 5. The single *Lasioceros aroa* and the two *Cleora injectaria* were taken here. *Chalciope alcyona* and *Trigonodes hyppasia* were unusually common, possibly coming from the extensive pasture by the road and across the river in a coconut plantation.

**Site 52.** *38*. 05238/76815. 100 m. Ste. Therèse. On spur of Mt. Tinbounon overlooking Amoa valley. Pasture dominated by niaouli with fragments of rainforest nearby; on ultramafic rock. 19.7.71, acl, weather as for 51. 198 individuals, 49 species. 79; 13; 8. Notable for the capture of five *Cerura amoa* and one of two *Plusiodonta theresae.*

**Site 53.** *39*. 05447/76357. 70 m. Probokonoro. Rich rainforest with patches of secondary vegetation on sedimentary rock, overlooking Houailou valley. 20.7.71, MV, warm, humid, overcast, light to moderate rain. 2923 individuals, 113 species. 94; 5; 1. The two *Calyptra argyrosigna* and the four *Hypena probokonora* were taken here, and also eight *Leucogonia amarginata.*

**Site 54.** *40*. 05467/76304. 340 m. Col des Roussettes. Amid niaouli savanna overlooking a broad arena of primary rainforest on the boundary of the ultramafic and sedimentary rocks. 21.7.71, acl, cool, overcast, light drizzle. 147 individuals, 50 species. 85; 9; 6.

**Site 55.** *41*. 05276/76375. 460 m. Mt. Krapé. Rainforest grading into *G. chamaecyparis* association in col at head of Kraoui valley as described on p. 114 and seen in Plates 36 and 37. The trap was sited on the edge of the road to the left of Plate 36. 22.7.71, MV, warm, overcast. 1256 individuals, 94 species. 73; 16; 11. One of two *Plusiodonta theresae* was taken here and, also at the next site, large numbers of *Scopula desita.*

**Site 56.** *42*. 05274/76373. 460 m. Mt. Krapé. Around spur from 55 in an area of almost pure *G. chamaecyparis* association as in Plate 35. 22.7.71, acl, warm, overcast. 168 individuals, 53 species. 65; 27; 8. One of the two *Caledophia pauli* was taken here and also the single *Chloroclystis callinda.*

**Site 57.** *43*. 05250/76490. 210 m. Col de Boa. Ultramafic outlier amid extensive niaouli savanna. The dense shrubby association with *A. spirorbis* an important component is described on p. 128 and illustrated in Plate 42, upper right. 23.7.71, acl, overcast but windy. 81 individuals, 26 species. 64; 28; 8. The single *Speiredonia mutabilis* was taken here.

**Site 58.** *44*. 05706/76444. 5 m. Houailou. On Poro road by a small valley on ultramafic rock. Some of the open bushy maquis containing both *Gymnostoma* and *A. spirorbis* can be seen in Plate 19, upper right, and is referred to on p. 93. 24.7.71, acl, overcast, humid, warm, with heavy rain at times. 68 individuals, 30 species. 50; 37; 13.

**Site 59.** *45*. 05531/76586. 190 m. Near Bâ. In isolated ultramafic hills on east coast. Rich, bushy maquis grading into low gully forest in places. 25.7.71, MV, clear, cool, 346 individuals, 59 species. 73; 25; 2.

**Site 60.** *46*. 04764/77257. 800 m. Mt. Panié. In zone of *Podocarpus* forest and areas of low maquis as in Plate 48, described on p. 128. 27.7.71, acl, partly overcast, warm. 112 individuals, 31 species. 50; 50; 0. One of the more unusual samples made. The two *Chlorocoma octoplagiata* were mentioned in connection with site 35 but there was also the only *Idaea paniensis* and relatively high catches of *Maceda savura, Dirades latibrunnea, Symmetroctena variegata, Cleora illustraria* and *Chloroclystis lunifera.* A number of ultramafic maquis species were represented. A transect sampled on Mt. Panié would probably prove most interesting.

**Site 61.** 04774/77265. 400 m. Mt. Panié. The trap was set beneath the canopy of the more typical sedimentary rainforest of the lower slopes. 28.7.71, acl, warm, partly overcast. 46 individuals, 13 species. 91; 9; 0. *Cleora immemorata* was much commoner than *illustraria* and *Miresa obscurimargo* was numerous, the latter perhaps characteristic of subcanopy samples.

**Site 62.** *47.* 05298/76594. 520 m. Mt. Aoupinié. Rich rainforest on sedimentary rock on the north slopes in the Napoué Areu valley, an area known as Haut Mou. The trap was set in a sawmill clearing; Plate 12 indicates the stature of the surrounding forest. 30, 31.7.71, MV, generally cool and clear. 289 individuals, 66 species. 89; 10; 1. The single *Anadevidia peponis* and the two *Elphos subrubida* were taken here as well as one of the more distinctive *Fabresema, catherinae.*

**Site 63.** *48.* 06334/75497. 80 m. Col Nassiboué. On coast between Nouméa and Anse Longue near Tiaré. Part of the outlook is shown in Plate 9. There is extensive *Leucaena,* some niaouli and small areas of leguminous thicket. 1.8.71, acl, clear, coolish. 138 individuals, 28 species. 57; 7; 36.

**Site 64.** *49.* 06497/75321. 100 m. Ouen Toro. This hill at the end of the Nouméa peninsula is clad in thicket that is a mixture of niaouli savanna, the *A. spirorbis* association, Anse Longue ticket, with introduced scrub (lantana, *Schinus*) and possibly a few of the maquis elements. 2.8.71, acl, cool, clear, moonlit later. 98 individuals, 37 species. 55; 10; 35. Samples from Ouen Toro (81 and 82 also) included both *Casbia albinotata,* the singleton *Leptotroga costalis* and *Crioa acronyctoides* and most of the *Idaea fruticeti* taken.

**Site 65.** *50.* 06701/75407. 220 m. Col Mouirange. Overlooking the broad valley to the north of the col. Mainly dry, bushy maquis with an area of *Gymnostoma* association in the valley bottom. 3.8.71, MV, mainly clear and moonlit. 147 individuals, 38 species. 41; 59; 0.

**Site 66.** *51.* 06728/75474. 540 m. Mt. Pouédihi. Trap set in *Baeckea*-sedge maquis near gully forest fringed with *Alphitonia* discussed on p. 100 and illustrated in Plate 21, top right. 4.8.71, MV, clear, cool, brightly moonlit. 75 individuals, 25 species. 44; 56; 0.

**Site 67.** 06984/75478. 60 m. R. Fausse Yaté. Trap overlooking the valley with dry, tall rainforest, sited in an area of *Baeckea*-sedge maquis. 5.8.71, acl, warm but with bright moon. 9 individuals, 4 species.

**Site 68.** *52.* 06955/75499. 120 m. Yaté Gorge. The *Arillastrum* and *Cocconerion* association discussed on p. 103 and illustrated in Plate 24. 6.8.71, MV, partly overcast but cloud brightened by moonlight. 57 individuals, 24 species. 48; 52; 0.

**Site 69.** 06880/75385. 300 m. Pic du Pin. Trap set in clearing amid *Nothofagus*. 7.8.71, acl, clear, cool, moonlit. 12 individuals, 7 species. 50; 33; 17. Two *Miresa obscurimargo* and two species of *Fabresema* were recorded.

**Site 70.** 06880/75386. 350 m. Pic du Pin. Trap sited in open area between *Nothofagus* forest and higher mixed forest (p. 100). 8.8.71, acl, overcast, warm but bright from moon. 4 individuals, 3 species, including one *Fabresema*.

**Site 71.** *4* (+4). 06980/75368. Grand Lac. 250 m. Amid herbaceous marsh association as for site 4. 9.8.71, acl, overcast but bright from moon. 16 individuals, 4 species, all of the ultramafic association.

**Site 72.** *3* (+3). 06965/75354. 270 m. Grand Lac. Near site 3 but somewhat further up the hillside with a better outlook. 11.8.71, acl, cool, clear. 72 individuals, 25 species. 42; 58; 0.

**Site 73.** *53*. 07039/75275. 15 m. Port Boisé. Trap was set at edge of rainforest overlooking the strand vegetation described on p. 86 and illustrated in Plates 6 and 10. Ultramafic rock. 12.8.71, MV, 13.8.71, acl, clear, cool. 63 individuals, 31 species. 69; 14; 17.

**Site 74.** *54*, 07039/75276. 20 m. Port Boisé. Further inland in tall rainforest as described on p. 100 and illustrated in Plate 20. 12.8.71, acl, 13.8.71, MV, clear, cool. 215 individuals, 63 species. 84; 9; 7. The singletons of *Eunoumeana fletcheri*, *Schrankia boisea* and *Chloeres boisensis* were taken here.

**Site 75.** *55*. 07030/75307. 40 m. R. Koué. Overlooking valley of Koué river where taller forest with much *Arillastrum*, largely burnt, grew amid extensive *Gymnostoma deplancheana* association. 14.8.71, MV, clear but warm. 65 individuals, 27 species. 52; 48; 0.

**Site 76.** 06903/75404. 250 m. Anna-Madeleine. On Plaine de Lacs road by Mine Anna-Madeleine. A deep erosion ravine in laterite is flanked by an open *Gymnostoma*/*Dacrydium* association with numerous *Agathis ovata*. Extensive *Baeckea*-sedge maquis is also present. 15.8.71, MV, refuelling overslept, overcast, warm. 38 individuals, 24 species. 50; 50; 0.

**Site 77.** *56*. 06713/75420. 220 m. Les Dalmates. Ultramafic rock clad mainly in bushy maquis (p. 93) with burnt *Arillastrum* and fragments of gully rainforest; the former is shown in Plate 14. 18.8.71, MV, clear at first then overcast, warmish. 124 individuals, 46 species. 57; 38; 5. One of two *Agrotis munda*, a single *Athetis tenuis* and the only specimen taken of the Australian pyralid, *Nomophila australica* were recorded in this sample. The first two were recorded as vagrant to Norfolk I. later in the year (HOLLOWAY, 1977) and the last is also vagrant there. Their co-occurrence in this sample away from more characteristic habitats could indicate that they too were vagrants from Australia.

**Site 78.** *57*. 06714/75424. 220 m. Les Dalmates. In next valley to east from site 77, with similar vegetation but the gully forest less extensive. 18, 19.8.71, acl, mainly warm and overcast. 110 individuals, 23 species. 45; 50; 5. Several *Fabresema* species were recorded here.

**Site 79.** *58*. 06550/75463. 500 m. Mt. Koghis. Rich rainforest on ultramafic rock. The trap was sited on a track running into the forest in the foreground of Plate 27. 20.8.71, acl, cool, strong wind; 21.8.71, MV, clear, cool, breezy. 116

individuals, 41 species. 71; 22; 7. One of two *Anoratha paritalis*, several *Schistophleps subtilis* and species of *Fabresema* were recorded here.

**Site 80.** *59.* 06556/75456. 600 m. Mt. Koghis. Ridge with poor *Baeckea*-sedge maquis dominated by niaouli (p. 104), with small valleys full of rainforest on either side. 21.8.71, acl, clear, coolish breezy, 94 individuals, 35 species. 67; 33; 0.

**Site 81.** *60.* 06494/75317. 10 m. Ouen Toro. As for site 64. 24.8.71, acl, clear, warm. 94 individuals, 35 species. 44; 9; 47.

**Site 82.** *61.* 06497/75318. 10 m. Ouen Toro. As for site 64. 25.8.71, acl, clear but warm; 26.8.71, acl, warm, overcast. 134 individuals, 35 species. 36; 10; 54.

The total sample included just under 21,500 individuals.

# V. THE ECOLOGY OF THE NEW CALEDONIAN MACROHETEROCERA

The samples of moths made during the survey of New Caledonia differ from those analysed for Norfolk I. by HOLLOWAY (1977) both in their smaller size and in their greater species-richness. Figure 9 illustrates the latter difference, showing plots of the relationship between size of sample and number of species taken in each area. The upper points represent amalgamations of samples, the ultimate being the total sample. Such amalgamations are mixtures of the three ecological associations of species identified later in this chapter, as are the majority of individual samples except some of those from the richer rainforest vegetation on sedimentary rock (solid triangles) where the west coast and ultramafic associations are virtually absent. In Figure 9 the samples in the west coast and ultramafic clusters identified in the classification of samples described later in the chapter are identified by solid squares and open circles respectively.

The curve is still rising steadily, suggesting that the total sample fell somewhat short of embracing the whole fauna. Indeed a number of species known to occur on the island were not taken and since the survey others have been recorded by ORSTOM entomologists. The Norfolk I. curve, included for comparison, is also still rising but this is due to continued recording of species vagrant from Australia (HOLLOWAY, 1977). The resident moth fauna of Norfolk I. totals 55 species and thus a curve for the same samples and groups of samples including only resident species would be expected to level out where indicated by the open triangle.

## The numerical methods

The nature of the sampling programme required that some modification be made to the numerical approach used in analysing the Norfolk I. data by HOLLOWAY (1977). The reasons for using certain methods in preference to others have been given in detail by HOLLOWAY and will only be summarised briefly here, though particular attention will be paid to a comparison of average-link cluster analysis with those methods.

Common to all methods is the initial stage of deriving a numerical measure that expresses the similarities or dissimilarities of all possible pairs amongst the objects to be classified (generally known as OTUs; operational taxonomic units) in terms of certain attributes. In the case of the New Caledonian survey it was necessary to obtain classifications both of the various samples in terms of their species composition and of the various species in terms of their distribution amongst the samples. The former was necessary in order to provide a better understanding of the latter, associations of species being ecologically more significant than associations of samples (HOLLOWAY, 1977).

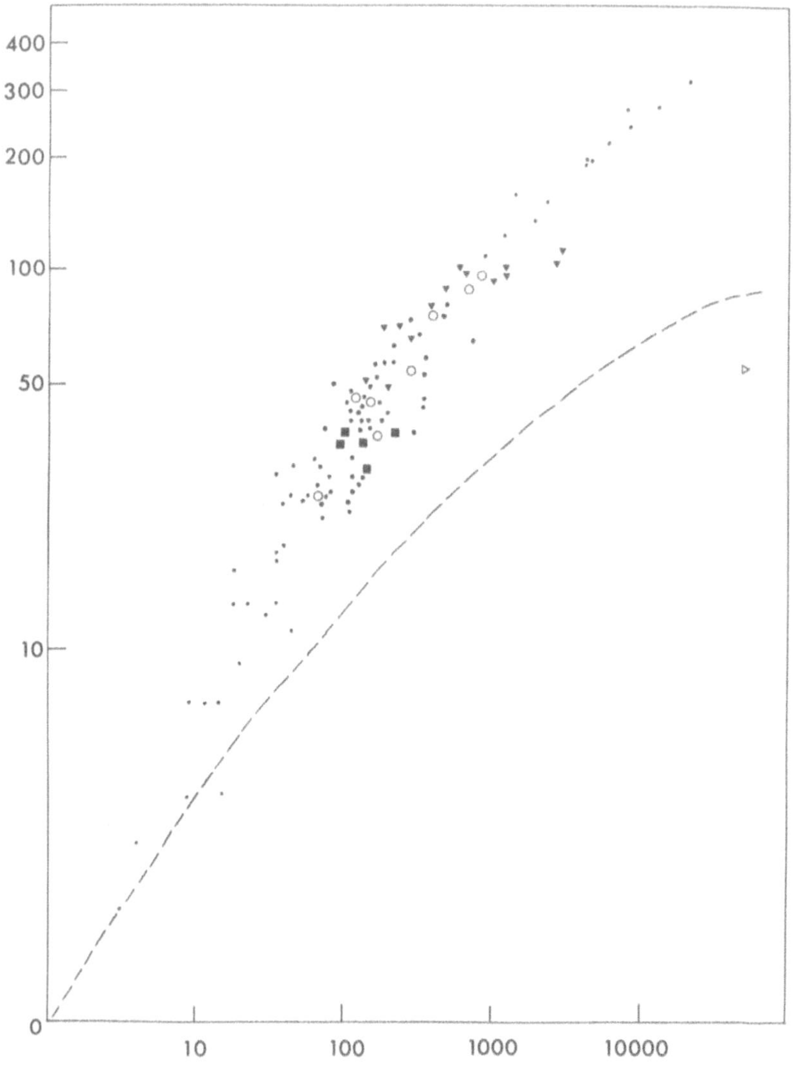


*Fig. 9.* A plot of sample size (horizontal axis) against species number (vertical axis) for New Caledonia. The broken line is a similar plot for Norfolk I. but including vagrants; the curve for resident species alone would level out where indicated by the open triangle. The other symbols indicate those samples grouped in the numerical analysis, namely ultramafic (open circles), west coast (solid squares) and forest and secondary (inverted solid triangles).

The taxonomic distance coefficient (described by CLIFFORD & STEPHENSON (1975) in their useful introduction to numerical methods of classification) was used for classifying the species according to their distribution amongst the samples. This was found in the Norfolk I. analyses to give the same results as the overlap coefficients used on such data previously (HOLLOWAY, 1970; ROBINSON, 1975); they both give a measure of the extent to which the distributions amongst the samples of a pair of species coincide or overlap in terms of relative abundance. Such species distributions are first expressed on a percentage or proportional basis to facilitate this comparison.

Before these coefficients are calculated it may be desirable to apply some system of weighting to the samples. For example, in order to give each sample equal emphasis, the samples may be normalised by expressing the representation of each species as a percentage or proportion. Such equal weighting of samples was used in analyses by HOLLOWAY (1970, 1977) and ROBINSON (1975). In these instances the samples were composites of several samples made at different times from the same locality. The New Caledonian data involve a larger number of samples of smaller size, mostly made on a single night or on a few consecutive nights. The only exceptions are those made in the Plaine des Lacs (Grand Lac and Pic du Pin) both at the beginning and at the end of the survey.

The New Caledonian samples and the proportions of their component species are therefore more likely to be affected by transient phenomena such as seasonal changes of abundance, localised population fluctuations, or even the emergence of the progeny of a single female from the neighbourhood of a foodplant on which oviposition took place (e.g., perhaps, *Rhesalides asphalta* at Sarraméa). Such transient effects would probably become subordinated to more general trends given the opportunity to sample at any locality more extensively through the seasons. They may also serve to obscure more general patterns. Indications from the Norfolk I. survey are that this is a valid assessment. There the general distinction of species into forest and widespread categories has become stronger, more clear cut as the samples from each site or area have become larger and as seasonal coverage has improved. A full seasonal coverage may be very important for, on Norfolk I., the two distributional elements are to a large degree seasonally segregated as well. Limited observations from the New Caledonian survey, discussed at the end of the chapter, suggest something similar may occur there as well.

Extremely localised population fluctuations are also more likely to be manifest in the smaller samples, as these are usually catches made on cool, clear nights when the flight range of the insects is lower than on warm nights. Catch size generally increases logarithmically with minimum nocturnal temperature (discussed for Norfolk I. by HOLLOWAY). The range of the trap is therefore much more limited on cool nights. Thus equal weighting of the samples is liable to enhance the effects of such localised fluctuations relative to more general trends.

Analyses were conducted both with the samples equally weighted and unaltered. In order to overcome some of the effects of localised fluctuations a third analysis was made with several groups of samples pooled together. These groups were those shown by a classification of samples to be relatively similar in terms of their overall species composition, but perhaps less so in terms of the proportions of those species. In this manner the 'noise' effects of localised fluctuations may be cancelled out to some extent, permitting more general patterns to be appreciated.

The samples were amalgamated on the basis of a classification in terms of species composition alone, though two further classifications incorporating information on the proportions of species were obtained for comparison. For the latter the BRAY-CURTIS coefficient was used (CLIFFORD & STEPHENSON, 1975: 57; HOLLOWAY, 1977: 165). This is calculated in a similar manner to the overlap coefficient for the comparison of species distributions but compares species proportions within samples. If the raw data are used the resulting classification is primarily determined by the relative proportions of the most abundant species. As these tend to be the species most prone to extreme fluctuations in population (HOLLOWAY, 1977) such classifications are liable to be of limited value.

A logarithmic transform of the data reduces but does not entirely eliminate this effect, giving a more equal weighting to the attributes (species) determining the classification. Such a transform is usually applied by ecologists using the BRAY-CURTIS measure (e.g. FIELD & MCFARLANE, 1968).

For the classification of samples in terms solely of species composition the coefficient of PRESTON (1962) was used. This has been criticised by CLIFFORD & STEPHENSON (1975) as being too complex and based on a premise (the log-normal distribution of species abundance and a related relationship between species number and area) requiring further examination. Such examination has, to a certain extent, been made. The log-normal distribution has been supported by recent theoretical and practical work (MAY, 1975; HOLLOWAY, 1977) and a clearer picture is emerging of the relationship between species and area (e.g. ABELE & PATTON, 1976; DIAMOND & MAYR, 1976). CLIFFORD & STEPHENSON recommended the use of simpler measures, such as the JACCARD coefficient, which involve no assumptions about underlying quantitative factors for comparing purely qualitative data on the species composition of samples or areas.

The problem with the JACCARD coefficient and related measures (discussed also by BALGOOY (1971)) is that, involving some representation of species shared between two areas in terms of the total number of species in the areas, they tend to give high measures of dissimilarity (or low measures of similarity) where there is great disparity in the number of species in the two floras, faunas or samples being compared. Where, for example, the faunas of small, offshore islands are compared with the mainland fauna of which they can be regarded as 'derived samples', or when, as in the case of the New Caledonian moth samples, difference in 'faunal size' are merely an artifact of the sampling procedure (undercollecting of island faunas is a similar instance) this property of such coefficients renders them unsatisfactory.

Various measures have been proposed to circumvent this problem. SIMPSON

(1943) suggested simply that the proportion of shared taxa in the smaller fauna or flora should be the measure, assuming that this proportion would remain constant should the smaller fauna be hypothetically enlarged to equal the larger, as in certain circumstances of low sampling or undercollecting it eventually might be, subsequent to the original analysis. The log-normal distribution of abundance amongst species and the fact that the more abundant insect species on the whole tend to be more widespread and dispersive mean that this assumption is not justified.

Ideally a measure should be sought that indicates whether the two faunas or samples under comparison have been drawn from the same source fauna in sampling or recent historical terms or alternatively from parent faunas differing from each other to a greater or lesser extent. It should also be indicative of phenomena such as the differentiation of endemic species. The faunas under comparison are effectively treated as samples and the measure will involve assumption about the distribution of abundance amongst the species of the source fauna.

WILLIAMS (1947) proposed a measure based on the log-series model of abundance distribution. The log-series is now regarded as only fitting small samples from a larger population where generally some 'humped' logarithmic distribution applies, such as the log-normal. The measure requires knowledge of the two areas being sampled, which should be unequal, and cannot readily be calculated in the majority of cases.

PRESTON's formula is based on the log-normal model and does not require knowledge of the areas concerned. The relationship between individuals, species and area is implicit in the log-normal model but the arguments used by PRESTON in deriving his measure leave much to be desired from a statistical point of view and could not be described as a sound foundation for the measure (R. SIBSON, *in litt.*). The formula requires further examination from a mathematical angle. From a more pragmatic viewpoint the results obtained using the PRESTON measure correlate closely with those gained from the application of the more simple measures, such as the JACCARD coefficient, when relatively equal floras or faunas are compared and are much more satisfactory when disparate biotas are examined. The measure gave an extemely rational assignment of the smaller New Caledonian samples in relation to a subjective assessment of vegetation type as will be seen below. No doubt more rigorous field tests could be devised to investigate the properties of the measure in detail.

One potential disadvantage of the PRESTON, WILLIAMS and SIMPSON measures is that they can lead to arrays of dissimilarity coefficients that are not metric. In a metric array involving comparison of three OTUs, A, B, and C, the dissimilarity coefficient between B and C, $bc$, will range between $ab+ac$ and $ab-ac$ where $ab$ is the larger. The above measures can give results where small fauna A has all its species in common with both larger faunas B and C, giving minimum dissimilarity values, yet faunas B and C are dissimilar to a greater or lesser degree. The array of coefficients would only be metric if $bc$ were likewise of minimum dissimilarity. CLIFFORD & STEPHENSON (1975: 78) have described at greater length the properties of metric measures of dissimilarity.

In general it is preferable to apply the various clustering or ordination methods of analysis to metric measures. An example of the potential effect of applying a measure that is not metric is provided by application of the PRESTON measure to BALGOOY's data on the plant genera of Pacific island groups (p. 41). In these data several island groups are listed which contain primarily small atolls supporting only a strand flora composed of genera ubquitous in the Pacific and fully represented on virtually all the larger archipelagoes as well. If such island groups are included in an analysis numerous instances occur where they have dissimilarity coefficients of zero with the larger islands. A single-link cluster analysis performed on such an array of coefficients would result in all the OTUs (island groups) clustering together at zero dissimilarity, despite there being significant differences in the floras of the groups with larger islands. Here there would be good biological reasons for excluding all predominantly strand genera from the analysis (effectively excluding all the atoll groups), as they form a general plant community that is distinct from the inland forest association that develops on larger islands. The strand flora could be the subject of a separate analysis at the species level. In the analysis on p. 41 the atoll groups have been excluded.

PRESTON suggested that his measure tended to be inaccurate where floras or faunas were more disparate than 4:1 but satisfactory results can still be obtained when the disparity is 5:1. Above this the dangers of non-metric arrays are probably much greater, though it is possible to check the metric properties of any array of coefficients by inspection.

After a suitable coefficient has been applied to the data a number of numerical methods can be used to identify significant structure or trends in the array of coefficients. Where an array is metric one can envisage the coefficients as distances (measuring dissimilarity), arranging points representing the OTUs in multidimensional space. Cluster analysis methods seek to identify agglomerations or associations of points within that space and ordination methods seek to identify the predominant trends in their dispersion.

Each method aims to present a simplified picture of the data and yet to preserve as much of their original structure as possible. In an ordination method the axes are ordered according to how much of the variation of the original data they successively recapture, measured as the percentage variance for each axis. Amongst cluster methods only the $B_k$ sequence of JARDINE & SIBSON (1968a, b) permits an analogous gradual recapture of the structure of the original data, this occurring as the value of $k$ increases from 1 (single-link) to $n-1$ where there are $n$ OTUs. The application of this sequence to biogeographical and ecological data was discussed in detail by HOLLOWAY (1977). Briefly, for any value of $k$, an OTU can only join an established cluster at a given level of dissimilarity if it has $k$ links at or below that level with OTUs already in the cluster. If there are less than $k$ OTUs in the cluster, then it must link with all of them. It is possible for up to $k-1$ OTUs to belong to two clusters at one time without bringing about the fusion of the clusters (overlapping clusters are shown in Figures 19 and 20, most graphically in Figure 30). Thus the resulting classification of the OTUs is non-hierarchic except when $k = 1$ (single-link).

The results of a single-link analysis can be presented as a dendrogram, as in Figure 10, where merely the levels of clustering of the OTUs are indicated, or as a minimum spanning tree (Figure 7), where the actual links bringing about the clustering are portrayed as distances connecting the two OTUs concerned. The minimum spanning tree thus incorporates only sufficient links such that all points shall be joined directly or indirectly to all other points, the sum of the lengths of the links being minimum.

The analyses for values of $k$ greater than 1 permit more and more of the structure of the data to be appreciated, providing information on the homogeneity or otherwise of clusters derived by a single-link analysis. The results, however, become increasingly complex, particularly where the clustering structure of the data is weak. There is often little gained from computing results for more than a few values of $k$ higher than 1. Rigorous computation of such results is very lengthy and such facilities were not available for this work. A more *ad hoc* approach was adopted, building up linkage diagrams (e.g. Figures 11 and 19) at ever increasing levels of dissimilarity and inspecting the clustering at each stage in relation to the conditions of the $B_k$ sequence. The mode of operation of this approach will become clearer when actual examples are discussed (pp. 41, 175, 195).

This rather cumbersome approach is more sensitive to structure in the data than the averaging and centroid methods of cluster analysis usually preferred by ecologists (e.g. CLIFFORD & STEPHENSON) and is thus less likely to give misleading classifications. These other cluster methods yield a single, cut-and-dried solution where the resultant dendrogram often suggests stronger clustering structure than is merited by the data. This will be illustrated later in the chapter when both the $B_k$ sequence and average-link analysis are applied to some of the New Caledonian data. In average-link, again proceeding gradually from a low to a high dissimilarity level, an OTU or cluster is joined to another cluster only if the average of the links between the two entities is at the level of dissimilarity being considered.

The various clustering methods give similar results where there is marked clustering structure in the data in the form of clear, isolated agglomerations of points. Where clustering structure is weak the averaging and centroid methods tend to 'impose' structure on the data by breaking up what are best regarded as continua amongst the points into what appear from inspection of the resultant dendrogram to be distinct clusters. This is exemplified by some of the New Caledonian data. It is under such conditions as these that the theoretical objection of discontinuity (when small changes in the data can lead to large changes in the resultant classification), stated by JARDINE & SIBSON (1968b) to be inherent in these methods, tends to be most serious.

The approach preferred here, the $B_k$ sequence combined with an ordination, is more reliable and more informative about the structure of the data. The average-link analyses gave useful results and were certainly more informative about the structure of the data than single-link alone. But the full approach demonstrated that some of the average-link clusters were at best only weakly defined in the data and reliance on this method alone could be misleading.

Mention must be made finally of another ordination method, non-metric multidimensional scaling. This was applied to the classification of samples using the PRESTON measure in order to obtain a two-dimensional plot of points representing the samples. On the plot the distributions of species and of the ecological associations of species could be portrayed clearly. The geographical map, with its complex mosaic of vegetation types, is not suitable for such portrayal.

The technique of principal coordinate analysis is unsatisfactory for this purpose as plotting of the first two, or even three axes usually only reveals major trends within the data. Single OTUs or small groups of OTUs that are isolated as outliers in a cluster analysis are often not segregated in principal coordinate analysis until several axes have been plotted (as shown on p. 177).

KRUSKAL's (1964a, b) technique of non-metric multidimensional scaling seeks a disposition of points in a selected low number of dimensions that is as close as possible to their original disposition in multidimensional space. In this chapter a two-dimensional solution is sought; other applications have been described by HOLLOWAY & JARDINE (1968) and JARDINE (1972). The measure of goodness-of-fit between the derived and original dispositions of points which is optimised by the method is a monotone regression measure (JARDINE, 1972). The results reflect the best possible summary of the data in the selected number of dimensions and thus will take account of any outliers (e.g. OTU 46 in Figure 12).

The goodness-of-fit, or stress, is recorded as a percentage, a low percentage indicating a good fit. In many instances of application it is advisable to obtain solutions for a number of dimensions, so as to select that one where the dimensions are fewest without the stress having risen so high as to render the results of unsatisfactorily low significance. The two-dimensional solution for the New Caledonian data was required to enable other results to be displayed but the stress was high (23.38%) and thus the results cannot be taken as significant in isolation. Their usefulness can, however, be gauged by comparison with the cluster analysis results using the same array of coefficients.

## The data

Most of the numerical analyses were performed on a matrix recording the distributions of 118 species amongst 61 samples, though the classifications of samples included information on more species, that involving the PRESTON measure including all the species recorded. The 118 species, listed in Table 5 where they are assigned numbers to enable them to be identified in the various figures of dendrograms and ordinations, are those where 20 or more individuals had been recorded. The only exception was *Prasinocyma albicosta*, OTU 116, with 19 individuals, which was included because of its proven association

*Table 5.* List of the 118 species classified according to their distribution amongst the samples in Chapter V, numbered as in the dendrograms, linkage diagrams and ordinations concerned.

1. *Eilema caledonica*
2. *Oeonistis lifuensis*
3. *Rhodogastria albivitrea*
4. *Spilosoma lifuense*
5. *Spilosoma lateritica*
6. *Agape lifuensis*
7. *Argina cribraria*
8. *Leucania pseudoformosana*
9. *Leucania latericia*
10. *Leucania scottii*
11. *Callopistria maillardi*
12. *Callopistria dimorpha*
13. *Craniophora fasciata*
14. *Hypoperigea lunulata*
15. *Athetis postpuncta*
16. *Leucocosmia onagrica*
17. *Platysenta illecta*
18. *Sasunaga tenebrosa*
19. *Spodoptera litura*
20. *Spodoptera mauritia*
21. *Molvena caledonica*
22. *Stictoptera tridentifera*
23. *Blenina donans*
24. *Characoma vallata*
25. *Achaea janata*
26. *Achaea serva*
27. *Ophiusa disjungens*
28. *Ophiusa discriminans*
29. *Mocis frugalis*
30. *Mocis trifasciata*
31. *Chrysodeixis eriosoma*
32. *Chrysodeixis illuminata*
33. *Anomis involuta*
34. *Ericeia hirsutitarsus*
35. *Lacera* spp.
36. *Oxyodes scrobiculata*
37. *Serrodes campana*
38. *Serrodes mediopallens*
39. *Hydrillodes norfolki*
40. *Progonia micrastis*
41. *Gnathothlibus erotus*
42. *Compsulyx cochereaui*
43. *Hippotion velox*
44. *Neola octofera*
45. *Teara circumfumata*
46. *Adeixis major*
47. *Oenochroma unifasciata*
48. *Nadagara serpentina*
49. *Aeolochroma hypochromaria*
50. *Anisozyga caledonica*
51. *Thalassodes pseudochloropis*
52. *Gnamptoloma rubra*
53. *Scopula undilinea*
54. *Chloroclystis lunifera*
55. *Euphyia perseverata*
56. *Horisme grisearia*
57. *Gonanticlea pulcherrima*
58. *Gymnoscelis acidna*
59. *Poecilasthena spp.*
60. *Syneora strixoides*
61. *Caledasthena montana*
62. *Casbia rectaria*
63. *Casbia serpentina*
64. *Cleora lacteata*
65. *Cleora immemorata*
66. *Cleora illustraria*
67. *Cleora acacicola*
68. *Oxymacaria insularis*
69. *Symmetroctena variegata*
70. *Eilema nitens*
71. *Nyctemera baulus*
72. *Athetis striolata*
73. *Chasmina basiflava*
74. *Blenina lichenopa*
75. *Maceda savura*
76. *Chalciope alcyona*
77. *Parallelia prisca*
78. *Parallelia redunca*
79. *Parallelia lateritica*
80. *Parallelia cuneilineata*
81. *Thyas miniacea*
82. *Avatha discolor*
83. *Hypospila bolinoides*
84. *Ophyx maculosus*
85. *Othreis fullonia*
86. *Pantydia capistrata*
87. *Rhesalides asphalta*
88. *Adrapsa ablualis*
89. *Bocana manifestalis*
90. *Dichromia quinqualis*
91. *Hypena laceratalis*
92. *Nodaria cornicalis*
93. *Simplicia caeneusalis*
94. *Euchromia cincta*
95. *Miresa obscurimargo*
96. *Euproctis dimorphissima*
97. *Agrius fasciatus*
98. *Macroglossum hirundo*
99. *Psilogramma menephron*
100. *Adeixis inostentata*
101. *Adeixis baeckeae*
102. *Epiplema lateritica*
103. *Thalassodes viridimargo*
104. *Pyrrhorachis pyrrhogona*

*Table 5 (Cont.)*

| | |
|---|---|
| 105. *Thalassodes baladensis* | 112. *Catoria camelaria* |
| 106. *Uliocnemis biplagiata* | 113. *Gonodontis clelia* |
| 107. *Urolitha bipunctifera* | 114. *Petelia trifascia* |
| 108. *Scopula desita* | 115. *Codalithia* spp. |
| 109. *Scopula homodoxa* | 116. *Egone bipunctalis* |
| 110. *Traminda mundissima* | 117. *Prasinocyma albicosta* |
| 111. *Horisme xylinata* | 118. *Orgyia sarramea* |

with *Acacia spirorbis*. Preliminary analyses using only the 69 species with 50 or more individuals did not give significantly different results, the less abundant species included in the larger analyses tending to fall into the groupings already identified by the smaller analyses.

The 61 samples were those with 25 or more species and 50 or more individuals. Several are amalgamations of samples from two sites that were close to each other, covering identical vegetation. Thus OTUs 3 and 4 include samples from sites 3+72 and 4+71 respectively, OTU 7 covers the R. Bleue rainforest sites 8+9, OTU 12 the coastal thicket sites 15+16, OTU 29 the Mt. Mou maquis slope sites 34+36, OTU 30 the Dome de Tiebaghi forest sites 41+42, and OTU 33 the Col d'Amoss maquis sites 45+46. As a result the numbering of the OTUs differs from that of the sites as indicated by *italic* in the list in the previous chapter.

## The classification of samples

The 61 samples were classified in three ways as described on p. 150. BRAY-CURTIS coefficients were derived from quantitative data on the 200 most abundant species, both with and without logarithmic transformation, and PRESTON coefficients were obtained from the complete qualitative data.

The analysis on untransformed quantitative data reflected patterns in the distribution of the commonest species. Thus OTUs 11, 28, 29 and 48 were brought together in a single-link analysis and persisted as a cluster when $k = 2$ by virtue of their high proportion of *Hydrillodes norfolki*, the commonest species encountered in the survey. Similarly OTUs 17, 33, 34, 37, 39, 41 and 42 were brought together through having a high proportion of *Leucocosmia nonagrica*. On the other hand the Plaine des Lacs *Baeckea*-sedge maquis samples 2, 3, 4 and 6 were united, as were those of the Mt. des Sources (9 and 10) and Port Boisé (53 and 54). The Ouen Toro samples (49, 60 and 61) formed a single-link cluster, uniting at a higher level with the previously paired samples from Anse Longue (12 and 27). Thus the resultant classification would appear to be partially determined by the abundance of the extremely common species and partially by vegetation type.

The analysis of logarithmically transformed data gave a classification where the influence of vegetation type was enhanced relative to that of the most abundant species. The single-link dendrogram is shown in Figure 10. Two

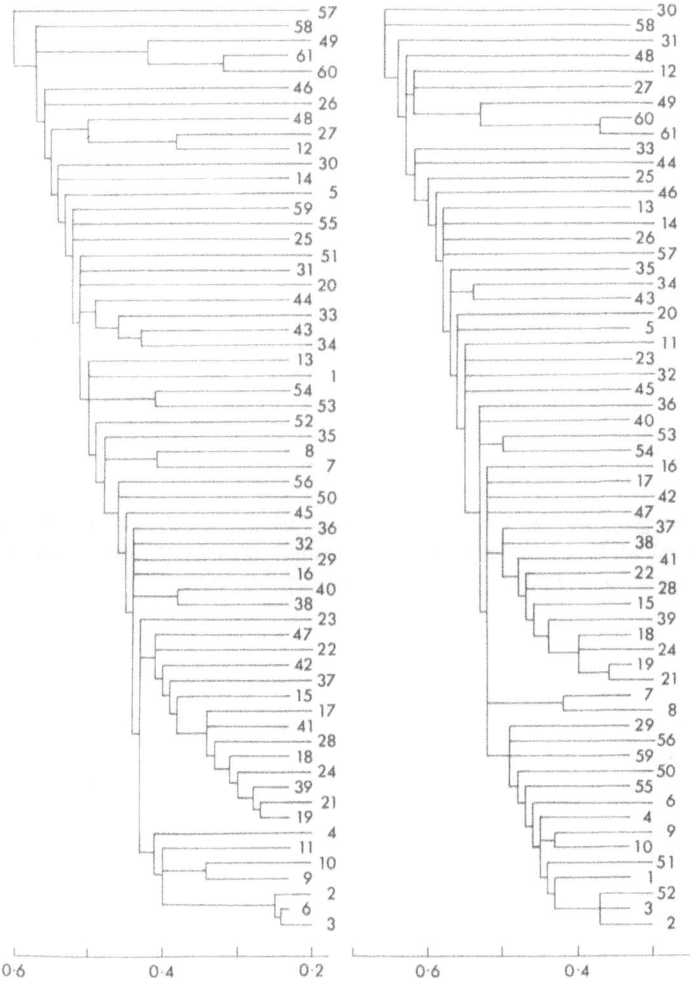


*Fig. 10.* Single-link dendrograms derived in the classification of samples from (left) an array of BRAY-CURTIS coefficients, the data transformed logarithmically, and (right) an array of PRESTON coefficients. The samples are numbered as indicated by *italic* in Chapter IV. The scales at the bottom indicate dissimilarity level.

moderate and several small clusters are apparent but there is no well-defined structure, suggesting there is much chaining in the data. This is only to be expected when samples have been taken from varying mixtures of different types of vegetation. Notable in this respect are samples from areas of valley rainforest amid maquis on the ultramafics (11, 28 and 41).

The largest cluster includes most of the larger samples taken from richly forested localities, including some of those on ultramafic rock (15, 17, 18, 19,

21, 22, 24, 28, 37, 39, 41, 42, 47). The other moderate cluster contains the larger samples from the Plaine des Lacs (2, 3, 4, 6) and the Mt. des Sources (9, 10, 11), all from areas where *Baeckea*-sedge maquis was extensive. The Port Boisé and Ouen Toro samples again form clusters but the latter are separated from a cluster containing the three west coast leguminous thicket sites (12, 27, and 48). The R. Bleue samples 7 and 8 are paired and there is also a rather weak cluster which includes the two Col d'Amoss samples (33, 34), the Col de Boa sample (43) and that from Houailou (44). The first three are grouped before 44 joins them and all are from rather unusual maquis habitats.

Examination of linkage diagrams for the situation at higher values of $k$ revealed the persistence of the two major clusters, though that of samples from ultramafic sites remained weak. At a dissimilarity level of 0.45 there was a cluster containing OTUs 17, 18, 19, 21, 24, 37, 39 and 41 when $k=6$. These are all samples from mixed rainforest and secondary vegetation on sedimentary rock with the exception of 37 (Amoa R.) and 41 (Mt. Krapé forest). When $k=5$ OTUs 15, 22 and 28 joined the above, as did 47 when $k=4$, 23 when $k=3$ and 36, 28 and 40 when $k=2$, all samples from similar vegetation. OTU 42, brought into the single-link cluster through its association with 41, was excluded at higher values of $k$. The sample concerned was taken from the *Gymnostoma chamaecyparis* association on Mt. Krapé. Only when $k=2$ at that level did a cluster of ultramafic maquis samples appear, containing OTUs 2, 3, 4, 6 and 9.

At higher levels of dissimilarity the ultramafic cluster merged with the forest and secondary cluster at low values of $k$ but tended to remain weakly distinct at higher values. OTUs such as 1, 8, 10, 29, 50 and 56 from areas of maquis on ultramafic rock were associated primarily with the ultramafic cluster and other such as 11, 28, and 45, from localities of mixed forest and maquis, served to bring the two major clusters together.

The purely qualitative analysis using the PRESTON measure produced a much clearer segregation of the samples into those associated with the ultramafic maquis on the one hand and those associated with areas of rainforest and secondary vegetation, mainly on sedimentary rock, on the other. The single-link dendrogram is shown in Figure 10 and the linkage diagram for dissimilarity level 0.55 is presented in Figure 11. The latter also indicates by dotted lines the links at higher levels of dissimilarity that bring the outlying OTUs into the main clusters in the single-link analyses; such links are part of the minimum spanning tree.

The ultramafic cluster is much enlarged in the dendrogram, now including many more of the Plaine des Lacs and other ultramafic samples, samples from localities that might be expected to have similar moth faunas in view of the overall uniformity of their vegetation. The forest and secondary cluster contains most of the OTUs referred to its counterpart in the previous analysis. The Port Boisé and R. Bleue samples form pairs as before and one of the Col d'Amoss samples is paired with the Col de Boa sample.

The west coast samples form a single-link cluster with the exception of OTU 48 (Col Nassiboué). This cluster is persistent at dissimilarity level 0.66 when

$k = 4$, though it then excludes OTU 12 and includes 48. OTU 12 is primarily associated with it (three links) and has one link with OTU 32 (Bambou R.) which in turn has two links with the coastal cluster, testimony perhaps of its similar, dry, rather open habitat. OTU 16 (Poupa; an *Acacia spirorbis* dominated site) is also associated (two links) but is overall closer to the main ultramafic and forest clusters which have more or less coalesced at that level through the same rather intermediate samples that brought them together in the previous analysis. The two other samples from localities where *A. spirorbis* is dominant (23, 25) join the main single-link cluster by virtue of their links with OTU 16.

At level 0.55 the two main clusters are united when $k = 1$ but separated though overlapping when $k = 2$. When $k = 5$ there is a forest cluster consisting of OTUs 15, 17, 18, 19, 21, 24, 28, 37, 39 and 41. It is these samples which are amalgamated in the third analysis of species distributions described later. There is also an ultramafic cluster with complete linkage consisting of OTUs 2, 4, 9, 10 and 56. When $k = 4$ this is joined by OTUs 1, 3 and 6 and the forest cluster is joined by 22, 38, 40 and 47. When $k = 3$, 29, 50, 51 and 52 join the former and 16 and 36 join the latter. When $k = 2$, 55 is brought into the former, 53 joins the latter and 28 joins both clusters so they overlap.

These general relationships amongst the sample OTUs are also apparent in the non-metric multidimensional scaling plot (Figure 12). The single-link clusters are enclosed by broken lines and the $B_4$ forest and ultramafic clusters are indicated by solid and enclosed open stars respectively. The $B_4$ west coast cluster at level 0.66 is indicated by large solid circles. Those samples made in vegetation associations on ultramafic rock can largely be segregated from those made off it by a single boundary (the finely broken line). The OTUs placed between the forest cluster and the coastal cluster are mainly samples from dry, open habitats where secondary vegetation predominates or samples from localities where *A. spirorbis* is dominant. The Port Boisé forest samples are also placed in this area. They stand apart somewhat from the rest of the forest and ultramafic samples in the cluster analysis and will be seen (e.g. Figure 26) to have a moderate representation of species characteristic of the west coast localities.

To the lower left of the plot are placed the smaller samples from the more distinctive ultramafic localities such as Tiebaghi (30) and Houailou (44), and on the lower right are the samples from rather unusual maquis vegetation mainly off the ultramafics. The latter include the samples from Mt. Panié (46), the Col d'Amoss (33, 34), the Plateau de Dogny (26) and the sample from the Ouambéo R. (20) made near the cliffs with *Araucaria*-dominated maquis shown in Plate 47.

In Figure 13 the preponderance of rainforest at each of the sample localities is indicated on the scaling plot, as are the localities where *Acacia spirorbis* was a prominent feature. The highly forested samples (solid squares) are seen to be mainly those falling into the forest cluster, the exceptions being the R. Bleue samples (7, 8) and that from Mt. Koghis (58), all relatively small samples from ultramafic areas and in the neighbourhood of extensive maquis.

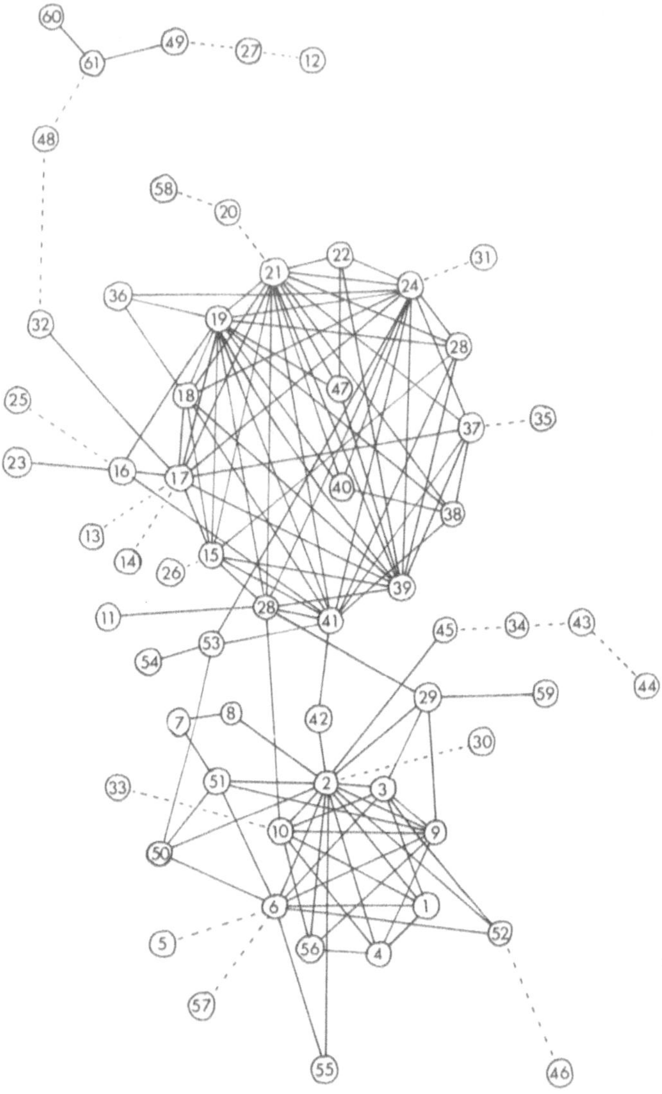


*Fig. 11.* A linkage diagram at level 0.55 for the samples from the array of PRESTON coefficients. The broken lines indicate the links at levels higher than 0.55 that bring in the outlying OTUs in the single-link analysis.

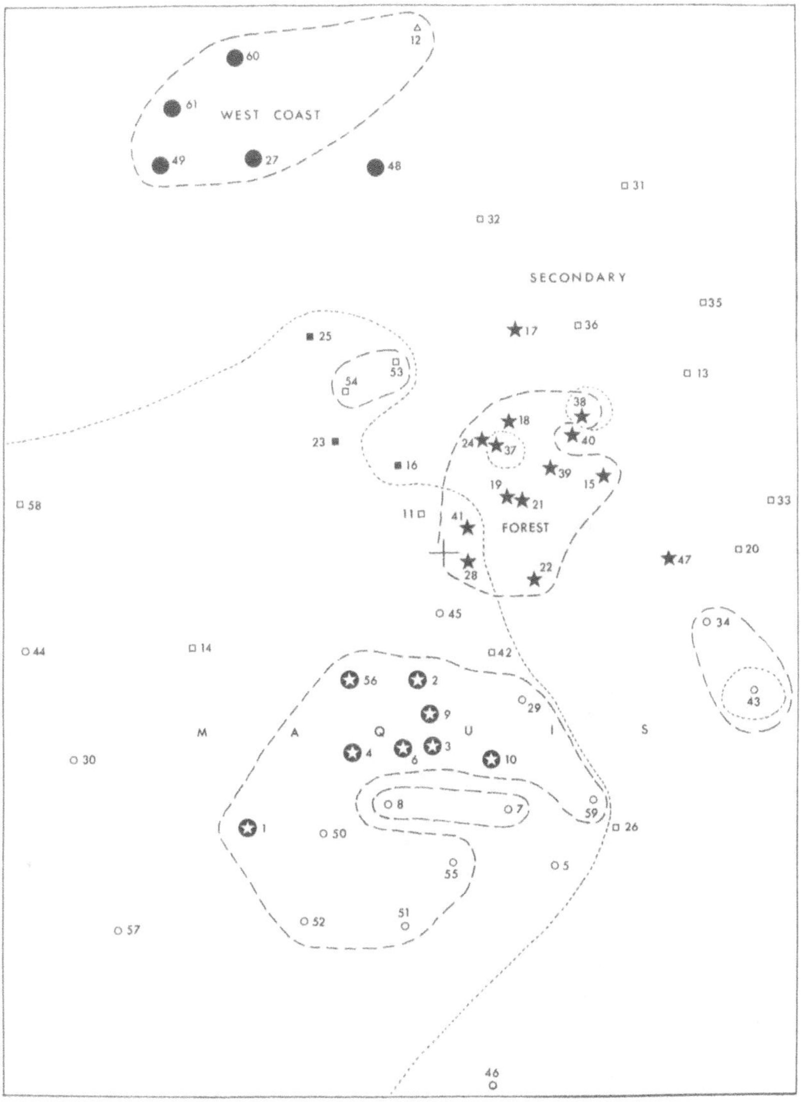


*Fig. 12.* Non-metric multidimensional scaling plot of the sample OTUs using the PRESTON measure. The dotted line separates the samples made in ultramafic localities from the rest. The broken lines enclose the OTUs of the single-link clusters, the large circles, the solid stars and the enclosed stars indicate the $B_4$ clusters discussed in the text including samples from west coast, rainforest and ultramafic localities respectively, and the smaller symbols indicate those OTUs associating with the ultramafic cluster (open circles) and the forest cluster (squares, the solid ones indicating a predominance of *Acacia spirorbis*) in the single-link analysis.

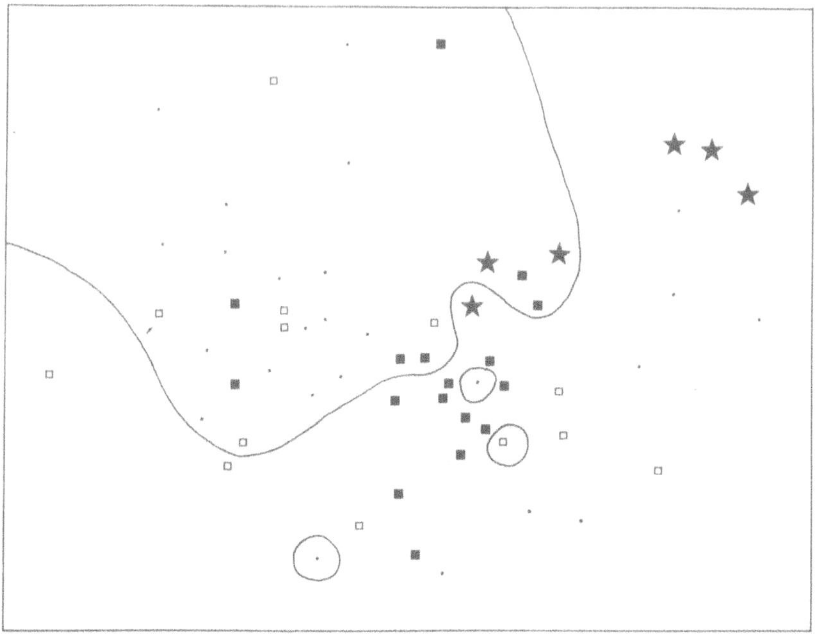

*Fig. 13.* The scaling plot showing the samples with rainforest significantly present (open squares) or predominant (solid squares) and those from localities with extensive *Acacia spirorbis* (stars).

These three analyses have shown that, for single night samples of the type made in New Caledonia, stressing the qualitative aspects of the data leads to a segregation of the samples according to the vegetation type in which they were made. Addition of quantitative information tends to obscure this relationship, mainly through the effects of what may be purely transitory fluctuations in the abundance of the commoner species.

On the basis mainly of the qualitative analysis it was decided to amalgamate as single samples the $B_5$ forest cluster samples and the $B_4$ ultramafic cluster samples as at dissimilarity level 0.55, the three Ouen Toro samples which were associated in all analyses, and the R. Bleue and Port Boisé pairs. This amalgamation was performed for the final analysis of species distributions described in the next section.

## The classification of species

Classifications of the 118 species (Table 5) according to their distribution amongst the samples were made firstly with the samples weighted equally as described on p. 149, secondly with no modification of the original data so the samples are weighted effectively according to their size, and thirdly with groups of samples amalgamated as described above.

Three of the OTUs included were later found to represent mixed series, each incorporating two species. OTU 35 included two species of *Lacera* (p. 500); the majority of the individuals proved to be *L. vinacea*, a new endemic, and the rest were *L. noctilio*. The distributions of both are primarily amongst the forest and secondary samples. Similarly both species of *Codalithia* (p. 427), included together as OTU 115, were distributed amongst the ultramafic samples. The two species of *Poecilasthena* (p. 317), included together as OTU 59, were assigned in the analyses to the ultramafic association of species. When separated, the commonest (*P. dimorpha*) proved to have a much more characteristic ultramafic distribution and the other species (*P. leucydra*) had a distribution more typical of the forest and secondary association of species.

Only a single-link analysis was performed on the equally weighted data. It produced many small, ill-defined clusters that appeared over a narrow range of dissimilarity values and fused together at a barely greater level. There was thus much chaining in the data, possibly arising through over-emphasis of the smaller samples and thus of the effects of localised population fluctuations. Three moderate clusters were of interest. Several smaller ones contained species largely restricted to the ultramafic maquis though these did not form a single cluster.

Of the three larger clusters one contained species largely restricted to samples from rich rainforest (OTUs 4, 38, 50, 51, 64, 68, 74, 77, 82, 84, 103, 104, 106, 112, 113, e.g. *Serrodes mediopallens*, *Thalassodes pseudochloropis*, *Parallelia prisca*, *Oxymacaria insularis*), another contained mainly species widely distributed amongst the samples, more particularly off the ultamafic rocks and possibly associated with secondary growth (OTUs 3, 8, 9, 11, 15, 16, 17, 19, 20, 29, 30, 34, 39, 41, 65, 78, 81, 118, e.g. *Leucania pseudoformosana*, *Platysenta illecta*, *Spodoptera mauritia*, *S. litura*, *Cleora immemorata*), and the third contained species concentrated in samples from west coast localities and those with *Acacia spirorbis* dominant (OTUs 10, 40, 53, 70, 72, 86, 92, 93, 96, 109, 110, 114, e.g. *Leucania scottii*, *Progonia micrastis*, *Scopula homodoxa*). The second cluster includes seven species that are widespread on Norfolk I. and probably characteristic of disturbed habitats (HOLLOWAY, 1977).

The nature of each association of species revealed by this and subsequent analyses can be displayed on the non-metric multidimensional scaling plot. Using the data transformed by equal weighting of the samples, the mean representation of the species in any association can be calculated for all the samples, yielding the distribution (expressed as a percentage amongst the samples) of what might be termed the 'average species' of the association. Figure 14 illustrates 'average species' for the first and second clusters described in the previous paragraph. The mean distribution of the species in the third cluster is not illustrated but would not be very different from that shown in Figure 25 which illustrates a grouping containing most of the species concerned but obtained in a subsequent analysis.

Both the unweighted data and the data with groups of samples amalgamated

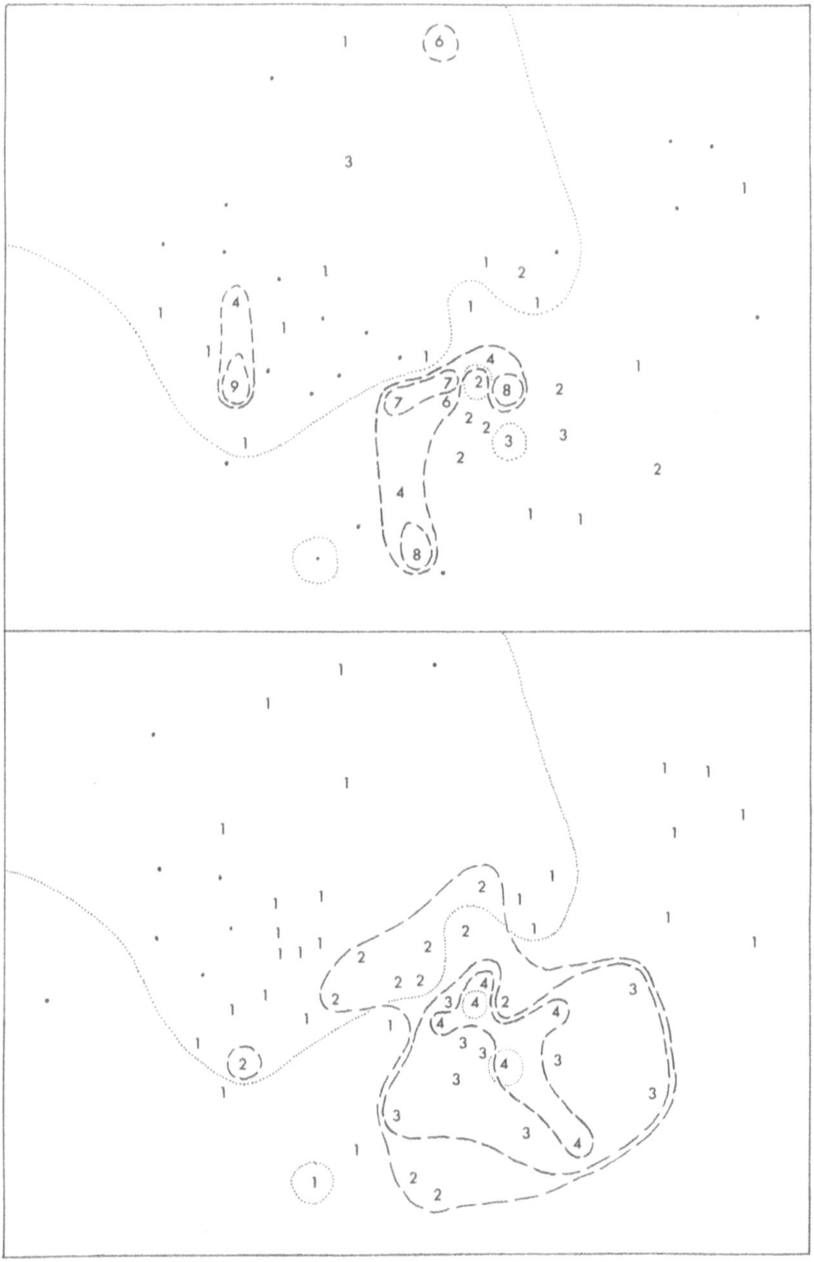

*Fig. 14.* 'Average species' for the forest (top) and secondary (bottom) clusters for the analysis of species distributions with samples equally weighted.

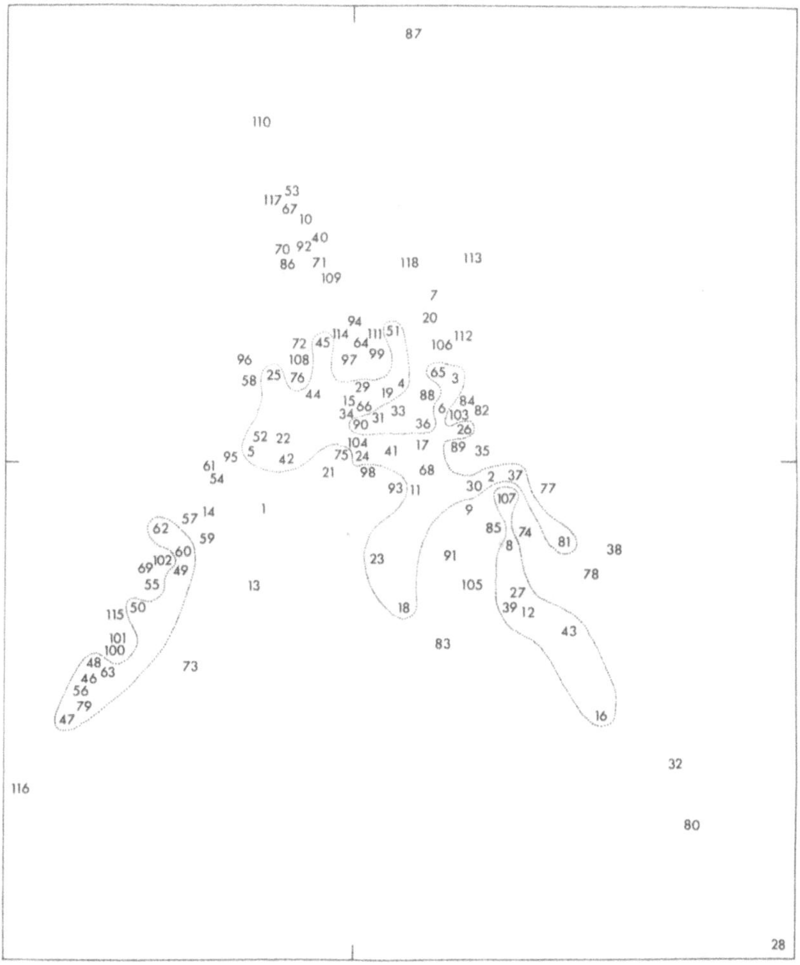


*Fig. 15.* A plot of the first (horizontal) and second (vertical) axes of the principal coordinate analysis of species with samples not equally weighted or amalgamated. The numbers are as in Table 5 and the major single link clusters are indicated by dotted lines.

were subjected to cluster analysis using both the $B_k$ sequence and average-link, and to principal coordinate analysis.

The ordination of the unweighted data suggests that no clear cut discontinuities occur within the distribution of the points representing the OTUs. The first two axes, plotted in Figure 15, recapture only 18.5% and 9.4% respectively of the variation of the original data but nevertheless are the only axes to produce significant dispersion of the majority of OTUs. Later axes, except the fourth, serve only to segregate one or two OTUs that are revealed as outliers by the cluster analyses.

The first axis sees the separation of species associated with samples from the ultramafic maquis (lower left 'arm' of the plot) from the rest. The second separates those species identified as associated with west coast and *A. spirorbis* localities in the equal weighted analysis, these being grouped near the top of Figure 15. The third axis (8.1%) segregates OTUs 100 and 116 (*Adeixis inostentata* and *Egone bipunctalis*) which were commonest in samples (sample OTU 4) from the marshy habitat of the Plaine des Lacs. The fourth (7.1%) sees a further slight segregation of the west coast species from the rest. At the extremes of the fifth axis (6.1%) are segregated, on the one hand, OTUs 53, 109 and 110 (Sterrhinae common in the Anse Longue samples) and, on the other, OTU 108 (*Scopula desita*), a species abundant at the two Mt. Krapé sites yet with only a single individual recorded elsewhere. The sixth axis (5.2%) segregates OTU 87 (*Rhesalides asphalta*), a species which was exceedingly abundant at Sarraméa where the trap was sited near a group of the *Albizzia* trees on which its larvae probably feed. In Fiji the species is characteristic of drier habitats with grassland or secondary growth (Robinson, 1975) and, apart from the Sarraméa sample, the species was found to be largely restricted to such habitats in New Caledonia.

The cluster analyses support the indications of the ordination, both of the largely continuous nature of the data and of the general segregations that can be made. The single-link analysis again indicates little structure and the probability of much chaining in the data. Two weak clusters are obtained that include species restricted to ultramafic maquis on the one hand and species most abundant in samples from rainforest and secondary vegetation on sedimentary rock on the other. A third small cluster incorporates some of the OTUs from near the extremity of the lower right 'arm' of the ordination, all characterised by particularly high numbers in the Probokonoro sample (sample OTU 39), the largest made during the survey. These three clusters are indicated in Figure 15.

At higher values of $k$ up to 4 (and possibly greater, though the clustering became too complex for accurate interpretation) the ultramafic species were partially distinguished from the large forest and secondary cluster, though the Probokonoro cluster proved far less persistent.

The same segregation was apparent in the average-link analysis (the dendrogram is shown in Figure 16), though this gave greater emphasis to the segregation of an enlarged Probokonoro cluster (solid square). The clusters are identified by symbols on the dendrogram which are then used to indicate the OTUs of these clusters as positioned in the ordination (Figure 17). The ultramafic cluster also encompasses more species than in the single-link analysis, including virtually all the OTUs in the lower left 'arm' of the ordination (solid circles). The main forest and secondary cluster was subdivided largely into four sub-clusters, also indicated by symbols on the dendrogram and in Figure 17. These clusters are not distinguished as aggregations of points by the first two axes of the ordination but appear to replace each other across a continuum. The two larger sub-clusters are segregated to some extent at higher values of $k$ in the $B_k$ sequence. For example, when $k = 2$ at a dissimilarity level

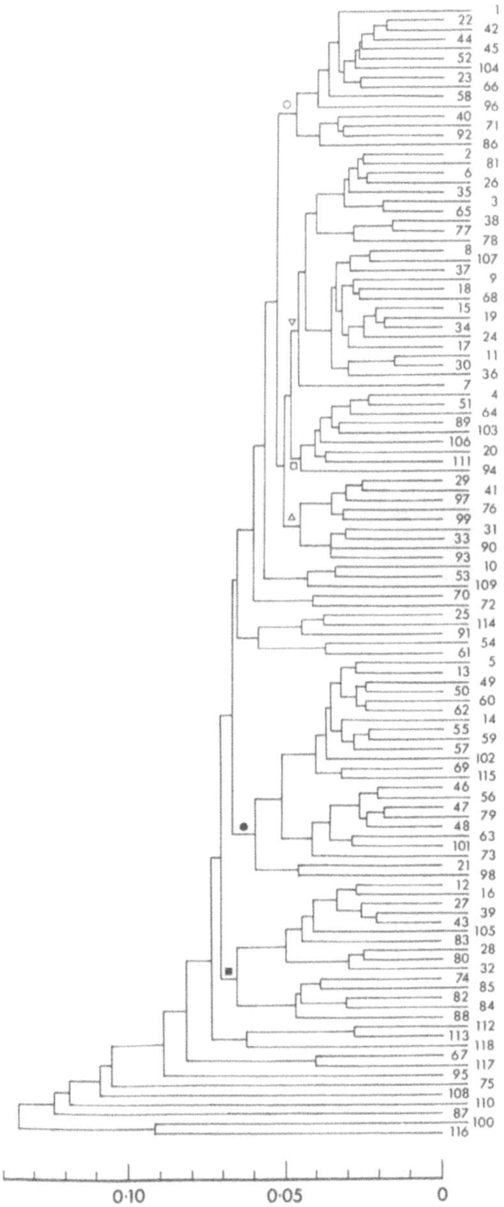


*Fig. 16.* The average-link dendrogram derived for the 118 species, the data not equally weighted or amalgamated. The scale indicates dissimilarity level and the numbers are as in Table 5. Various clusters are marked with symbols for cross-reference with Figure 17.

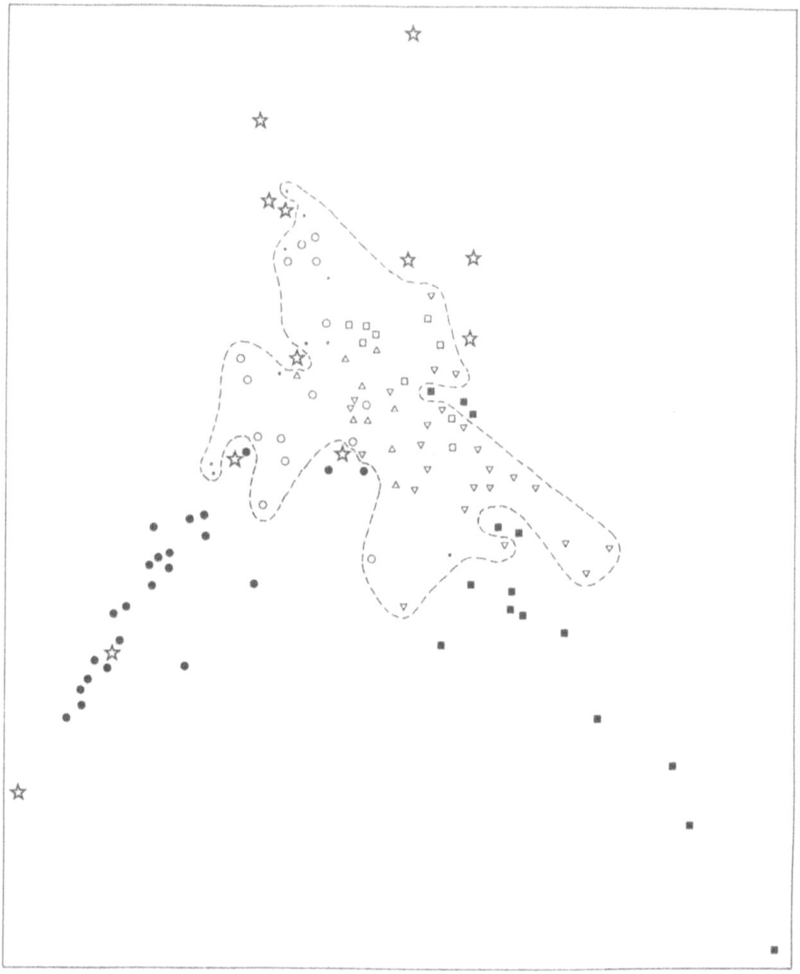

*Fig. 17.* The principal coordinate plot of Figure 15 with the OTUs of the average link analysis of Figure 16 indicated by symbols. Stars indicate outlying OTUs. The broken line encloses the large forest and secondary cluster.

of 0.0256 there are two clusters with OTU 34 in an overlap between the two. One consists of OTUs 6, 15, 18, 19, 24, 26 and 34, and the other of OTUs 5, 22, 42, 44, 45, 52 and 34. Associated with each (by single links) are a number of other OTUs and these associations broadly make up the two average-link clusters, the first indicated by open inverted triangles and the second by open circles. They both contain species well represented generally in samples from rainforest areas but the latter is more strongly represented in samples from ultramafic rainforest (including species such as *Stictoptera tridentifera*,

*Spilosoma lateritica* and *Compsulyx cochereaui*). This grouping of species features to some extent in the next analysis. The other two average-link sub-clusters were not particularly apparent when clustering structure at higher values of *k* was examined but they generally consist of species taken in rainforest samples off the ultramafics on the one hand (open squares; e.g. *Uliocnemis biplagiata, Thalassodes pseudochloropis, Cleora lacteata*) and species of drier, open habitats with secondary vegetation on the other (open triangles; e.g. *Mocis frugalis, Dichromia quinqualis, Gnathothlibus erotus*).

The cluster analyses did not segregate in entirety the group of west coast and *Acacia spirorbis* sample species identified in the equally weighted analysis and, to some extent, in the ordination, but the average-link analysis identified three groupings of the component species. OTUs 40, 71, 86 and 92 formed a sub-cluster of the 'open circle' cluster, 10, 53 and 109 formed a triplet and 70 and 72 were paired, coming into the forest and secondary cluster at a high level. The foursome is generally associated with dry woodland and thicket, especially that dominated by *A. spirorbis* (*Progonia micrastis, Pantydia capistrata, Nodaria cornicalis* and *Nyctemera baulus*). *N. baulus* was recorded by day also and the diurnal records indicate a more typically forest and secondary distribution. The triplet is perhaps more characteristic of dry, open habitats with grassland (*Leucania scottii, Scopula homodoxa, S. undilinea*) and the pair (*Athetis striolata, Eilema nitens*) is generally similar in character to the foursome apart from high representation at site 17 (OTU 13).

OTUs 67 (*Cleora acacicola*) and 117 (*Prasinocyma albicosta*) are paired as they were in the single-link analysis and in the analysis of equally weighted data. In the average-link analysis they join the main cluster at a high level. In their distributions they are strongly associated with *Acacia spirorbis* (Figure 18) and the *Prasinocyma* was reared from it during the survey.

The effect of amalgamating groups of samples of similar species composition can be illustrated best by examining classifications of species of the families Geometridae, Uraniidae and Lymantriidae only. These families are weaker in flight and generally less mobile and dispersive than the other families such as the Noctuidae and Sphingidae. The difference between the Noctuidae and Geometridae is illustrated by aspects of their dispersal to, and colonisation of Norfolk I. (HOLLOWAY, 1977). Poorly dispersive species are less likely to be recorded away from suitable habitats, so distribution patterns are liable to be more clear cut.

Figure 19 shows single-link dendrograms for the data with and without amalgamation and a linkage diagram for the data without amalgamation at a dissimilarity level of 0.039. The dendrogram for data without amalgamation has no marked, isolated clusters though at approximately 0.030 dissimilarity there is some distinction between forest and ultramafic species. The linkage diagram indicates a much clearer segregation with two large $B_4$ clusters distinguished, though overlapping (OTU 52; *Gnamptoloma rubra*). These largely persist even when $k = 6$ though several OTUs such as 54 and 101 are excluded from the larger cluster.

Amalgamation of samples serves merely to accentuate this segregation. The

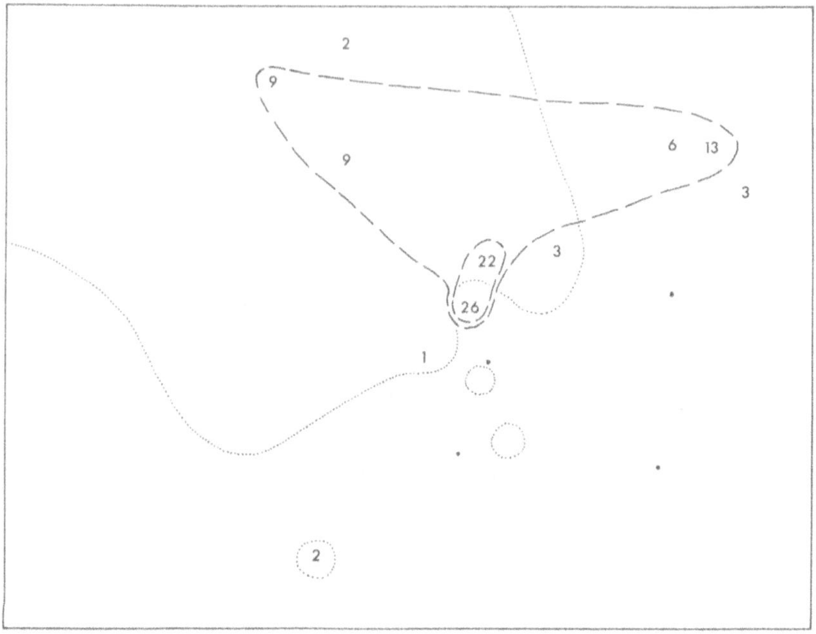


*Fig. 18.* The 'average distribution' of *Prasinocyma albicosta* and *Cleora acacicola*, two species associated with *Acacia spirorbis*.

two major clusters of the linkage diagram are separated in the single-link dendrogram for amalgamated data, OTU 52 falling within the ultramafic cluster. At dissimilarity level 0.047 the two clusters unite when $k = 1$ but for higher values of $k$ they remain as distinct, homogeneous clusters, in fact showing virtually complete linkage for the majority of their component OTUs. The single-link dendrogram distinguishes a pair cluster with OTUs 67 and 117 and a triplet (53, 109, 110). The former is discussed above and the latter contains the three species of Sterrhinae associated mainly with the dry west coast habitats.

The effects of amalgamation of samples on the classification of all 118 species can be appreciated by examining Figures 20–22, where single-link and average-link dendrograms and the principal coordinate analysis results are presented. Aspects of the cluster analyses are illustrated on the ordination plots.

The arrangement of the OTUs with regard to the first two axes of the ordination (Figure 20; rotated 90° relative to Figure 15) is very similar to that for the unaltered data. However, there is much greater segregation of the three broad groups of OTUs suggested by the first two analyses. The extent to which the structure of the data has been simplified by the amalgamation is illustrated by the fact that the first and second axes recapture 63% and 15% respectively

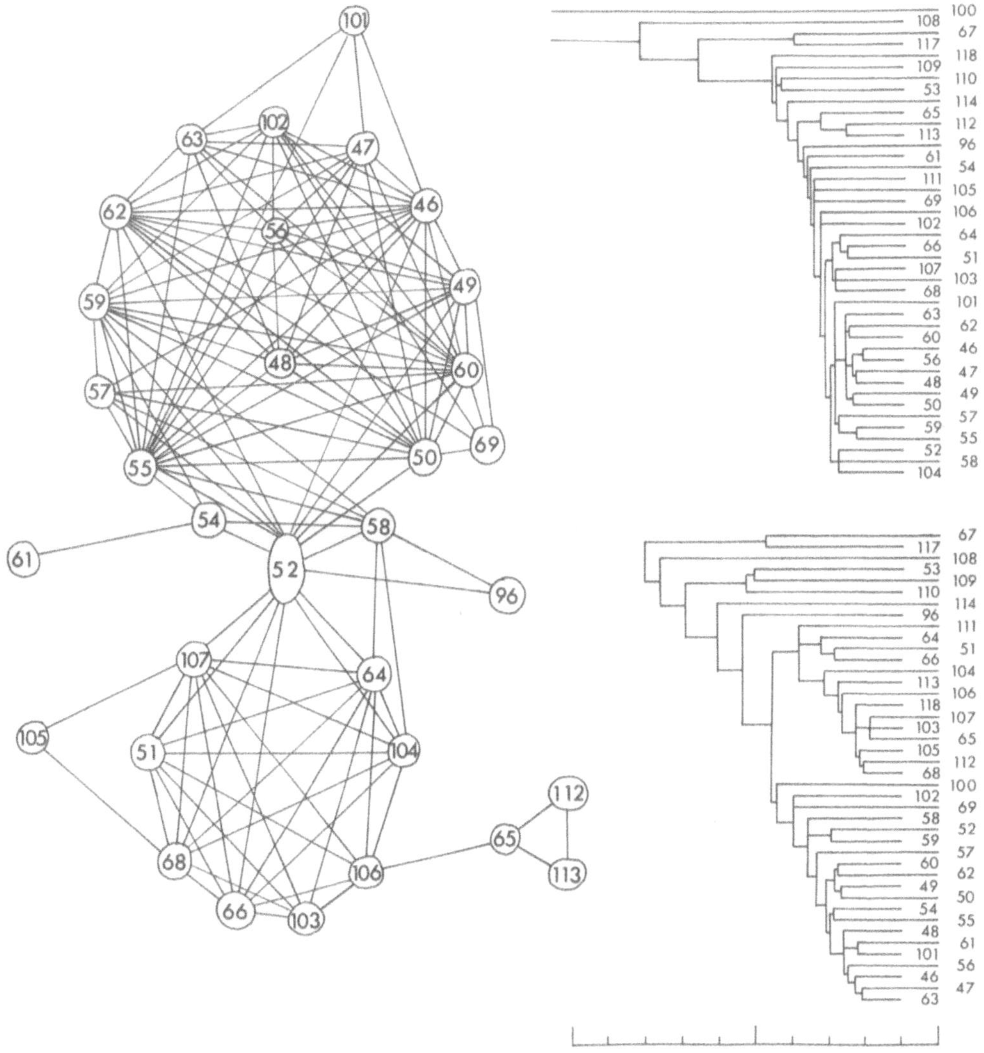

*Fig. 19.* Analyses of Geometridae, Uraniidae and Lymantriidae only, indicating the effects of amalgamating samples. The upper dendrogram is from data without amalgamation, the lower from data with samples amalgamated. The linkage diagram on the left is for unamalgamated data to a dissimilarity level of 0.039.

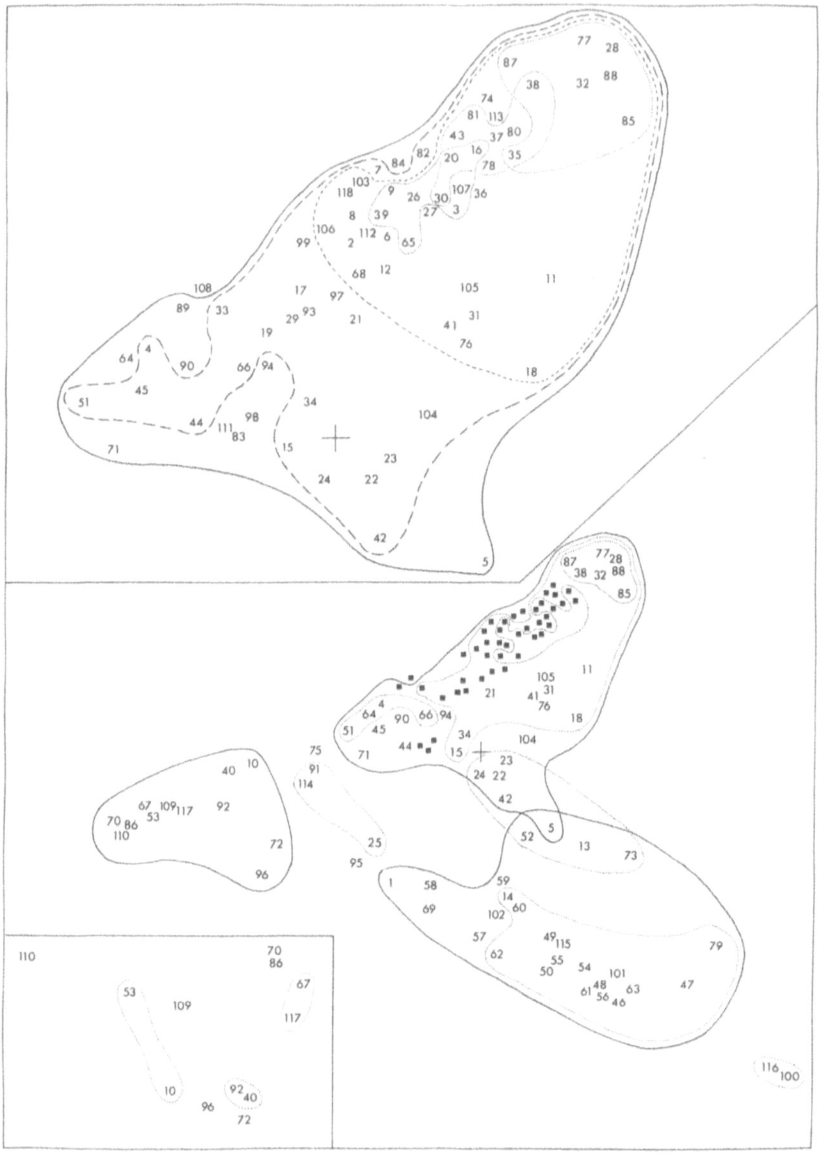


*Fig. 20.* The first (vertical) and second (horizontal) axes of the principal coordinate analysis of species with samples amalgamated. The lower section illustrates the single-link clusters (dotted lines), the major ultramafic and forest associations recognised from the $B_k$ analyses and the loose association of west coast and *A. spirorbis* species (all enclosed by solid lines). The dispersion of the last group by the second (vertical) and third (horizontal) axes is indicated at bottom left, single-link pairings being indicated by dotted lines. The upper diagram shows the forest association on a larger scale and illustrates how various $B_k$ clusters build up as described in the text.

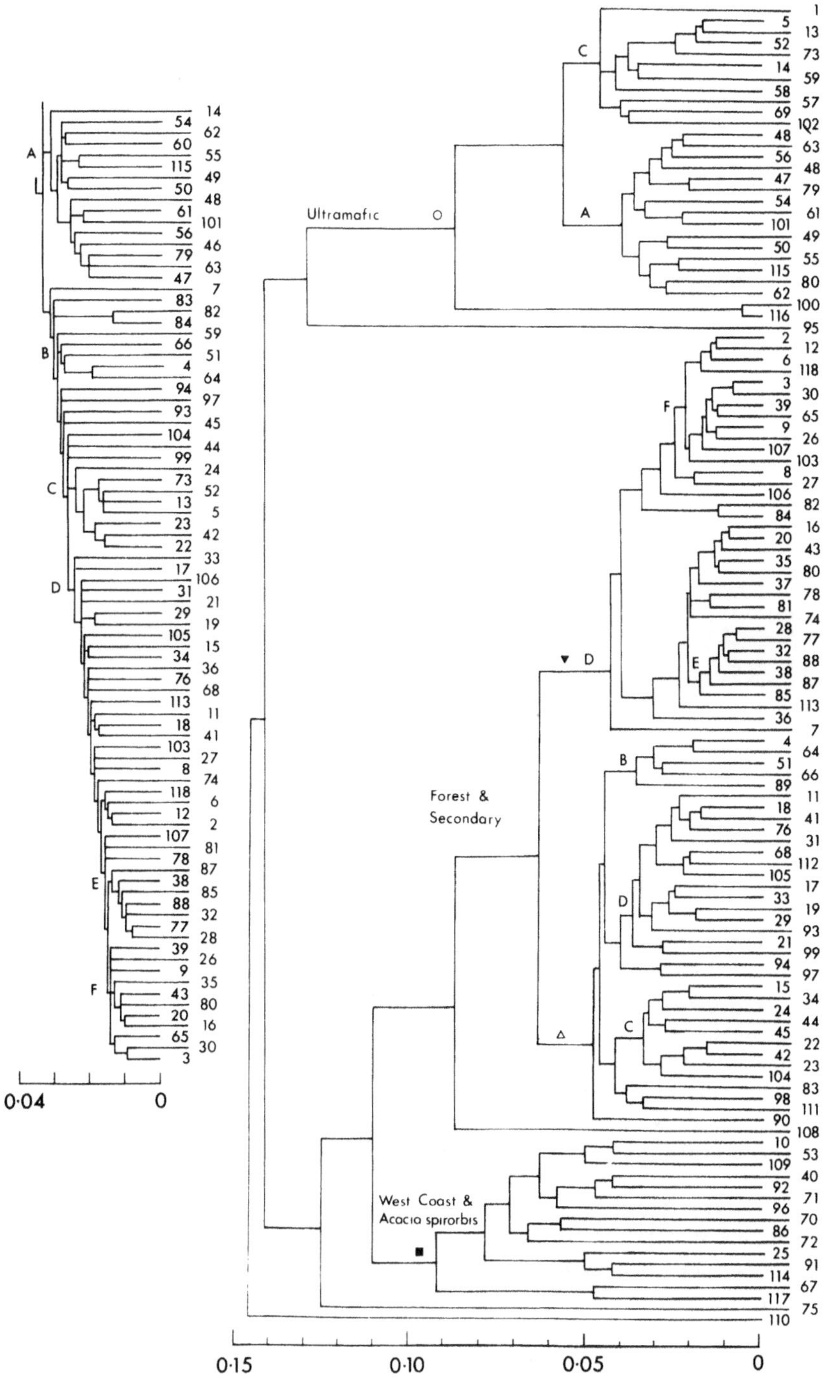


*Fig. 21.* Single-link (part only, left) and average-link (right) dendrograms for the classification of species using data with samples amalgamated. Cross-reference from the former to the latter is facilitated by the letters and the symbols on the latter cross-refer to Figure 22.

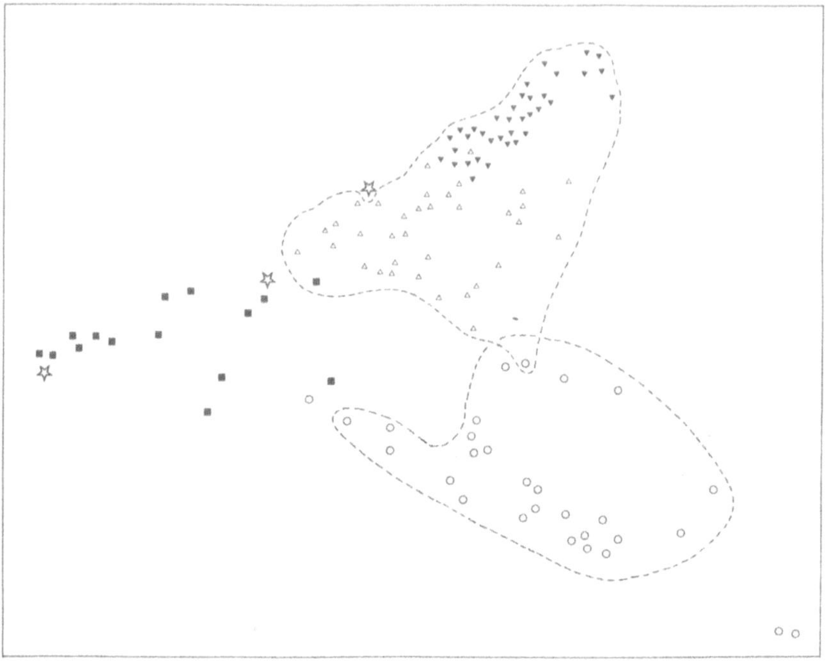

*Fig. 22.* The principal coordinate plot of Figure 20 with the major $B_k$ clusters of Figure 20 (broken lines) and the major average-link clusters of Figure 21 (symbols as on the dendrogram) indicated. OTUs outlying the major clusters 1are indicated by stars.

of the variation of the original data. The next two axes add only another 3% each.

The single-link dendrogram (Figure 21, left) is only drawn up to the dissimilarity level where the major forest and ultramafic clusters become linked. As in the analyses performed on equally weighted and non-amalgamated data, the clustering occurs mainly over a narrow range of dissimilarity values and there is probably much chaining in the structure of the data. Any agglomerations in the data must be sought through analyses at higher values of *k*. The clusters mentioned above embrace the majority of OTUs in the upper right and lower right 'arms' of the ordination respectively. Cluster A (ultramafic) and cluster D (the major part of the forest and secondary cluster) are indicated by dotted lines in Figure 20, as are subclusters B and C of the forest cluster and E and F, subclusters of D. Cluster C appears intermediate between the forest and ultramafic clusters on the ordination and contains several OTUs in common with the 'open circle' cluster of the previous average-link analysis.

At higher levels of dissimilarity the remaining OTUs cluster in to the major cluster by degrees, only the odd pairs and triplets such as (53, 10), (67,117), (92, 40), (116, 100) and (25, 91, 114) being indicative of any further structure.

Analysis at higher values of *k* broadly confirms these groupings, though there are some differences. The forest and ultramafic clusters gradually build up to those indicated by unbroken lines (excluding the coastal group) in Figure 20 at a dissimilarity level of 0.045 when $k = 5$. OTU 5 (*Spilosoma lateritica*) falls within an overlap between the two; it is part of the single-link cluster C. Cluster C does not persist at higher values of $k$, its OTUs segregating roughly equally between the two major clusters. The mean distribution of species in C is shown in Figure 24.

Some structure is evident within the forest and secondary cluster. At dissimilarity level 0.019 three overlapping $B_5$ clusters occur as illustrated by dotted lines in the upper part of Figure 20. They replace each other roughly linearly in the ordination and represent a continuum within the data from species largely restricted to the major samples from rainforest on sedimentary rock to those somewhat more widespread though still centred on those major samples. The distributions of the two extreme clusters, comparable in their component OTUs to the single-link clusters E and F, are shown in Figure 23. $B_4$ clusters at levels 0.028 and 0.035 are also illustrated on the upper part of Figure 20, showing how the forest cluster builds up to include the more dispersed OTUs in the ordination. Within this dispersion there is some indication of weak segregation of OTUs 15, 22, 24, 34 and 44. They form a $B_3$ cluster at 0.030 and have complete linkage at 0.033, though are united with the main cluster at that level when $k = 4$. Associated with this group are OTUs 23, 42, 45, 66 and 104, and typical of it are such species as *Athetis postpuncta*, *Characoma vallata*, *Ericeia hirsutitarsus* and *Neola octofera*. They are species generally widespread in the island, found in a variety of mainly wooded or forested habitats and are more strongly represented in the ultramafic samples than many species of the forest and secondary cluster. They form a sub-cluster in the average-link analysis.

There is no marked segregation of the west coast and *Acacia spirorbis* species. At level 0.061 the OTUs 40, 86, 92 and 96 form a $B_3$ cluster with 92 and 40 in an overlap with the large forest cluster. Weakly associated with them are OTUs 10, 53, 70, 109, 114 and, more peripherally, 72 and 110. They all form what might be termed a loosely associated peripheral appendage to the forest and secondary cluster.

The third axis of the ordination brings about further dispersal of these coastal species though not of the OTUs of the other clusters. It is plotted for these OTUs (horizontal axis) against the second axis at the bottom left of Figure 20. It separates at the extremes those species (e.g. 67, 117) characteristic of dense, low, dry forest, mainly dominated by *A. spirorbis*, from those characteristic of more open savanna habitats (53, 110). The mean distribution of the coastal association OTUs enclosed by an unbroken line in Figure 20 is shown in Figure 25.

The average-link dendrogram (Figure 21, right) contains a cluster that is referable to this west coast and *Acacia* group, though it excludes OTU 110 and includes the triplet of 25, 91 and 114 that also forms a single-link cluster (*Achaea janata*, *Hypena laceratalis* and *Petelia trifascia*). The levels of clustering are generally much higher than in the other major clusters, indicative both

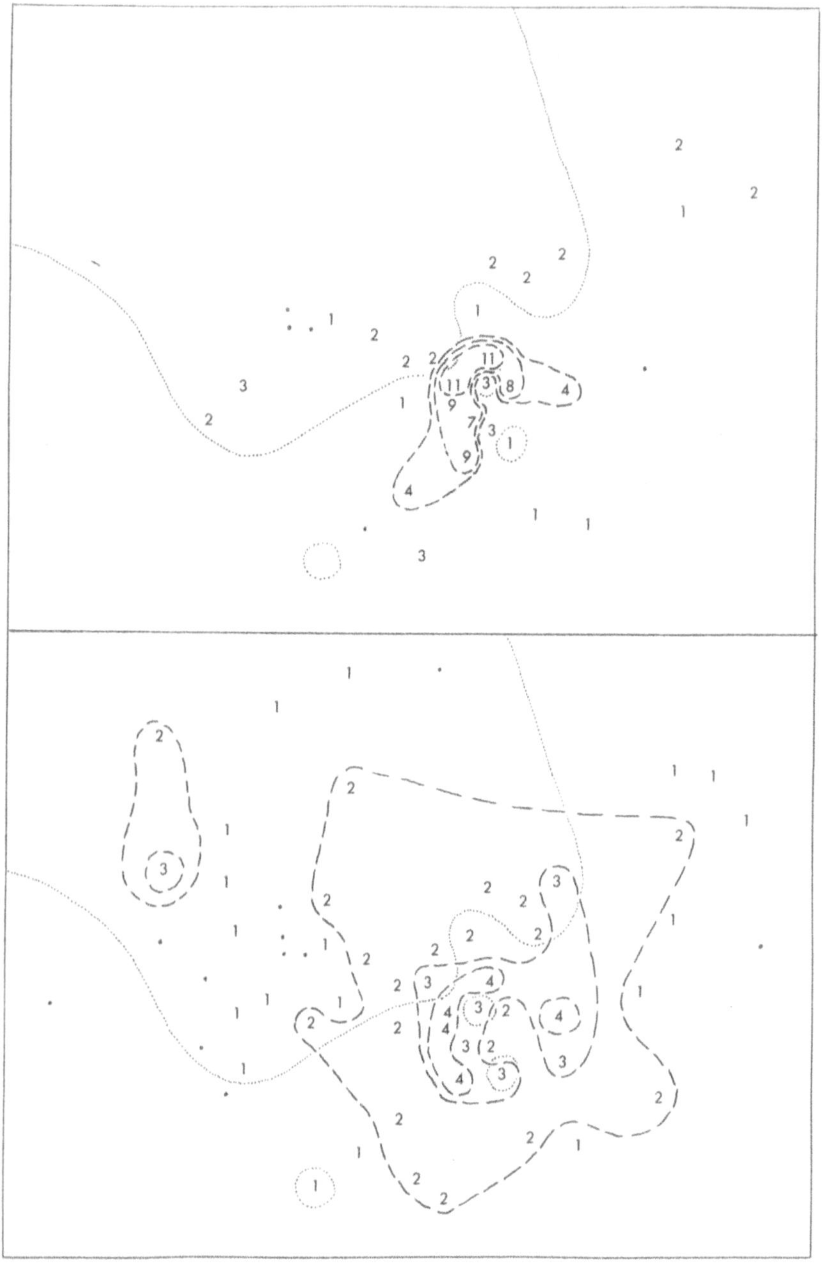

*Fig. 23.* 'Average species' for the two extreme clusters of the three overlapping ones at level 0.019 when $k = 5$ (p. 175); samples amalgamated.

of the more tenuous ecological association between the species and, perhaps, of the undersampling of these dry west coast and *Acacia* habitats during the survey. They would repay further investigation.

The average-link analysis resulted in the derivation of three major clusters virtually identical to the three major segregates of the ordination and corresponding well with the structure suggested by application of the $B_k$ sequence, including the assignation of OTUs of the single-link cluster C. The distribution of OTUs in the various single-link clusters amongst the average-link clusters is conveyed as far as possible by the lettering in Figure 21. Several OTUs placed as outliers in previous analyses are assigned to one or other of the major clusters, only three OTUs being unplaced.

The three outlying OTUs are *Traminda mundissima* (110), *Scopula desita* (108) and *Maceda savura* (75). The first was taken almost exclusively at the west coast sites at Anse Longue and is segregated from other members of the west coast group by axis 3 of the ordination. The second, segregated on axes 8 and 9 of the ordination, was exclusive to the two Mt. Krapé samples (sample OTUs 41 and 42) excepting a single individual from the Kopéto sample (25). The species has thus only been recorded from the central western ultramafic massifs. *M. savura*, segregated on axes 4 and 5 of the ordination, was particularly abundant in the rather unusual Mt. Panié sample (sample OTU 46). This rather localised concentration of the species may be more than an artefact of the sampling method, as a large number of specimens were taken by Miss CHEESMAN on Mt. Tinchialit further north in the Panié range, possibly from a similar habitat.

OTU 95 (*Miresa obscurimargo*) outlies the ultramafic cluster and is segregated on axes 4 and 5 of the ordination. It would appear to characterise sub-canopy samples from rainforest, particularly that on ultramafic rock.

The average-link analysis divided both major clusters into apparently well-defined sub-clusters. The ultramafic cluster is divided into those OTUs of the single-link cluster A on the one hand and those more widespread in the samples, tending towards the forest cluster and including members of the single-link cluster C on the other. The forest cluster has a primary subdivision into a group of OTUs restricted to, or commoner in, rainforest samples (solid, inverted triangles in Figure 22) and one of OTUs generally widespread in their distribution amongst the samples (open triangles). The former is generally referable to single-link cluster D and to the three overlapping $B_5$ clusters mentioned on p. 175; the latter segregate to some extent at a lower level in the average-link analysis as do the similar single-link clusters E and F. The second group includes the part of single-link cluster C assigned to the forest cluster at higher values of $k$ and also the group of OTUs mentioned as associating weakly at level 10.030 when $k = 3$ (p. 175).

The dispersion in the ordination of OTUs of the second forest and secondary sub-cluster and the indications from the $B_k$ analyses suggest that its marked segregation in the average-link analysis is somewhat misleading. The component OTUs are best regarded as representing trends in distribution that are peripheral to, but continuous with, the nuclear grouping of OTUs in the forest cluster that contains the three overlapping $B_5$ clusters. These trends are characterised by reduced concentration in the main rainforest samples and greater representation in samples from more open habitats or the ultramafic vegetation. Similar trends were indicated by the analysis where the samples were equally weighted.

The average-link analyses produced useful groupings of the OTUs but it is perhaps wiser in general to accept such groupings with confidence only when other methods have been used in conjunction that are more informative about the overall structure of the data. Average-link is perhaps most useful as a method for assigning outlying OTUs to major clusters. An averaging method was used by HOLLOWAY (1977) to assign such OTUs to the widespread or forest categories on Norfolk I. once this distinction became apparent from applying the $B_k$ sequence and ordination to the data.

From the data available for the New Caledonian fauna it is possible to recognise with confidence two major associations of species. One is characteristic of open maquis vegetation on ultramafic rock and the mean distribution of its species is shown in Figure 25. The other is associated with rainforest and secondary vegetation, mainly on sedimentary rocks. The mean distribution of its more typical species (the OTUs in the $B_4$ cluster at level 0.028) is illustrated in Figure 24. It is also possible to recognise a much looser grouping of species associated with drier habitats, particularly of the west coast, such as the *Acacia spirorbis* associations and the leguminous thicket and grassland at Anse Longue.

There are indications that further discrimination of species associations within these groupings may be possible, particularly within the forest and west coast ones. Such discrimination will only be possible when a much more intensive sampling programme has been carried out.

On the basis of the numerical analyses the three basic associations of species are recognised as listed below. Species with less than 20 individuals in the samples have been placed subjectively by inspection of their recorded distributions. In most cases such placings were obvious but some species did not fit easily in any category and are listed separately. All singletons taken are also listed separately though these are grouped to some extent according to the habitat type in which they were taken. For each association the species are ranked according to abundance, the number of individuals taken being indicated in brackets.

The percentage of species referable to each association in each sample is given for each locality in Chapter IV and is indicated for the 61 samples used in the analysis on the non-metric multidimensional scaling plot in Figure 26.

**The west coast and *Acacia spirorbis* association:** *Leucania scottii* (241); *Cleora acacicola** (158); *Progonia micrastis* (72); *Scopula undilinea* (50); *Nodaria cornicalis* (48); *Athetis striolata* (41); *Pantydia capistrata** (38); *Traminda mundissima* (37); *Eilema nitens* (29); *Euproctis dimorphissima* (28); *Scopula homodoxa* (20); *Prasinocyma albicosta** (19); *Lysimelia litoralis* (17); *Leucania loreyimima, Maliattha inconcisoides* (10); *Luceria cooki, Amyna punctum* (9); *Amyna natalis* (8); *Westermannia poupa** (7); *Idaea fruticeti* (6);

*Parallelia propyrrha**, *Acrapex exsanguis*, *Eublemma cochylioides* (5); *Epiplema spatulata*, *Eois tiebaghi**, *Mythimna consanguis*, *Leucania loreyi* (4); *Earias perhuegeli*, *Comostola thalassias** (3); *Casbia albinotata*, *Eublemma anachoresis*, *Digama marmorea*, *Helicoverpa assulta* (2). Asterisked species are those that appear to be strongly associated with *A. spirorbis*. Singletons recorded from the west coast group of samples were of *Helicoverpa armigera*, *Crioa acronyctoides*, *Leptotroga costalis*, *Idaea trypheropa* and the two indeterminate ophiderines and single indeterminate hypenine described in Chapter IX.

**The ultramafic association:** *Euphyia perseverata* (493); *Poecilasthena dimorpha* (200); *Chloroclystis lunifera*, *Eilema caledonica* (193); *Spilosoma lateritica* (189); *Gymnoscelis acidna* (157); *Nadagara serpentina* (156); *Horisme grisearia* (151); *Syneora strixoides* (150); *Casbia serpentina* (118); *Caledasthena montana* (104); *Adeixis major* (86); *Casbia rectaria* (85); *Symmetroctena variegata*, *Anisozyga caledonica* (80); *Oenochroma unifasciata* (79); *Hypoperigea lunulata* (69); *Gnamptoloma rubra* (64); *Tiracola circularis* (60); *Aeolochroma hypochromaria* (55); *Craniophora fasciata* (51); *Chasmina basiflava* (46); *Gonanticlea pulcherrima* (43); *Adeixis baeckeae* (37); *Egone bipunctalis* (33); *Parallelia lateritica*, *Epiplema lateritica* (25); *Adeixis inostentata* (23); *Codalithia quincuncialis* (17); *Scopula caledonica* (12); *Thalassodes microchloropis*, *Corgatha figuralis* (11); *Codalithia subtilis* (10); *Anisodes lateritica* (9); *Parosteodes grisea*, *Colnettia brinoni* (7); *Lophoptera hemithyris*, *Scopula quinquefasciata*, *Encryphia frontisignata*, *Chloroclystis scintillata* (4); *Idaea lateritica*, *Anisodes dalmatensis*, *Cofimpacia luteata*, *Stictoptera describens* (3); *Avitta ophiusalis*, *Caledophia pauli* (2). Singletons recorded from typical ultramafic maquis sites were of *Adeixis montana*, *Derambila pellucida*, *Chloeres boisensis*, *Hemithea acaudata*, *Chloroclystis callinda*, *Chloroclystis macroaedeagus*, *Neogyne elongata*, *Micronia cynthiae*, *Epiplema desistaria*, *Thumathoides fumosa*, *Eublemma miasma*, *Gyrtona divitalis*, *Cofimpacia violacea*.

**The forest and secondary association:** *Hydrillodes norfolki* (2381); *Leucocosmia nonagrica* (1972); *Tiracola plagiata* (1400); *Ericeia hirsutitarsus* (1165); *Platysenta illecta* (655); *Cleora immemorata* (598); *Rhodogastria albivitrea* (452); *Spodoptera mauritia* (387); *Gnathothlibus erotus* (363); *Leucania latericia* (328); *Chrysodeixis eriosoma* (266); *Oeonistis lifuensis* (256); *Serrodes mediopallens* (217); *Sasunaga tenebrosa* (209); *Leucania pseudoformosana*, *Callopistria maillardi* (201); *Mocis trifasciata* (198); *Spilosoma lateritica* (189); *Teara circumfumata* (164); *Agape lifuensis* (159); *Achaea janata* (155); *Cleora lacteata*, *Cleora illustraria* (151); *Compsulyx cochereaui* (145); *Ophiusa disjungens* (144); *Spilosoma lifuensis* (142); *Callopistria dimorpha* (124); *Spodoptera litura* (123); *Oxyodes scrobiculata* (120); *Thalassodes pseudochloropis* (117); *Stictoptera tridentifera* (116); *Mocis frugalis* (113); *Achaea serva* (96); *Molvena caledonica* (92); *Athetis postpuncta* (87); *Poecilasthena leucydra* (85); *Characoma vallata* (81); *Neola octofera* (76); *Ophiusa discriminans*, *Oxymacaria insularis*, *Hippotion velox* (67); *Serrodes campana*, *Blenina donans* (62); *Argina cribraria* (56); *Chrysodeixis illuminata* (52); *Anomis involuta*

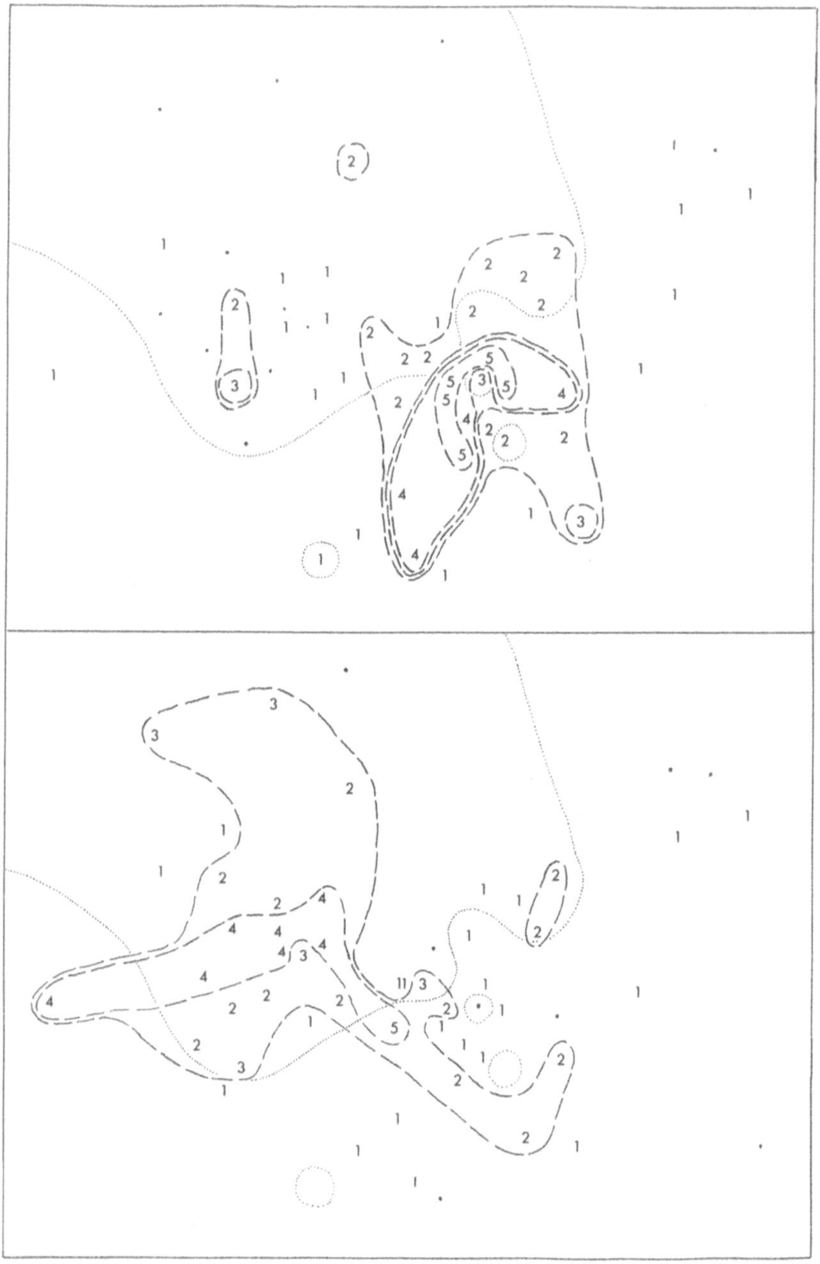

*Fig. 24.* 'Average species' for the forest and secondary cluster at level 0.028 when $k = 4$ (top) and for single-link cluster C, for data with samples amalgamated.

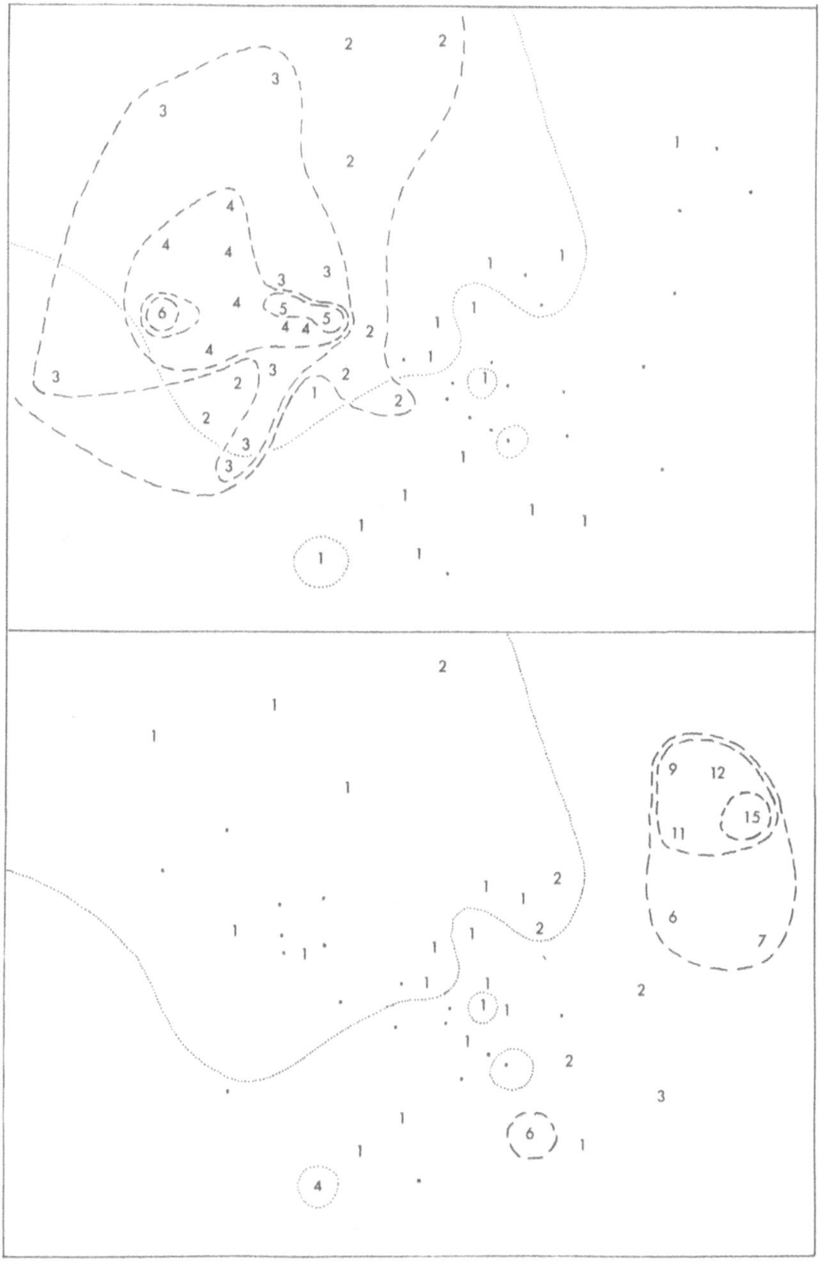


*Fig. 25.* 'Average species' for the ultramafic association (top) and the west coast and *Acacia spirorbis* association (bottom).

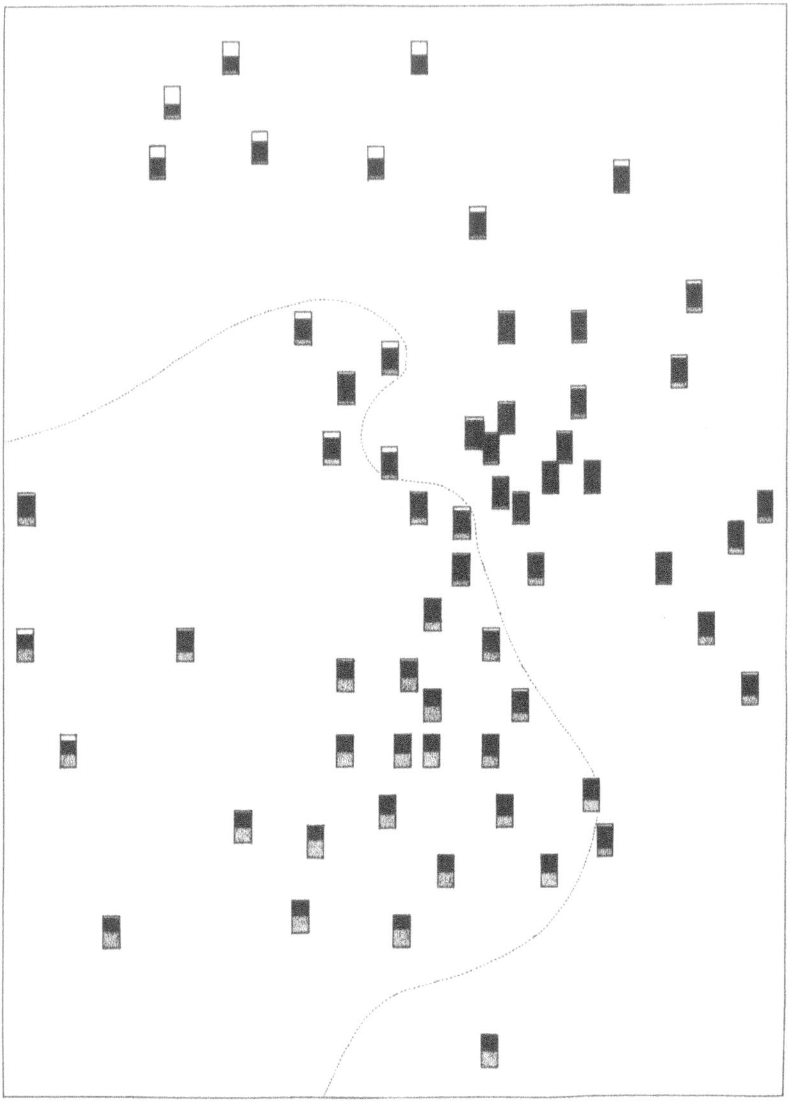

*Fig. 26.* The scaling plot of Figure 12 showing the percentage of species in each sample OTU referable to the ultramafic (stippled), forest and secondary (black) and west coast (white) associations.

(50); *Orgyia sarramea* (49); *Blenina lichenopa* (45); *Psilogramma menephron* (44); *Urolitha bipunctifera* (43); *Parallelia prisca, Macroglossum hirundo* (42); *Agrius fasciatus* (37); *Catoria camelaria* (35); *Pyrrhorachis pyrrhogona, Simplicia caeneusalis* (34); *Thalassodes viridimargo, Bocana manifestalis* (31); *Parallelia cuneilineata* (30); *Hypena laceratalis, Uliocnemis biplagiata, Rhesalides asphalta* (29); *Thyas miniacea, Avatha discolor* (27); *Nyctemera baulus, Hypospila bolinoides* (26); *Euchromia cincta, Thalassodes baladensis, Chalciope alcyona* (25); *Parallelia redunca, Anomis nigritarsis* (24); *Anomis lyona, Dichromia quinqualis* (23); *Othreis fullonia, Horisme xylinata, Leucogonia amarginata* (22); *Ophyx maculosus, Gonodontis clelia, Adrapsa ablualis* (21); *Daphnis placida* (20); *Polyacme dentata, Trigonodes hyppasia, Harita nebulosa* (19); *Pataeta carbo* (18); *Hippotion scrofa, Comostola leucomerata, Comibaena mariae, Lacera vinacea, Hypocala deflorata, Ericeia inangulata* (16); *Cerura amoa, Ctenoplusia placida, Oruza kavatcha, Platysenta dolorosa* (15); *Eublemma variochrea* (14); *Schistophleps subtilis* (13); *Plecoptera violacea* (12); *Pseudocoarica caledonica, Penicillaria maculata, Agathia asterias* (11); *Ophiusa hituensis* (10); *Gymnoscelis pseudofluctuosa, Amyna octo, Eublemma rufogalactea, Parallelia solomonensis, Argyrogramma signata, Polydesma boarmioides* (9); *Anisodes paragloria, Sauris curvicosta, Utetheisa lotrix, Anomis flava, Arcte coerula, Pantydia metaspila, Rivula caledonica* (8); *Ziridava xylinaria, Amyna flavirena, Calathusa basirufa, Parallelia vitiensis, Ctenoplusia albostriata, Lacera noctilio, Harita irregularis* (7); *Anisodes niveopuncta, Eutamsia quadrispina, Ophiusa trapezium, Harita nodyna* (6); *Collix ghosha, Chasmina tibialis, Maliattha ritsemae, Avitta australis* (5); *Eumelea rosalia, Eupithecia eupitheciata, Scotocyma legalis, Cleora dothionis, Athetis reclusa, Calogramma picta, Paectes cristatrix, Lophoptera squammigera, Earias luteolaria, Hyperlopha cristifera* (4); *Anisodes samoana, Gymnoscelis chlorobapta, Callopistria falcata, Platysenta praesecta, Nigramma polionota, Earias flavida, Anticarsia irrorata, Pantydia sparsa, Veia pectinata, Cofimpacia chusaroides, Agrius convolvuli* (3); *Gymnoscelis sara, G. tristrigosa, Cleora injectaria, Elphos subrubida, Eugnesia correspondens, Leucania venalba, Corgatha concavata, Eublemma baccalix, Eublemma pudica, Calathusa brehoa, Nycteola pseudonilotica, Nycteola canoides, Phyllodes imperialis, Aedia leucomelas, Anomis revocans, Avitta pectinata, Calyptra argyrosigna, Anoratha paritalis, Hypena probokonora, Simplicia toma, Hippotion celerio* (2). Singletons recorded from typical rainforest sites were of *Ozola plana, Prasinocyma ocyptera* (?), *Anisodes kohensis, Chloroclystis pyrsodonta, Eoasthena sarramea, Eunoumeana fletcheri, Epiplema yoshimotoi, Oeonistis entella, Utetheisa salomonis, Lasioceros aroa, Acrapex pacifica, Pseudaletia separata, Eublemma glaucochroa, Eublemma rivula, Oruza cariosa, Penicillaria jocosatrix, Ophiusa tirhaca, Ophisma gravata, Trigonodes cephise, Anadevidia peponis, Othreis materna, Prionopterina grammatistis, Speiredonia mutabilis, Hippotion boerhaviae.*

**Possible vagrants** (as recorded during survey; some may be resident as well): *Agrotis ipsilon, Agrotis munda, Athetis tenuis, Xanthorhoe sodaliata, Euproctis aliena.*

**Anomalous distributions:** *Chlorocoma octoplagiata, Scopula desita, Idaea scintillans, Idaea paniensis, Eupithecia rigida, Phrissogonus laticostata, Clepsimelia phryganeoides, Aplochlora vivilacha, Petelia trifascia, Scardamia eucampta, Dirades latibrunnea, Miresa obscurimargo, Calathusa uniformis, Calathusa mesospila, Maceda savura, Maurilia iconica, Plusiodonta theresae, Rhesala irregularis, Speiredonia zamis, Theretra nessus.*

**Diurnal species:** *Eutrichopidia macchia* and *Ctimene nocturnignis* are associated with ultramafic maquis and *Macroglossum corythus* (dusk flying) and *Euchromia creusa* are probably forest species.

**Uncounted:** Nolidae, species of *Fabresema, Macaduma, Thallarcha multifasciata, Araeopteron canescens, Trissernis greeni, Schrankia boisea, Trigonistis toroensis, Hypena fijiensis* (Malaise trap capture).

**Not recorded during the survey** (but noted elsewhere as from New Caledonia): *Thalassodes pilaria, Scopula rubraria, Micrulia pacifica, Micrulia tenuilinea, Xanthorhoe finitima, Ctimene rubropicta, Ctimene fulvimacula, Euchromia rubricollis, Theretra pinastrina, Cephonodes lifuensis, Asota caricae, Asota alienata, Asota caledonica, Syntheta nigerrima, Spodoptera exempta, Neogalea esula, Xanthodes congenita, Ophiusa coronata, Ophiusa parcemacula, Grammodes oculicola, Eudocima salaminia, Hypena masurialis. Chasmina candida, Platysenta conducta, Eublemma ragusana, Phlegetonia delatrix.*

## The composition and diversity of the species associations

Table 6 illustrates the percentage of species of each association referable to various distributional types (described in the next chapter), to various degrees of endemism and to major taxonomic groups.

Whereas Noctuidae are two to three times more numerous than Geometridae in the west coast and forest associations, the proportions are reversed in the ultramafic association. Noctuidae generally predominate in other Melanesian faunas but Geometridae, particularly the Oenochrominae, Larentiinae and Ennominae, feature much more strongly in the Australian fauna. Hence it is not surprising to find that the ultramafic association contains a relatively high proportion of Australian species or endemics of Australian affinity.

The ultramafic association contains three times the proportion of endemics (74% versus 25% in the other associations) and 37% of its total species are taxonomically isolated endemics, these latter containing roughly equal numbers of species of Indo-Malayan and Australian affinity. Thus, as with the flora, the ultramafic association contains those elements most peculiar to New Caledonia and which are probably the oldest members of the macrolepidoptera fauna.

Surprisingly, though sampling from these habitats was not extensive enough for a confident assertion, the floristically more diverse associations of the base-rich serpentine soils, such as the *Gymnostoma*-dominated low forests, did not support a correspondingly more diverse moth fauna than the *Baeckea*-sedge maquis. Indeed the reverse would appear to hold. If this is confirmed by

*Table 6.* Percentages of various distributional and taxonomic categories in the three major New Caledonian species associations and in the group of species strongly associated with *Acacia spirorbis;* these latter are bracketed and also are included in the west coast group.

| | *WEST COAST* | *(ACACIA)* | *FOREST & SECONDARY* | *ULTRAMAFIC* |
|---|---|---|---|---|
| *Total species assigned* | 33 | (7) | 172 | 46 |
| Widespread element | 34 | (0) | 49 | 9 |
| Australian elements (wide, Coral Sea, savanna) | 39 | (43) | 8 | 15 |
| Melanesian elements | 0 | (0) | 10 | 0 |
| S.W. Pacific elements | 3 | (0) | 7 | 2 |
| Endemics | 24 | (57) | 26 | 74 |
| | 100 | (100) | 100 | 100 |
| Endemics of Australian affinity | 6 | (14) | 6 | 28 |
| Endemics of Indo–Malayan affinity | 9 | (14) | 18 | 33 |
| Endemics of uncertain affinity | 9 | (29) | 2 | 13 |
| | 24 | (57) | 26 | 74 |
| Endemics that are taxonomically isolated (including members of isolated endemic species complexes) | 3 | (0) | 4 | 37 |
| Species shared with Fiji | 36 | (14) | 42 | 11 |
| Geometridae | 27 | (57) | 19 | 63 |
| Noctuidae | 61 | (43) | 67 | 30 |

further investigation then the explanation could be biochemical. It may be that adaptation of plants to the more demanding mineral regimes of the base-rich ultramafic soils has rendered them biochemically unsuitable for the larvae of foliage-feeding Lepidoptera. Nickel accumulation by some plants and the production of resins by others (e.g. *Arillastrum* and *Cocconerion*) would all contribute to such an effect. The only moth species that could conceivably be associated with the richer flora is *Caledophia pauli*, a species in a monotypic, endemic genus of Indo-Malayan affinity. Only two individuals were taken but both were in samples from *Gymnostoma*-dominated associations (in the Plaine des Lacs and on Mt. Krapé).

The valley rainforests on ultramafic rocks are, like the maquis, associated with base-poor soils but with much organic material and a high moisture content. They too have a relatively rich moth fauna but this appears to be an attenuation of the sedimentary rainforest association, though a number of species such as *Compsulyx cochereaui*, *Neola octofera* and *Stictoptera tridentifera* might be considered more particularly associated with ultamafic rainforest.

The ecology of other phytophagous insects, sap-suckers as well as defoliators, in relation to ultramafic vegetation types would repay investigation, especially if coupled with further research into the biochemistry of the flora.

The west coast group of species also has a high representation of Australian elements but the lowest proportion of endemics. Several species are shared with Fiji where they also characterise drier, more open habitats (talasinga grassland and secondary bush (ROBINSON, 1975)). Just over half the species are drawn from three noctuid subfamilies (Hadeninae, Acontiinae and Hypeninae) and one geometrid subfamily (Sterrhinae). All these subfamilies are strongly represented throughout the Old World in open herbaceous communities, including more arid ones.

The forest and secondary association is by far the largest of the three and is the one most comparable with other south-west Pacific faunas. Species shared with Norfolk I. are almost entirely within this group. It contains a high proportion of widespread Indo-Australian species and the majority of those with more restricted Melanesian and south-west Pacific distributions. The endemic species are also primarily of Indo-Malayan affinity.

The total sample was not large enough to permit the sort of analysis of diversity that has been possible on Norfolk I. Nevertheless the separation of the three species associations enables their species/abundance curves to be examined independently. These are plotted in Figure 27, the logarithm of abundance of each species being plotted against its abundance rank within the association concerned (the method of WHITTAKER (1965)). Curves 1, 2 and 4 are those for the west coast, ultramafic and forest associations respectively, 3 is that for all Norfolk I. resident species (from HOLLOWAY, 1977) and 5 is that (incomplete) for the lower montane association from Mt. Kinabalu (HOLLOWAY, 1973a), representing the richest tropical rainforest community so far examined quantitatively. Singletons for the three New Caledonian associations have been assigned roughly on the basis of the lists at the end of the previous section.

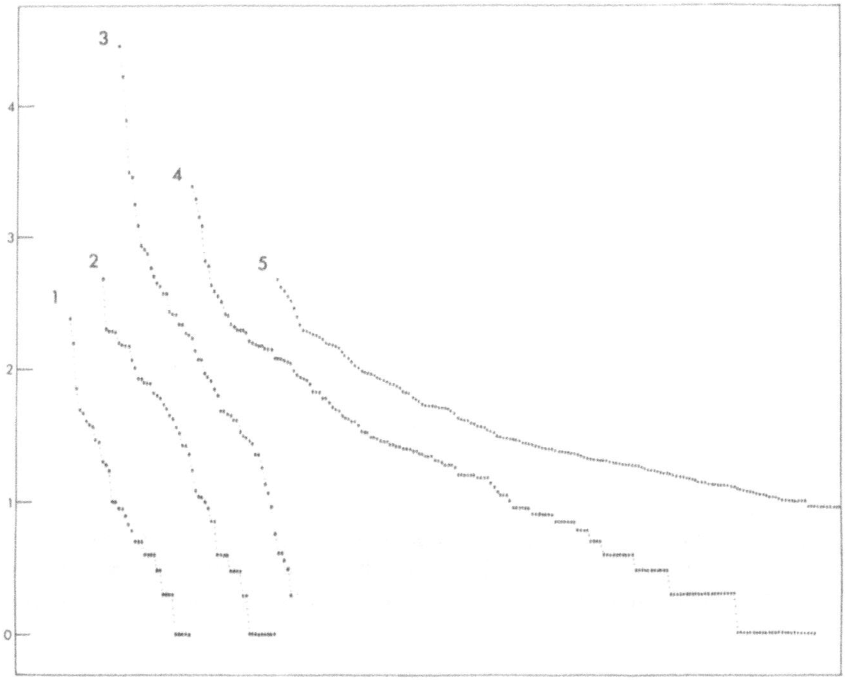


*Fig. 27.* Species/abundance plots for the west coast (1), ultramafic (2) and forest and secondary (4) associations in New Caledonia, with that for Norfolk I. resident species (3) and the upper part of that for the Mt. Kinabalu lower montane associaton for comparison. For each association the logarithm of abundance of each species (vertical axis) is plotted against its abundance rank (horizontal axis), the species being ordered according to rank.

The characteristic, complete, sigmoid log-normal curve, exemplified by that for Norfolk I., has not been obtained for any of the three associations. A total sample of five to ten times the size would probably be needed to obtain these. At a purely subjective level of comparison it might be said that the curves for the coastal and ultramafic associations resemble that for Norfolk I., whilst that for the forest association approaches the Kinabalu curve in shape.

The above comparison reflects primarily the richness of the three associations. With such a small sample it is not possible to separate reliably the richness (number of species) and equitability (evenness of abundance) components of diversity for each curve. The various statistical measures currently applied to incomplete samples are being criticised increasingly on several grounds, such as their tendency to vary with sample size (Robinson, 1975). A more empirical examination of the species/abundance curve has been recommended (Williamson, 1972, 1973).

Examination of the shapes of the curves in Figure 27 suggests that the west coast association may be similar in diversity parameters to the Norfolk I.

resident community, both in richness and equitability. Historically, as with Norfolk I. itself, it may be a relatively young association. It has a low level of endemism consistent with this hypothesis. The drier habitats it favours may have been much less extensive prior to the Pleistocene trend of climatic cooling, if the factors producing increasing aridity in Australia also had influence in New Caledonia. The vegetation types concerned are somewhat fragmented and limited in extent relative to the other associations.

The forest association, on the other hand, relates to by far the greatest extent of suitable habitat in New Caledonia and is correspondingly rich in species. It is not as rich as the Kinabalu association illustrated and has a greater spread of species at the more abundant extreme of the curve which could be indicative of lower equitability and looser packing of species. A larger sample, providing a more complete assessment of the curve, will be necessary to confirm this.

The ultramafic association curve has its points relatively bunched compared with the Norfolk I. curve, though the richness is roughly of the same order. If this bunching, indicative of greater equitability, is confirmed with a much larger sample, then it might be testimony to the greater age and history of development of the New Caledonian ultramafic community, the species number having approached more closely the maximum feasible for the environmental resources available. The evidence for a tendency for equitability to increase to a maximum under such circumstances was discussed with reference to Norfolk I. by HOLLOWAY (1977).

The ultramafic association includes most of the examples of endemic radiation in New Caledonia. There are three *Adeixis*, two *Codalithia*, two related *Scopula* and two out of three *Cofimpacia* included in the association. Seven of the *Fabresema* species were also recorded primarily from ultramafic localities.

It is surprising that the richness of the ultramafic association is only of the order of the resident fauna of Norfolk I. The size of the total moth fauna of New Caledonia is close to that expected for its area in relation to other Pacific islands (ROBINSON, 1975), given that there is linear correlation between the logarithm of area and the logarithm of faunal size for all but very small islands. But comparing the relative areas of ultramafic and sedimentary rocks leads to the expectation of a much larger fauna associated with the former unless the 'volume' of that environment available for exploitation by phytophagous insects such as Lepidoptera is only a small proportion of the whole. The indications are that the truly ultramafic species are largely restricted to the floristically poor *Baeckea*-sedge maquis. They may be excluded from the richer associations on base-rich soils for biochemical reasons and from the rainforests by competition from species of the forest association. The same species occur, somewhat depleted, on sedimentary and metamorphic rocks in association with elements of the maquis such as on the Plateau de Dogny and parts of the Panié range.

Lack of suitable larval foodplant may not be the only possible limiting factor. Two species of the ultramafic association were reared from larvae beaten from *Baeckea ericoides* in the Plaine des Lacs, namely *Adeixis baeckeae* and *Syneora strixoides.* They readily accepted *Baeckea virgata,* which is found mainly on sedimentary rock over the central part of the island, and the *Adeixis* accepted *B. pinifolia,* a localised riverine species noted at the Cascade de Ciu and at Cap Colnett during the survey (the *Syneora* had pupated before this latter species was encountered). *S. strixoides* was recorded as an adult from most areas where *Baeckea* was abundant, but was most numerous in ultramafic areas. *A. baeckeae* was only recorded from the southern ultramafic massif, being particularly numerous in the Plaine des Lacs. If further sampling confirms this restriction, then presumably the distribution of *baeckeae* is limited by factors other than the availability of *Baeckea.* Its close relative and possible competitor, *A. major,* a larger species, was found more abundantly and occurred away from the southern massif on the Dome de Tiebaghi, on the ultramafic outlier near Bâ (site 59) and even on sedimentary rock amid the hillcrest association of Col Toma where maquis elements were present in the vegetation.

A lot more information on the biology of the ultramafic species must be collected before the reasons behind their apparent low numbers overall and their low abundance in, or exclusion from, the more highly adapted floristic associations of the base-rich soils can be elucidated. Extensive beating for larvae in the Grand Lac *Baeckea*-sedge maquis was entirely fruitless apart from the yield from *Baeckea.* Beating and sweeping in other maquis areas (mainly for Psocoptera) also produced only one or two larvae. A geometrid larva was discovered feeding on *Tristania guillainii* on the Dome de Tiebaghi but was not reared out.

*Acacia spirorbis* and *Alphitonia neocaledonica* grow on a variety of substrates throughout the island. It might be possible to investigate the degree of defoliation of these species by larvae in relation to the base content of the soils in which they grow. *Alphitonia* trees on the Col de Petchécara were observed during the survey to be much more heavily defoliated than those growing in some ultramafic areas, though the single species reared from *Alphitonia, Casbia rectaria,* was found as a larva on the Dome de Tiebaghi and was assigned to the ultramafic association in the numerical analysis. Beating of the Petchécara *Alphitonia* yielded five different types of lepidopteran larva. As well as the *Casbia* there were two more Geometridae, one of which was probably a geometrine, a noctuid and what was possibly a small lymantriid. The others were not reared out.

One point of interest is that many of the taxonomically isolated endemics associated with the ultramafics, presumably some of the earlier lepidopterous colonists of the island, are of Indo-Malayan affinity. Noctuids such as *Parallelia lateritica* and *Chasmina basiflava* and geometrids such as *Nadagara serpentina* and *Gonanticlea pulcherrima* belong to polytypic genera that are widespread throughout the Indo-Australian tropics, though mostly centred in the Oriental Region. Among the butterflies *Austroypthima petersi* is another taxonomically isolated ultramafic species of Indo-Malayan derivation. Such Indo-Malayan species must, in their original colonisation, either have followed their Indo-Malayan host plants in establishing on New Caledonia or have been able to shift to a new host on arrival, assuming they were moderately host-specific in the first place. Somehow during this process the insects or the plant-insect associations became restricted to the ultramafic rocks, yet today the more typically Malesian vegetation occurs on the sedimentary rocks.

It was suggested in Chapter I that the ultramafic rocks were probably more extensive, contributing a far greater proportion of the total land area of New Caledonia, soon after their emplacement. Subsequently erosion has steadily reduced their area to its present extent. In such circumstances, if the original area of sedimentary rock associated with the ultramafics was very small, there would have been a premium on adaptation by colonists to conditions on the ultramafics, be they insect-host complexes with the host plant adapting to a more demanding mineral regime or insects adapting to a new host. Such pressures must have been most stringent immediately following emplacement of the ultramafics and it may thus be reasonable to date the arrival of those early Indo-Malayan colonists predominantly at some time from the later Oligocene to the middle of the Miocene, coinciding with a period of increasing contact between the Oriental Region and Australasia (p. 24) yet whilst the ultramafics were still extensive. There may have been much greater competition at all times amongst Indo-Malayan taxa in establishing on the sedimentary rocks and their associated flora and adaptation to the ultramafics could have been correlated with reduced viability in the face of such competition, the species concerned thus becoming restricted to the ultramafics.

Once the ecological barrier of adaptation to the ultramafics had been broached by a species it would appear from present day evidence that there was potential for radiation, for, as stated earlier, instances of endemic radiation are largely concentrated in the ultramafic association. Interspecific competition and perhaps the volume of immigration were possibly too high to permit similar radiation in the sedimentary rock associations. In a few instances of radiation such as in *Cofimpacia* and the *Idaea* triplet (*fruticeti*, *paniensis* and *lateritica*) there would appear to be some ecological segregation of the species. This may be incipient in the various ecotypes of *Teara circumfumata.*

It may be significant that the ultramafic association Indo-Malayan species, indeed all the species, are associated most strongly with the *Baeckea*-sedge maquis on poor soils with relatively undemanding mineral regimes; the challenge of adaptation by foliage feeding insects to such vegetation or of its penetration by potential host plants might have been similarly undemanding. Even Australian moth species, to some extent preadapted to sclerophyll vegetation and to many of the plant taxa widespread on the ultramafics, are largely restricted to the *Baeckea*-sedge maquis.

A similar penetration of a less demanding regime may have occurred with the Indo-Malayan rainforest elements now found partially on the ultramafic rocks, such forests being associated with base-poor humic soils. Several of the forest association moth species more particularly associated with ultramafic rainforest, such as those falling into the intermediate single-link cluster C (p. 174), are also taxonomically isolated Indo-Malayan derivatives. Amongst the moths there are *Compsulyx cochereaui*, *Stictoptera tridentifera* and *Spilosoma lateritica*, and amongst the butterflies *Delias ellipsis* and *Nacaduba deplorans*.

Only two Indo-Malayan ultramafic species have been associated with host plants, one only tentatively. HOLLOWAY (1973b) suggested that *Austroypthima petersi* appeared to be associated in its adult distribution with species of the endemic bamboo, *Greslania*, which is related to the Papuan genus *Nastus*. Gramineae are virtually excluded from the ultramafics, Cyperaceae predominating. *Greslania* is the major exception. *Austroypthima* belongs to the *Ypthima* group of genera which are generally grass and bamboo feeders, so the association of *Austroypthima* with *Greslania* is likely, though not yet confirmed by observation of oviposition or rearing of larvae. If confirmed, this might be an instance where Indo-Malayan insect and host have penetrated the ultramafics more or less together. However, *Nastus* has an odd distribution that includes Madagascar and excludes the Oriental Region, so it may have had a longer history in Australasia than the butterfly. The butterfly has no specific relatives elsewhere, though one may await discovery in New Guinea, perhaps in areas where *Nastus* is present.

The second species is *Anisodes lateritica*, endemic, not highly isolated taxonomically but belonging to a widespread Indo-Malayan genus. It was reared by M. A. DELOBEL from *Codia discolor*. The genus *Codia* is restricted to E. Australia and New Caledonia. In the latter its species are important components of various associations on the ultramafics. It belongs to the predominantly Australasian family Cunoniaceae (p. 66–7). The closest relatives of the moth are firstly in Queensland and New Guinea (*A. compacta*) and secondly, more distantly, in Fiji (*A. lautokensis*). *A. lautokensis* has been reared from *Myrtus* in Fiji (ROBINSON, 1975), so the complex may be flexible in its larval foodplant requirements.

Akin to this *Anisodes* complex, several other of the older Indo-Malayan elements in New Caledonia have their nearest taxonomic association with distinctive Queensland or New Guinea species, both being more distantly related to Oriental forms. This Inner Melanesian Arc association suggests that the ancestors of these pairs may have first colonised and adapted to typical Inner Arc rainforest conditions in New Guinea and Queensland, an adaptation which could have facilitated penetration of the more demanding New Caledonian ultramafic environment at a later date. These Inner Arc associations will be discussed in detail in the next chapter.

More recently *Craniophora fasciata*, a member of the widespread Indo-Australian element yet associated with the ultramafics, appears to have colonised New Caledonia from Australia, being absent from the rest of eastern Melanesia. The genitalia differ slightly from those of Oriental specimens and here too there may have been a degree of preadaptation. Similarly *Parallelia propyrrha* appears strongly associated with *Acacia spirorbis*, prominent mainly on the ultramafics, and is essentially an Australian species. However, it is closely related to Oriental taxa such as *P. feneratrix*.

The key to the colonisation of New Caledonia by Lepidoptera will lie in studies of the biology of these complexes. Unfortunately little is known of their life histories in various parts of their ranges.

## Seasonal trends

As the survey progressed it became apparent that a number of species were much less abundant in what had previously been productive habitat types, suggesting that some sort of seasonal decline might have been occurring. Particularly notable in this respect were the noctuids *Achaea janata, Chrysodeixis eriosoma* and *Callopistria maillardi* and the geometrid *Gnamptoloma rubra.* The last named may have been assigned to the ultramafic species association in the analyses largely because of this. The early samples when it was more common were made mainly in the Plaine des Lacs. *Achaea janata* (OTU 25) also does not fall readily into any of the three associations. It is perhaps most characteristic of secondary vegetation and areas of grassland. It was observed to be numerous both by day and at light at Anse Longue.

It would not be possible to assess overall trends without sampling through a whole year. However, comparison of samples made at the beginning and end of the survey period in the same localities in the Grand Lac and Pic du Pin areas indicates that, whilst the abundance levels of the ultramafic species on the whole changed little, those of the forest and secondary species almost all showed a marked decline.

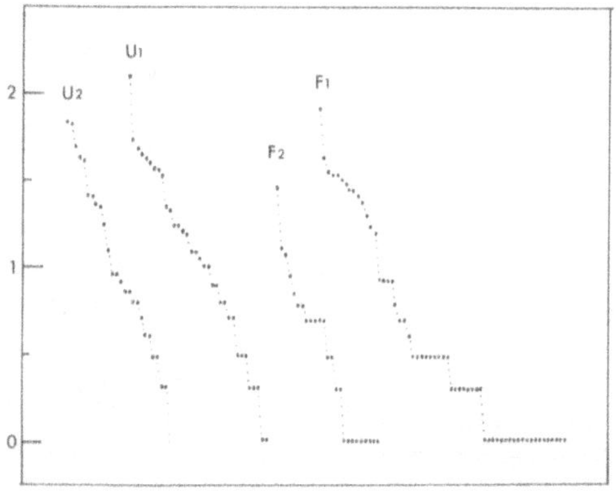


*Fig. 28.* Species/abundance curves for the ultramafic (U) and forest (F) associations in samples from Plaine des Lacs localities made at the beginning (1) and end (2) of the survey period, illustrating seasonal changes.

Figure 28 shows the species/abundance curves for the ultramafic (U) and forest and secondary (F) associations at the beginning (May: suffix 1) and end (August: suffix 2) of the four-month survey period. In the initial period sites 2, 3, 4 and 7 were sampled, 3 and 4 being represented by samples from the very similar sites 72 and 71 respectively in the second period. All were in habitats with extensive *Baeckea*-sedge maquis around the trap or nearby. The difference in behaviour of the two associations is obvious from the figure but it can be illustrated numerically also. Expressing all abundances as logarithms, the difference between the two periods in overall sample size is 0.45. Comparisons of the abundances of the first twelve species in the abundance rank order of the ultramafic association during the first period with those of equivalent rank in the second period showed a mean fall, rank by rank, of 0.13, ranging from an increase of 0.09 to a decrease of 0.28. A similar comparison for the forest association showed a mean fall of 0.63, ranging from a decrease of 0.46 to a decrease of 0.75.

These rank comparisons were made irrespective of species identity. Individual species often showed a much more extreme decline. With very few exceptions the forest and secondary association species were represented by fewer individuals in the second period. Notable decreases amongst the more abundant species were shown by *Rhodogastria albivitrea* (36:5), *Callopistria maillardi* (32:0), *Leucocosmia nonagrica* (35:2), *Platysenta illecta* (26:5), *Achaea janata* (16:1), *Chrysodeixis eriosoma* (28:0), *Hydrillodes norfolki* (35:3), *Compsulyx cochereaui* (24:6) and *Cleora immemorata* (17:5). Species showing a much smaller decline were *Sasunaga tenebrosa* (28:13), *Stictoptera tridentifera* (31:12) and *Ericeia hirsutitarsus* (83:29), species placed in the trend towards the ultramafic association in the ordination in Figure 20 (OTUs 18, 22 and 34). Two species, *Neola octofera* (0:7) and *Teara circumfumata* (0:6), showed a definite increase. These last are both Australian derivatives and therefore somewhat unusual in the forest association; the latter has a distinct ecotype restricted to the ultramafic maquis.

A much greater range of variation in abundance is evident for the ultramafic species. Several of these also showed a marked decline, such as *Spilosoma lateritica* (44:9) which is shared with the forest association, *Adeixis inostentata* (17:6), *Anisozyga caledonica* (36:6), *Gnamptoloma rubra* (15:0), *Chloroclystis lunifera* (36:3) and *Casbia rectaria* (39:2). Most remained relatively constant such as *Parallelia lateritica* (11:7), *Chasmina basiflava* (10:12), *Adeixis major* (23:17), *Oenochroma unifasciata* (33:41), *Euphyia perseverata* (127:65) and *Horisme grisearia* (42:48), and several showed a marked increase, notably *Eilema caledonica* (2:42), *Nadagara serpentina* (7:66) and *Symmetroctena variegata* (3:25).

On Norfolk I. the two major ecological associations of species are also largely segregated seasonally (Holloway, 1977). The preliminary indications are that a similar segregation may occur in New Caledonia, the ultramafic species generally maintaining their adult populations through the cooler, drier winter and the forest and secondary species suffering an overall decline. The peak abundance of the latter may therefore coincide with the period of heaviest rainfall, as does that of most species of the widespread association on Norfolk I. Indeed the associations concerned have many species in common. On Norfolk I. the overall population level of the widespread species association was shown to be correlated with irregular fluctuations in rainfall as well as the general pattern (Holloway, 1977), most of the species being continuously brooded.

# VI. THE GEOGRAPHY OF THE NEW CALEDONIAN LEPIDOPTERA

The discussion in this chapter is based mainly on a numerical analysis of patterns of geographical distribution shown by Lepidoptera species in the south-west Pacific. This analysis includes the majority of species recorded for New Caledonia here (records of species such as *Ophiusa parcemacula* coming too late for inclusion) and for Fiji by ROBINSON (1975). Also included are those species collected in the New Hebrides in 1971 by ROBINSON that have been reliably identified. A number of small and insignificant species are excluded (mainly nolids and acontiine and hypenine noctuids) that have probably been overlooked by most collectors and are therefore under-recorded.

Information on the distributions of these species was gained mainly from reference to the collections in the British Museum (Natural History) but reference was also made to sources in the literature such as VIETTE (1949, 1951), PROUT (1928), TAMS (1935), ROBINSON (1975) and MADDISON (1976), and to an unpublished collation of records for the Pacific by ROBINSON. Unpublished records for Tonga made by the FAO/UNDP pests and diseases survey were included and Dr I. F. B. COMMON kindly provided information on the distribution of Australian species.

The distribution amongst the areas listed in Table 7 was recorded for each species, providing the basic data for the numerical analysis. The following dissimilarity coefficient was calculated for each pair of species except when it appeared likely to exceed 0.50:

$$1-\frac{\text{Number of areas in which both species occur}}{\text{Number of areas in which either or both species occur}}$$

It is a measure of the overlap of the distributions of the two species and has been used previously by HOLLOWAY & JARDINE (1968) and HOLLOWAY (1973b, 1974a, 1977).

Single-link cluster analysis was used to derive the dendrogram in Figure 29 (the very large widespread cluster is excluded) and a linkage diagram (Figure 30) was drawn to study the situation at values of $k$ greater than 1 in the $B_k$ sequence (p. 152). In the figure solid lines indicate links to dissimilarity level 0.33, broken lines links to 0.38 and dotted lines links to 0.40.

The largest cluster obtained in the single-link analysis contained 133 species distributed widely through the Indo-Australian area. With few exceptions the same species formed a $B_3$ cluster at level 0.40. The distribution of this widespread element is shown in Figure 31 where the numbers of species referable to the element found in each area are indicated and the fall-off of

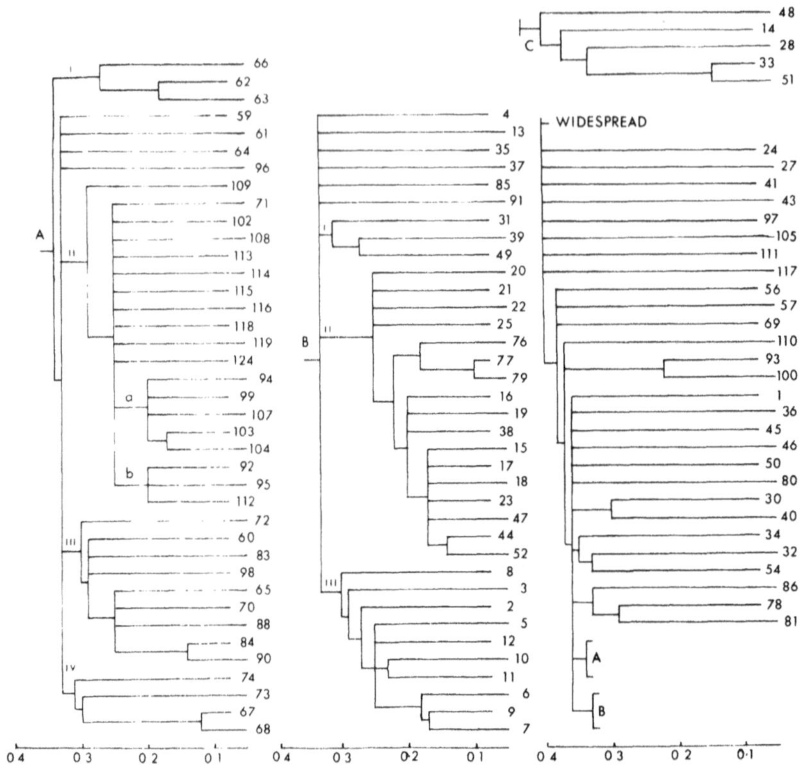

*Fig. 29.* Single-link dendrogram derived in the classification of species according to their geographical distributions as discussed in the text. The numbers are as in Table 7. Species of the widespread element are excluded.

representation of the element from its centre is illustrated by contours at regular intervals of species number.

At dissimilarity levels lower than 0.40 the widespread cluster showed some segregation into a main cluster of species generally referable to the Oriental and Melanesian tropics and a smaller cluster of more ubiquitous species, extending far out into the Pacific. This segregation was only revealed when the situation for values of $k$ greater than 1 was examined. The majority of species central to the main cluster were Rhopalocera or Sphingidae which have probably been much better recorded in the area. This applies to some extent to the smaller cluster and it could be that when the distributions of the more insignificant species are known more accurately the segregation will prove to be more clear cut.

At dissimilarity level 0.30 the smaller cluster was represented by a $B_5$ cluster of 10 OTUs, two of which were also included in the main cluster. The ten species concerned are *Agrius convolvuli*, *Achaea janata*, *Chrysodeixis eriosoma*, *Spodoptera litura*, *Spodoptera mauritia*, *Amyna octo*, *Zizina otis*,

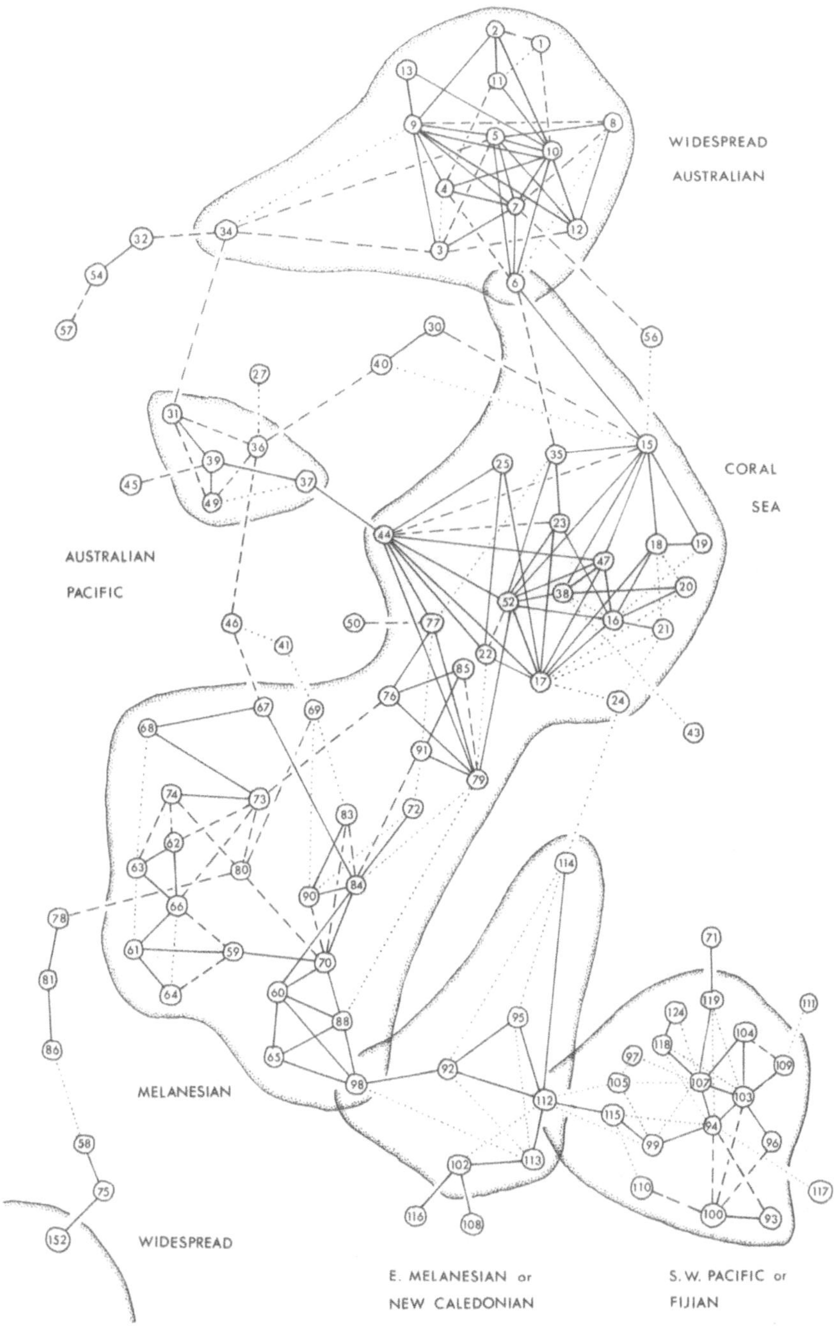


*Fig. 30.* Linkage diagram for OTUs representing species of the several Australasian elements discussed in the text. $B_2$ clusters at level 0.40 are enclosed by stippled lines. Links to level 0.33 are indicated by solid lines and subsequent ones to 0.38 then to 0.40 are indicated by broken and dotted lines respectively.

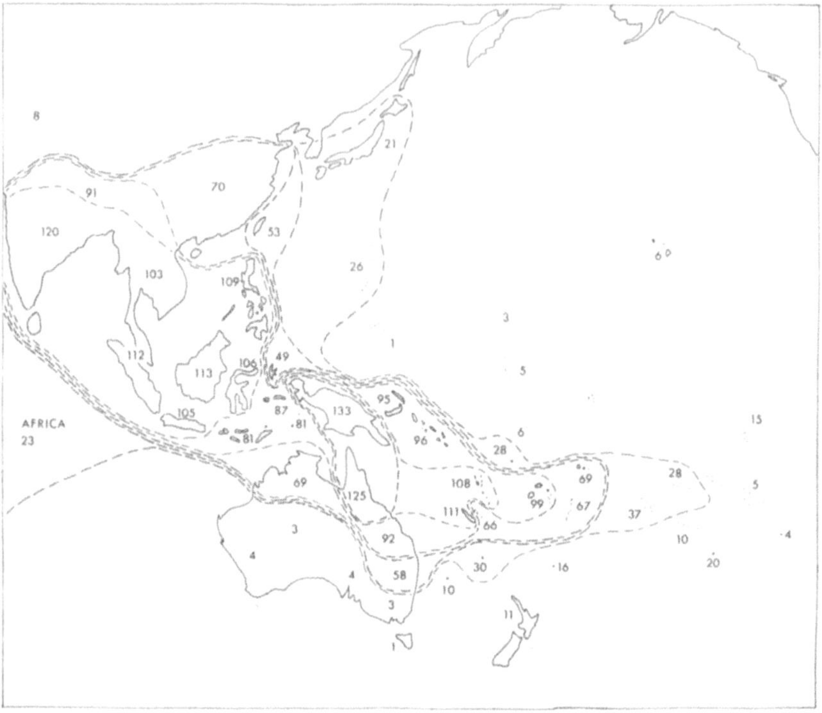


*Fig. 31.* The distribution of species of the widespread element.

*Hypolimnas bolina*, *Melanitis leda* and *Platysenta illecta*, the last two being the two OTUs in the overlap with the main cluster. A number of other species associated mainly with this cluster rather than with the main cluster, namely *Gnathothlibus erotus*, *Hippotion celerio*, *Anomis involuta*, *Simplicia caeneusalis*, *Chasmina tibialis*, *Tiracola plagiata*, *Anticarsia irrorata*, *Othreis fullonia*, *Helicoverpa armigera*, *Helicoverpa assulta*, *Mocis frugalis* and *Lampides boeticus*.

Apart from being extremely widespread in the Indo-Australian and Pacific tropics and subtropics, about half the species occur in Africa and others have close relatives there. When the Pacific island groups have been better worked there are other species that might correctly be included with the element, having been recorded from areas far into the Pacific such as Rapa I. and the Marquesas, such as *Utetheisa pulchelloides*, *Callopistria maillardi*, *Amyna natalis*, *Phlegetonia delatrix*, *Avatha discolor* and *Gymnoscelis imparatalis*.

These 30 species comprise the waif element of the Pacific. Many are migratory: 22 are recorded from Norfolk I. and of these 17 are definite or suspected vagrants (HOLLOWAY, 1977). They include many of the dominant species (e.g. the two *Spodoptera*) of the widespread community of species characteristic of disturbed habitats on Norfolk I. and are generally those associated with the smaller Pacific islands. There are 19 on Rapa I. and 18 on Rotuma I., 58% and 25% of the faunas respectively. They include many of those species distributed over a wide range of habitats in New Caledonia (p. 163). Many are agricultural pests.

The distribution of species in the total widespread element shows some degree of segregation into 'Melanesian' or 'Coral Sea' patterns in Australasia, patterns characteristic of more localised distributional elements discussed later on. The former extend strongly through Melanesia with little extension into Australia, often excluding New Caledonia; the latter are largely absent from E. Melanesia, reaching the south-west Pacific via Australia. These distributions may reflect the dispersal routes of what were probably Oriental species originally, perhaps those more characteristic of rainforest favouring the Melanesian route and those tolerant of drier conditions being more successful in Australia. Examples of the former are *Ericeia inangulata, Felinia precedens, Anisodes obliviaria, Symmacra solidaria* and *Jamides bochus.* Only the first occurs in New Caledonia, though *Jamides carissima* may prove to be merely a form of *bochus.* Examples of the latter are *Catopsilia scylla, Mythimna consanguis, Eublemma cochylioides, Eublemma pudica, Lophoptera hemithyris, Blenina donans, Ophiusa tirhaca, Ophisma gravata, Hypospila bolinoides, Anadevidia peponis, Ctenoplusia placida, Calyptra argyrosigna, Harita nebulosa, Traminda mundissima, Ziridava xylinaria* and *Cleora injectaria.* Many of these species do not occur outside New Caledonia in the Pacific.

Most of the remaining non-endemic species fall into one of a number of Australasian elements. Figure 29 shows the single-link dendrogram derived for such species which mostly cluster together as a broadly Australasian element before uniting with the widespread cluster at level 0.40, and Figure 30 shows the linkage diagram as described earlier, the species being identified by numbers as listed in Table 7. Interpretation of the linkage diagram in terms of the $B_k$ sequence of cluster analyses led to the recognition of aggregations of OTUs (species) somewhat different from those indicated by the single-link ($B_1$) clusters, aggregations that are generally more homogeneous and therefore more satisfactory.

The widespread Australian cluster (*Biii* in Figure 29) links with the Coral Sea cluster (*Bii*) at level 0.33 in the single-link analysis but at level 0.40 it forms a distinct $B_2$ cluster, though sharing OTU 6 with the Coral Sea cluster. This $B_2$ cluster is enclosed in Figure 30 and is recognised as containing species of a widespread Australian element, fourteen in number. The element is illustrated in Figure 32 and contains species widely distributed through the south and east of Australia that extend out into the Pacific, primarily to New Caledonia and Norfolk I. This element was recognised to be a major component of the Norfolk I. fauna by HOLLOWAY (1977). It includes species such as *Agrotis munda, Athetis tenuis, Cynthia kershawi* and *Bassaris itea* that may be wholly or partially vagrant to New Caledonia and have been recorded as vagrant to Norfolk I. Others included are definitely resident such as the ecologically localised *Adeixis inostentata* and the butterfly *Nacaduba biocellata.* Six species are found in New Zealand and the species may generally be

*Table 7.* List of primary areas used in the numerical analysis of Lepidoptera geographical distribution, together with a list of species numbered as in Figures 29 and 30 to facilitate cross-reference.

Africa
S. Palaearctic
N. Palaearctic
Japan
China
Taiwan
Himalaya
India + Sri Lanka
S.E. Asia
Malaya + Sumatra
Borneo
Java
W. Lesser Sundas
Philippines
Sulawesi
N. Moluccas
S. Moluccas
Kei + Tenimber
New Guinea
W. Australia
Eyrean (central)
N. Territory
N. Queensland
S. Queensland
New South Wales
Victoria
Tasmania
S. Australia
New Zealand
Norfolk I.
Lord Howe I.
Chatham Is.
Bismarck Is.
Solomon Is.
New Hebrides
New Caledonia
Loyalty Is.
Fiji
Rotuma
Samoa
Tonga
Kermadec Is.
Cook Is.
Society Is.
Austral Is.
Tuamotus
Marquesas
Rapa
Pitcairn
Mariana Is.
Caroline Is.
Marshall Is.
Gilbert Is.
Tuvalu Is.
Phoenix Is.

1. *Cynthia kershawi*
2. *Bassaris itea*
3. *Agrotis munda*
4. *Athetis tenuis*
5. *Syntheta nigerrima*
6. *Pantydia capistrata*
7. *Pantydia sparsa*
8. *Adeixis inostentata*
9. *Hippotion scrofa*
10. *Scopula rubraria*
11. *Phrissogonus laticostata*
12. *Xanthorhoe sodaliata*
13. *Diarsia intermixta*
14. *Xanthodes congenita*
15. *Eilema nitens*
16. *Aeolochroma hypochromaria*
    *Calathusa mesospila*
    *Encryphia frontisignata*
    *Acrapex exsanguis*
    *Casbia albinotata*
    *Neogyne elongata*
17. *Casbia rectaria*
18. *Characoma vallata*
19. *Euproctis aliena*
20. *Lysimelia litoralis*
    *Araeopteron canescens*
21. *Egone bipunctalis*
    *Calathusa stenophylla*
22. *Delias nysa*
23. *Corgatha figuralis*
24. *Parallelia propyrrha*
25. *Urolitha bipunctifera*
26. *Chloroclystis pyrsodonta*
27. *Gymnoscelis acidna*
28. *Cephonodes janus*
29. *Idaea scintillans*
30. *Prasinocyma albicosta*
31. *Acraea andromacha*
32. *Anaphaeis java*
33. *Catopyrops florinda*
34. *Nacaduba biocellata*
35. *Digama marmorea*
36. *Leucania scottii*
37. *Athetis striolata*
38. *Corgatha concavata*
39. *Oruza cariosa*
40. *Pataeta carbo*
41. *Lophoptera abbreviata*
42. *Grammodes oculicola*
43. *Parallelia hicanora*
44. *Crioa acronyctoides*
45. *Rhesalides asphalta*
46. *Agathia asterias*
47. *Comibaena mariae*
48. *Scopula desita*
49. *Scopula homodoxa*
50. *Scopula oppilata*
51. *Gnamptoloma rubra*
52. *Horisme xylinata*
53. *Catoria camelaria*
54. *Precis villida*
55. *Agrotis ipsilon aneituma*
56. *Leucania loreyimima*
57. *Earias perhuegeli*

*Table 7. (Cont.)*

58. *Macroglossum hirundo*
59. *Euploea nemertes*
60. *Euploea treitschkei*
61. *Cyrestis telamon*
62. *Hypolimnas alimena*
63. *Eurema candida*
64. *Danis schaeffera*
65. *Luthrodes cleotas*
66. *Borbo impar*
67. *Euchromia creusa*
68. *Platysenta praesecta*
69. *Anigraea ochrobasis*
70. *Nigramma polionota*
71. *Penicillaria dinawa*
72. *Blenina lichenopa*
73. *Ophiusa hituensis*
74. *Chalciope alcyona*
75. *Mocis trifasciata*
76. *Parallelia redunca*
77. *Parallelia solomonensis*
78. *Parallelia vitiensis*
79. *Phyllodes imperialis*
80. *Thyas miniacea*
81. *Chrysodeixis illuminata*
82. *Serrodes mediopallens*
83. *Mniothripa bradleyi*
84. *Harita nodyna*
85. *Comostola leucomerata*
86. *Thalassodes pilaria*
87. *Anisodes niveopuncta*
88. *Cleora lacteata*
89. *Clepsimelia phryganeoides*
90. *Nadagara irrectracta*
91. *Cleora illustraria*
92. *Hypolimnas octocula*
    *Euploea boisduvalii*
93. *Anisodes samoana*
    *Euploea lewinii*
94. *Appias albina subspecies*
95. *Badamia atrox*
96. *Nola samoana*
97. *Nola transversata*
98. *Utetheisa salomonis*
99. *Maceda savura*
100. *Parallelia prisca*
101. *Aedia sericea*
102. *Ericeia hirsutitarsus*
103. *Hypospila similis*
104. *Hydrillodes surata*
105. *Speiredonia simplex*
106. *Cephonodes armatus*
107. *Thalassodes chloropis*
108. *Trissernis greeni*
109. *Chloroclystis lepta*
110. *Chloroclystis scintillata*
111. *Gymnoscelis concinna*
112. *Scardamia eucampta*
113. *Agape lifuensis*
    *Parallelia cuneilineata*
    *Harita irregularis*
    *Cephonodes lifuensis*
    *Cleora immemorata*
    *Danaus pumila*
    *Calathusa basirufa*
114. *Jamides carissima*
115. *Rhesala irregularis*
    *Gymnoscelis sara*
116. *Sauris curvicosta*
    *Papilio ilioneus*
    *Cepora perimale subspecies*
117. *Poecilasthena leucydra*
118. *Apothripa vailima*
    *Rivula dipterygosoma*
    *Chloroclystis bosora*
    *Collix lasiospila*
119. *Ericeia leichardtii*
    *Nacaduba samoensis*
    *Anomis samoana*
    *Cleora samoana*
    *Rivula polynesiana*
152. *Maliattha ritsemae*

more characteristic of temperate climates than those of the Coral Sea element. The species *Xanthorhoe sodaliata* and *Syntheta nigerrima* may also only be vagrant to New Caledonia.

A few other species are associated in the analysis with this element but do not cluster with it, being somewhat intermediate with other clusters or of somewhat anomalous distribution. *Leucania loreyimima* (56) has a slightly more northerly distribution in Australia than is typical for the Australian widespread element, trending towards the Coral Sea element. The distribution

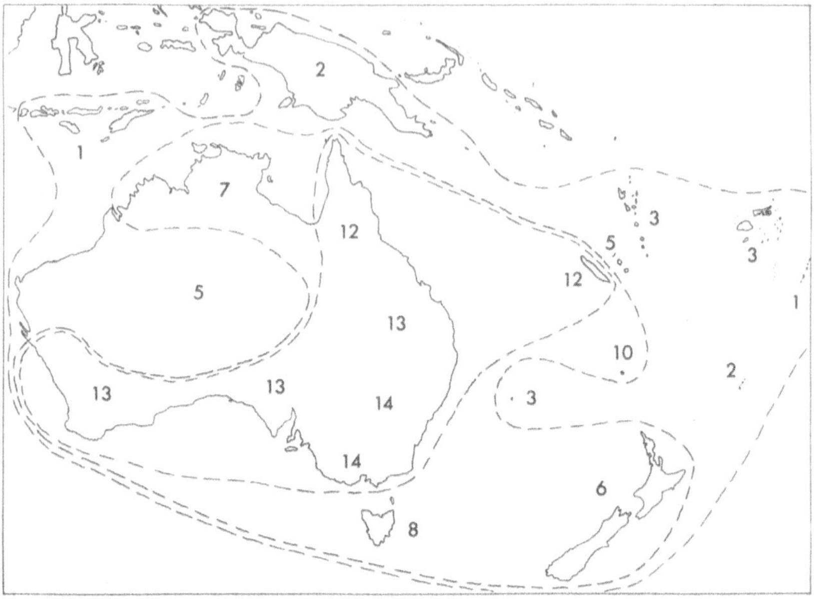


*Fig. 32.* The distribution of species of the widespread Australian element.'

is also somewhat anomalous as the species has been recorded from scattered localities in the Oriental Region. It is a grassland feeder as a larva and could have been spread to these localities during the transport of animals with fodder.

A loose group of four species includes *Nacaduba biocellata*, a member of the widespread Australian element (OTU 34). But this forms a triplet in the single-link analysis with two other OTUs, *Anaphaeis java* (32) and *Precis villida* (54). The last links with the fourth member of the group, *Earias perhuegeli* (57). These are all species widespread in Australia that extend strongly into the Pacific. *A. java*, *P. villida* and *E. perhuegeli* have been recorded from Norfolk I., the last two probably only vagrant. At dissimilarity levels higher than 0.40 links occur with the Australian Pacific element described below as well as with the widespread Australian element.

The Australian Pacific element contains only five taxa in the strict sense in that these form a $B_2$ cluster (indicated in Figure 30) at level 0.40, three OTUs (31, 39, 49) having formed a single-link cluster (*Bi* in Figure 29). The five species involved are *Acraea andromacha*, *Leucania scottii*, *Athetis striolata*, *Oruza cariosa* and *Scopula homodoxa*. They are generally distributed through N. Australia and New Guinea, extending strongly into the south-west Pacific. Three were assigned to the west coast/*Acacia spirorbis* association in the previous chapter, the *Acraea* is also associated with the western side of the island (HOLLOWAY & PETERS, 1976) and *Oruza cariosa*, apparently rare in

New Caledonia, is associated with the dry talasinga habitats in Fiji (ROBINSON, 1975).

At higher levels of dissimilarity than those shown in Figure 30 several other OTUs come into loose association with the Australian Pacific group which thus, in a broader sense, could be considered to include *A. java*, *N. biocellata*, *Grammodes oculicola* (42) and *Rhesalides asphalta* (45). *G. oculicola* is rare both in New Caledonia and Fiji but *R. asphalta* is a talasinga species in Fiji and, apart from a high catch at Sarraméa (p. 139), is mainly found in the west coast habitats in New Caledonia. Both the first two are perhaps commoner on the drier western side of the island and the *Nacaduba* may feed as a larva on *Acacia spirorbis* flowers (HOLLOWAY & PETERS, 1976).

The distributions of all these species and *Earias perhuegeli* are mapped summed in Figure 33. *Earias perhuegeli* was also only taken in New Caledonian west coast localities. The overall distribution and ecology of these species suggests that the element generally could be considered typical of dry tropical grasslands and associated shrubby species.

Also associated exclusively with this element in the analysis are *Gymnoscelis acidna* (27) and, at a higher dissimilarity level than 0.40, *Anisodes niveopuncta* (87) and *Digama marmorea* (35). The first, though assigned to the ultramafic ecological association in the previous chapter, is common in west coast localities and the last was also recorded (two localities) only from the west coast. The *Anisodes* is a species of coastal vegetation in Fiji (ROBINSON; as *decolorata*) and appears to be characteristic of open habitats with secondary vegetation in New Caledonia, though it has not as yet been extensively recorded.

The loose single-link cluster C in Figure 29 does not persist at higher values of *k*, tending to unite with the Coral Sea cluster. It includes five OTUs, these being the species *Xanthodes congenita*, *Cephonodes janus*, *Catopyrops florinda*, *Scopula desita* and *Gnamptoloma rubra*. Their joint distribution is mapped in

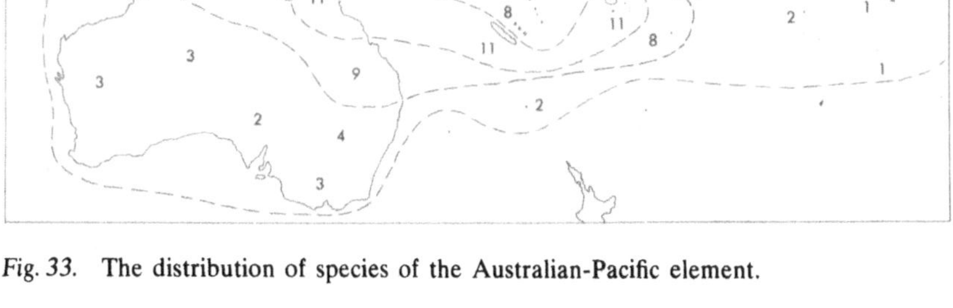

Fig. 33. The distribution of species of the Australian-Pacific element.

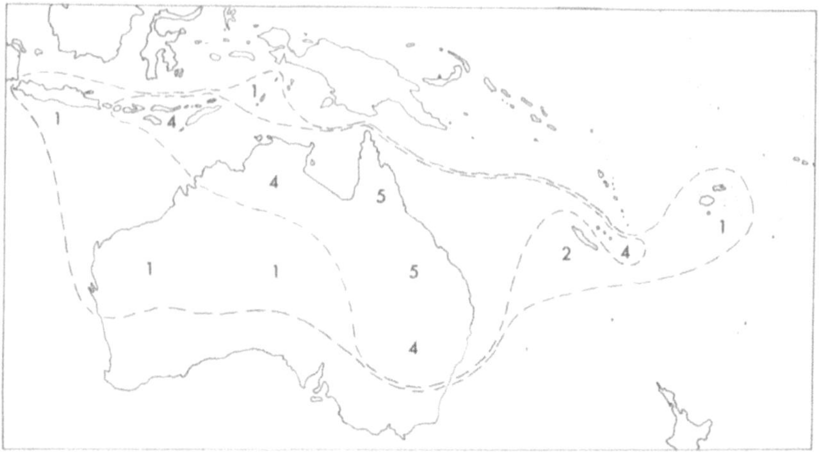

*Fig. 34.* The distribution of species of single-link cluster C.

Figure 34. They are found in N. Australia but are absent from New Guinea. Four are found in the Lesser Sunda Is. and the Loyalties, two in New Caledonia and one in Fiji. The first three are rare in the south-west Pacific and apparently restricted to the Loyalties in the New Caledonian island group. The last two are common in New Caledonia, *S. desita* having an anomalous distribution restricted to the western ultramafic massifs.

The Coral Sea element also appears as a single-link cluster (*Bii* in Figure 29), though this includes a triplet of OTUs (76, 77, 79) that are intermediate between the Coral Sea element and the Melanesian element. At level 0.40, however, the Coral Sea element is manifest as a $B_3$ cluster, though uniting with the Melanesian group of species when $k = 2$. It includes 22 species distributed amongst 15 OTUs (some OTUs represent several species with identical distributions), that is 15–23, 25, 38, 44, 47 and 52; 77 and 79 are also part of the cluster but for reasons discussed below will be regarded as intermediate with the Melanesian element. Associated more or less exclusively with the Coral Sea cluster are *Parallelia propyrrha* (24), *Prasinocyma albicosta* (30) and *Pataeta carbo* (40) which form a pair cluster, and *Parallelia hicanora* (43).

The summed distributions of the 22 species are mapped in Figure 35. The element is strongly represented in north and east Australia, the species all being recorded in southern Queensland and New Caledonia. Exchange between these two areas, probably primarily from the former to the latter, would no doubt have been facilitated by the exposure of the Chesterfield Reefs and other intervening areas during the Pleistocene glaciations. The species concerned occur in all three ecological associations in New Caledonia. For example, *Aeolochroma hypochromaria* and *Egone bipunctalis* are ultramafic species, *Urolitha bipunctifera* and *Comibaena mariae* are forest species and *Eilema nitens* and *Lysimelia litoralis* are west coast species.

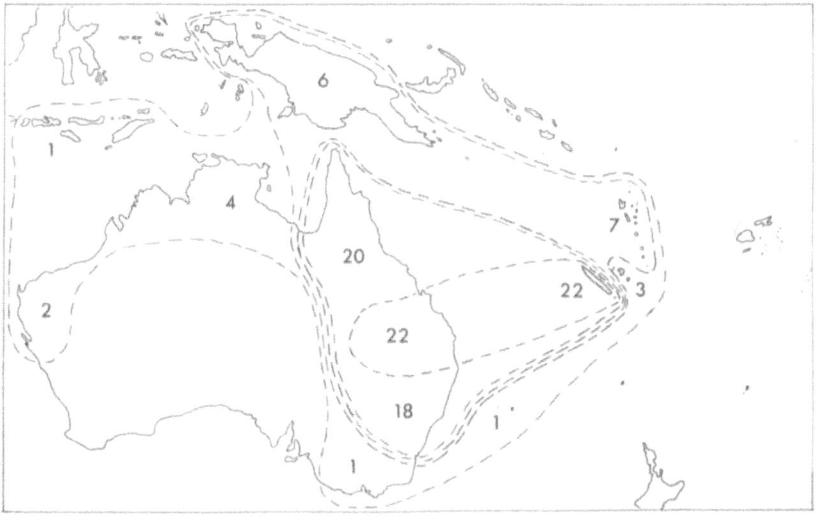


*Fig. 35.* The distribution of species of the Coral Sea element.

The element extends in moderate strength to the New Hebrides but none of the species occurs in Fiji except the associated species *P. carbo, P. hicanora* and possibly *P. propyrrha. P. hicanora* appeared in numbers during a very dry spell in Fiji and has not been recorded since (ROBINSON, 1975); it has not been recorded from New Caledonia.

The Coral Sea element is united with a more loosely defined Melanesian element or group of elements at level 0.40 when $k = 2$. This association occurs primarily through a group of 6 OTUs that form a $B_2$ cluster themselves at level 0.38, sharing OTU 44 with the Coral Sea element. The species involved in these OTUs are *Parallelia redunca* (76), *Parallelia solomonensis* (77), *Phyllodes imperialis* (79), *Comostola leucomerata* (85), *Cleora illustraria* (91) and, in the overlap with the Coral Sea cluster, *Crioa acronyctoides* (44). In general these species extend further south in Australia than the more characteristically Melanesian species but, unlike the Coral Sea species, are represented in the Bismarck and Solomon Is. *C. leucomerata* and *C. illustraria* have not been recorded from the New Hebrides and are both represented by distinct subspecies on New Caledonia, that of the former closest to the Australian race. Thus they probably reached New Caledonia from Australia. The other species are perhaps more likely to have arrived through extension eastwards in Melanesia.

Associated exclusively with these species at level 0.40 in the analysis is *Scopula oppilata* (50), recorded from the Loyalty Is. but not from New Caledonia or the New Hebrides.

The Melanesian element is only partially evident in the single-link analysis. Cluster *Aiii* persists as a $B_2$ cluster of 8 OTUs at level 0.38 which divides into two $B_3$ clusters of 4 OTUs, (60, 65, 88, 98) and (70, 83, 84, 90). The species concerned are firstly *Euploea treitschkei, Luthrodes cleotas, Cleora lacteata* and *Utetheisa salomonis* and secondly *Nigramma polionota, Mniothripa bradleyi, Harita nodyna* and *Nadagara irretracta.* The distributions of all eight are summed in Figure 36. The element extends from New Guinea eastwards as far as Samoa but is poorly represented in Australia. The first subgroup includes species excluded from Fiji, though *C. lactata* extends to Samoa (possibly from the New Hebrides via the chain of islands including Rotuma that was exposed during the glaciations (ROBINSON, 1975)), and the second contains species that are all absent from the Loyalty Is., two of which (*Mniothripa, Nadagara*) are also absent from New Caledonia. The second subelement has most links with the group of OTUs intermediate with the Coral Sea element discussed earlier.

The other subelement of the Melanesian element is not so clearly defined. In the single-link analysis it appears as two separate small clusters, *Ai* and *Aiv*, but at level 0.38 it includes two overlapping $B_2$ clusters from which two OTUs (67, 88) of *Aiv* are excluded, that is (59, 61, 64, (66) 62, 63, 73, 74, 80). At level 0.40 these all are united with the Coral Sea and Melanesian cluster when $k = 2$ but are not segregated when $k = 3$ except for the group of four OTUs. The summed distributions of all the species in these OTUs are shown in Figure 37. The species are *Euploea nemertes, Cyrestis telamon, Danis schaeffera,* then *Borbo impar* in the overlap, *Hypolimnas alimena, Eurema candida, Anua hituensis, Chalciope alcyona* and *Thyas miniacea.*

These nine species are much more widespread in the Papuan Region than those of the other Melanesian subelement, being well represented in the Moluccas and extending further west into the Oriental Region and into northern Australia. Only three of them have been recorded from the New Hebrides and Fiji but all occur in New Caledonia, though the occurrence of *E. candida* and *E. nemertes* needs confirmation (HOLLOWAY & PETERS, 1976). The majority are butterflies, their distributions possibly more thoroughly recorded. Taxa of the other subelement may prove to occur in the Moluccas

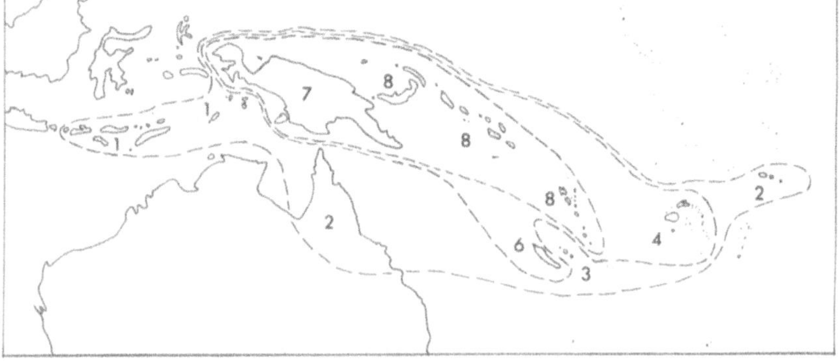


*Fig. 36.* The distribution of species of the eastern Melanesian element.

*Fig. 37.* The distribution of species of the more widespread Melanesian element.

which would appear to be undercollected when one examines the representation of the widespread element there relative to other areas (Figure 31). The distinction of the two subelements might then become blurred. The Bismarck group may also be undercollected, and the paired OTUs 67 and 68 (*Euchromia creusa* and *Platysenta praesecta*) of cluster *Aiv* mentioned earlier may owe their exclusion at higher values of *k* to their absence from the Bismarcks in the record.

It may therefore be justified to consider the Melanesian element as a whole, including the two subelements described and OTUs 67 and 68. Furthermore, in a broader sense, it might be taken to include at least some of the species in the group intermediate with the Coral Sea element and also several OTUs associated primarily with it such as *Lophoptera abbreviata* (41), *Agathia asterias* (46), *Anigraea ochrobasis* (69) and *Blenina lichenopa* (72). *L. abbreviata* only extends as far east as the New Hebrides.

This overall Melanesian element coincides broadly with the Papuan Region as defined by HOLLOWAY & JARDINE (1968). It resembles the Melanesian element derived for a larger sample of butterfly species by HOLLOWAY (1973b).

Also associated with the Melanesian element (via OTU 80) is the triplet in the single-link analysis consisting of *Parallelia vitiensis* (78), *Chrysodeixis illuminata* (81) and *Thalassodes pilaria* (86). These species have a distribution extending from the N. Moluccas to Samoa, a wide Melanesian distribution, and for the purposes of subsequent discussion will be included with the Melanesian element. They form a chain of links, connecting via *Macroglossum hirundo* (58) and *Mocis trifasciata*, with very widespread Melanesian and Pacific distributions, to *Maliattha ritsemae* (152), part of the widespread element proper.

The remaining two elements recognised are centred on the south-west Pacific, the eastern part of Melanesia. In the single-link analysis they form a single cluster *Aii* with their 'nuclear' OTUs partially separated as subclusters, *Aiia* (Fijian) and *Aiib* (New Caledonian). At level 0.40 they form larger $B_2$ clusters with a shared OTU, 112 (Figure 30). The New Caledonian or East Melanesian element shares OTU 98 with the Melanesian element, a link that brings the two elements together in the single-link analysis; for higher values of *k* the Melanesian element associates more strongly with the Coral Sea element as described earlier.

*Utetheisa salomonis* (98) extends from the Bismarck Is. to the New Hebrides, Loyalty Is. and New Caledonia. Together with *U. clareae* from Fiji and an undescribed species from Niue it forms a complex of species derived from *U. pulchelloides* (ROBINSON, 1971a) that possibly radiated originally from a centre in Melanesia, probably New Guinea.

The distributions of the components of the two $B_2$ clusters are mapped summed in Figures 38 and 39. The elements may be taken to indicate general trends of speciation in the south-west Pacific in that they centre on New Caledonia and Fiji respectively, suggesting that these have been the major centres of speciation and radiation amongst the island groups concerned.

Some of the central OTUs of the New Caledonian cluster, however, contain species found in both the centres. *Hypolimnas octocula* and *Euploea boisduvalii* (92) extend from the Solomons to Fiji and New Caledonia, *Badamia atrox* (95) is recorded from New Caledonia to Fiji with an outlying subspecies in the Marquesas (a single female only so its status requires further investigation) and *Scardamia eucampta* (112) is distributed *as atrox*, though absent from the Marquesas. Similarly *Jamides carissima* (114) is shared between New Caledonia and Fiji, occurring also on the Loyalty Is., though this species may prove to be merely a race of the widespread *bochus*. Endemics and species

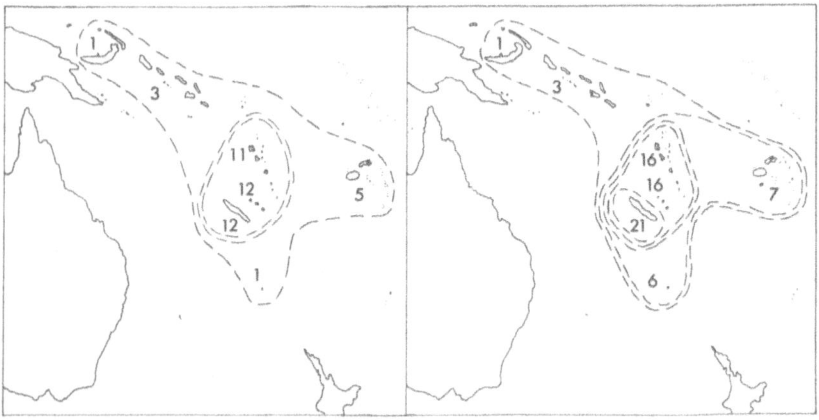


*Fig. 38.* The distribution of species of the New Caledonian element as strictly (left) and loosely (right) defined in the text.

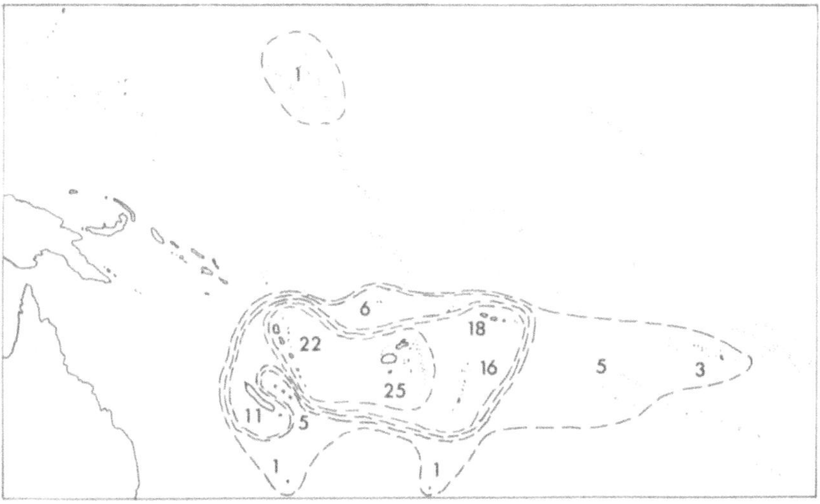


*Fig. 39.* The distribution of species of the Fijian element.

recorded from only two areas were excluded from the analysis as they yield only a limited range of coefficient values and are best discussed independently. A few more species found in both centres fall within the Fijian element in the classification and will be discussed later.

The other OTUs included in the New Caledonian $B_2$ cluster are 102 (*Ericeia hirsutitarsus*) and 113 which contains seven species found in New Caledonia, the Loyalties and the New Hebrides. Associated with 102 are two OTUs, 108 and 116, which contain species extending to Norfolk I. Both the New Hebrides (ROBINSON, 1976) and Norfolk I. (HOLLOWAY, 1977) have faunas derived very recently from other areas, so it may be justified as a general rule to treat the species mentioned in this paragraph as 'extended New Caledonian endemics', precinctive species whose original divergence from ancestral stock occurred in isolation in New Caledonia and whose current wider range is a result of recent dispersal. These species, together with *Asota alienata* shared between New Caledonia and the New Hebrides, could be included in a more loosely defined New Caledonian element as mapped on the right hand side of Figure 38. New Caledonian endemics and species shared only with the Loyalty Is. are not included but historically, in terms of their original site of isolation and divergence, they could also be included with this element.

The final element, the south-west Pacific or Fijian, also contains a few species common to Fiji and New Caledonia apart from *Scardamia eucampta* which occurs in the overlap with the New Caledonian cluster. The OTUs concerned occur mainly in the zone of interaction between the two clusters, several linking with OTU 112. These OTUs are 93, 94, 96, 99, 100, 105, 110, 115, 117. A further species that should possibly be mentioned in this context is *Nola transversata* (97) which occurs on Norfolk I. and may have been overlooked in the survey of New Caledonia (nolids were not sampled quantitatively).

Several of these species common to New Caledonia and Fiji are members of extensive Indo-Australian species complexes where the member species are usually contiguous but allopatric in their geographical distributions. Examples are the *Appias albina* subspecies group, *Euploea lewinii* and *boisduvalii*, *Parallelia prisca*, *Speiredonia simplex* (recorded only from the Loyalties in the New Caledonian island group), and possibly also *Jamides carissima*, *Maceda savura* and *Anisodes samoana*. *Badamia atrox* may have resulted from a previous invasion of the south-west Pacific, possibly from Australia, of the widespread and mobile *B. exclamationis*. This may parallel the more ill-defined situation in the races or forms of *Euploea tulliolus* and *Hypolimnas bolina* described by HOLLOWAY & PETERS (1976) where Australian stock may have reinvaded the south-west Pacific after an earlier invasion had already undergone divergence there. All these species are widespread in the south-west Pacific, the majority being recorded from Samoa and Tonga and some from further east. *Nola samoana* is of interest in that it extends from New Caledonia to Samoa via Rotuma, excluding Fiji where the closely related *fijiensis* flies. These species probably represent an earlier invasion by *Nola insularum* stock. The Rotuma distribution pattern is discussed further on p. 214.

The remaining five shared species are taxonomically more isolated. *Chloroclystis scintillata* and *Poecilasthena leucydra* show no evident subspeciation but the latter has a sibling species in New Caledonia. *Rhesala irregularis* and *Gymnoscelis sara* have distinct subspecies in New Caledonia and Fiji. As with *Maceda savura*, it is the Fijian subspecies of *R. irregularis* that flies in the New Hebrides; material of *G. sara* from the New Hebrides is too limited to enable its subspecific status to be determined. *Hypolimnas octocula* is closest to New Guinea species. It occurs as several subspecies from the Solomons to Fiji and New Caledonia and there are similar, but imperfectly known taxa flying in the Mariana and Palau archipelagoes.

The summed distributions of species in the Fijian element are mapped in Figure 39. The central OTUs of the cluster form a $B_3$ cluster at level 0.40, being 94, 103, 107, 118, 119. These include species that are generally distributed through the New Hebrides, Fiji, Samoa and Tonga, an association of island groups that is also apparent within the overall element in Figure 39. A similar association has already been noted for the floras of these areas in Chapter II.

Other OTUs associated with the cluster are 71 (*Penicillaria dinawa*), 111 (*Gymnoscelis concinna*), and 117 (*Poecilasthena leucydra*) which has already been discussed. With the exception of the first which occurs in New Guinea and may therefore be extinct or unrecorded in intermediate localities, these, together with species shared between Fiji and one other of Samoa, Tonga and the New Hebrides not included in the analysis, could be added to the $B_2$ cluster Fijian element. This broader Fijian element will be used as a basis for further discussion. The species shared between pairs of island groups are:

*Nacaduba dyopa, Hasora chromus, Stictoptera vitiensis, Garella rotundipennis, Mocis vitiensis, Chloroclystis linda, Sauris elaica* and *Heteropan dolens* (Fiji-New Hebrides); *Oriens augustula, Phlegetonia barbara, Gyrtona hopkinsi,* and *Anomis vulpina* (Fiji-Samoa).

Nine species were included in the analysis that did not associate with any cluster at the 0.40 level. *Chloroclystis pyrsodonta* occurs in N. Queensland, New Caledonia and Fiji and associates primarily with the Coral Sea element, as does *Idaea scintillans* (S. Queensland, New Caledonia, New Hebrides). *Grammodes oculicola* has been mentioned already in association with the Australian-Pacific element. *Catoria camelaria* is probably closest to the Coral Sea element. *Agrotis ipsilon aneituma* is closest to widespread Melanesian and Pacific species such as *Mocis trifasciata. Serrodes mediopallens* is closest to the more widespread Melanesian subelement but has not been recorded from the Bismarcks and Solomons; *Clepsimelia phryganeoides* is similar. *Aedia sericea* ranges from New Guinea to the Austral Is. and is perhaps best regarded as an extended central and eastern Melanesian species. *Cephonodes armatus* is best assigned to the south-west Pacific element though it may be conspecific with *picus* and would then be a widespread element species.

The numerical analysis has indicated a dichotomy of speciation in the south-west Pacific based primarily on New Caledonian and Fijian centres. In Table 8 the percentage endemism in the four major south-west Pacific island groups is given, together with percentages of species from the seven broad categories of geographical distribution pattern. Fiji and New Caledonia (including the Loyalties in the table) have been well collected, Samoa less so. The

*Table 8.* The numbers of Lepidoptera species recorded from the archipelagoes of the south-west Pacific and the percentages of various geographical elements in each. Each geographical element is treated in its broadest sense as described in the text and species unassigned in the numerical analysis have been assigned subjectively to the most likely element. The last line indicates the approximate percentage of species that could be attributed to early episodes of colonisation in that they are taxonomically isolated or belong to groups of species that have radiated within the island group concerned.

| | *New Caledonia & Loyalty Is.* | *New Hebrides* | *Fiji* | *Samoa* | *Tonga* |
|---|---|---|---|---|---|
| Macroheterocera | 380 | 304 | 353 | 157 | 91 |
| Rhopalocera | 64 | 60 | 42 | 26 | 20 |
| Total species | 444 | 364 | 395 | 183 | 111 |
| Endemics | 31 | 29 | 46 | 27 | 1 |
| Widespread | 37 | 45 | 36 | 53 | 74 |
| Wide Australian | 3 | 1 | 1 | – | 1 |
| Australian-Pacific | 3 | 2 | 3 | 3 | 7 |
| Coral Sea | 9 | 3 | – | – | – |
| Melanesian | 8 | 10 | 3 | 4 | 2 |
| New Caledonian | 6 | 4 | 2 | – | – |
| Fijian | 3 | 7 | 9 | 13 | 15 |
| Old endemics and endemic radiations | 12 | 1 | 15 | 5 | – |

UNDP/FAO survey has recently added to our knowledge of the Tongan fauna.

The New Hebrides collections made by G. S. ROBINSON in 1971 have only been subjected to a preliminary analysis (ROBINSON, 1976) and the figures presented, based on ROBINSON's calculations, may be subject to change. For example, the percentage endemism is liable to be reduced when the fauna of the Solomon Is. is better known.

Tonga has only one endemic species, *Gymnoscelis erymna*, the rest being species recorded from Fiji (ROBINSON, 1975). The Australian-Pacific element is proportionately better represented than in Fiji. The more southerly position of the group brings it more within the range of influence of the westerly dispersal of Australian species discussed for Norfolk I. by HOLLOWAY (1977). The Tongan fauna is obviously a very recent one, three-quarters of the species being in the widespread element.

The proportions of the different elements in the New Hebrides, Fiji and Samoa are moderately similar. However, the New Hebrides has a higher proportion of Coral Sea and New Caledonian species, Fiji has a greater proportion of endemics and Samoa has few New Caledonian element species and a high proportion of Fijian ones. These differences are what would be expected from present geographical disposition of the areas, though the New Hebrides might have been expected to have a much stronger relationship with New Caledonia.

Proportions in New Caledonia are very different. The widespread element representation is similar but there is a much greater representation of Australian elements, particularly the Coral Sea element, only weakly represented elsewhere. In the previous chapter it was shown that these Australian elements were most strongly represented in the ultramafic and west coast associations, whereas the other elements are more characteristic of, or restricted to, the forest and secondary association.

Even species of the widespread element show a degree of dichotomy in their Australasian distribution patterns as mentioned earlier. New Caledonia has a number of species, such as *Blenina donans* and *Ophisma gravata*, that are not represented elsewhere in the Pacific, show a Coral Sea element type of distribution in Australasia and have obviously spread to New Caledonia from Australia. Conversely the other island groups have widespread species that are not recorded from New Caledonia, such as *Sasunaga oenistis*, *Parallelia arctotaenia*, *Ercheia kebea*, *Chrysodeixis acuta*, *Maceda mansueta*, four *Nanaguna* species, *Leucania yu*, *Eublemma crassiuscula*, *Felinia precedens*, *Hypena iconicalis*, *Anisodes obliviaria* and *Symmacra solidaria*. The majority of these are frequent or common in these areas and thus may be genuinely absent from New Caledonia.

The dichotomy of New Caledonian and Fijian influence in the south-west Pacific has probably only recently been blurred by interaction through the New Hebrides. These islands have only presented a significant land area from the end of the Pliocene and previous to that what land there was would have been closer geographically to Fiji than to New Caledonia. The New Hebrides fauna has a specific endemism only slightly lower proportionately than that of New

Caledonia (Table 8). But, given that it will not be reduced by a large amount when the Solomons fauna is known, it is generally a much younger endemism than that of New Caledonia, most endemics having close relatives in neighbouring land areas. There are no endemic species complexes except possibly for a quartet of *Schrankia* described by HOLLOWAY (1977) and several undescribed *Callopistria* that appear to be related to *dimorpha* in New Caledonia and the widespread *maillardi*. There is only one potentially endemic genus, an undescribed catocaline noctuid. Thus the New Hebrides fauna would appear to be of predominantly recent derivation and the group has probably only been of major significance in the development of the south-west Pacific fauna over the past two or three million years.

The development of the Fijian lepidopterous fauna within the context of the Pacific as a whole has already been discussed by ROBINSON (1975). The situation in New Caledonia can now be related to this general picture.

Excepting Hawaii, there is little endemism or radiation of macrolepidopterous species in Polynesia, the major part of the fauna being made up of the more ubiquitous species of the widespread element. Such endemism as there is occurs primarily in eastern Polynesia, in the Marquesas, the Society Is. and Rapa I. But many of the island groups have not been studied and the picture may change. The endemism is mainly concentrated in the Hypeninae, the *Chloroclystis*/*Gymnoscelis* complex of the Larentiinae, and in the Ennominae, primarily *Cleora*. These endemics are probably all of Indo-Malayan origin, apparently mainly recent isolates from more widespread species. An exception to this is the noctuid on Rapa I. attributed to the polytypic but otherwise rather restricted Indo-Australian genus *Euplexia* (CLARKE, 1971). No *Euplexia* have been recorded from the south-west Pacific.

The Hypeninae have *Hydrillodes* and *Hypena* species extending to eastern Polynesia, often endemics or of restricted distribution. The endemic genus *Fautaua* occurs in the Society Is. with two species. There is endemism at the generic level in Fiji and Samoa, with marked radiation in three apparently related genera in the former. Further work is needed but these Pacific endemic genera in the Hypeninae do not appear from superficial examination to form a related group.

The ennomine genus *Cleora* has also been extremely successful in the Pacific, often with more than one species endemic to an island group. There are two on Rapa and three in the Marquesas. There is a derived genus in Hawaii and endemic species in New Zealand. *Cleora* is much richer in species in Fiji than in New Caledonia. A detailed study of the Pacific *Cleora* as a whole might prove rewarding from a biogeographical point of view.

The Samoan group has one or two distinctive elements in its moth fauna which may be comparable in antiquity of origin to those in Fiji. There are five endemic genera and several more species that are taxonomically isolated within the genera to which they have been assigned. Otherwise the level of endemism is similar to that in the New Hebrides. Figures for the percentage of species attributable to early colonisation or endemic radiation within the five south-west Pacific island groups are given at the bottom of Table 8. They can only be approximate and refer to species in endemic genera, species in complexes largely within the island group concerned (e.g. *Eoasthena* in Fiji) and species apparently taxonomically isolated within their genera.

The genera endemic to Samoa are all monotypic. There are two Hypeninae (*Machaeropalpus fasciatus* and *Mormecia lachnogyia*), two Catocalinae (*Anomocala hopkinsi* and *Cymodegma buxtoni*) and one genus of the Lithosiinae (*Chrysaeglia samoensis*). The four noctuids, described by TAMS (1935), have not had their affinities investigated thoroughly yet but the genus *Chrysaeglia* bears at least superficial resemblance to a group of Oriental genera that includes *Oeonistis* and *Monosyntaxis*.

The taxonomically isolated species are two *Schrankia* discussed by HOLLOWAY (1977), an undescribed *Callopistria*, two *Asura* species, remarkable outliers if they do indeed belong to that largely Oriental genus, and *Ziridava dysorga*, a larentiine also discussed by HOLLOWAY. A third Samoan *Asura*, *uniformeola*, is recorded from the Solomons and further west but the relationship of the endemic species to it and to other species in the genus has so far been investigated on the basis of external characters only (TAMS, 1935). None of these taxonomically isolated species have parallels in Fiji or New Caledonia.

The explanation of these distinctions in the Samoan fauna from that of Fiji may lie in the suggestion by ROBINSON (1975: 340) that some species extending from Melanesia may have by-passed Fiji via a chain of islands including Cherry, Mitre, Tikopia, Rotuma and Wallis and extending along the northern margin of the Fiji Plateau that would have been much more extensive during periods of low sea level (Figure 6). The age of the Rotuma volcanics is early Pliocene and, if such vulcanism is common to all these islands and seamounts along the shear zone of the Vityaz Trench, then this route could have enabled spread to occur both at the Pliocene inception of the chain when the volcanics were little eroded and during subsequent periods when sea level has been lower than today, particularly during the Pleistocene glaciations. The island chain connects the St. Cruz Is. to Samoa, the former providing an entry point for species from the Solomons and the New Hebrides. The most recent sea level minima may have led to the distributions recorded for *Asura uniformeola* and *Tiracola rufimargo*, both extending through Melanesia to the Solomons with outliers in Samoa, the occurrence of a *Callopistria* in Samoa resembling various New Hebrides species and *dimorpha* in New Caledonia, and the distributions of *Nola samoana*, recorded from New Caledonia, the New Hebrides, Rotuma, Samoa and Tonga, and *Cleora lacteata*, extending from New Guinea to the New Hebrides and New Caledonia with a subspecies in Samoa. None of these species has a counterpart in Fiji apart from the *Nola*. Only the *Nola* occurs on Rotuma so the role of the island chain in the dispersal of these species can only be surmised.

Several genera are notable for having speciated quite considerably in the south-west Pacific. Apart from *Cleora* there are several *Macaduma* species, *Callopistria* and *Thalassodes*. *Tiracola* has an endemic species in New Caledonia and one, possibly two, in the New Hebrides as well as the widespread *plagiata* and the Melanesian *rufimargo*. *Anisodes* is well represented, as is *Sauris*, though, as with *Cleora*, the main centre of species richness is Fiji, where taxonomically isolated species are present and endemic radiation (in *Sauris* at least) may have occurred. New Caledonia has shared in some of this speciation though it supports only one *Sauris* and fewer *Cleora*, *Thalassodes* and *Anisodes* than Fiji. The origins of the respective New Caledonian and Fijian species in these genera also differ to a large extent. *Cleora illustraria*, *lacteata* and *acacicola* have no parallels in Fiji, the first two being Melanesian species and the last possibly derived from Australia. *C. immemorata* and *C. dothionis* may be associated primarily with Fijian species but only *C. injectaria* is hared. There are ten *Cleora* in Fiji (described by ROBINSON (1971b, 1975)) but only six in New Caledonia.

In *Anisodes* there are parallels between the two areas in the *niveopuncta* and *compacta* derivatives and *A. samoana* is shared. The two remaining New Caledonian species, *dalmatensis* and *kohensis*, are probably of Australian origin. Fiji supports five more species, the widespread *obliviaria* and *monetaria* and three endemics, one of which is undescribed.

The marked difference between the Fijian and New Caledonian faunas can be illustrated further by reference to the endemic species and their origins, and to the few instances where there appears to have been parallel development in the two areas.

ROBINSON (1975) stated that the bulk of the Fijian fauna is Papuasian in origin, very few species being derived unequivocally from Australia. In contrast the proportion of New Caledonian endemic species of Australian affinity is almost as great as that of species of Melanesian affinity (Table 6), generally referable to wider Indo-Australian or Indo-Malayan species complexes. The non-endemic Australian element in Fiji consists of species shared with New Caledonia apart from *Parallelia hicanora* and *Scopula sublinearia*. Some of the Fijian Australian-derived endemics also show parallels as discussed for the *Calathusa*, *Poecilasthena* and *Anisodes* species and the *Chlorocystis insigillata* complex later. However, many have no close parallels in New Caledonia. The endemic pair of *Casbia*, probably most closely related to each other, show no close relationships with New Caledonian species. *Austrocarea albipicta* is related to the Australian species *iocephala*, but no counterpart has been recorded from New Caledonia. Nor has one for the high moss forest dominant *Chloroclystis nina*, related to the Australian *C. testulata* which also occurs in New Zealand and, with *A. iocephala*, on Norfolk I., an indication of sufficient mobility to reach New Caledonia. *Polyclysta gonycrota* and *Gathynia yasawa* also have no parallels in New Caledonia.

It is possible that a *C. testulata* derivative may await discovery at high altitudes in New Caledonia, such as on the Panié range, but by and large there is not a moss forest element comparable to that described by ROBINSON for Fiji. Of eight species more or less characteristic of this habitat only one is shared with New Caledonia where it has a somewhat less restricted ecology, namely *Poecilasthena leucydra*. This species and *Sasunaga tomaniiviensis* are not so restricted to high altitudes in Fiji as the others. *Chloroclystis nina* has already been mentioned and the remaining five belong to old endemic or Fijian centred species groups, namely *Lophocoleus mirabilis*, *Chloroclystis rubicunda*, *Sauris victoria*, *Eoasthena quilla* and *Cleora ochricollis*. In New Caledonia only *Chlorocoma octoplagiata*, *Caledasthena montana* and possibly *Adeixis montana* could be considered primarily montane though *Chloroclystis lunifera* is perhaps commoner at high altitudes.

Before discussing endemic radiations within the two areas it is helpful to consider the Fijian endemic taxa that are morphologically rather isolated. None are of Australian affinity, being of Papuan (*Stenopterygia nausoriensis*, *Tibiocillaria magnifica*, *Echanella hirsutipennis*, *Ruttellerona presbytica*, *Xois sesara*) or wider Indo-Malayan affinity (*Chasmina viridis*, *Dyrzela trichoptera*, *Prospalta caerulea*, *Barasa triangularis*, *Parallelia koroensis*, *Diomea fenella*, *Speiredonia strigiformis*, *Anisodes harrietae* and *Anisodes prionodes*), apart from the enigmatic *Chloroclystis mariae* and *Horisme chlorodesma*. The genera *Mesurodes* (Geometrinae), *Levuana* (Zygaenidae) and *Acritocera* (Cossidae) are monotypic endemic genera, the last two in families not recorded from New Caledonia. The majority of these are characteristic of undisturbed rainforest. There are also eight indeterminate species, mainly noctuids, that may also belong to this category.

Several related species pairs may have arisen through endemic radiation, namely *Chloroclystis bosora* and *rubicunda* (the first has spread to the New Hebrides or could even have speciated there), *Idaea bathromyses* and *dicenea*, *Casbia alphitoniae* and *aedoea* and possibly *Poecilasthena leucydra* and *inhaesa* though the former is not endemic and the status of the latter is uncertain, one specimen only being known.

There are several larger groups of interrelated species. The genus *Sauris* contains a group of four, *Dasychira* three, *Beggina* six (with another species in the Solomons), and *Hypenagonia* five (a New Hebrides species may prove to be related). The largest group includes three hypenine genera, all endemic, that may be interrelated. *Lophocoleus* contains six species and *Palaeocoleus* and *Tholocoleus* are monotypic.

The larentiine genus *Eoasthena* has its greatest development in Fiji with five species. Two occur in the New Hebrides, one in both Samoa and New Caledonia. It could be regarded as an endemic Fijian genus that has recently radiated more widely in the south-west Pacific.

The New Hebrides, Fiji and Samoa also differ from New Caledonia in having an abundance of small grey sarrothripine noctuids in genera such as *Apothripa*, *Nanaguna*, *Microthripa*, *Mniothripa* and *Giaura*. None of these are represented in New Caledonia but there are five species of the Australian genus *Calathusa* (compared with two in the New Hebrides and one common to Fiji and Tonga), two *Nycteola* and *Characoma vallata*, a Coral Sea species. Conversely New Caledonia has a number of typically Australian Oenochrominae that do not occur elsewhere in the south-west Pacific.

New Caledonia has about fifty taxa that appear morphologically isolated, somewhat less than Fiji as can be seen from the last line of Table 8. There are twelve endemic genera, namely *Paratisiphone*, *Austroypthima*, *Fabresema*, *Thumathoides*, *Codalithia*, *Colnettia*, *Cofimpacia*, *Pseudocoarica*, *Compsulyx*,

*Caledophia, Caledasthena* and *Eunoumeana. Fabresema* has radiated considerably, having at least 13 species, *Cofimpacia* has three, *Codalithia* two and the rest one each. The three new species attributed to *Adeixis* probably represent a radiation within New Caledonia as may the two *Spilosoma* species (though one extends to, or has a relative in, the New Hebrides) and the three related *Idaea.* There are also a number of other species taxonomically isolated within their genera, such as *Delias ellipsis, Elodina signata, Parallelia lateritica, Chlorocoma octoplagiata, Gonanticlea pulcherrima* and *Symmetroctena variegata,* bringing the total of species arising from putative old colonisations to about fifty.

In very few instances are there parallels either in these old endemics or in more evidently recent ones between Fiji and New Caledonia. Apparently recent parallels are as follows: *Maliattha inconcisoides* and *M. melanesiana* are probably allied to the Melanesian *M. inconcisa; Parallelia cuneilineata* (also in the New Hebrides) and *P. duplicata* are part of the *joviana* complex, possibly most closely allied to the Melanesian *solomonensis* which also extends into the south-west Pacific; *Lacera vinacea* and *L. contrasta,* together with the New Hebridean *L. violacea,* are derivatives of the widespread *L. noctilio* which has now reinvaded the area as far as Tonga; *Plusiodonta theresae* and *P. dimorpha* appear closely allied to each other and to Papuan species, as do *Thalassodes pseudochloropis* and *T. chloropis,* though the latter is found from the New Hebrides to Tonga; *Hydrillodes norfolki* and *H. surata* are allopatric members of the same species complex precinctive to the New Caledonian and Fijian areas respectively; *Anisodes paragloria* and *A. gloria* are derived from the Australian-Pacific species *A. niveopuncta; Anisodes lateritica* and *lautokensis* are related to the New Guinea and Queensland species *A. compacta,* but the Fijian species has extremely divergent male genitalia and thus stands apart from the other two; *Chloroclystis lunifera* and *C. encteta* are allied to the Australian *C. insigillata* (recorded as vagrant to Norfolk I. by HOLLOWAY (1977)), *encteta* extending from the New Hebrides to Samoa and the Cook Is; *Scotocyma legalis, S. miscix* and an undescribed species from Samoa may have had a common history; *Petelia trifascia* and *P. aesyla* are part of the widespread *medardaria* complex which also has a representative in the New Hebrides that is related to more westerly forms; *Papilio ilioneus* (shared with Norfolk I.), *P. schmeltzi* and the Samoan *P. godeffroyi,* together with the distinctively tailness *P. aegeus* (Australia, New Guinea) could be the divergent fragments of an earlier invasion of the area by the widespread *P. fuscus*; a closely related group of subspecies of *Cepora perimale* occurs in the New Caledonian archipelago and Norfolk I, and a race with very reduced markings, *inopinata,* flies in Fiji, both of which, but especially the former, may have been derived from the Australian race rather than the rather constant chain of Melanesian races that extend as far east as the New Hebrides (HOLLOWAY & PETERS, 1976); *Polyura gamma* and *P caphontis* form an eastern Melanesian complex with *P. epigenes* from the Solomons.

These pairs comprise only about 3% of their respective faunas and are almost all derivatives of widespread genera or even widespread species, therefore powers of dispersal are likely to have been good. Species common to Fiji and New Caledonia comprise only about a third of the faunas, these also being predominantly species of the widespread element.

There are only six instances where pairs of species may represent more ancient episodes in the colonisation of the south-west Pacific by Lepidoptera. *Asota caledonica* and *A. woodfordi* are very similar in wing pattern but show marked differences in their male genitalia. The genus *Asota* is centred broadly over the Indo-Australian tropics with the highest number of species in New Guinea. *Chasmina basiflava* and *C. viridis* are very different in external structure, especially the disposition of coremata which are abdominal in *viridis* and at the base of the hindwing in *basiflava*. The latter may be a more ancient isolate within the genus than *viridis*, as may be *Stictoptera tridentifera* as compared with Fijian species of the same section of the genus. The third possible pair consists of *Parallelia lateritica* and *P. koroensis* which have similar wing pattern and colouring but, like the *Asota* species, differ markedly in the structure of their male genitalia. *Poecilasthena dimorpha* and *P. leucydra* may be parallel derivatives of their largely Australian-centred genus though in male secondary sexual characters and in genitalia they are markedly different. *P. leucydra* also flies in New Caledonia and Samoa and there is a second species, *inhaesa*, in Fiji, though the status of the latter is somewhat enigmatic (ROBINSON, 1975). *Eunoumeana fletcheri* is close to the genus *Ruttellerona* of which there is a distinctive species, *presbytica*, in Fiji and another as yet undescribed in the New Hebrides. Again the genitalia show greater differences than the facies. The relationship of *Austroypthima petersi* and *Xois sesara* has been treated in detail by HOLLOWAY (1974b). *Austroypthima* is an aberrant member of the *Ypthima* group of genera but *Xois* includes a second species, *arctous*, found in New Guinea and Australia.

Insufficient material of the *Asota* and *Eunoumeana* exists from New Caledonia to convey much about their ecology but the rest are all members of the ultramafic association. Their Fijian counterparts are associated with rainforest apart from *Xois* which is more widespread. All belong to genera that contain at least some mobile, widespread species or are related to genera that do so.

Thus New Caledonia and Fiji appear overall to have has a very different history of colonisation by Lepidoptera except where more mobile elements were involved. This was indicated by the floristic patterns discussed in Chapter II and correlates with the geological history of the two areas. In the next chapter it will be seen that this difference is general throughout the fauna.

In the discussion of the flora in Chapter II an attempt was made to determine whether there were any indications from the distributions of generic or subgeneric groups of the possible geological heterogeneity of Melanesia, the development, followed by partial fusion, of separate Inner and Outer Melanesian Arcs. The systematics of the moths is not advanced enough to enable a similar detailed analysis though there are indications from the distributions of butterfly genera of a similar heterogeneity in the form of a dichotomy into genera restricted to New Guinea and genera widespread through Melanesia. The precinctive subgroups of the genus *Delias* recognised by TALBOT (1928–1937) segregate readily into one or other category as do the sister groups *Ornithoptera* and *Schoenbergia* of the *Troides* complex discussed by ZEUNER (1943). The heterogeneity was also indicated by the more general analysis of HOLLOWAY & JARDINE (1968).

Two major groupings of the butterflies are restricted to Australasia. The Hypocystini of the Satyridae have their centre of diversity in Australia, with genera in New Zealand, New Caledonia (*Paratisiphone*) and New Guinea. Only one genus occurs in the Outer Arc, the rather aberrant *Argyronympha* in the Solomons. The most recent account of this group was by HOLLOWAY (1974b) but further work is needed to determine whether it is a natural grouping and to discover its affinities outside Australasia. These latter could possibly be with South America, a possible Antarctic connection.

The second grouping is the subfamily Trapezitinae of the Hesperiidae. This may have resulted from an early radiation following invasion by Indo-Malayan elements of the Hesperiinae, possibly of the *Taractrocera* group. It contains 16 genera, 12 restricted to Australia, one shared with New Guinea but centred in Australia, and three restricted to New Guinea (EVANS, 1949; COMMON & WATERHOUSE, 1972). The *Taractrocera* group itself has radiated extensively in Melanesia, containing eight genera centred on New Guinea but with little radiation into Outer Arc island groups (EVANS). The degree of radiation in the Trapezitinae suggests that this subfamily may represent an early Tertiary colonisation of Australasia by Indo-Malayan elements.

The ancestors of the lycaenid tribes centred in Australasia were probably also early Indo-Malayan colonists as they have no relatives in South America (ELIOT, 1973). The thecline tribe Ogyrini is almost confined to Australia, only a few species occurring in New Guinea. The genus *Ogyris* resembles *Delias* in feeding on Loranthaceae primarily but with one species feeding on Santalaceae. In the tribe Candalidini seven genera are restricted to, or strongly centred on, Australia. The eighth, *Holochila*, contains four species groups and is a large genus shared between Australia and New Guinea but not occurring in Outer Arc areas (ELIOT). COMMON & WATERHOUSE referred species of this genus to *Candalides*.

The lycaenid tribe Luciini contains a section almost confined to Australia and another includes several genera restricted to New Guinea or (*Hypochrysops* and *Philiris*) centred in both Australia and New Guinea. *Hypochrysops* has extended weakly into the Outer Arc and there is one Asian species, distributed in Sundaland.

The tribe Zesiini contains a monotypic section restricted to southern India and another of three genera, including the polytypic *Jalmenus*, restricted to Australia. This connection, together with the occurrence of association with ants in many lycaenid groups and with Leguminosae, particularly *Acacia* in Australia, suggests another possible origin for the Australian lycaenid fauna, or parts of it. They could have been derived from Africa, in parallel with the Proteaceae and possibly *Acacia*, early in the Cretaceous before the major break-up of Gondwanaland has proceeded very far and India could still have participated in the exchange. Both the Indian *Zesius* and the Australian *Jalmenus* are attended by ants as larvae (WYNTER-BLYTH, 1957; COMMON & WATERHOUSE). There is only a scanty fossil record for the butterflies but they probably had evolved by the middle of the Cretaceous and possibly occurred as far back as the Jurassic (ELIOT).

None of these Australian lycaenid groups are represented in New Caledonia, possibly because Leguminosae are only poorly represented in the more ancient floristic associations of the ultramafic rocks (JAFFRÉ, 1974). No Trapezitinae have been recorded and only one hypocystine satyrid, *Paratisiphone*, which is largely restricted to the *Baeckea*-sedge maquis of the ultramafics and is probably one of the oldest elements of the New Caledonian macrolepidopterous fauna. Only the hypocystines among these old Australasian groups are represented in New Zealand.

There is one South American connection in the Australian butterfly fauna and that is the monotypic genus *Cressida* (Papilionidae), distributed widely over north and east Australia and therefore primarily a tropical species. It is related to the South American genus *Euryades* (COMMON & WATERHOUSE).

Returning to the Lycaenidae, the rest of the genera largely restricted to Melanesia are all derived more recently from the Oriental Region. The majority have more characteristically Outer Arc distributions, for example *Arhopala* s. s., *Hypochloris* (*Pseudonotis*), *Danis* and *Epimastidia*. The species of these genera show a slight tendency to be concentrated in the northern part of New Guinea and the Vogelkop.

*Danis* and *Epimastidia* are placed in the *Danis* section of the tribe Polyommatini by ELIOT, the third member of which is *Psychonotis*, a genus that includes the rather aberrant New Caledonian species *purpurea* (also present on the Loyalties). *Psychonotis* and *Danis* are separable on genitalic characters but not all the species have been dissected. Most *Psychonotis* species determined so far are in New Guinea, with one in Sulawesi (ELIOT, *pers comm.*). It is possible, therefore, that *Psychonotis* may be an Inner Arc genus. *Psychonotis purpurea* is a rainforest species and not restricted to the ultramafics. HOLLOWAY & PETERS assigned *Danis schaeffera* to *Psychonotis* but the validity of this must be tested by dissection. *Delias ellipsis* may also be more referable to Outer Arc sections of the genus than to those restricted to New Guinea, though its affinities are obscure and an Inner Arc relationship cannot be discounted. Amongst the older endemics (the taxonomically isolated ones) Inner Arc relationships predominate, a phenomenon that could perhaps be predicted from the working hypothesis of the geological history of the area presented in Chapter I. Only in the *Ruttellerona* and *Eunoumeana* complex and perhaps in *Scotocyma* are there potentially Outer Arc relationships but these may be more referable to the more recent speciation events shared by south-west Pacific archipelagoes described on p. 217. Many of the Inner Arc associations involve taxa from northern Australia as well as New Guinea and the New Caledonian member of the association could well have been derived from Australia in some instances. About fifteen examples are known so far and are described below. Some other New Caledonian endemics of uncertain affinity may also prove to be of this category when the faunas of northern Australia and, more particularly, New Guinea are better known. A further six New Caledonian endemics have their closest relatives in New Guinea, namely *Epiplema lateritica*, *Anisozyga caledonica*, *Chloroclystis macroaedeagus*, *Agrius fasciatus*, *Nycteola canoides* and *Veia pectinata*, but these are perhaps not so isolated taxonomically.

*Danaus pumila* is a New Caledonian precinctive species (it occurs in the New Hebrides) whose external features are most similar to the *D. schenkii* group of species in New Guinea. *Elodina signata*, rather distinct taxonomically, belongs to a genus with several species in New Guinea and northern Australia and could have been derived from either source. *Austroypthima petersi* has no close relatives in New Guinea but its putative foodplant has; a relative may wait to be discovered in New Guinea where the bamboo *Nastus* is found. The closest relative of *Miresa obscurimargo* would appear to be a New Guinea species. The endemic *Caledophia* is part of an Australasian complex that includes *Callipotnia* (two New Guinea species), *Physetostege* (New Guinea, Queensland) and *Encryphia* (one species, Queensland and New Caledonia). All are probably derived from the African and Oriental genus *Conolophia*; *Caledophia* is perhaps closest to *Encryphia*. *Gonanticlea pulcherrima* is isolated within the genus as is the Melanesian *sublustris*, possibly its closest relative. The latter may be an expanded New Guinea species as it shows little variation throughout its range, though obviously this can only be included tentatively as an Inner Arc association. The genus *Papuanticlea* has three species in New Guinea, one in Queensland and one in New Caledonia. The New Caledonian species is closest to the Australian one. *Nadagara serpentina* is very isolated within the genus and can only be related tentatively to New Guinea and Queensland species, similarly isolated taxonomically. *Ctimene nocturnignis* is also closest to New Guinea species. *Polyacme dentata* has its nearest relatives in New Guinea (two species). *Compsulyx cochereaui* is an early offshoot from two Oriental genera but its closest geographical relatives in either of these are a number of possibly interrelated species of *Oxyambulyx* in New Guinea. *Cerura amoa* is also closest to New Guinea species but not perhaps as taxonomically isolated as some of the others discussed here. *Molvena caledonica* is also quite close to an Australian species but the genus itself is distributed in New Guinea, Queensland and New Caledonia, with distinct species in each area. *Leucogonia amarginata* is congeneric with two New Guinea species only but the monotypic Queensland genus *Metaxanthiella* is also closely allied and there is a case for regarding all four species as congeneric. *Ophyx maculosus* is closest to a Queensland species but related taxa occur in New Guinea.

There is one connection with New Guinea at the specific level that excludes the rest of Melanesia and that is *Elphos subrubida*.

These Inner Arc associations in New Caledonia are almost entirely of ultimate Indo-Malayan affinity and include the majority of the oldest elements in the New Caledonian macrolepidoptera. A large number are members of the ultramafic association, such as the species of *Austroypthima*, *Caledophia*,

*Gonanticlea*, *Nadagara* and *Miresa*, and the rest are restricted to rainforest, some of the latter even favouring rainforest on ultramafic rock such as *Compsulyx*.

The restriction to ultramafic rock suggests that many of the species concerned colonised New Caledonia in the middle of the Tertiary when ultramafic habitats may have predominated, perhaps to the virtual exclusion of all others. This would also have been a time when Inner Arc New Guinea was well separated from the Outer Arc and rainforest was still extensive in northern Australia. Under such conditions the development of taxonomic groupings associated with the three areas would be expected.

Unlike some other faunal groups Australian elements of apparent long standing in New Caledonia are few. *Paratisiphone lyrnessa* has already been mentioned. The related group of three *Adeixis* species is another instance though this is likely to have been a result of a colonisation after the opening of the Tasman Sea, as the putative ancestral stock would have also been ancestral to *Adeixis inostentata*, a species of the widespread Australian element. The final definite old Australian connection is *Chlorocoma octoplagiata*, a species that appears to be associated with high altitude *Podocarpus* forests. The genus is characteristic of temperate regions of Australia.

There are no old connections with New Zealand, shared species belonging to either the widespread element or the widespread Australian element. The New Zealand macroheterocera fauna includes numerous instances of genera that have radiated *in situ* or polytypic endemic sections of more widespread genera. Examples of the former are *Pseudocoremia* (Ennominae) and a few genera of arboreal feeding Hadeninae such as *Graphania*, and the latter are exemplified by *Pasiphila* and *Helastia* species, these taxa being perhaps most informatively placed as subgenera or sections of the more widespread *Chloroclystis* and *Xanthorhoe* (*sodaliata* is a *Helastia*).

The taxonomy of the New Zealand fauna is still being worked out and it has not been related fully to the faunas of other areas. Unlike New Caledonia there are a number of apparently austral, Antarctic or even Gondwanan groups. Literature consulted included the works of HUDSON (1928, 1939) and DUGDALE (1961, 1964, 1971a, 1975).

The ennomine genus *Declana* has eleven species which include amongst their larval foodplants some of the oldest elements of the New Zealand flora such as *Myrtus*, *Laurelia*, the araliad *Pseudopanax* and *Nothofagus*. The genus is probably of S. American affinity, as is the larentiine *Tatosoma*, which also has eleven species, some of which feed on *Podocarpus*, *Nothofagus*, *Weinmannia* and *Coriaria*. *Gargaphia*, *Sestra* and *Azelina* are interrelated endemic genera, the larvae feeding on ferns. They also may be related to South American taxa. The genus *Xyridacme*, also endemic, may be related to the South American genus *Phellinodes* and includes *Hebe*, *Pittosporum* and the araliad *Nothopanax* amongst its larval hosts.

There is an interrelated group of six species of the Australian genus *Dichromodes* (Oenochrominae), the larvae probably lichen feeders. The endemic oenochromine genera *Theoxena* (one species) and *Samana* (two species, the larvae feeding on the myrtaceous genus *Leptospermum*) resemble slightly the New Caledonian *Adeixis*, particularly as regards the central black streak to the forewing, but the male genitalia are somewhat different. There are three species of *Poecilasthena*, two feeding on *Leptospermum*, but none are closely related to New Caledonian species. The genus *Microdes* (Larentiinae) is also shared with Australia.

Though the Indo-Malayan plant genera invading New Zealand in the early and middle Tertiary often exhibit significant radiation or even generic subcentres in New Caledonia today, the moths do not show a comparable pattern. The *Helastia* and *Pasiphila* groups in New Zealand, probably derived from north temperate and tropical montane faunas (HOLLOWAY, 1977), do not have counterparts in New Caledonia. Nor are New Zealand genera such as *Austrocidaria* and *Pseudocoremia* represented, though they include elements mobile enough to have colonised Norfolk I.

Apart from the three Australian connections mentioned, New Caledonia has not participated in any old austral radiation of macrolepidoptera. Families identified as archaic or largely restricted to Australia by COMMON (1970), such as the Hepialidae, Cossidae, Castniidae, Anthelidae, Cyclotomidae and Carthaeidae, are mostly absent, the Hepialidae being represented by one species of probably recent derivation from Australia. The Thaumetopoeidae contain large blocks of genera endemic to Australia but the single New Caledonian species again appears to be of recent derivation. The Oenochrominae also have large blocks of endemic genera but only the *Adeixis* group in New Caledonia has any semblance of antiquity. The Larentiinae have radiated significantly in Australia, though the systematics of the species involved has not been worked out to any great extent; again the New Caledonian relatives of these species are of recent derivation.

The oldest elements in the New Caledonian fauna are mainly taxonomically isolated species of Indo-Malayan derivation (probably Miocene). There is a much stronger Australian influence in the fauna than elsewhere in the Pacific but this appears to be relatively more recent and may have arisen mostly when exchange between the two areas was facilitated by exposure of the Chesterfield Reefs during the Pleistocene.

There are perhaps several explanations for the lack of old austral macrolepidoptera, contrasting with the situation in the flora and several other phytophagous groups. Firstly the major radiation of these groups may have occurred subsequent to the opening of the Tasman Sea. This can largely be discounted, as it would apply also to New Zealand where there are several genera that could be termed ancient and austral. Also some of the Australian archaic groups have affinities with southern African taxa, such as the Thaumetopoeidae generally which are also represented strongly in Madagascar and have a few genera in India and the Mediterranean Region. The Australian group is homogeneous and has morphological features intermediate between the African genera and the Mediterranean *Thaumetopoea* (KIRIAKOFF, 1956, 1970). N. MCFARLAND (*in litt.*) has reared the larvae of Australian *Teara* from *Acacia*, *Cassia* and *Eucalyptus*.

The second possible explanation is that any austral groups present in New Caledonia in the early Tertiary would have been ousted subsequently by competition from Indo-Malayan invaders. But the Lepidoptera are generally a rather specifically phytophagous group and austral Lepidoptera could well have been conserved with the austral plant taxa, particularly given the effect of the ultramafics in this conservation, largely isolating the austral plant taxa from competition. Also there are a number of non-phytophagous insect groups such as the Trichoptera (not specifically or restrictedly phytophagous anyway) where austral elements coexist with more modern ones.

This leads on to a third possible explanation. The predominance of the ultramafic habitats in the middle Tertiary may have had a selective effect on austral elements present from before the time of emplacement. Some adapted, some became restricted or died out. In the early stages too there may have been fewer deposits of eroded ultramafic sediments where the desaturation permitted less adapted forms to thrive. The austral Lepidoptera may have been feeding on plants that were less versatile biochemically and thus were less able to respond to the demands of adaptation to the ultramafics and more prone to extinction or, through restriction to sedimentary rocks, more liable to competition from later invaders. The biochemically more versatile plants would be those most likely to have evolved chemical defences against defoliation by insect larvae prior to the emplacement of the ultramafics and later, by toleration or even accumulation of toxic minerals in their sap during adaptation to the ultramafics, would have become even more unpalatable to such insects. Today the plants of the most extreme soil types, floristically diverse, do not support a correspondingly diverse Lepidoptera fauna.

The microlepidoptera, as far as they have been worked, and other families such as the Pyralidae and Hepialidae generally resemble the macrolepidoptera in their New Caledonian biogeography. The more qualitative collection of Pyralidae and microlepidoptera made during the 1971 survey has only just begun to be studied; only a little of the material has been included in recent generic revisions (WHALLEY, 1973, 1976; GASKIN, 1974a).

The most interesting discovery was that of four species of the primitive micropterigid genus *Sabatinca*, unusual in its family in that the species come to light instead of being exclusively diurnal. The genus today is otherwise known only from New Zealand (18 species) and Australia (six species) and is thus the only old austral element so far known in the New Caledonian fauna. The Australian species differ from the New Zealand ones in characters of wing venation and palps (COMMON, 1970). Investigation of the New Caledonian species has not proceeded far enough to determine their affinities precisely but they show much diversity in wing pattern and are possibly closer to New Zealand taxa. They were all taken in ultramafic localities, three being recorded at Pic du Pin. Though the present day distribution of the genus is Australasian there are several fossil records from the northern hemisphere. A *Sabatinca*-like moth has been recorded from Lebanese amber (−130 my approximately) and the genus has been recognised in the Baltic amber (Eocene, Oligocene) and the Burmese amber (Late Miocene, about −7 my) (WHALLEY, 1977).

Only one species of the Hepialidae has been recorded from New Caledonia, *Aenetus* (*Oenetus*) *cohici*, described by VIETTE (1961) from two females taken on Mt. Koghis. VIETTE related this species to *marginatus* from the Trobriand Is., off south-east New Guinea but if a male in the Bishop Museum collection, also from Mt. Koghis, is that of *cohici* then the relationship is closest to the Queensland species *mirabilis*, both species having males with pale green forewings and a diagnostic pair of coremata, one large one each side of the base of the abdomen. This species-to-species association suggests that the New Caledonian species is probably of recent derivation and not an old austral element.

DUMBLETON (1966) made a detailed study of the New Zealand Hepialidae which fall into two subfamilies. One is the austral Oxycaninae with one genus in South America, four in New Zealand and several in Australia. The other is the Hepialinae which includes *Aenetus*, with one species in northern New Zealand, several in tropical Australia and New Guinea as well as New Caledonia, *Aoraia*, a New Zealand genus of Australian affinity with four species, six Australian genera and two restricted to the south-west Pacific. These last are *Phassodes* (Fiji, Samoa) and *Toenga* (Rarotonga in the Cook Is.), both monotypic. This is an intriguing instance where Fijian area representatives of a taxonomic group appear to be more ancient than those in New Caledonia.

A cursory examination suggests that the Pyralidae are generally of a broad Indo-Australian tropical character with no obvious endemic radiation. However, a characteristic species of the ultramafic maquis is a large crambine with grey wings, the forewing bisected longitudinally by a broad white streak. It has been determined by Mr. M. SHAFFER as a new species of the otherwise Australian genus *Corynophora* (two species in eastern Australia, one in the west). The New Caledonian species is morphologically distinctive and is probably one of the older members of the fauna.

GASKIN (1974b) described a group of less specialised and related species of the crambine genus *Pareromene* found in New Guinea, Rennell I., New Caledonia and New Zealand (only one species in New Caledonia). This Inner Melanesian Arc distribution he dated tentatively as arising in the Upper Miocene. The large number of New Zealand species in this genus are generally related to less specialised Papuan Region forms. Here there may be parallels with the *Chloroclystis-Pasiphila* and *Xanthorhoe-Helastia* type of distribution, but where New Caledonia is included, albeit weakly.

A preliminary examination of some of the Tortricinae has been made by Mr. J. S. DUGDALE who found no strong relationships with any other area but suggested there was a general relationship in the fauna with New Guinea and Queensland but not New Zealand (DUGDALE, *in litt.*).

There is one group of species which appears to have radiated within New Caledonia and spread to other areas of the Pacific. The species are assigned at present to the closely related tortricine genera *Xenothictis* and *Xeneda* and are being studied by Mr. K. TUCK (BMNH) who has suggested they might be congeneric. The two genera together contain at least six species in New

Caledonia, two in the Loyalties and one in each of the New Hebrides and Fiji.

A preliminary sorting of the other microlepidoptera collected during the survey has revealed several general features of interest (Dr. G. S. ROBINSON and Dr. K. SATTLER, *pers. comms.*). As in Australia (COMMON, 1970) there is a high proportion of Xylorictidae and, to a lesser extent, Oecophoridae. The fauna is generally Australian in aspect and endemism appears to be high. Whereas Fiji and the rest of the Pacific have numerous species of Cosmopterygidae in their faunas, New Caledonian representatives of the family are few. Similarly the Immidae have many species of the genus *Imma* in Fiji but only one, shared with Norfolk I. (HOLLOWAY, 1977), occurs in New Caledonia.

On pp. 220–1 it was remarked that New Caledonian endemics of uncertain affinity might prove on further investigation to show Inner Arc relationships. A case in point, *Graphium gelon*, arises from the study published in Japanese (1977) of the subgenus *Graphium* by SAIGUSA, NAKANISHI, SHIMA & YATA. Their diagrams of phylogenetic relationships show *gelon* to be related primarily to *macleayanum* (E. Australia), *weiskei* (New Guinea) and *stresemanni* (Ceram), thence to a more general grouping that is completed by two further related pairs: *empedovana* (Sundaland) and *codrus* (wide Melanesian–Philippines and Sulawesi to Solomons); *sarpedon* (widespread element) and *cloanthus* (China, Himalaya, Sumatra).

# VII. THE ZOOGEOGRAPHY OF OTHER GROUPS.

The knowledge of the New Caledonian fauna differs markedly from group to group and the following review only attempts to deal with those where there has been some attempt at a zoogeographic synthesis. In many cases the synthesis is incomplete because of undercollecting in neighbouring island groups.

### Oligochaeta

In the Oligochaeta New Caledonia supports the genus *Acanthodrilus*, shared only with the Kermadecs. It belongs to the Gondwanan family Acanthodrilinae (KEAST, 1973). Terrestrial oligochaetes were stated by KEAST to have very poor powers of dispersal but the subfamily Acanthodrilinae includes some coastal strand species that can tolerate salt water, and the presence of species on the Kermadecs, isolated, recent volcanic islands, and on several of the recently glaciated Antarctic islands is indicative of long distance dispersals within the group in recent times.

### Mollusca

SOLEM (1958) defined three main zoogeographic elements in the Pacific mollusc fauna: a southern relict element, involving southern Australia (including Tasmania), New Zealand and New Caledonia; a Polynesian element, an endemic development of mainly Oriental derivatives; and a Palaeo-Oriental (= Indo-Malayan) element derived from the Oriental Region from the late Tertiary onwards. These broad categories are applicable to most other taxonomic groups but the proportions of each in each island group vary, a variation that may be related to dispersal powers, and for most other groups the second category is more Melanesian than Polynesian. PEAKE (1969), in supporting these categories for the molluscs, also noted the similarity of molluscan geography to that of the birds and insects.

The Palaeo-Oriental element predominates in the Solomons for the molluscs, the Polynesian in Fiji and the southern relict in New Caledonia. The New Hebrides contain more of a mixture: two Palaeo-Oriental families and two Polynesian families do not reach New Caledonia but occur in the New Hebrides; endemics occur in 24 genera and of these genera eight (including four widespread ones) occur in New Caledonia and fourteen in Fiji.

SOLEM (1961), referring to previous accounts of the New Caledonian fauna, suggested that revision would reduce the total number of species to about 160 or 170. He noted (1959) that most taxonomy of Pacific taxa had been based on shell sculpture and the crucial work of dissecting the living parts was only then beginning. The results of such dissection may indicate that much revision of molluscan biogeography is necessary. The New Caledonian list he presented included 11 freshwater species, 23 land prosobranchs, 20 bulimulids, 49 endemic endodontids, 24 paryphantids (these last three groups totalling 57% of the fauna), 19 miscellaneous endemic pulmonates and 18 introduced species.

In comparison with the New Hebrides the fauna is rather limited and restricted as regards the taxonomic groups represented. Endemism is high. In overall character the fauna is generally closest to that of New Zealand, less so to that of Australia. Endodontids dominate the Australian and New Zealand faunas and are very important in Melanesia and Polynesia. The Paryphantidae are represented in New Caledonia by one taxonomically isolated group, one with affinities with New Zealand and Australia and one, possibly polyphyletic, that has a broad Melanesian distribution from New Guinea eastwards.

The Bulimulidae are represented in New Caledonia by 20 endemic species (though a few are found in the Loyalty Is.) that appear to have resulted from a single radiation. They have all the characteristics as a family of an ancient austral group. They occur in Central and South America with Eocene fossils recorded in Patagonia, there is a possible bulimulid genus, *Prestonella*, in South Africa but this needs further study, and in Australasia there are the *Bothriembryon* and *Placostylus* groups (SOLEM, 1959). SOLEM discussed these last in some detail.

*Bothriembryon* is centred in south-west Australia with one species in Tasmania. The *Placostylus* group occurs in New Zealand, Lord Howe I., New Caledonia, the New Hebrides, the Solomons and Fiji. Most of the species are placed in *Placostylus* but in the New Hebrides occurs the genus *Diplomorpha* which includes *Quiros* as a distinct subgenus. SOLEM indicated that affinities within *Placostylus* are between New Zealand, Lord Howe and the New Hebrides on the one hand and Fiji, the Solomons and, more distantly, New Caledonia on the other. If the New Hebrides and New Caledonia were exchanged in these groupings the pattern would fit the general biogeographic trends in the area; otherwise it is somewhat anomalous. These affinities are based for the most part on shell characters and may have to be revised when the living parts have been studied. Most of the island groups have supported significant endemic radiation of *Placostylus*. The Lord Howe species are at present included in one of the New Zealand subgenera, *Maoristylus* (CLIMO, 1975).

CLIMO stated that land mollusca on the whole are evolutionarily conservative and many modern genera date from the Cretaceous, including most of the New Zealand fauna. The distribution of many New Zealand species today appears to reflect Pliocene geography, suggesting that dispersal in some molluscan taxa at least is very limited. The general resemblance between the New Caledonian and New Zealand faunas as distinct from the Australian may, as with much of the flora, only be a reflection of the response of the organisms to increasing climatic differentiation in Australasia. More detailed taxonomic studies are needed before it can be determined whether there is any more precise association between the two areas.

LADD (1960) suggested from molluscan distribution patterns that the central Pacific archipelagoes are of long standing. Many guyots may have been subaerial in the past, especially during periods of low sea-level. Drilling on several of the deeper guyots has produced rounded basalt pebbles and shallow water fossils dating back to the Cretaceous

### Arachnida

BERLAND (1934) wrote on the geography of Pacific spiders. The Pacific faunas are generally imperfectly known but the New Caledonian fauna is richer than those of Fiji, Samoa and Tonga (which have generally similar faunas). The Loyalties fauna is almost entirely in common with that of New Caledonia, though less rich. Endemism in New Caledonia exceeds 60% and the strongest connections of the fauna are with Australia. BERLAND referred to the Norfolk I. fauna as being predominantly Australian which would indicate that the spiders had generally good powers of dispersal (HOLLOWAY, 1977). They would be passively dispersed in air currents on threads of gossamer and might be wafted readily from Australia after lifting into the westerly upper winds.

The arachnid fauna of New Zealand has strong connections with that of Australia and there are also some probable ancient groups present with wider austral distributions involving southern Africa, Madagascar and South America (FORSTER, 1975). The connections with Australia are primarily amongst the Tasmanian fauna and to a less extent with rainforest taxa up the east coast of the mainland. The Australian sclerophyll forest fauna is generally not present in New Zealand. As with the flora, there are a number of spider and harvestman genera more characteristic of warmer latitudes that are today restricted to the extreme north of New Zealand, and FORSTER particularly mentioned the harvestman genus *Triregia* which has three species restricted to the region north of Auckland and Three Kings I., and is allied to the New Caledonian genera *Triconobunus* and *Diaenobunus*.

### Insecta

The geography of Pacific insects has been discussed generally by GRESSITT (1956, 1958, 1960a, 1961, 1974) and MUNROE (1965), the former mainly from the standpoint of the Coleoptera. Most Pacific faunas are oceanic, progressively more disharmonic towards the east and largely derived from Indo-Malayan and Papuan sources, akin to SOLEM's Palaeo-Oriental element. The subdivision of this element into Indo-Malayan and Papuan elements is important, the latter representing a generally more ancient phase of colonisation from the Oriental Region, as with the flora. The borderline between these two subelements is indefinite, as there has undoubtedly been a more or less continuous and probably increasing west to east expansion and dispersal of both Indo-Malayan and Papuan elements into the Pacific (GREENSLADE, 1969), the newer arrivals tending to displace the older colonists. The turnover rate of this process will depend largely on the dispersal powers of the taxonomic group in question and thus, when there have been subtle changes in the juxtaposition of land areas as suggested for the south-west Pacific in Chapter I, distribution patterns reflecting older events will be more likely to be preserved in the less mobile groups. This continuous displacement of older colonists by newer has been discussed in detail by DIAMOND for Melanesian birds and will be referred to when the New Caledonian avifauna is discussed (p. 249).

Both GRESSITT and MUNROE regarded New Caledonia as an exception to the general Melanesian pattern, having elements of Australian and New Zealand affinities in its insect fauna, varying in proportion from group to group. GRESSITT particularly drew attention to extended Inner Melanesian Arc relationships in the Coleoptera involving links between New Guinea, New Caledonia and New Zealand that excluded Australia and eastern Melanesia.

GRESSITT (1961) described the New Caledonian fauna as disharmonic but rich, with indications of great age. This he contrasted with the New Hebrides fauna which is largely young, derived by dispersal across water gaps. The disharmonic nature of the New Caledonian biota and the variation in proportions of various geographical elements in different taxonomic groups are both indicative of a predominantly oceanic history.

New Guinea, despite its proximity to Australia, has a rich fauna with much endemic radiation in which Oriental elements are predominant. GRESSITT (1958) suggested from biological evidence that there must have been land in the New Guinea area throughout most of the Tertiary, possibly more isolated from Australia and more archipelagic.

The Fijian fauna is richer than that of Samoa and the New Hebrides and as rich as that of New Caledonia in some groups. It is disharmonic compared with that of New Guinea and the Solomons, suggested by GRESSITT (1961) to be the result of isolation and thus colonisation by long distance dispersal, a conclusion supported by ROBINSON (1975).

Examination of individual insect Orders largely confirms this general picture but, as stated above, the emphasis varies from group to group.

## Dermaptera

The New Caledonian Dermaptera have been treated in detail by BRINDLE (1976a), who has also described the New Hebrides fauna (BRINDLE, 1976b). The New Caledonian fauna of 18 species contains 13 endemics, one species with a Melanesian distribution pattern that extends far into the Pacific, and four cosmopolitan ones. The degree of endemism is very much higher than in other south-west Pacific archipelagoes and this is mainly due to the presence of a group of seven interrelated *Brachylabis* species and two large species of *Anisolabis*. Both genera are very widespread and there are as yet no indications of the relationships of these endemics with other faunas. Each genus is represented in the New Hebrides by a single small species only. New Caledonia has two endemic *Nesogaster*, one of possible New Zealand affinity, though the genus is widespread in the Indo-Australian area from Borneo eastwards.

There is a marked contrast with the New Hebrides fauna which is perhaps more characteristic of the other south-west Pacific areas. There are 16 species that fall into the lepidopteran distribution categories roughly as follows. There are five endemics, one Fijian element species, and six Melanesian species of variable range including *Hamaxas nigrorufus*, the species in New Caledonia that extends far into the Pacific, and another species shared only with the Solomons. Four cosmopolitan species complete the list. Apart from the differences in the genera mentioned in the previous paragraph there are representatives of three genera that do not occur in New Caledonia, namely *Auchenomus*, *Sphingolabis* and *Marava*.

## Ephemeroptera

EDMUNDS (1972) reviewed the biogeography of the Ephemeroptera. They have generally poor powers of dispersal. Three subfamilies are restricted to Australia, New Zealand and South America (Ameletopsinae, Coloburiscinae, Oniscigastrinae) and in each case the Australian and South American forms are more closely related to each other than either are to those from New Zealand. In New Caledonia only the family Leptophlebiidae is found, related most immediately to New Zealand members of the family and more generally to a broad 'Gondwanan' grouping within the family, with representatives in Australia and South America, South Africa, Madagascar and Sri Lanka. The biogeography of the Order in these instances shows close correlation with what is known of the dispersal of Gondwanaland.

The larvae of the New Caledonian mayflies are found in the upper reaches of the streams, above 600 m. The lower reaches are dominated by Odonata, Chironomidae and Crustacea. EDMUNDS related this to the northward drift of New Caledonia during the Tertiary affecting a group unaccustomed to tropical conditions but it could in part be due to competition from more recent colonists such as the Odonata, which are generally of a more tropical character.

## Odonata

The New Caledonian Odonata are being studied at present by LIEFTINCK who has already published descriptions of the adult stages (LIEFTINCK, 1975). Forty species have been recorded from New Caledonia and the Loyalties and they can be referred fairly readily to the geographical elements recognised for the macrolepidoptera in the previous chapter. Sixteen are endemic (40%), six are shared with the New Hebrides (New Caledonian element) with one extending to Fiji (15%), six are referable to the widespread element, six are broadly Melanesian and seven (17.5%) can be assigned to the Australian-centred elements (widespread, Coral Sea and Australian-Pacific). Compared with the macrolepidoptera the endemism is greater, the number of widespread species much less, the Australian elements representing a similar percentage and the New Caledonian and Melanesian elements a higher one. These differences, especially the proportion of endemics to widespread species, could be indicative of poorer dispersal powers in the Odonata of Australasia overall, though some of the larger species are known to be extremely mobile.

In the Argiolestinae there are three genera represented, two endemic and one shared with the New Hebrides, containing a total of four species. They belong to a generally Australo-Papuan group of genera and could have colonised New Caledonia initially from either source. There is an interrelated group of six endemic *Isosticta*, a genus shared with Australia (two species) in an Australo-Papuan family. A taxonomically isolated species of *Metaphya* is related to species from New Guinea and Borneo.

There are six endemic *Synthemis*, a primarily Australian genus with two species in New Guinea and one in Fiji. The genus is part of a tribe that consists otherwise of three genera restricted to Australia (CAMPION, 1921).

Five Oriental-Papuan zygopterous families do not reach New Caledonia nor is the genus *Nesobasis* represented, though this has approximately 20 species endemic to Fiji and two shared between Fiji and the New Hebrides. Thus the older elements of the fauna have most affinity with Australia and there are also indications of Inner Arc relationships; there is a discontinuity with areas to the north and east. The small New Zealand fauna (11 species with six endemic) is also related to that of Australia (McLELLAN, 1975).

**Phasmatodea**

Amongst the orthopteroid orders only the Phasmatodea have been treated in any detail (NAKATA, 1961). In New Caledonia there are five subfamilies represented by 12 genera and 24 species. Sixteen species belong to the subfamily Eurycanthinae which is restricted to Melanesia and lacking in Fiji and eastwards. There are five eurycanthine genera virtually restricted to New Caledonia (one extends to the New Hebrides). Only one other species of the sixteen occurs outside New Caledonia, a species of the New Guinea-centred genus *Eurycanthus* that is shared with New Guinea and the Bismarcks. The subfamily shows strong Inner Arc association. The Lord Howe monotypic endemic genus *Dryococelus* is a eurycanthine.

The tribe Macranthini is shared between Australia and New Zealand. Three genera occur in New Zealand, two endemic and one, *Clitarchus*, shared with New Caledonia (including a shared species).

The Loyalties have eight species in as many genera, three shared with New Caledonia but the rest, including two genera, being endemic. This is rather different from the usual close relationship between the two areas but the situation could have arisen through undercollecting on the main island.

In the New Hebrides endemism is low and the fauna appears to have been derived from sources all around. The Fijian fauna is very different from that of New Caledonia, even shared genera being represented by different species. The situation on Lord Howe I. is of great interest for, as well as three Australian species as might be expected from its geography, there is the eurycanthine already mentioned and another old endemic, *Parasosibia australica*, belonging to an otherwise Papuo-Oriental subfamily Necrosciinae. Both of these appear to be Inner Arc relationships that exclude Australia.

The Phasmatodea are probably poorly dispersive, suggested by NAKATA to be rafted as eggs. None have been recorded from Norfolk I. which is consistent with the hypothesis of poor dispersal. The *Clitarchus* species shared between New Caledonia and New Zealand is therefore most unusual.

## Psocoptera

The New Caledonian psocids are in the process of being worked by I. W. B. THORNTON and C. N. SMITHERS. So far they have published accounts of the Myopsocidae (SMITHERS & THORNTON, 1974) and the Philotarsidae (THORNTON & SMITHERS, 1974). These publications are part of a systematic analysis of south-west Pacific Psocoptera, concentrating first on the Inner Melanesian Arc and then on the Outer, working to a broad geological hypothesis similar to that outlined in Chapter I. As yet the Outer Arc faunas are little known but those of Australia, New Zealand and New Guinea have begun to be studied.

Psocids are very variable as regards powers of dispersal, some being recorded frequently in Pacific aerial plankton and others, including both winged and flightless groups, being much less mobile and ecologically more restricted. The Myopsocidae of New Caledonia, three *Phlotodes* species, may belong to the more dispersive extreme. The genus has a wide distribution and an Australian species has been recorded from Norfolk I. and the Kermadecs, an indication of mobility. The affinities of the New Caledonian species are not clear. One is part of a widespread Palaeotropical group and another is part of a smaller group recorded from Madagascar, New Guinea and New Caledonia, a possible Inner Arc relationship.

The generic placements of the New Caledonian Philotarsidae are provisional and may have to be revised when the New Guinea and New Zealand faunas have been better studied. At present the New Caledonian fauna has been divided amongst three genera. There are five *Haplophallus* species, the genus being widespread in the Old World tropics. One is shared with the Loyalties, Fiji and Samoa with a close relative on Norfolk I., and the other four form an interrelated group of Australian affinity.

The other two genera, *Austropsocus* and *Zelandopsocus*, share several characters that set them apart from other Philotarsidae. New Caledonia supports seven *Austropsocus* and 27 *Zelandopsocus* compared with five and two respectively from the New Zealand area. Apart from a dubious and unconfirmed record of *Zelandopsocus* from Australia and the possibility that the New Guinea fauna may contain these two genera, it is likely that they are restricted to New Caledonia and New Zealand

## Hemiptera

Relatively little is known of the Hemiptera. Both DUMBLETON's (1961) account of the Aleyrodidae and FENNAH's (1969) account of the Fulgoroidea stress that little is known of the overall Melanesian fauna and therefore biogeographical conclusions must be tentative. Amongst the New Caledonian Aleyrodidae there are several links with Australia but little affinity with New Zealand.

FENNAH described 16 new fulgoroid genera but these may only be endemic by default because of under-collecting elsewhere. He noted the presence of a number of mainly Australian genera that had not been recorded further to the north and east. Much of the fulgoroid fauna is of widespread Indo-Australian character. FENNAH observed a marked discontinuity between the Solomons and New Hebrides on the one hand and New Caledonia on the other, transgressed by apparently recent dispersals. He suggested that the geographical discontinuity between the two areas might have been greater in the past.

The Australasian cicadas are of great interest, with many groups seeming to reflect fairly old radiations. The general patterns of distribution observed and the high degree of endemism suggest only limited powers of dispersal.

Two cicada species were taken in the *Baeckea*-sedge maquis of the Plaine des Lacs during the 1971 survey, distinguishable in the field by song differences. They have been determined by Mr. J. S. DUGDALE as species of *Notopsalta*, one being *N. melanesiana*. DUGDALE (1971b) referred two more New Caledonian species to this genus (previously placed in *Cicadetta*), and there is one in each of New Zealand and eastern Australia. *Notopsalta* belongs to a large and somewhat heterogeneous group of cicadas with a trifid aedeagus, discussed at length by DUGDALE. The group is almost entirely centred in Australia and New Zealand, there being an outlying genus in South Africa. If this is a natural assemblage it may well be of ancient austral character; several of the subgroups recognised by DUGDALE are well represented in both Australia and New Zealand. New Zealand and Australia also share a few other groups of cicadettines.

New Caledonia is thus the centre of species richness for *Notopsalta*; seven other species of *Cicadetta* in New Caledonia mentioned by DUFFELS (1977) need examination and may prove to be referable to *Notopsalta* or these other groups. Only the family Tibicinidae is represented in New Caledonia as in New Zealand, though in the latter the representation is restricted to the tribe Cicadettini (DUGDALE). Several genera in the family are very heterogeneous and there is much need for revisional work, so indications of biogeographical affinities within the genera as they stand now must be treated with caution. Such revision will no doubt prove extremely valuable from a biogeographical point of view as illustrated by the work of DUFFELS (1977) which will be discussed further below.

DUFFELS, in his treatment of the genus *Diceropyga*, reviewed the zoogeography of Melanesian cicadas, referring to New Caledonia. Apart from the *Notopsalta* there are two species of *Abricta*, an otherwise Australian genus with two outliers on Mauritius, *Kanakia typica* which is congeneric with a single Queensland species only, the endemic and monotypic genus *Mouia* and four endemic species of *Ueana*, a genus that has one further species in the Bismarck Is. *Ueana* was shown by DUGDALE to have affinities with the

Australian genus *Pauropsalta*. The outlier in the Bismarcks requires examination with reference to the New Caledonian species. It may occur in New Guinea as well and might prove to be an Inner Arc relationship but this is merely surmise. The New Caledonian fauna is thus predominantly Australian with *Notopsalta* being more generally austral, though DUGDALE suggested that the New Zealand species may have been derived from Australia relatively recently by long distance dispersal. The New Caledonian fauna is largely distinct from that of the rest of Melanesia.

Cicada distribution east of Sundaland generally shows intriguing patterns relative to the geological discussion of Chapter I. The Lesser Sundas have a rather impoverished Malaysian fauna. The Philippines have a richer, more endemic fauna with some Malaysian invaders. Sulawesi supports a member of the *Diceropyga* group and also *Prasia* with four species, one other in New Guinea, and *Lembeja* which has one species in the Philippines, three in Sulawesi (one extending to the Lesser Sundas and the Philippine species is recorded from Sumba), two restricted to Sumba, three in New Guinea and one more shared with Queensland, and one restricted to Queensland. If confirmed by revision, this is an interesting Sulawesi/Inner Arc connection. The species distributions are taken from METCALF (1963).

Apart from *Diceropyga* and related genera there are a few more tibicinid genera restricted to Melanesia, two of which, *Gymnotympana* and *Thaumastopsaltria*, are centred on and largely restricted to New Guinea. A third, *Baeturia*, has a wider Melanesian distribution, occurring in New Guinea (the centre), the Moluccas and Kei, with three species in Queensland, one extending from the Moluccas to Samoa and a single outlier in the Ryukyu Is. Samoa also supports the monotypic endemic genus *Moana*. Fiji has species attributed to *Tibicen* and *Cicadetta* as well as endemic radiation in the *Diceropyga* group.

The genus *Diceropyga* and relatives appear to represent a clear instance of colonisation of Melanesia by Oriental forms that has resulted in the evolution of two sister groups referable one to each of the Inner and Outer Melanesian Arcs. The nearest allied genera to these two sister groups in the Oriental Region are *Orientopsaltria*, *Platylomia* and *Dundubia*. The Inner Arc (here excluding New Caledonia) group consists of three interrelated genera virtually restricted to New Guinea, *Fatima*, *Sawda* and *Cosmopsaltria* s. s. A single Fijian species is included in *Sawda* but this was only a tentative assignation and may be unjustified.

The Outer Arc group is represented by *Diceropyga*, divided by DUFFELS into three closely related genera. *Dilobopyga* is restricted to Sulawesi, apart from a species endemic to the S. Moluccas in one of the two species groups represented; here is an apparent association of Sulawesi with the Outer Arc. *Aceropyga* is an eastern Melanesian genus with two species groups (one monotypic) endemic to Fiji and a third with species in the Bismarcks and Solomons, one in Tonga and one shared between the New Hebrides and Kusaie in the E. Carolines. The genus *Diceropyga* contains four species groups. One occurs in the Moluccas, one in the Vogelkop, Salwatti and Waigeu and one in the Bismarck Is. The fourth and largest group has fifteen species in New

Guinea and the Solomons, one of those in New Guinea extending into the Bismarck group. The seven Solomons species are placed in a distinct subgroup by DUFFELS who also speculated on the phylogeny of the eight New Guinea ones. The Solomons group is allopatric within the archipelago, species usually being restricted to a single island. The New Guinea group initially divides into two lines. One is restricted to the northern coast of New Guinea and includes the species shared with the Bismarck Is. (*gravesteini*). The other includes a species restricted to the islands to the south-east of New Guinea on the one hand and a group of three, one also restricted to such islands (Goodenough, Normanly), one restricted to the extreme south-east and the third distributed widely over southern New Guinea and recorded from the Aru Is. and N. Queensland. The development of this second group could be relatively recent, following the approach and fusion of the two arcs and it is interesting to observe the way the species appear to replace each other with very little overlap around the coast of south-east New Guinea.

EVANS (1974) stated that the known cicadellid fauna of New Caledonia was very small. He described a new monobasic tribe of the Cicadellidae, the Selenomorphini, with at least three species, all restricted to New Caledonia and the Loyalties.

## Thysanoptera

The New Caledonian Thysanoptera were studied by BIANCHI (1945, 1952). He found a predominance of Australian affinities, a fact that has been confirmed by more recent studies by L. A. MOUND (*pers. comm.*), though again the faunas of neighbouring areas are not well known. Some of the species may prove to be more widespread, for example MOUND has sunk *Cryptothrips dimidiatus* to *Nesothrips propinquus* and *Gastrothrips noumeae* to *G. acuticornis*, both of which are widely distributed through introduction along the Europe to New Zealand trade route.

*Adrothrips* (*Scopaeothrips*) *intermedius* occurs in Australia and is associated with *Casuarina*, as may be *Ensiferothrips primus* which has a similar distribution. *Aelothrips inauditus* is to be transferred by MOUND to a new genus, together with two further undescribed species from New Zealand.

## Coleoptera

Study of the Coleoptera has again been patchy but the New Caledonian fauna was reviewed and listed at the beginning of the century by FAUVEL (1903–1907). From this account one gains a general impression of a fauna of oceanic character yet with significant generic endemism and approximately equal representation of links with Australia and New Zealand. GRESSITT (1960a) contrasted the Micronesian, New Hebrides and Samoan faunas with the New Caledonian, the former being characteristic oceanic faunas derived by aerial dispersal from primarily Indo-Malayan and Papuan sources. The Fijian fauna is generally associated with these.

The New Caledonian cerambycids and chrysomelids generally show more

Inner Melanesian Arc affinities than Australian ones, though one strictly Australian cerambycid genus extends to New Caledonia (GRESSITT, 1956). GRESSITT also illustrated the distribution of the weevil tribe Pachyrrhynchini which is of great interest as it extends from islands north of the Philippines through a centre in Luzon to New Guinea and the Solomons, a distribution that virtually excludes Sundaland and Sulawesi. It thus follows the putative extension of the Outer Melanesian Arc westwards along the boundary of the Pacific tectonic plate that was suggested to have occurred in the middle of the Tertiary in Chapter I, the old northern route of faunal exchange between the Oriental Region and Melanesia proposed by HOLLOWAY & JARDINE (1968).

The Cerambycidae of the south-west Pacific have been relatively well studied, for example the Fijian fauna by DILLON & DILLON (1943) and the New Caledonian by BREUNING (1951, 1953). The proportions of affinities within the Norfolk I. cerambycid fauna were suggested by HOLLOWAY (1977) to indicate relatively poor dispersal abilities and hence on New Caledonia the family might be expected to show up the older biogeographical associations of the island.

HAYASHI (1961) reviewed the geography of the New Caledonian cerambycids. The high percentage of endemic genera and tribes suggests a long history of isolation. Endemic genera in primitive groups such as the Piesarthrini, Rhagiomorphini and Spintherini have affinities with Australia. New Caledonian members of the Enicodini and Tmesisternini, tribes of the Lamiinae, indicate old links with New Zealand and New Guinea. The Enicodini have an endemic genus in New Guinea but the principal centres are in New Zealand and New Caledonia; a species of *Enicodes* has recently spread from the latter to Norfolk I. The Tmesisternini occur mainly in New Guinea, the Bismarcks and Solomons, with three genera endemic to New Caledonia.

In the Chrysomelidae there are six hispine genera in New Caledonia, five of which are endemic (GRESSITT, 1960b). Affinities are with New Guinea and the Solomons rather than Australia; the subfamily does not occur in New Zealand. Hispines are closely associated with palms, *Pandanus* and *Freycinetia.*

WENZEL (1955) found the New Caledonian Histeridae to be predominantly oceanic in character except for an endemic species of *Sternaulax*, a primitive genus represented otherwise by a single species in New Zealand.

## Diptera

The New Caledonian Diptera are poorly known, detailed studies being more or less restricted to the Tipulidae, Culicidae and Tabanidae.

The Tabanidae were described by MACKERRAS (1961). New Caledonia lacks *Tabanus* but supports four *Philoliche*, a curious African connection. There are ten *Cydistomyia*, five from three different Papuan species groups and five of Australian affinity. One of three *Dasybasis* is of Australian affinity but the other two species relate to New Zealand. Overall the tabanids appear to reflect the more ancient biogeographical associations of the island. The Fijian fauna is predominantly Melanesian in character.

The Pacific mosquitos have been studied intensively by BELKIN (1962). Their medical importance means that collecting has probably been thorough in most areas and thus the biogeographical patterns may be fairly reliable. The twenty-eight more widespread species and some of the more localised ones correspond in their distribution patterns to the elements derived for the macrolepidoptera in the previous chapter. There are 155 endemic or precinctive species as follows: Solomons (81); New Zealand (20); Fiji, Tonga and Samoa (23); St. Cruz and New Hebrides (15); Solomons-New Hebrides (1); New Caledonia (9); New Caledonia + New Hebrides (1); E. Polynesia (5).

The majority of these species represent an attenuation of Indo-Australian tropical elements and generally the more 'primitive' taxa occur nearest the periphery of this attenuation. The oldest elements are in New Zealand where the fauna is distinctive though poorly known. It is generally not related to the New Caledonian fauna but one species of the group (*Culex iyengari*) occurs in New Caledonia and one (*C. pacificus*) in the New Hebrides. The affinities of this New Zealand group are with southern Australia and, more distantly, South America. Two culicids of New Zealand affinity occur on Norfolk I. (FERRAR, STANDFAST & DYCE, 1975) and the colonisation of New Caledonia and the New Hebrides by this group might thus be a recent event, perhaps a parallel of the spread of the parakeet, *Cyanorhamphus novaeseelandiae*, through most of the same areas.

The New Caledonian culicid fauna is distinct from that of the rest of the Pacific, endemism standing at 45%. Australian elements are conspicuous, particularly on the drier west coast. The endemic element may be a relict of an earlier fauna and shows a great deal of affinity with that of south-east Australia and Tasmania. The Loyalties fauna resembles that of New Caledonia.

The New Hebrides fauna has been derived from many sources, the endemics and more widespread elements probably having resulted from several different dispersals from various directions. The St. Cruz Is. have links with both the Solomons and the New Hebrides, and one connection with Rotuma. The rich Solomons fauna is mostly of Papuan affinity.

The group of Fiji, Samoa and Tonga contains the most highly differentiated fauna in the Pacific, excepting that of New Zealand. The larger part is of Oriental derivation but there is also a small Australian element. The largest number of species is on Fiji but there are distinctive elements on Samoa as well and general affinities between the two. The Tongan fauna is not well known but is probably poorer with low endemism. The situation overall in these islands resembles that found for the Lepidoptera, though there are far fewer widespread species.

In dispersal the culicids may be transported passively in wind currents as adults, thus the south-east trades would bear them from New Zealand to New Caledonia and westerly airstreams would transport colonists from Australia into the Pacific. Dispersal of taxa from the north-west might be more dependent on cyclones. The Tipulidae, of similar build and weak flight, might also be dispersed in this way but VANE-WRIGHT (1967) has postulated marine dispersal, rafting of eggs, larvae or pupae in plant material, to explain the presence of a few species of the large tropical South American genus *Zelandotipula* in New Zealand.

The Tipulidae of New Caledonia have been described in a series of papers by ALEXANDER (1929, 1934, 1945, 1948a, b). Several tipuline genera thought to be primitive and having today an austral distribution occur in New Caledonia and the author was encouraged by Mr. R. I. VANE-WRIGHT to make a collection during the moth survey. The family comes readily to light and many species are conspicuous during the day or can be swept, so it was possible to make a large collection and obtain much information on local distribution. It was thought possible during the planning of the survey that New Caledonia would prove to support species of either *Ischnotoma* or *Zelandotipula*, but none were obtained. These are basically southern genera reviewed by VANE-WRIGHT (1967).

*Zelandotipula* has its centre of diversity in northern South America but four species occur in New Zealand, forming a distinct, interrelated group. Because of their close monophyly and the lack of greater radiation in New Zealand VANE-WRIGHT considered that, despite their overall distinctiveness, the possible rafting in circumpolar ocean currents might have occurred relatively recently. The low radiation in New Zealand contrasts with the extensive speciation that has occurred in the widespread genera of the Limoniinae, *Gynoplistia* and *Austrolimnophila*. Present day ocean currents would be as likely to carry propagules to Australia (especially the circumpolar current) and Melanesia generally (Southern Equatorial Current) as to New Zealand. The northerly distribution in South America today suggests that the genus may favour tropical conditions and hence may be at a competitive disadvantage in New Zealand. The glaciations there might have reduced a hypothetical large *Zelandotipula* fauna to a single species at one time, the ancestor of the four modern ones.

*Ischnotoma* occurs in southern South America and south-east Australia, a distinct and polytypic subgenus in each area. Though not recorded in New Caledonia it is of interest to the general debate on the origins and radiation of the older elements of the southern biota. A fossil species, together with one of *Macromastix*, has been recorded from the Oligocene of the Isle of Wight in England (VANE-WRIGHT). These genera thus resemble other plant and animal groups that have been recorded, albeit sparsely, as fossils in the northern hemisphere in the Tertiary or earlier; the lepidopteron *Sabatinca* was mentioned in this context in the previous chapter. Perhaps the lepidopterous families Hepialidae and Thaumetopoeidae provide instances where, despite predominantly austral distributions, northern hemisphere relictual elements have not yet died out. The possible penetration of elements from a drifting India into Asia after contact in the early Tertiary has also to be considered; the Mediterranean and Asian Thaumetopoeidae may be a case in point.

ALEXANDER recorded 56 tipulid species from New Caledonia. Four further species in the BMNH collection were taken by Miss L. E. CHEESMAN and, of the 35 recorded during the 1971 survey, 19 were new, giving a total of 79 species. Preliminary determination of material collected during the moth survey has been made by Mr. P. S. CRANSTON and a representative selection has been sent to Professor C. P. ALEXANDER. Apart from an *Elephantomyia* no further genera have been recorded, the additions being mainly to *Gynoplistia*, the various subgenera of *Limonia* and to *Macromastix*.

Though *Zelandotipula* and *Ischnotoma* were not discovered, there are a number of other primitive tipuline genera represented, all with probable Australian affinities. *Dolichopeza austrocaledonica*, though belonging to a cosmopolitan genus, has its closest affinities with Australian species. There are two *Phacelodocera* species, *herroni* and *margaritae*, with congeners only in Tasmania and South America. One species of the subgenus *Plusiomyia* of the Australian genus *Ptylogyna* (*neocaledonica*) was described by ALEXANDER from high altitude near St. Louis. A further species was taken at Grand Lac in 1971 where it appeared to be associated with the herbaceous marsh vegetation by the lake. The majority of specimens of the species mentioned above were taken in association with the ultramafic rocks.

The largest tipuline genus on the island is *Macromastix*, now placed as a subgenus of *Leptotarsus*. *Macromastix* is today restricted to Australia, New Zealand and South America apart from a single species, *errans*, in Sri Lanka. The fossil record from the English Oligocene mentioned above suggests that the distribution of the genus in the past was probably wider than its present day 'Gondwanan' range.

ALEXANDER noted six species. The 1971 survey recorded four of these (the missing pair being the closely related duplex of *repleta* and *risbeci*) together with four new species. All the species except *risbeci* have been recorded frequently in ultramafic localities, mostly in the south where they predominate in samples made. Six have not been recorded away from the ultramafics and those that have (*cockerellae*, *novocaledonica*, *risbeci* and a new species) are restricted to localities with rainforest. *M. novocaledonica* was taken in numbers flying by day over wet, reedy pastures near Sarraméa. *M. caledoniana* and two of the new species were taken in numbers flying soon after dawn over the montane maquis at 760 m on the Mt. des Sources.

All the Tipulinae are endemic. At present 85% of the rest of the fauna is endemic, consisting mainly of species from very widespread genera or subgenera. This endemism may prove to be lower when the faunas of neighbouring areas are better known and the affinities of the endemics can be defined more precisely. Of the nine species recorded outside New Caledonia one is worldwide (*Erioptera* (*Trimacra*) *pilipes*) and the remainder, all *Limonia* species, are more localised, one shared with the New Hebrides, two Australian, four shared with the Fiji-Samoa area and one widespread in the Indo-Australian tropics. Where affinities of the endemics are known these are predominantly with Australia. The only endemic with New Zealand affinities is *Limonia* (*Discobola*) *caledoniae*, belonging to an Indo-Australian subgenus with much radiation in New Zealand.

About two thirds of the non-tipuline endemics have only been recorded from ultramafic localities, particularly the more distinctive ones and most of the several species duplexes. However, a number of these are known from very few specimens and may prove to occur in a wider range of habitats. Eight of the nine more widespread species have been recorded more away from the ultramafic rocks.

### Plecoptera, Megaloptera and Mecoptera

The Plecoptera, Megaloptera and Mecoptera all contain groups with purely Antarctic (S. America, Australia, New Zealand) distributions (ILLIES, 1965; MCLELLAN, 1975) but a search of the literature has failed to locate any reference to the occurrence of these Orders in New Caledonia.

### Trichoptera

Most adult Trichoptera come readily to light and, during the course of the moth survey, a selection from each locality was collected so as to include all species present with some indication of their relative abundance. The material is now with Professor H. H. ROSS who also has adult and larval material collected by a recent Austrian expedition (STARMÜHLNER) and by several American collectors (ROSS, 1975, 1976). He is at present working on this material and the references just given represent the first of the results.

These publications describe the twelve New Caledonian Helicopsychidae. They are all *Helicopsyche* species but a diverse assemblage in which ROSS recognised four species groups. The *vallonia* group contains two species with Australian affinities. The other three groups are more distinctive, possibly the result of endemic radiation, but can also be related generally to Australian and New Zealand members of the genus. Detailed study of members of the genus from other parts of the world may lead to modification of this picture (ROSS, *in litt.*).

Professor ROSS has kindly commented on some of the other families present in New Caledonia in correspondence but has indicated that for many of them a survey of the world fauna would be necessary before reliable conclusions could be drawn. The examples cited below are those where there is some degree of confidence. Information has also been drawn from ROSS (1956) and MCLELLAN (1975).

Several trichopteran genera, particularly of the more primitive families discussed by ROSS (1956), have appeared in the Baltic (early Tertiary) amber, suggesting that evolution has been somewhat conservative and radiation of some of the complexes had begun in the Mesozoic. Several families and subfamilies have an Antarctic distribution today or are restricted to Australasia, such as the Pycnocentrellidae, Helicophidae and Philorheithridae (MCLELLAN).

The New Caledonian fauna generally relates to the Australian and New Zealand faunas. In the Hydrobiosinae (of the Rhyacophilidae) and Kokiriidae (Rhynchopsychidae) the affinities of the New Caledonian fauna are with New

Zealand, the Kokiriidae being otherwise only found in Chile. The Hydrobiosinae are, with the exception of a probably early offshoot *Apsilochorema*, restricted to Central and South America, Australia and New Zealand. Eight genera are endemic to New Zealand and the New Caledonian *Xanthochorema caledon* is related to them. *Apsilochorema* is distributed through the Indo-Australian tropics and has three species in Fiji but none in New Caledonia.

The *Hydrobiosella* of New Caledonia now show affinities with both Australia and New Zealand, though in ROSS (1956) only the New Zealand connection was apparent. The situation in *Hydropsyche* has also been modified by further work since that paper was written. The New Zealand genus *Orthopsyche* is dominant in New Caledonia but there are other genera that appear to be endemic. In the Hydroptilidae there is a distinctive endemic genus, *Caledonotrichia*, of uncertain affinities that contains two species (MARSHALL, 1977).

The faunas of New Guinea, the New Hebrides and Fiji have not been studied in depth but these areas have strong affinities amongst themselves and with the Oriental Region but little with New Caledonia (ROSS, *in litt.*).

## Hymenoptera

The Pacific Hymenoptera are perhaps the least known of the insect Orders. MICHENER (1965) commented that the New Caledonian bees had not been studied to any great extent, only seven genera having been recorded. These included both Australian and Indo-Malayan elements.

The ant subfamily Myrmeciinae includes the most primitive living species and is restricted today to two genera in Australia and New Caledonia. However, the subfamily is represented by extinct Oligocene genera from the Baltic amber and Argentina (BROWN & TAYLOR, 1970).

## Vertebrates

The isolated position of New Caledonia has meant that few vertebrates have been able to colonise it apart from birds. There is one freshwater fish of interest, no native amphibia, a relatively large reptile fauna that includes a group of extemely interesting geckos, a poor but diverse bird fauna and a few native bats. Various frogs and mammals (rats, pigs, deer, cats, dogs) have been introduced and have naturalised.

## Fish

The freshwater fish belongs to the endemic genus *Nesogalaxias*, the status of which was discussed by MCDOWALL (1968). It is taxonomically isolated but related to the *Galaxias* of Australia and New Zealand. MCDOWELL suggested Australia as the origin. Though probably restricted to freshwater in New Caledonia, its ancestors were probably diadromous and colonised the island through marine dispersal; present day currents favour colonisation from New

Zealand though the circumpolar current of Oligocene times and earlier (p. 28) would have favoured colonisation from Australia if it flowed as suggested.

The single species inhabits the lakes of the Plaine des Lacs, Lac-en-Huit and Grand Lac. It was still present in Grand Lac in 1971 but introduced *Tilapia* and projected mining operations in the lakes have placed its future in jeopardy.

## Amphibia

Though no Amphibia occur in New Caledonia it is worth mentioning the presence of two frog species of the genus *Platymantis* on Fiji. This genus has been discussed by BROWN (1965) and GORHAM (1965, 1968) and, in its broadest sense (embracing both *Platymantis* and *Cornufer*, the latter shown to be an inappropriate name by GORHAM), it is distributed along the islands of the Outer Melanesian Arc and its supposed extension north-west into the Philippines. There are eight species in the Philippines, one on Waigeu, five in New Guinea, one (possibly introduced) in the Palau group, two in the Bismarcks, nine in the Solomons and two in Fiji.

## Reptiles

Fijian reptiles also show a marked contrast to those of New Caledonia. *Ogmodon vitianus* is an endemic monotypic genus of elapid snake related to Solomons species that at present rest uneasily in the genera *Denisia* and *Micropechis*. *Ogmodon* has characters in common with the Asiatic genus *Bungarus* (MCDOWELL, 1967, 1969). The Fijian iguana and its Tongan relative have their closest allies in South or Central America. Their presence in the south-west Pacific is enigmatic. Rafting from the Americas without colonisation of intermediate localities or restriction as relicts from a much wider Indo-Australian range are possible hypotheses (AVERY & TANNER, 1970). The Fijian gekkos and skinks represent extensions of genera with wide Indo-Australian tropical ranges.

The New Caledonian reptiles were described in detail by ROUX (1913) and listed briefly by VIROT (1956). More recent treatments of various elements of the fauna may be found for the Gekkonidae in WERMUTH (1965) and KLUGE (1967a, b) and for *Leiolopisma* (Scincidae) in GREER (1974). The New Zealand reptiles have been described by BULL & WHITTAKER (1975) and those of the New Hebrides by MEDWAY & MARSHALL (1975).

The boa *Candoia bibroni* has been recorded from the Loyalty Is. but not the New Caledonian mainland. The species has a Melanesian distribution that extends to Fiji and Samoa. It has been recorded in the New Hebrides on the major islands from Santo to Erromanga. The genus has three species that range broadly through the Papuan Region, two extending to the Tokelau Is. This may be another example of an Outer Arc distribution that excludes New Caledonia (virtually). The 'flowerpot snake', *Typhlops braminus*, was recorded for the first time in both New Caledonia and the New Hebrides in 1971, in both cases almost certainly a horticultural introduction. No recent reference has been located for *Typhlops willeyi*, mentioned by ROUX as being endemic to the Loyalties with affinities with the Solomons; it might prove to be *braminus*.

GREER (1974) recognised three generic groupings within the *Leiolopisma* complex. The first contains five genera, a subgroup of two extending through Europe and Asia and recurring in the south-east U.S.A., and a subgroup of three Indo-Australian tropical genera, two largely restricted to New Guinea and one Indo-Australian widespread genus. The last, *Lipinia*, includes *L. noctua* which extends into the Pacific to Hawaii, occurs in the New Hebrides but has not been recorded from New Caledonia.

The second group contains all the New Caledonian skinks. A subgroup of three genera is broadly Melanesian with *Eugongylus* extending from the Moluccas to Cape York and the St. Cruz Is., *Phoboscincus* with two species (*bocourti* and *garnieri*) restricted to New Caledonia and the Loyalty Is., and *Tachygia*, monotypic, endemic to Tonga.

Of the other genera in the second group two (*Morethia* and *Proablepharus*) are restricted to Australia, the monotypic *Cryptoblepharus* extends throughout the Malagasy, Indo-Australian and Pacific tropics, including the New Caledonian group, and *Anotis* has two species in Australia and an interrelated trio (*gracilis*, *mariae* and *slevini*) in New Caledonia.

Two genera, *Emoia* and *Leiolopisma* s. s., complete the second group. *Emoia* is a large genus extending throughout the Indo-Australian tropics and into the Pacific. MEDWAY & MARSHALL recorded seven species in the New Hebrides. Three of these are only found in the more northerly islands of the group and two, *cyanura* and *samoensis*, extend to the Loyalties but not New Caledonia. *E. samoensis* also occurs in Fiji and Samoa and *E. cyanura* is much more widespread. It would appear that *Emoia* is still in the process of colonising the New Hebrides and has only recently extended to New Caledonia.

*Leiolopisma* contains 38 species centred mainly in southern Australia, especially Tasmania, with subcentres in New Caledonia and New Zealand. There is a single outlier on a small island near Mauritius. There are six species in New Caledonia of which two extend to the Loyalties and one (*austrocaledonica*, placed in *Lampropholis* by MEDWAY & MARSHALL) to the more southerly islands of the New Hebrides. Relationships within the genus have not been determined fully. The New Zealand species (including *Sphenomorphus*, lumped with *Leiolopisma* by GREER) are probably derived from between one and three invasions followed by radiation, and there are some cytological indications of relations with Australia; the cytology of the New Caledonian species has not been investigated and might lead to a change in this picture when known (BULL & WHITTAKER, 1975). One New Zealand species appears to be closely related to the Lord Howe species, a parallel with the plant genus *Dracophyllum*.

The third subgroup of the *Leiolopisma* complex contains eight genera, three African and five Australasian, mostly Australian but with one genus extending into Melanesia and one a monotypic endemic on Bougainville in the Solomons.

Two gekkonid subfamilies are represented in New Caledonia. The Gekkoninae contain large numbers of warm temperate and tropical genera and those represented in the New Caledonian group are all widespread in the Indo-Australian tropics. *Gymnodactylus* (*Cyrtodactylus* in MEDWAY & MARSHALL) *pelagicus* has a widespread Melanesian distribution, *Hemidactylus garnotii* is recorded from India to Hawaii but not from the New Hebrides, and *Lepidodactylus lugubris* is widespread in the Indo-Australian tropics. These three species are found in New Caledonia and the Loyalties but *Gehyra oceanica* (*vorax* in ROUX), similarly widespread, occurs only in the Loyalties. ROUX referred also to *Lepidodactylus crepuscularis*, endemic to the New Caledonian group; other endemic species occur one in each of Fiji and Rotuma.

The Diplodactylinae are biogeographically much more interesting. The subfamily is restricted to Australasia and could be taken in many ways to epitomise the effects of climatic differentiation on elements of the ancient and general Australasian biota. One tribe, the Diplodactylini, contains five genera endemic to and widespread in Australia with species that are primarily terrestrial and adapted to more arid habitats. The other tribe, the Carphodactylini, are primarily arboreal and in Australia are centred in the tropical forests of the north with only minor penetration of southern and arid zones. This tribe has three interrelated genera in New Caledonia, *Rhacodactylus* with six species, *Eurydactylodes* with two species and *Bavayia* with two species. The *Bavayia* species are coastal and occur also on the Loyalties. The other genera are more strictly characteristic of rainforest. Some are montane and two *Rhacodactylus* species may be restricted to the southern ultramafic massif. In New Zealand there occurs another trio of closely related genera, *Heteropholis*, *Hoplodactylus* and *Naultinus*. *Naultinus* and *Heteropholis* species produce fertile intergeneric hybrids (BULL & WHITTAKER, 1975). The New Zealand genera are most closely allied to the New Caledonian ones. In Australia there are four genera. Three are closely related and distinct within the tribe (*Carphodactylus*, *Phyllurus* and *Nephrurus*), and the fourth, *Pseudothecadactylus*, restricted to the Cape York area, tends towards the New Caledonian genera in morphology.

The amphibian and reptile faunas of the south-west Pacific illustrate markedly the difference between New Caledonia, where austral elements, probably ancient, are predominant, and areas to the north where Indo-Malayan and Melanesian elements predominate. There are indications from the local distributions of species within the islands of the New Hebrides and the New Caledonian group that segregation of the two elements has only recently begun to break down and is still in the process of doing so.

## Birds

The avifauna of New Caledonia, excluding marine and migrant species, totals 68 species distributed amongst 57 genera. Fifteen of those genera are not found elsewhere in the Pacific, being endemics or derived from Australia or

New Zealand (MAYR, 1940a, b, 1965). MAYR subdivided the fauna as follows: 31 species are of Australian derivation and not endemic, nine more of that provenance having evolved into endemic species or genera. *Cyanorhamphus novaezeelandiae* is derived from New Zealand, presumably recently and possibly via Norfolk I. where it also occurs. Eighteen species are derived from the Papuan Region and are undifferentiated and nine species from the same source have become endemic to species or genus.

Thus Papuan and Australian influence is roughly equal in the older, more differentiated elements but the latter is almost twice as strong among more recent colonists. The Loyalties fauna is different, more like that of the southern New Hebrides with the Australian element reduced relative to New Caledonia and endemism much lower. BERLIOZ (1962) also made this distinction. MAYR suggested that the distribution of birds indicated that the Loyalties are much more recently emergent than New Caledonia and never linked with it by dry land.

Land birds on the whole appear to be much less dispersive than insects (HOLLOWAY, 1977) and the increase in the Australian influence in more recent colonists could have been facilitated by recent exposure of the Chesterfield Reefs and other stepping stones between the two areas. The difference noted for the Loyalty Is. suggests that the sea barrier required to exert some filtering effect on faunal exchange may be quite small.

The kagu (*Rhynochetos jubatus*) is endemic to family level, related to the cranes and of uncertain origin. There are four endemic genera. *Eunymphicus* and *Megalurulus* are of Australian affinity and *Drepanoptila* and *Leptomyza* are of Papuan affinity, possibly derived from New Guinea. Single species of *Ducula* and *Corvus* are well differentiated from Papuan relatives and could be accorded subgeneric status.

In New Caledonia it is the rainforests that are most rich in species (BERLIOZ, 1962), though the endemic genus *Megalurulus* is more characteristic of open habitats. During the moth survey a multispecific flock of insectivorous species was observed working through taller *Baeckea*-sedge maquis at Grand Lac. Some attempt is being made to determine the status of rare and threatened species, particularly the endemics (BÉLAND, 1975).

The birds of the rest of eastern Melanesia from the Solomons to Fiji and Samoa are derived mainly from 16 families found in New Guinea where they make up 59% of the total number. The further eastwards from New Guinea the higher is the proportion of species that belong to characteristic colonising genera (MAYR, 1965).

Fiji has 54 species with eight endemic genera, including the dove genus *Chrysoena* with three species. There is a further genus shared only with Samoa, which has 33 species and two endemic genera. The Tongan fauna shows little endemism but there are four genera endemic to E. Polynesia (MAYR, 1940a, b). The overall picture for the birds is generally similar to that for the moths including the higher representation of taxonomically isolated endemics in Fiji, but the widespread element is almost absent, perhaps a reflection of the poorer dispersal powers of birds.

In the New Hebrides (DIAMOND & MARSHALL, 1976) there are 56 breeding land bird species, only five of which have differentiated to the level of endemic species and only one to endemic genus. Eighteen species are colonists from Australia, mostly via New Caledonia as fifteen are shared with New Caledonia and two others have relatives there. Thirteen have been derived from New Guinea via the Bismarcks and Solomons and are absent from Australia. Conversely seven New Hebrides species extend to San Cristobal in the Solomons and a few more to species-poor communities such as mountains and isolated islands such as Rennell where ten species of New Hebrides origin occur. There are nine widespread species where it is uncertain whether they were derived via the Australian or Melanesian route. There are 22 species derived from Australia and/or New Guinea in the first two categories extending also to Fiji, with 17 in New Caledonia and 13 in Samoa, thus resembling the widespread Melanesian and Australian-Pacific elements derived for the macrolepidoptera. One species is definitely derived from Fiji and thirteen with distributions generally restricted to the south-west Pacific, only five extending to the Solomons outliers. These species are shared with, or have allospecies in, New Caledonia (seven), Fiji (eight) and Samoa (five). They are probably referable to the New Caledonian and Fijian geographical elements as postulated for the macrolepidoptera. They all belong to wider superspecies or genera that arose in New Guinea, or to genera derived from New Guinea genera. It is not clear whether they arose directly from an eastward wave of colonisation or whether the eight shared with Fiji differentiated on Fiji first and then invaded the New Hebrides.

The factors governing the dispersal, establishment, coexistence and extinction of bird species in Melanesia are extremely complex (DIAMOND, 1975). Many are insectivorous; the frugivorous ones do not show the same degree of specificity in relation to the flora as is observed in the Lepidoptera. Frugivorous birds aid the dissemination of many plant propagules and hence any coevolution between plants and birds will be based on cooperation rather than antagonism such as determines the relationship between plants and defoliating larvae; extreme specificity might not be expected in the former case but has evolved frequently in the latter. This difference may have resulted in the relationship of bird diversity primarily with the structural diversity of the vegetation and that of lepidopteran diversity to a great extent with the floristic diversity of the vegetation. Nevertheless, as seen above and in the analyses of HOLLOWAY & JARDINE (1968), the overall biogeographic patterns are very similar in the two groups. Common factors behind this similarity may be the geographical disposition of land areas and the effects of climate and seasonality on vegetation type generally. The effects of these ecological differences are perhaps most evident on a more local scale; the checkerboard distribution patterns, species guilds, incidence functions and assembly rules noted for Melanesian birds by DIAMOND appear to have no close parallels in the Lepidoptera, though further observations of patterns on small islands as opposed to large ones would be valuable in this respect.

## Mammals

The only native mammals in New Caledonia apart possibly from the Polynesian rat are bats. There are four fruit bats and three microchiroptera. The most comprehensive account is by SANBORN & NICHOLSON (1950) but supplementary information was obtained from FELTEN (1964), WALKER (1964) and, for the New Hebrides, from MEDWAY & MARSHALL (1975). In the New Hebrides there are three fruit bats and eight microchiroptera. With the exception of *Pteropus anetianus* only those four species shared with New Caledonia are recorded from the three southern islands of Erromanga, Tanna and Aneityum. *P. anetianus* is endemic. The six not recorded from New Caledonia apart from *anetianus* fall one into the Fijian element, two into the widespread element and three into the Melanesian element.

Ther are only four bats on Fiji. There is one endemic *Pteropus* and another, *P. tonganus*, shared with New Caledonia, the New Hebrides, Tonga and Samoa, and recorded as far west as Karkar I. off New Guinea. *Notopteris macdonaldi* is a third fruit bat that also occurs in New Caledonia and the New Hebrides. *Notopteris* is monotypic, its affinities being with *Dobsonia* which extends from the Philippines and Sulawesi through Melanesia to the Solomons, a possible Outer Arc distribution.

New Caledonia supports two endemic *Pteropus* (*vetulus* (= *macmillani*) and *ornatus* which has a distinct subspecies in the Loyalties) as well as *tonganus* and *N. macdonaldi.* Two of the microchiroptera, *Miniopteris australis* and *M. medius*, are Indo-Australian widespread species shared with the New Hebrides. The third, *Chalinolobus neocaledonicus*, is endemic but belongs to an Australian genus of three species, one extending to Norfolk I. (*gouldii*); a further species is endemic to New Zealand.

# VIII. DISCUSSION

The survey of the biogeographic relationships of New Caledonia presented in preceding chapters revealed a plethora of different patterns. Nevertheless there are a number of salient features common to many or most of the groups studied, each emphasised to a varying degree from group to group.

Some of this variation is due to factors such as the antiquity of the various groups and their richness in the different source areas involved, but overall the patterns are consistent with what is known to date about the complex geological and climatic history of the area. Against this must be set the trend for older groups to be replaced slowly by younger invaders in island colonisation. MAYR (1965) wrote about New Caledonia as follows:

'About two-thirds of the 68 species in the land bird fauna are only subspecifically different from their Australian and Papuan relatives, or even entirely identical with them. We thus find that in an island which is presumably more than 60 million years old most of the species seem to have arrived within the last 50,000 years. There is no evidence that would suggest that the island was not always able to accommodate at least 30 or 40 species of land birds, and we must therefore conclude that there was a continuous cycle of colonization and rapid extinction'.

MCDOWALL (1969) came to the same conclusion in comparing the New Caledonian fauna with that of New Zealand, where the greater isolation and, to some extent, the larger area may have served to slow the progress of the extinction of older elements by competition from younger invaders; New Zealand generally has a much older land bird fauna. MCDOWALL suggested there was a bias towards extinction of older elements in the equilibrium of immigration and extinction proposed for islands by MACARTHUR & WILSON (1967). The reason for this may be a general loss of fitness and competitiveness in small island populations relative to large island ones due to impoverished gene pools, reduced selection pressure and other factors (DIAMOND, 1970). MCDOWELL observed that the older elements of the New Zealand avifauna appeared to be increasingly more prone to extinction, illustrated by extinctions recorded in recent times since the advent of man. This pattern is consistent for both main islands and for Chatham and Stewart Is.

It follows that the turnover rate of this process must depend on dispersal powers from group to group and on the relationship to the source or sources (isolation, relative size) from island to island (DIAMOND, *in press*). The extent to which endemic radiation can occur will also be dependent on such factors and others such as population size. In general this turnover rate will be much slower for less dispersive groups and older elements will survive for much longer and radiate more extensively, particularly on isolated islands; these older elements will tend to reflect older dispersals and older geographical associations.

The effects of geographical isolation in this process are apparent in the greater conservation of older elements in New Caledonia relative to New Guinea, illustrated particularly in the floras of the two areas (Chapter II). The effects of differences in dispersal powers are illustrated by the situation on Lord Howe I.

The present day geography of Lord Howe I. leads to the prediction that there will be a predominance of Australian elements in the biota for groups with widely different dispersal powers, given that elements from the three possible source areas (Australia, New Zealand and New Caledonia) have an equal chance of establishment. On Norfolk I. the Australian influence would be expected to decline relative to that of New Zealand and New Caledonia, with decrease in dispersal powers as the greater proximity of the latter areas becomes a more important factor than that of the much greater area of Australia as a source. Such predictions for Norfolk I. were shown to hold for various elements in the flora and fauna by HOLLOWAY (1977). Norfolk I. is geologically very young and thus all colonists can be considered as recent. The predictions also appear to hold for the similarly young Kermadec group, where the New Zealand influence is increased relative to that of the other two areas, and a slight Fijian element replaces the New Caledonian (for example *Hydrillodes surata* amongst the moths).

On Lord Howe I., geologically very much older, Australian elements predominate in dispersive groups such as the muscid flies and Lepidoptera but in the less dispersive groups such as the flora and land birds New Caledonian and New Zealand elements are almost on a par with the Australian ones (HOLLOWAY, 1977). This is highly suggestive of an old pattern of affinity being progressively ousted by new colonists, occurring more rapidly in the more dispersive groups. The possible nature of this old pattern of affinity will be discussed further below.

The old austral elements in New Caledonia are most evident in the groups with generally poor powers of dispersal such as the Ephemeroptera, Cerambycidae and Gekkonidae, but much less so in the more dispersive ones such as the Lepidoptera and, to some extent, the birds.

These old connections are predominantly with Australia or with Australia and New Zealand together, particularly in the flora; exclusive connections with New Zealand are few except in some insect groups such as the Ephemeroptera, Trichoptera and Cerambycidae. Phytogeographic patterns, such as illustrated by several generic groupings in the Proteaceae, by *Dracophyllum* and, in a different way, by the *Nothofagus brassii* group, suggest that many, if not all exclusive links between New Caledonia and New Zealand could have arisen through the extinction of Australian representatives of the groups concerned during the increase of aridity and the consequent restriction of rainforest on the Australian continent throughout the later Tertiary. Such extinction would have been most likely in groups dependent on tropical rainforest conditions. It may be imminent in the gecko tribe Carphodactylini where the Australian relative of New Caledonian and New Zealand genera is found only in the Cape York area of Queensland. Australian counterparts of many apparently exclusive links may be similarly localised, rare, and awaiting discovery in the rainforests of Queensland.

These old austral elements in New Caledonia, New Zealand and Australia may owe their origins to dispersals prior to the opening of the Tasman Sea. At that time exchange between the three areas was not precluded for less dispersive organisms by the large water gaps of today. Subsequent to the opening it is possible that the geographical isolation of the three areas relative to each other has remained more or less constant. More recent exchanges would thus have been over extremely long distances or via stepping stone routes, a few of which are geologically possible.

Modern New Zealand taxa have spread north into the Pacific such as the parakeet genus *Cyanorhamphus* and the phasmid *Clitarchus*; the New Zealand geometrid genera *Pseudocoremia* and *Austrocidaria* have reached Norfolk I. Other instances were cited by FLEMING (1976). During the millions of years of the Tertiary other taxa may have crossed the gap between New Caledonia and New Zealand, the shared Trichoptera groups for example. Furthermore, the Indo-Malayan plant taxa that colonised New Zealand in the Tertiary must either have crossed the Tasman Sea from Australia or have been derived by dispersal down the Inner Melanesian Arc via New Caledonia or even via scattered atolls along the Outer Melanesian Arc. One must conclude that exchange was and is possible between New Caledonia and New Zealand even without stepping stones, though the establishment of a colonist must be an extremely rare event. It is therefore not necessary to invoke a closer geographical association of the two areas through the Tertiary, or at least the earlier part, to explain their present biogeographical connections.

The position of Lord Howe I. in this scheme of things is somewhat puzzling. The strong connection with New Caledonia and New Zealand amongst the older and probably less dispersive elements is contrary to what might be expected from present day geography. Explanation of this paradox, as it was termed by PARAMONOV (1958-1963), might provide a key to much of the biological history of the Inner Melanesian Arc relative to Australia in the latter half of the Tertiary, possibly earlier if land persisted from the first volcanic episode (−30 my) to the second. The presence of an endemic species of *Dietes* (Iridaceae), an otherwise South African genus, might indicate the latter though there is a school of thought that it may be an introduction (BALGOOY, 1971).

Amongst New Caledonian connections are the birds *Tricholimnas sylvestris*, *Janthoenas godmanae* and *Turdus xanthopus*. Amongst those with New Zealand are the *Leiolopisma* skink, the *Dracophyllum* species, the *Placostylus*, species of *Coprosma* of a complex shared with Norfolk I., the Kermadecs and Chatham Is. but centred on New Zealand, and the parakeet *Cyanorhamphus novaeseelandiae*. These last two are possibly fairly recent colonists, so access from New Zealand may have been potentially fairly constant over the latter half of the Tertiary. Associations with both New Caledonia and New Zealand occur in the palms and in the Gesneriaceae (*Negria*) and there are also more general associations with Inner Arc areas to the north in the phasmid fauna and again in the palms.

The native vegetation consists of subtropical rainforest in the lowlands and a sort of cloud forest at higher altitudes (700–800 m approximately). It is the latter that contains most of the older elements (PARAMONOV). This suggests that the origins of the paradox may be ecological. At some time in the past conditions on Lord Howe may have been such as to favour the establishment of colonists from New Caledonia and New Zealand rather than those from Australia. The contrast of Lord Howe rainforest with the extensive dry sclerophyll associations of the nearest sector of the Australian coast would suggest that something of the sort would be operative today. Nevertheless Australian elements predominate in the most dispersive groups, so perhaps the contrast is not so great that all niches cannot be filled by selective establishment of the predominantly Australian immigrants. Alternatively, for these dispersive groups, Lord Howe may be differentially more accessible to ecologically more compatible colonists from the tropical north of Australia than for groups with poor dispersal powers.

But in the past, during the most recent glaciations when Australia was more arid overall, during the Messinian Regression of the Miocene (−5 my) and its associated Antarctic glaciation, and perhaps even during the Oligocene cold period when the first volcanics may have presented a significant land area, the ecological contrast between Lord Howe and neighbouring areas of Australia may have been much greater. Such glacial periods would also have seen low sea levels and consequent exposure of stepping stones. These would have served to increase potential contact between Lord Howe and New Zealand and New Caledonia (Figures 4, 6) but not with Australia. This might have applied particularly in the Miocene when the Lord Howe volcanics and possibly those of atolls and seamounts further north on the Lord Howe Rise towards the Chesterfield Reefs were less extensively eroded and of much greater area. SUTHERLAND & RITCHIE (1974) suggested that the present island is only a fortieth of the area of the original volcanic shield. The situation could have been similar in the Oligocene. The period immediately following eruption, with the island area at a maximum, would also be the time most favourable for the establishment of colonists, arrivals at a later date meeting with more competition from the established biota, particularly as the area available was being progressively reduced by erosion. Lord Howe today has a flora richer in genera than would be expected from its area and isolation (BALGOOY, 1969).

It is possible that a more extensive chain of islands on the Lord Howe Rise have provided a route for interchange between New Caledonia and New Zealand, but considerations of the relative areas involved suggests that Lord Howe would have acted mainly as a sink for elements dispersing from both. Where groups are shared between all three areas and detailed taxonomic investigations have been made, the Lord Howe member of the complex usually appears to be close to taxa in one or other of its neighbours (e.g. the New Zealand affinities of the *Dracophyllum*, *Placostylus* and *Leiolopisma*). It is also possible that such a distribution pattern could have arisen through extinction of counterparts in Australia much as similar extinction may have produced apparently exclusive links between New Zealand and New Caledonia. The gesneriads with *Coronanthera* in Australia and the *Archontophoenix* alliance in the palms may provide instances where the process is incomplete. Lord Howe could have interacted with tropical rainforest elements in Australia much as New Caledonia may have done through stepping stones provided by the Barrier Reef and the Chesterfield Is.

Exchange between New Caledonia and Australia through the Chesterfield Is. may have occurred on several occasions in the past. It almost certainly did during the most recent glaciations; the increase in Australian influence in the more recent bird colonists relative to the older ones might be taken as evidence. The Coral Sea element in the Lepidoptera and other organisms may well have been augmented by the emergence of these stepping stones, especially as regards those species of the west coast association.

The presence of these stepping stones is not essential for exchange in the most dispersive groups, as indicated by the large number of vagrant Australian species recorded on Norfolk I. (35 out of 90 macroheterocera species) over a period of six years (HOLLOWAY, 1977). Two of these Norfolk vagrants, *Chloroclystis insigillata* and *Hypoperigea tonsa*, have close endemic relatives in New Caledonia. Exchange would no doubt have been augmented by stepping stones, especially for the less dispersive groups such as the geometrids where a number of the ultramafic element species are also found in Australia or have close relatives there. Elements of the flora of Australian affinity, the *Baeckea* and *Grevillea* species perhaps, may have reached New Caledonia by this route some time during the Tertiary.

Palynological investigations on the Chesterfield Reefs might be of value in elucidating their role in floral exchange between New Caledonia and Australia. Their biology today, described by COHIC (1959), offers no clue. There are atolls of calcareous sand with a flora of 20 species in as many genera and 16 families, Gramineae having five species. They are all strand and atoll species, including coconuts and shrubs such as *Sophora* (Leguminosae), *Scaevola* (Goodeniaceae; this genus contains a widespread Pacific species but also has a centre of radiation in New Caledonia) and *Colubrina* (Rhamnaceae). The insects collected by COHIC were indicative of a waif oceanic fauna. The only lepidopteron taken was the extremely widespread pyrale, *Hymenia recurvalis*, which was suggested by COHIC to have been feeding as a larva on *Achyranthes aspera* (Amaranthaceae).

Exchange between Australia and New Caledonia, probably mainly from the former to the latter, is likely to have been moderate, if sporadic, throughout the Tertiary. Individuals of five moth species taken during the survey may have been vagrants from Australia, borne across the Tasman Sea in frontal systems. The predominance of Australian elements in the New Caledonian Tipulidae may be due to passive transport of the weakly flighted but aerially buoyant adults in a similar manner in the past.

The Inner Melanesian Arc association between New Guinea and New Caledonia, perhaps in many instances involving in the past a more extensively forested and tropical Queensland, appears to be of Oligocene and Miocene origin. Most of the taxa showing it in New Caledonia are taxonomically distinct and, in the Lepidoptera at least, largely associated with the ultramafics, suggesting a period of colonisation after their emplacement and before they were eroded away to expose a significant extent of other, underlying rock types.

Biological evidence of the decreasing separation and eventual fusion of the two Melanesian Arcs can be found in many groups but further investigation of ambiguous cases and increasing taxonomic knowledge of poorly dispersive groups akin to the cicadas is needed. Bird families such as the Columbidae and Psittacidae might repay examination in this light. Data in PETERS' *Checklist of the Birds of the World*, Vol. III, suggest that *Ptilinopus* and *Gallicolumba* of the former and *Domicella* and *Charmosyna* of the latter might be typical Outer Arc genera, *Ptilinopus* and *Charmosyna* being somewhat richer in species in the northern part of New Guinea than in the south.

Identification of evidence of independent evolution in the two arcs will be difficult, both because of intermingling of elements in the two arcs following fusion and because of possible alternative explanation of the patterns in terms of differential expansion and radiation of New Guinea centred genera into outlying archipelagoes. For example, bird genera with a high proportion of mobile species of the tramp and supertramp categories of DIAMOND (1975) might be expected to have 'Outer Arc' distributions and not be so successful in large areas such as New Guinea, whereas genera with species of the high-*S* category predominant (poorly dispersive species characteristic of species-rich areas) might be expected to be more strongly centred in New Guinea. In such cases a lot will depend on the taxonomic relationships of the species concerned, i.e. whether there are indications of overlapping but relatively independent centres of development along the chain of the Outer Arc such as in the cicadas discussed, or whether there is evidence of waves of dispersal into the Outer Arc areas from New Guinea.

The most crucial evidence will come from instances where sister groups can be recognised and related one to each arc as in the cicadas, one or two plant genera, and the butterflies of the genus *Delias* and of the *Troides* and *Danis* (perhaps) complexes.

There is more or less a gradation from the Coral Sea type of distribution to the Melanesian type in the Lepidoptera today; this may apply more generally as well. Such continuity in the past would have been more broken, especially for the less dispersive groups, given the geography of Figure 3. Independent evolution and radiation of the more successful species into genera in each of the more fragmented distribution types that would be expected from such geography could well have led to the development of Inner and Outer Arc genera.

The concept of two Melanesian Arcs and their effect on the biology of the area requires more rigorous investigation. Other factors need to be considered, such as the probability of a gradation of ecological conditions from the drier rainforests of Queensland to the insular conditions of the Outer Arc, and, conversely, the relative success of Inner and Outer Arc genera in Australia subsequent to the fusion of the Arcs, assuming that the geological hypothesis is supported by further evidence and that genera attributable to one or other are in their early evolution can be reliably identified.

There is much stronger correlation between the geological reconstructions of Chapter I and the biological evidence as regards the relationships between New Caledonia, the New Hebrides and Fiji. New Caledonia and Fiji are floristically and faunistically very distinct, any parallels being found mostly amongst the more dispersive elements. The New Hebrides relate more strongly in their biota to Fiji than to New Caledonia, and exchange with the latter appears to be extremely recent, exemplified particularly by the Reptilia. Such a pattern is much more consistent with the geological hypothesis of Chapter I than with the alternative of the extrapolation of present day geography back into the past. The strong Fijian influence in the New Hebrides biota indicates that the archipelago has probably been interacting biologically with neighbouring areas over five or more million years, despite having only presented a significant area over the last one or two. The replacement of older colonists by more viable new ones in an island discussed at the beginning of this chapter may not occur so rapidly when that island or archipelago is rapidly increasing in size through geological uplift as in the New Hebrides. Thus plants and animals involved in earlier exchanges with Fiji when the group was of limited area may have been preserved despite the considerable influx of widespread Indo-Australian taxa in such groups as the Lepidoptera.

This widespread element in the New Caledonian fauna is the most modern in derivation and is only really predominant in the most dispersive groups, notably the Lepidoptera amongst those discussed here. A considerable number of such widespread species in New Caledonia have probably been derived by dispersal from Australia, whereas Fiji and the New Hebrides have probably received more of their widespread species through dispersals eastwards in Melanesia.

The role of the ultramafics in the evolution of the New Caledonian biota, particularly as regards conservation of the older elements, has already been discussed at length in previous chapters (pp. 68, 188, 224). The suggestion that they may have had a contrary, adverse effect on the survival of older elements in the phytophagous insect fauna through the adaptation of the older elements of the flora to the bizarre mineral regimes of the soils requires further investigation. The relationships of other phytophagous groups such as the piercing Hemiptera and Coleoptera to the vegetation of the base-rich soils needs investigation. It might be possible to investigate evidence of defoliation in various plant species and contrast its extent with the mineral content of the plant material concerned. This might prove particularly fruitful in such instances where a plant species grows on a variety of substrates such as *Alphitonia neocaledonica*. Several moth species feed on it (p. 189) and this fauna could be studied in detail both through sampling from *Alphitonia* in the field in various conditions and in laboratory experiments where larvae of the species concerned could be offered foodplant from a variety of soil types. The detritus feeding *Hydrillodes norfolki* might also prove a useful experimental animal in this respect as *Pteridium*, which it found palatable in the author's mattress (p. 532), also grows on a range of substrates.

Sampling of Lepidoptera prior to and during any seasonally determined flush of new growth in the vegetation of base-rich ultramafic soils might reveal whether there are any species missed during the survey that have evolved a precise univoltinism to capitalise on a period when the toxicity of such vegetation might be low. Estimates of the overall population level of Lepidoptera in extensive areas of the rich ultramafic flora could be gained by relating samples to the minimum nocturnal temperature and comparing the general relationship obtained (the relationship between the logarithm of the number of individuals caught and the temperature is linear as shown for Norfolk I. by HOLLOWAY, 1977) with that for samples from other vegetation types.

Another observation may be relevant to the question of the relationship of the Lepidoptera to the ultramafic vegetation. A number of endemic noctuids have pectinate antennae in the male. The genera *Colnettia* and *Pseudocoarica* are good examples but more significant are the species *Veia pectinata*, *Avitta pectinata* and the ctenuchid *Euchromia cincta*, these belonging to genera where such pectination is unusual or otherwise absent. All these species number amongst the more taxonomically isolated of the New Caledonian endemics, though the *Veia* and *Avitta* are less isolated than the rest. Such pectination in the male antennae is often an adaptation for the location of females by scent and is possibly more advantageous where populations are low and individuals of a species sparsely scattered. Could it be that the pectination in these New Caledonian species evolved in the past in response to general low levels of population in ultramafic habitats when these were much more extensive? Most of the species cited are relatively rare today, either associated with rainforest or the ultramafics. Only the *Euchromia* is common in light-trap samples but the majority (17 vs.8) were taken on the ultramafics, often where rainforest is present. Rainforest on the southern ultamafic massif today is often very limited in extent, scattered in small patches in valleys. If such an 'insular' situation occurred in the past, dispersal and exchange between forest patches might have been advantageous for species restricted to them, particularly for genetic reasons, and the need for an effective means of location for pairing may have arisen in consequence.

Apart from the interesting absence of Lepidoptera predominantly associated with the richest ultramafic associations, there was generally a good correlation between lepidopteran distribution and vegetation type. The indications are that greater precision would be gained from a bigger sample overall. Only then will it be possible to test rigorously whether there is good correlation between floristic and lepidopteran diversities. The relationships on the ultramafics suggest that such correlation will not be general, but possibly in their case the circumstances are exceptional.

The size of the overall sample obtained means that the survey failed to a large extent to achieve the original aim of testing the value of moth sampling as a means of environmental surveying and monitoring. Nevertheless the indications were not discouraging and several species such as the *Baeckea* feeders and those associated with *Acacia spirorbis* could be used as biotic indicators; no doubt more will present themselves as the fauna becomes better known.

Inevitably, this survey has only scratched the surface of many of the intriguing questions concerning the origins, development and ecology of the New Caledonian biota. It is to be hoped that the work necessary will be done before the unique associations of the ultramafics are irretrievably degraded or lost through mining operations. Indeed the results of such work could prove sufficiently important from an ecological or medical aspect to ensure their conservation.

# IX. A SYSTEMATIC ACCOUNT OF THE NEW CALEDONIAN MACROHETEROCERA

The butterflies of New Caledonia were described in detail by HOLLOWAY & PETERS (1976). This chapter deals with the remainder of the 'macrolepidoptera'. The systematic order is that currently in use in the British Museum (Natural History), largely based on that of COMMON (1970), and runs as follows: Zygaenoidea (Limacodidae); Geometroidea (Geometridae, Uraniidae); Sphingoidea (Sphingidae); Notodontoidea (Thaumetopoeidae, Notodontidae); Noctuoidea (Lymantriidae, Arctiidae, Ctenuchidae, Hypsidae, Nolidae, Noctuidae, Agaristidae). Only those families represented in New Caledonia have been listed in the above order. The subfamilies of the Noctuidae follow the arrangement of NYE (1975).

The following treatment is intended to be as full as possible without duplicating accounts of various species already published for the Fijian fauna by ROBINSON (1975) and for that of Norfolk I. by HOLLOWAY (1977). This applies particularly to synonymy and original references for each species, a cross-reference to these works being given. The geographical distribution of each species is given in full however, gained from the sources mentioned on p. 195.

Unless otherwise stated material listed is deposited in the British Museum (Natural History) (BMNH), though a representative collection will be deposited in the Museum National d'Histoire Naturelle in Paris (MNHN). Other material was examined on loan from Paris (MNHN), the Bernice P. Bishop Museum in Honolulu (BPBM), and the collection of the California Academy of Sciences (CAS). Dr. H. INOUE kindly arranged for me to examine several specimens taken by Mr. H. SUZUKI in the Loyalty Is.

Measurements are from the centre of the thorax to the apex of the forewing. These are not indicated on the plates for new species and subspecies as they are given in the text.

## LIMACODIDAE

### **Miresa obscurimargo sp. n.** (Plate 65: 11, 12, Figure 69: 1)

♂♂ 9–16 mm, ♀♀ 14-20 mm. The male antennae are bipectinate, the pectinations becoming gradually shorter towards the apex. In *M. incredibilis* Hering (New Guinea; type examined), to which the New Caledonian species bears a general resemblance, the pectinations cease abruptly one third from the base of the antenna. The thorax and forewings vary from pale ochreous, through ochreous speckled with reddish brown, to reddish brown; the hindwings and abdomen are more uniform, less speckled, the same colour as the forewings but paler. There is a longitudinal dark streak down the centre of the thorax above. The fringes along the centre of the forewing dorsum and on the margins of both wings are dark brown. On the forewing there is a dark discal spot and a slightly sinuous, oblique dark postmedial as illustrated; these markings tend to be obscured in the darker specimens. *M. incredibilis* is similar in appearance to the dark reddish brown forms but has rufous patagia and rufous lateral patches at the posterior of the thorax; the abdomen is basally much broader than in *obscurimargo*, darker than the hindwing and with a rufous apex; the forewing lacks a postmedial and the discal spot is more prominent, transversely elongate; the fringes are not darker. The male genitalia of *obscurimargo* are similar to those of typical *Miresa* (*albipuncta* (Herrich-Schäffer)). The valves are set more ventrally relative to the saccus than in *incredibilis* and the fused part of the gnathi is relatively broader and of even width instead of broader subapically; the aedeagus of *incredibilis* is straight instead of basally downcurved and differs in having the apex trumpet-shaped and the vesica scobinate.

The females of *obscurimargo* are larger, always medium reddish brown, more uniform than in the male, lacking the speckling of the forewing. The forewing discal spot and postmedial are as in the male but two large specimens from the R. Bleue have no discal spots, a few dark ones occurring instead at the posterior margin of the cell.

HOLOTYPE ♂. New Caledonia: Pic du Pin, 300 m. Site 69. 06880/75386 (J. D. HOLLOWAY) 7.8.1971, BM 1971–507. BM limacodid slide 123.

PARATYPES: 33♂♂, 10♀♀ New Caledonia: various localities. BM limacodid slides 124 (♂), 125, 126 (♀♀).

The species was recorded in New Caledonia as follows: Grand Lac site 2 (1); Pic du Pin site 7 (5), site 69 (2); R. Bleue site 8 (4), site 9 (16), site 10 (4); Mt. des Sources site 11 (1); Kopéto (1); Mt. Mou site 33 (2); Mt. Krapé site 55 (1); Mt. Panié site 61 (6); Mt. Aoupinié (4); Mt. Koghis site 79 (1). These are almost all rainforest sites, the majority on the ultramafics. The species appears commonest where the trap was sited below the canopy, surrounded by forest. It may therefore restrict its flight to below the canopy.

## GEOMETRIDAE: OENOCHROMINAE

### *Adeixis inostentata* (Walker) (Plate 52:7, Figure 40: 4, 5, 41:3)

*Panagra inostentata* Walker, 1861, *List Specimens Lepid. Insects Colln Br. Mus.*, 23: 1012.

Found throughout Australia, including Tasmania and excepting arid areas, *inostentata* was taken at only two localities within 100 m of each other near Grand Lac: site 4 (17), site 71 (6). It was abundant in this area, mainly vegetated by the herbaceous marsh association. In New Zealand occurs the closely related *griseata* Hudson which is characteristic of 'pakihi' country, poorly drained acid areas with sedges, some rushes, *Gleichenia* and scattered low shrubs (dwarf podocarps, epacrids, *Pittosporum*, *Coprosma*), occurring in

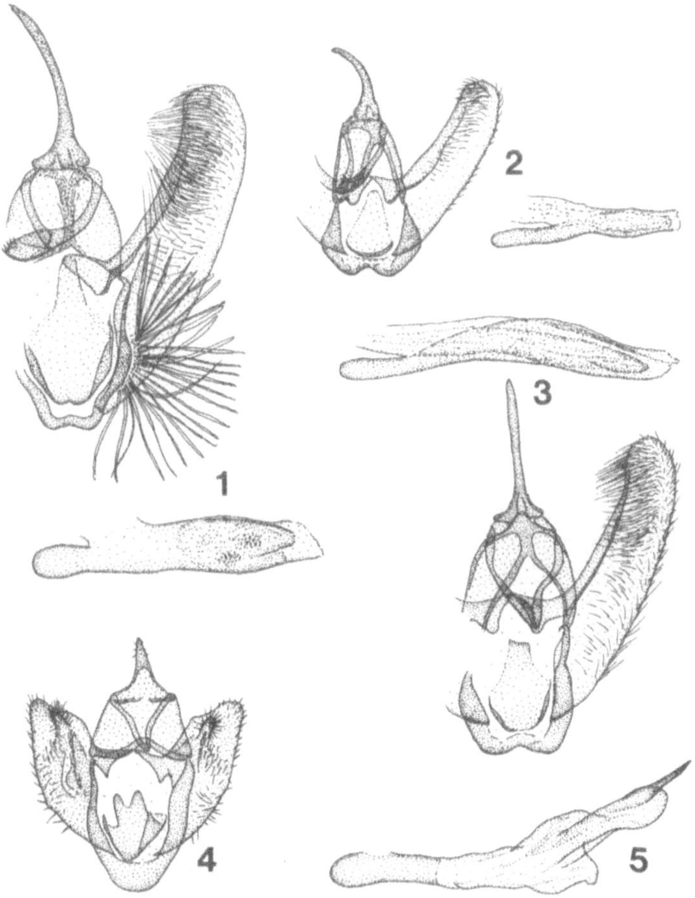


*Fig. 40.* The male genitalia of *Adeixis* species: *montana* (1), *baeckeae* (2), *major* (3) and *inostentata* (4, and aedeagus, 5).

such localities from sea level to 1000 m from Invercargill to N. Auckland (J. S. DUGDALE, *in litt.*).

**Adeixis baeckeae sp. n.** (Plate 52: 5, 6, Figure 40: 2, 41: 4)

♂♀ 9–11 mm. The whole insect is medium grey, the forewing with an oblique. darker postmedial, diffuse basad and serrate distad, the teeth running along the veins; there is a black discal spot. Females have the postmedial better developed and it is often almost absent in male specimens. In several specimens there are two black flecks just basad from the postmedial, opposite the cell. In a few specimens there is a black streak as illustrated in Plate 52: 5. The hindwings are slightly paler than the forewings and the discal dot is faint.

This and the next two species form an endemic, interrelated group of the genus and have probably resulted from a previous invasion by *inostentata*-like stock. The only other possible genus to which they might be referred is the Australian *Dichromodes* Guenée, also present in New Zealand where the larvae may be lichen feeders. The New Caledonian species have similar unipectinate antennae whereas *inostentata* has bipectinate ones but no *Dichromodes* examined has the uniform grey forewing with the serrate, diffuse postmedial that characterises the New Caledonian *Adeixis*. The structure of the male and female genitalia is also consistent with that of *inostentata*, though the three interrelated species lack the setose lobe to the valve and the aedeagus cornutus of *inostentata*. All four species have the apex of the fused gnathi spined.

In *baeckeae* the postmedial is more distal than in *inostentata* and less oblique. *Inostentata* does not have forms with a black streak on the forewing.

HOLOTYPE ♂+pupa. New Caledonia: Grand Lac, 250 m 06955/75355 (J. D. HOLLOWAY) May 1971, BM 1971-507. Reared from larva on *Baeckea ericoides*. Emerged 2.7.1971. BM geometrid slide 8502.

PARATYPES: 5♂♂, 2♀♀ as holotype; 15♂♂, 17♀♀, New Caledonia, various localities, BM geometrid slides 8551 (♂), 8420 (♀). Larval and pupal material was also preserved.

The species was recorded as follows: Grand Lac site 2 (14), site 3 (1), site 4 (2); Plaine des Lacs site 5 (2); Pic du Pin site 7 (2); Mt. des Sources site 11 (2), site 12 (2); Dothio site 18 (1); Mt. Mou site 33 (2), site 36 (3); Col Mouirange (1); Mt. Pouédihi (4); Les Dalmates site 77 (1).

Several larvae were beaten from *Baeckea ericoides* in the Grand Lac area. In the penultimate instar they are about 1 cm long, orange-green, darker dorsally. In the final instar they are 1.5 cm long, apple-green with a narrow, paler stripe laterally. When feeding, usually at night, they sit looped at the ends of the *Baeckea* twigs. A single *Apanteles* was obtained from one larva and retained. To pupate they spin up loose cocoons in the terminal leaves of the *Baeckea*, and emerge about 25 days after pupation.

These larvae would also thrive on *Baeckea virgata* and *B. pinifolia* but the adults were never recorded away from the southern ultramafic massif where *B. ericoides* predominates.

**Adeixis major sp. n.** (Plate 52: 1–3, Figures 40: 3, 41: 1)

♂♀ 13–15 mm. This is very similar to the previous species but distinctively larger. The forewing postmedial is more concave basad, less definitely serrate distad; in the female it is often tinged rufous, with an ochreous tinge extending basad posterior to the cell. Where a full black streak occurs on the forewing (in perhaps 5–10% of specimens as in *baeckeae*) it is more sharply angled towards the dorsum basally and, instead of tapering away anteriorly along the postmedial, it is broken up into separate spots; compared with *baeckeae* it is relatively narrower, especially in the male. The hindwing discal dot is generally slightly more prominent. The male genitalia are half as large again as in *baeckeae*, the

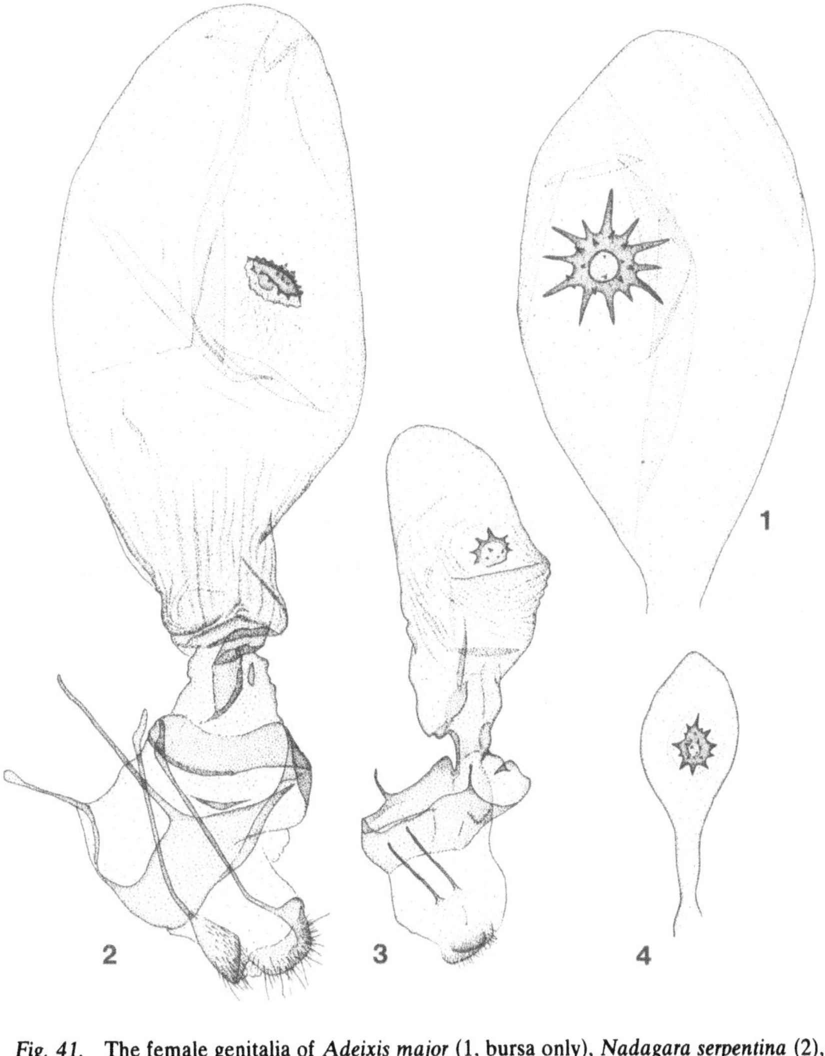


*Fig. 41.* The female genitalia of *Adeixis major* (1, bursa only), *Nadagara serpentina* (2), *Adeixis inostentata* (3) and *A. baeckeae* (4, bursa only).

valves slightly more elongate and the saccus more parallel-sided. The female genitalia are twice as large as in *baeckeae*, the signum much smaller relative to the bursa.

HOLOTYPE ♂. New Caledonia: Grand Lac, 250 m Site 2 06975/75356 (J. D. HOLLOWAY) 16.5.1971, BM 1971-507. BM geometrid slide 8503.

PARATYPES. 47 ♂♂ (slides 8552–5), 34 ♀♀ (slide 8421) New Caledonia, various localities.

The species was recorded as follows: Grand Lac site 1 (4), site 2 (22), site 3

(4), site 4 (3), site 72 (1); Pic du Pin site 7 (21); Mt. des Sources site 11 (3), site 12 (2); Dothio site 18 (1); Col Toma (8); Mt. Mou site 36 (1); Tiebaghi site 41 (1); Mt. Krapé site 56 (1); near Bâ (4); Col Mouirange (4); Yaté Gorge (4); Anna-Madeleine (1); Les Dalmates site 78 (1). It is more widespread in ultramafic localities than *baeckeae* and was taken in large numbers on Col Toma where elements of *Baeckea*-sedge maquis occur on sedimentary rock.

**Adeixis montana sp. n.** (Plate 52: 4, Figure 40: 1)

♂ 13 mm. A species similar in colour to *major* but with the forewing postmedial very diffuse, much more oblique, straight, extending from the apex to half way along the dorsum; there is also a faint, paler submarginal fascia parallel to the margin. In the male genitalia the saccus is more narrowly bilobed apically, more even, the juxta is broader and the valves bear a basal tuft of long, blade-like setae.

HOLOTYPE ♂. New Caledonia: Mt. des Sources, 760 m Site 11a 06650/75512 (J. D. HOLLOWAY) 29.5.1971, BM 1971–507. BM geometrid slide 8550.

## CALEDOPHIA Gen. n.

This genus is a Papuan Region relative of the African and Oriental genus *Conolophia* Warren, type species *conscitaria* Walker (tropical Africa). Other genera related are *Encryphia* Turner (type species *frontisignata* Walker, *q.v.*), *Physetostege* Warren (type species *miranda* Warren, New Guinea, Queensland) and *Callipotnia* Warren (type species *multicolor* Warren, New Guinea, and containing another New Guinea species). They are characterised (*Encryphia, Physetostege*) in TURNER'S key to Australian oenochromines (1929, *Proc. Linn. Soc. N.S. Wales*, 54: 470) by forewings with $R_1$ anastomosing with Sc and $R_2$; forewings with an areole; hindwings with Sc not anastomosing with the cell, being separate from it basally; the male antennae bipectinate. The genus *Noreia*, primarily Indo-Australian tropical, is separated in the key by the filiform antennae of the male but the tegumen of the male genitalia shows the same peculiar structure as in the other genera (male genitalia of species of *Conolophia* and *Noreia* are illustrated by HOLLOWAY (1976)). *Caledophia* keys out to *Encryphia*.

In addition to these external characters all these genera (*Noreia* not examined) have, in the male, lateral patches of sparse deciduous setae on the third sternite of the abdomen; the tegumen extends well below the top of the vinculum, thence turning up dorsally to join it on each side; a membrane extends between the valve costa and the tegumen, incorporating the transtillae.

*Caledophia*, *Encryphia* and *Conolophia* have a pair of small lateral lobes at the distal end of the third abdominal sternite, though in *Conolophia* there is a complex array of small, hirsute patches. *Caledophia* has basal tufts of long black scales on the male hind tibia and, in common with *Encryphia*, has a fan of black scales on each side of the sixth abdominal segment. *Encryphia* is distinguished by two brushes of long hairs on the fourth sternite and *Caledophia* by a prominent pair of coremata dorsally between the seventh and

*Plate 52.* 1 (♂), 2, 3 (♀♀). *Adeixis major*; 4. *A. montana* ♂; 5, 6 (♀♀). *A. baeckeae*; 7. *A. inostentata* (8 mm); 8. *Caledophia pauli* ♂; 9. (♂), 10 (♀). *Oenochroma fasciata*; 11 (♂), 12 (♀). *Encryphia frontisignata acornuta*.

eighth segments. Both these genera have a marked indentation subanteriorly, ventrally, in the fourth sternite. *Physetostege* is distinguished by two moustache-like pairs of hair-pencils between the sixth and seventh, and seventh and eighth abdominal sternites. On the wings *Conolophia* and *Caledophia* have approximately straight postmedials, *Physetostege* and *Callipotnia* have angled ones, whereas in *Encryphia* they are angled in the male and straight in the female. *Physetostege* has a subbasal fovea to the male forewing.

In the male genitalia the valves have a complexity of structure; those of *Caledophia* and *Encryphia* are illustrated in Figure 42. *Caledophia* and *Conolophia* have similar valves, uncus and juxta, but the aedeagus of *Caledophia* is distinguished by a prominent cornutus on a bulbous base. *Callipotnia* has two projections from the centre of the valve, one setose, and a projection from the base of the valve costa; there is a distinctive tuft of fine, hairlike setae from the base of the uncus; the gnathi are prominent, fused, serrate above; there is a very prominent, long, bifid anterior projection from the base of the juxta; the aedeagus has two prominent cornuti. In *Physetostege* there is an unequally bifid, ventrally directed projection from the valve costa as well as a dorsal flange; the juxta is strongly 'v' shaped and the aedeagus, though very short, has two large cornuti.

*Caledophia* has characteristics in common with all these genera but is perhaps closest to *Encryphia* (abdominal structures) and *Conolophia* (genitalia, straight postmedials). It fits readily into none of them and is therefore separated here. It is diagnosed by the dorsal coremata to the abdomen and the bulbous cornutus on the vesica. The type species is *pauli*, described below.

**Caledophia pauli sp. n.** (Plate 52: 8, Figure 42: 3)

♂♂ 21 mm. Wings with black speckles and markings on a pinkish grey ground as illustrated. Other characteristics have already been described in the generic treatment above.

The species is named for M PAUL COCHEREAU who showed the author much kindness during his stay in New Caledonia, but who was reluctant to let his family name yield a tonguetwisting epithet for a new species.

HOLOTYPE ♂. New Caledonia: Mt. Krapé, 460 m Site 56 05274/76373 (J. D. HOLLOWAY) 22.7.1971, BM 1971–507.

PARATYPE ♂. New Caledonia: Plaine des Lacs, 240 m Site 5 07020/75362 (J. D. HOLLOWAY) 21.5.1971, BM 1971–507. BM geometrid slide 8209.

**Derambila pellucida sp. n.** (Plate 53: 1)

♀ 10 mm. The insect is completely white except for the palps, fore- and mid-legs (excepting the tarsi) which are dorsally dark brown. The wings are semi-transparent, the forewings with a submarginal row of blackish brown dots on the veins, diffusing slightly towards each other across the spaces; two similar spots occur antemedially, one on the cell at the origin of Culb and the other posteriorly on vein 2A. The species is possibly closest to *catharina* Prout

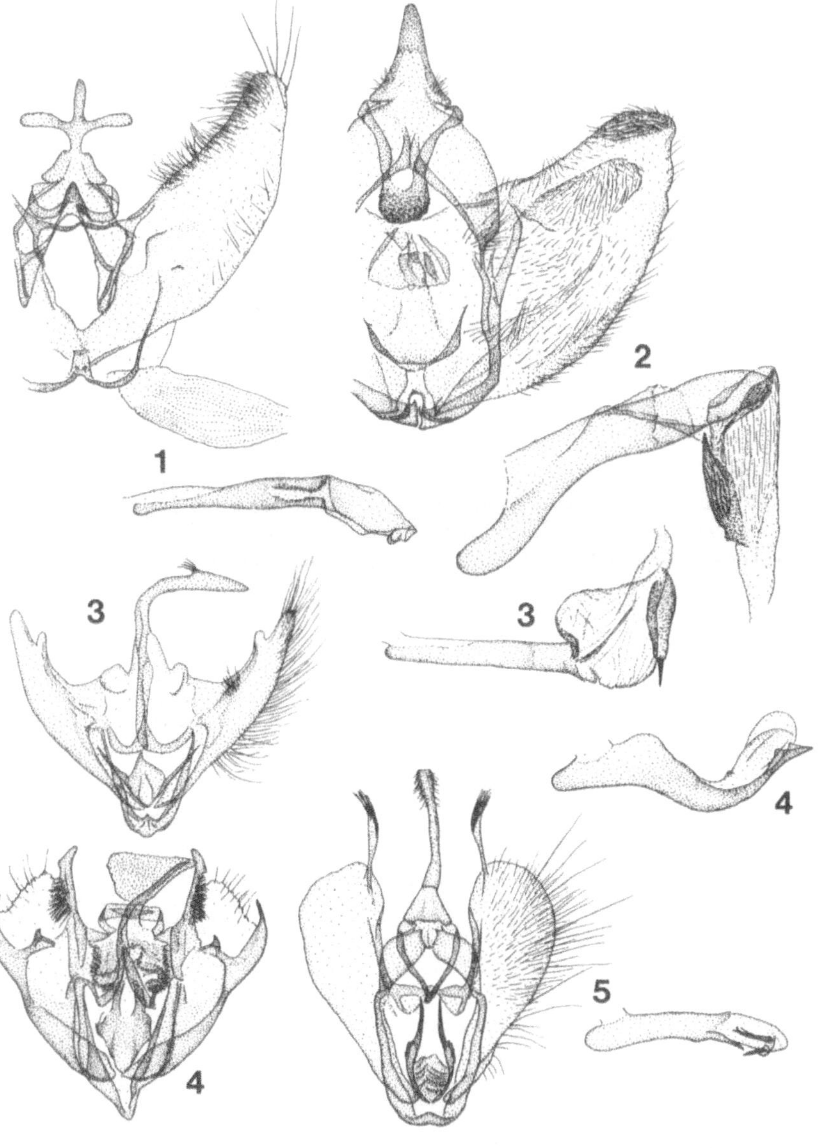


*Fig. 42.* The male genitalia of *Eumelea rosalia degener* (1), *Oenochroma unifasciata* (2), *Caledophia pauli* (3), *Encryphia frontisignata acornuta* (4), and *Nadagara serpentina* (5).

(Queensland, type examined; Lesser Sundas) but *catharina* is smaller (8 mm), a more opaque white, with discal spots on both wings, a more complete forewing antemedial, more sinuous submarginals and a marginal row of darker dots on the veins; discal, submarginal and marginal dots occur also on the hindwing, which is immaculate in *pellucida*.

HOLOTYPE ♀. New Caledonia: R. Bleue, 180 m Site 10 06704/75544 (J. D. HOLLOWAY) 27.5.1971, BM 1971–507. BM geometrid slide 8423.

**Encryphia frontisignata acornuta ssp. n.** (Plate 52: 11, 12, Figure 42: 4)

*Tacparia*? *frontisignata* Walker, 1863, *List Specimens Lepid. Insects Colln Br. Mus.*, 26: 1523, type examined.

♂♀ 15–16 mm. Ground colour of both sexes pinkish grey, marked and speckled black. The fasciae of the wings are pale yellow as illustrated, with a browner band adjacent and diffusing away towards the medial area. The subspecies is externally indistinguishable from the nominotypical race from Queensland. The male genitalia differ, the ventro-apical process to the valve being more curved and the dorso-apical process being much narrower and more thickly setose. The aedeagus lacks the cornutus in the vesica of the typical race, is more sinuous and is sagittate at the apex.

HOLOTYPE ♂. New Caledonia: Grand Lac, 250 m Site 4 06980/75367 (J. D. HOLLOWAY) 19.5.1971, BM 1971–507. BM geometrid slide 8210.

PARATYPES. 3♀♀ New Caledonia: Grand Lac site 2 (1); Ouen Toro site 82 (2).

*Eumelea rosalia degener* Warren **stat. n.** (Plate 53: 3, 4, Figure 42: 1)

*Geometra rosalia* Stoll, in Cramer, 1781, *Uitlandsche Kapellen*, 4: 152.

*Eumelea degener* Warren, 1894, *Novit. zool.*, 1: 375, type examined (Lifu).

The similarity of *degener* in facies and male genitalia to *E. rosalia australiensis* Warren (Queensland) indicates that it should be placed as a subspecies of *rosalia* (Stoll). The New Caledonian race is smaller than *australiensis*, the post- and antemedials of the forewing being more distinct in the male, the reddish striations less numerous in the female giving a paler yellow insect; the submarginal dark spots in both sexes are somewhat more prominent. Typical *rosalia* (Burma, Philippines, Palawan, Sula Is., Moluccas) has similar male genitalia but the male has a marked deep pinkish sheen in most examples and the submarginals are straighter, less broken. Subspecies *aurigenaria* Warren (Java, Lesser Sundas) is intermediate in appearance between *rosalia* and *australiensis*; typical *rosalia* has the females heavily striated red. Subspecies *attenuata* Warren occurs in New Guinea, the males resembling the typical race but the females being yellower. In the Solomons subspecies, *ditona* Prout, the pinkish sheen in the male is restricted to the margins of the wings, the rest being red-striated on yellow. Specimens from the Bismarcks are intermediate between *ditona* and *attenuata*.

Subspecies *degener* occurs in the Loyalty Is. and has been recorded in New Caledonia as follows: Mt. des Sources site 14 (1); Ouambéo R. (1); Col d'Amieu (1); Mt. Aoupinié (1). ORSTOM: Forêt de Thi. Nouméa (VIETTE, 1951). BPBM: Col des Roussettes; St. Louis valley.

**Oenochroma unifasciata sp. n.** (Plate 52: 9, 10, Figure 42: 2)

♂♀ 23–25 mm. The ground colour of the male is pale grey, tinged pink in a few specimens, and lightly flecked with black scales; the ground colour of the female is vinous pink. In both sexes the thorax, postmedial fascia and fringes are darker, the postmedial edged paler basad, pale grey in the male and orangeish in the female. The species is perhaps closest in general appearance to *decolorata* Warren (Queensland, type examined) but the forewing is less produced apically and less falcate. *O. decolorata* has additional diffuse, dark medial fasciae on both wings and a pale discal spot on the forewing.

The male genitalia have the valves simple but in *decolorata* the apex is cleft and there is a prominent, scobinate, convoluted harpe; the saccus is narrower in *unifasciata* and the fused gnathi are apically broader. The aedeagus vesica has a single cornutus; in *decolorata* this is basally scobinate, there is an additional sclerotised plate in the vesica and the aedeagus apex is spined.

HOLOTYPE ♂. New Caledonia: Pic du Pin, 240 m Site 7L 06879/75383 (J. D. HOLLOWAY) 7.8.1971, BM 1971–507. BM geometrid slide 8207.

PARATYPES: 30♂♂, 6♀♀ New Caledonia, various localities; 1♂ New Caledonia, La Coulée 28.10.1976 (A. DELOBEL) MNHN.

The species was recorded as follows during the survey: Grand Lac site 2 (34), site 3 (7), site 4 (3), site 72 (3); Plaine des Lacs site 5 (1); Pic du Pin site 7 (27); R. Bleue site 10 (3); Mt. des Sources site 11 (1); Mt. Mou site 33 (9); Mt. Krapé site 55 (2); Houailou (2); Col Mouirange (2); Mt. Pouédihi (1); Yaté Gorge (1); Anna-Madeleine (4); Les Dalmates site 77 (3), site 78 (6). The species is a member of the ultramafic association and is the only representative of its genus found outside Australia. In Australia the larvae of *Oenochroma* species have been recorded feeding on Proteaceae (COMMON, 1970).

*Ozola plana* (Warren) (Plate 53: 2, Figure 43)

*Desmobathra plana* Warren, *Novit. zool.*, 1: 380, type examined (Lifu).

This species, with the two discussed below from the New Hebrides, belongs to a rather isolated section of *Ozola* Walker (type species *microniaria* Walker). The wings are entirely orange-yellow instead of fawn or dark grey on white. The valves of the male genitalia have a characteristic diagonal crease not seen in the type species of the genus or those illustrated by HOLLOWAY (1976). The female bursa has one spined band as compared with the two of *microniaria*.

The two named forms from the New Hebrides completing this section of the genus are *acrophaea* Meyrick (type examined) and *hesperias* Meyrick (type examined). The former differs from the latter in the greyish apices to both wings, but otherwise they are very similar and may prove to be merely variations of one species. They differ from the New Caledonian *plana* in having a fine reddish speckling to the wings; the forewing fasciae of *plana* are obscure or absent in these species. A male *acrophaea* dissected had a shorter aedeagus than *plana* with diagnostic prominent scobination in the vesica. The valve in *plana* has diagnostic triangular processes at the apex of the costa and at the distal end of the diagonal crease.

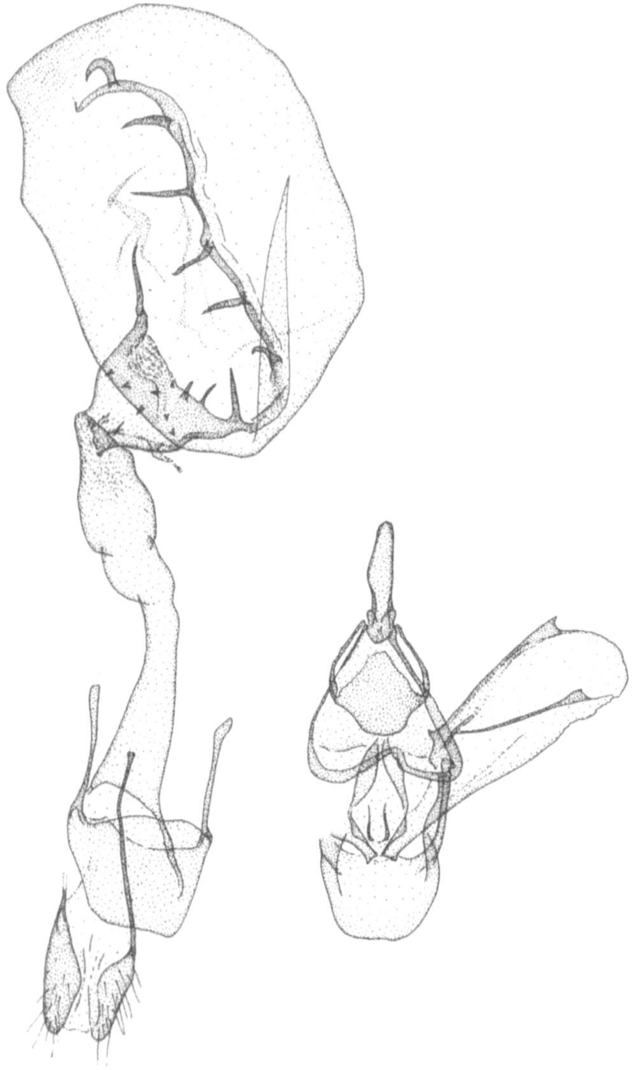

*Fig. 43.* Female and male genitalia of *Ozola plana.*

There is a long series from Lifu in the BMNH and one specimen from New Caledonia (no locality). VIETTE (1949: 327) also recorded *plana* from New Caledonia and a single female was taken at the Grottes de Koum during the survey.

**Aeolochroma hypochromaria caledonica ssp. n.** (Plate 53: 12, Figure 44: 2, 3)

*Cleora? hypochromaria* Guenée, 1857, *Hist. nat. Insectes* (*Lépid*), 9: 234, type examined.

♂♀ 16–19 mm. Ground colour straw, stippled bright green and marked and fasciated darker, the fasciae made up of brownish fawn and black scales. This dark fasciation and blotching is heavier than in typical *hypochromaria* (Queensland, New South Wales). In the male genitalia the uncus is shorter and slightly thicker, the subapical angle of the valve is less obtuse and the portion apicad from the angle is longer. The cornuti of the aedeagus are longer, more curved, and are not set on the broadly squarish sclerotised base that characterises the typical race.

HOLOTYPE ♂. New Caledonia: Mt. Koghis, 600 m site 80 06556/75456 (J. D. HOLLOWAY) 21.8.1971, BM 1971-507. BM geometrid slide 8202.

PARATYPES: 31♂♂, 5♀♀ New Caledonia, various localities; 1♂ Nouméa, Nov 20, 1940 (FXW) at light, BPBM.

The species was recorded during the survey as follows: Grand Lac site 2 (4), site 3 (5), site 4 (3); Pic du Pin site 7 (13), site 69 (1); R. Bleue site 9 (2), site 10 (1); Mt. des Sources site 11 (2), site 12 (1); Dothio site 18 (1); Bréhoa R. (1); Col d'Amieu (2); Col Toma (7); Mt. Mou site 33 (1), site 34 (1); Col des Roussettes (1); Mt. Aoupinié (2); Mt. Pouédihi (1); Mt. Koghis site 79 (3), site 80 (3). It is a member of the ultramafic association.

*Agathia asterias dimota* Prout (Plate 53: 5, 6, Figure 44: 1)

*Agathia asterias dimota* Prout; ROBINSON (1975: 251).

This is the Fijian subspecies of a species that has been recorded from the Kei Is., New Guinea, Australia (N. Territory, Queensland, New South Wales), Norfolk I., the Solomons, New Hebrides and Fiji. ROBINSON discussed the subspecies in detail and HOLLOWAY (1977: 118) described a new subspecies, *jowettorum*, from Norfolk I.

The species was recorded as follows in New Caledonia: R. Bleue site 10 (1); Dothio site 18 (1); Poupa (1); Col Toma (2); Mt. Mou site 33 (1); R. Amoa (1); Probokonoro (1); Mt. Krapé site 56 (1); Houailou (2). These are all rainforest or ultramafic localities.

*Anisozyga caledonica* (Thierry-Mieg) (Plate 53: 9-11, Figure 48: 1)

*Felicia caledonica* Thierry-Mieg, 1915, *Miscnea ent.*, 22: 40.

*Anisozyga* has its centre of speciation in New Guinea. The New Caledonian species is endemic but probably most closely allied to New Guinea species. The males are relatively constant in facies but the females vary from those with extensive white areas to those with the same areas suffused dark grey as illustrated.

The species was recorded as follows: Grand Lac site 1 (1), site 2 (14), site 3

*Plate 53.* 1. ***Derambila pellucida***; 2. ***Ozola plana*** (11 mm); 3 (♂, 18 mm), 4 (♀, 22 mm). ***Eumelea rosalia degener***; 5 (♂), 6 (♀). ***Agathia asterias dimota*** (19 mm); 7. ***Chloeres boisensis***; 8. ***Comostola thalassias***; 9 (♂), 10, 11 (♀♀). ***Anisozyga caledonica*** (14–16 mm); 12. ***Aeolochroma hypochromaria caledonica***.

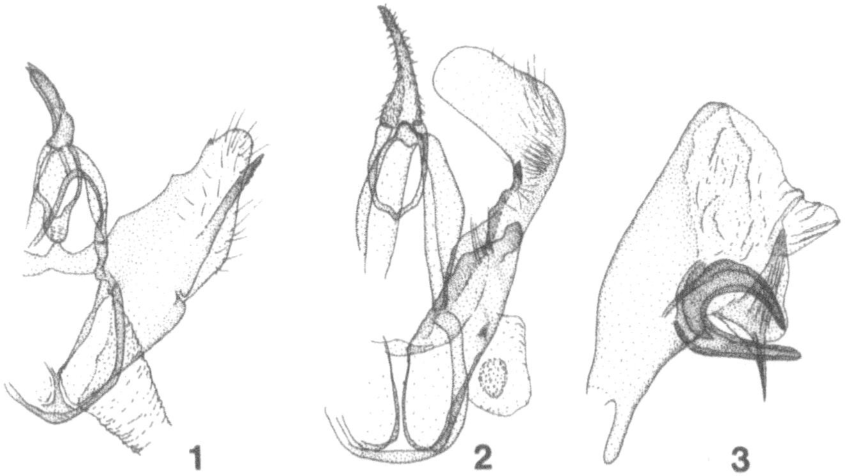


*Fig. 44.* Male genitalia of *Agathia asterias dimota* (1) and *Aeolochroma hypochromaria caledonica* (2, and aedeagus, 3).

(10), site 4 (1); Pic du Pin site 7 (17); R. Bleue site 8 (2), site 9 (4), site 10 (6); Mt. des Sources site 12 (1); R. Kuenthio (1); Ouambéo R. (2); Col Toma (5); Petit Gouli (1); Tiebaghi site 41 (1); Bambou R. (2); Col d'Amoss site 45 (1); Col des Roussettes (1); Mt. Aoupinié (5); Houailou (1); R. Koué (1); Anna-Madeleine (1); Les Dalmates site 77 (1); Mt. Koghis site 79 (1). MNHN: La Coulée. It is a member of the ultramafic association.

**Chloeres boisensis sp. n.** (Plate 53: 7, Figure 45: 5)

♂ 10 mm. Wings apple green with a paler costa to the forewing; the abdomen is also green with a paler dorsal stripe. The wing fringes are lemon yellow and the antennal shafts white. In appearance the species resembles *C. citrolimbaria* Guenée (type examined; Queensland, New South Wales) but is smaller (cf. 11–13 mm) and differs in the male genitalia. The valve has a diagnostic basal corema, is asymmetric apically (more curved ventrally than in the more symmetric valve of *citrolimbaria*), and has a less setose and shorter harpe which does not connect by a narrow flap from its apex to a subbasal ampulla near the ventral margin of the sacculus, a characteristic of *citrolimbaria*.
HOLOTYPE ♂. New Caledonia: Port Boisé, 20 m Site 74 07039/75276 (J. D. HOLLOWAY) 12.8.1971, BM 1971-507. BM geometrid slide 8530.

**Chlorocoma octoplagiata sp. n.** (Plate 54: 11, Figure 47: 3)

♂♂ 15–16 mm, ♀ 18 mm. Ground colour pale green, the wings fasciated paler green; fringes, base of antennae and bar between latter white; forewing costa

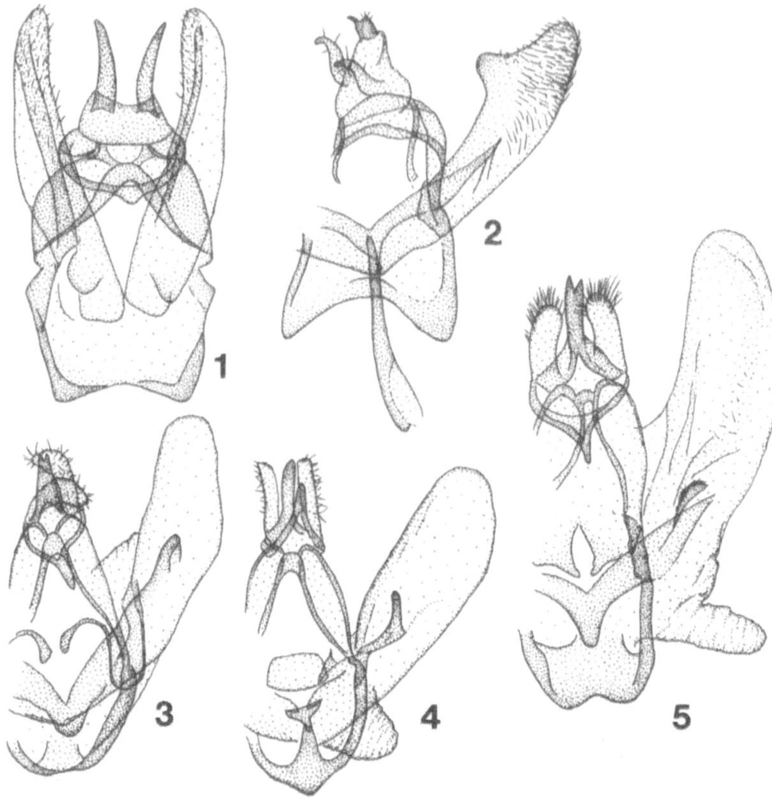


*Fig. 45.* Male genitalia of *Uliocnemis biplagiata reducta* (1), *Comibaena mariae* (2), *Comostola thalassias* (3), *Hemithea acaudata* (4), and *Chloeres boisensis* (5).

and, narrowly, wing margins red; spots in discal area of wings pinkish red, edged rufous; legs and rest of antennae pale reddish brown. Compared with other members of the genus (almost entirely from temperate regions of Australia) the wings are less bluish, more apple green. The fasciae are more diffuse, broader. The discal red spots on both wings are joined by similar spots in the space immediately posterior (two of the males lack this on the hindwing). The male genitalia are similar to those of *melocrossa* Meyrick (type examined; New South Wales, South Australia) but the projection from the centre of the ventral margin of the valve is very much more developed and the juxta has a pair of diagnostic blade-like processes.

Montane species of *Prasinocyma* from New Guinea, such as *votiva* Prout and *geminipuncta* Warren, are externally very similar but the male genitalia of these are different.

HOLOTYPE ♂. New Caledonia: Mt. Mou, 1100 m Site 35 06385/75594 (J. D. HOLLOWAY) 3.7.1971, BM 1971–507. BM geometrid slide 8531.
PARATYPES. 1♀ as holotype: 2♂♂ Mt. Panié site 60

This would appear to be a strictly montane species, perhaps associated with cloud forest dominated by *Podocarpus*.

*Comibaena mariae* (Lucas) (Plate 54: 1, Figure 45: 2)

*Iodis mariae* Lucus, 1888, *Proc. Linn. Soc. N. S. Wales* (2), 3: 1266.

This species, described from Queensland, occurs also in New South Wales and New Guinea. It was recorded as follows in New Caledonia: Mt. des Sources site 14 (2); Poupa (3); Ciu (1); Koh (3); Bréhoa R. (2); Petit Gouli (3); Kopeto (1); Ste. Therèse (1). BMNH: Pouébo (CHEESMAN). It is relatively widespread, mainly off the ultramafics, though showing some affinity with *Acacia spirorbis* localities.

**Comostola leucomerata caledonica ssp. n.** (Plate 54: 2, Figure 46: 1)

*Chlorocoma leucomerata* Walker, 1866, *List Specimens lepid. Insects Colln Br. Mus.*, 35: 1609, type examined.

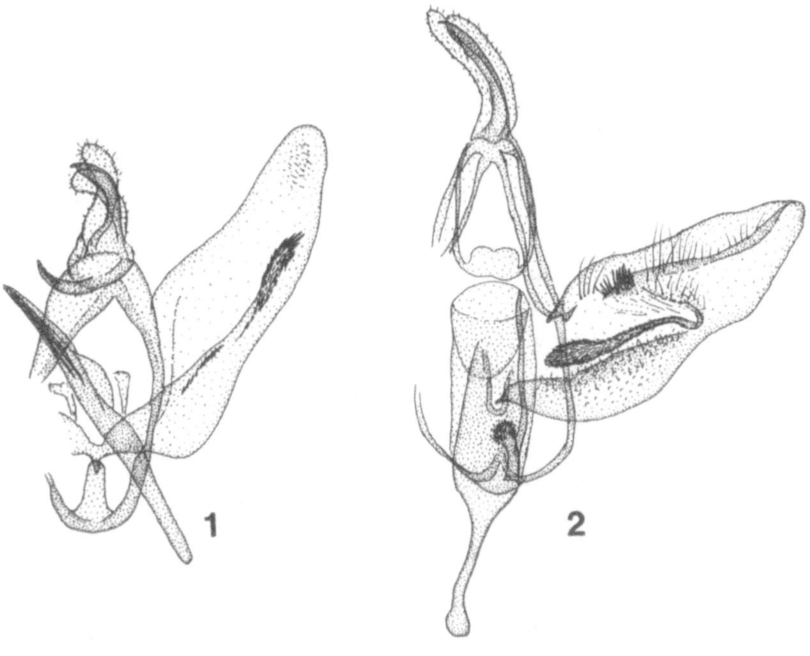


*Fig. 46.* Male genitalia of *Comostola leucomerata caledonica* (1) and *Urolitha bipunctifera* (2).

♂♀ 9–10 mm. Ground colour apple green marked with red as illustrated; fringes yellowish green; hindwing red discal spot is variable in size as in typical *leucomerata* (Queensland, New South Wales; *haplophanes* Turner, 1910, *Proc. Linn. Soc. N. S. Wales*, 35: 567, may prove to be synonymous). The subspecies is distinguished by its male genitalia which have valves narrower than in the typical race, lacking the angle to the ventral margin and the prominent lobe at the base of the dorsal margin; the scobination of the harpe is comparatively reduced over the basal half.

In New Guinea, extending to the S. Moluccas and Sulawesi, occurs ssp. *flavifimbria* Warren (1906, *Novit. zool.*, 13: 86, type examined) **stat. n,** with genitalia as in typical *leucomerata* but with the hindwing discal spot relatively invariable and intermediate between the extremes exhibited by both the typical and the New Caledonian race. In the Bismarcks and Solomons occurs ssp. *rubeibasis* Prout (1934, *Gross-Schmett. Erde*, 12: 129, type examined) **stat. n.,** slightly smaller but with broader marginal red chequering and larger discal spots to the wings.

HOLOTYPE ♂. New Caledonia: R. Bleue, 180 m Site 10 06704/75544 (J. D. HOLLOWAY) 27.5.1971, BM 1971–507. BM geometrid slide 9635.

PARATYPES. 10♂♂ (slide 8511), 3♀♀ New Caledonia, various localities.

The species was recorded as follows in New Caledonia: Pic du Pin site 7 (1); R. Bleue site 10 (2); Koh (1); Col d'Amieu (2); Col Toma (4); Sarraméa (1); R. Amoa (1); Probokonoro (2); near Bâ (1); Port Boisé site 74 (1). It is possibly a general rainforest species.

### **Comostola thalassias sp. n.** (Plate 53: 8, Figure 45: 3)

♂♀ 8 mm. Ground colour pale bluish green. Antennae, costa of forewing and fringes yellowish white. There are diffusely and faintly darker lunules at the discs of both wings but otherwise the species stands alone in the genus in lacking prominent wing markings. The general structure of the male genitalia is similar to that of the previous species and of *Chloeres boisensis.*

HOLOTYPE ♂. New Caledonia: Sarraméa, 160 m Site 29 05888/76065 (J. D. HOLLOWAY) 22.6.1971, BM 1971–507. BM geometrid slide 8527.

PARATYPES. 1♀ Petit Gouli, 1♂ Ouen Toro site 81.

The species may be associated with *Acacia spirorbis*; Sarraméa is not far from the extensive stands of the species at Petit Gouli.

### **Hemithea acaudata sp. n.** (Plate 54: 5, Figure 45: 4)

♂ 10 mm. Wings medium to pale, slightly olivaceous green with the distal half of the fringes deep pink. Postmedials on both wings consist of a row of white dots on the veins parallel to the margins. On the forewing the antemedial is represented by white dots on veins 1A and M. There are slightly darker, diffuse, transverse discal dashes on both wings. The species is somewhat similar to *H. stuhlmanni* (Prout) from Fiji, but that species is larger, darker olivaceous

*Plate 54.* 1. *Comibaena mariae* (10 mm); 2. *Comostola leucomerata caledonica;* 3. *Thalassodes microchloropis;* 4. *Thalassodes pseudochloropis;* 5. *Hemithea acaudata;* 6. *Pyrrhorachis pyrrhogona augustata* (9 mm); 7. *Thalassodes viridimargo;* 8. *Prasinocyma albicosta rufitincta* (13 mm); 9. *P.* near *ocyptera* (13 mm); 10. Urolitha bipunctifera (12 mm); 11. *Chlorocoma octoplagiata;* 12. *Uliocnemis biplagiata reducta.*

green, with more irregular fasciae on the wings; the fringes are more chequered and the hindwing tail much more acute. The male genitalia of *stuhlmanni* have the valves prominently bifid but in *acaudata* they are entire with a central harpe. The species is only tentatively placed in *Hemithea* but no closer match could be found in the BMNH collection.

HOLOTYPE ♂. New Caledonia: Tiebaghi, 330 m Site 41 04156/77366 (J. D. HOLLOWAY) 10.7.1971, BM 1971-507. BM geometrid slide 8526.

*Prasinocyma albicosta rufitincta* Warren (Plate 54: 8, Figure 47; 1)

*Geometra albicosta* Walker, 1861, *List Specimens lepid. Insects Colln Br. Mus.*, 22: 529, type examined.

*Prasinocyma rufitincta* Warren, 1897, *Novit. zool.*, 4: 44, type (Lifu) examined.

The typical race flies in Queensland, the Northern Territory and New Guinea. Subspecies *rufitincta* is found on Lifu, Uvea, the Isle of Pines, and was recorded from the following localities in New Caledonia: Poupa (6); Petit Gouli (6); Sarraméa (1); Kopéto (1); Tiebaghi site 39 (1), site 40 (1); Mt. Krapé site 55 (1); Ouen Toro site 82 (2). All these localities except Sarraméa are characterised by abundant *Acacia spirorbis*.

The larva was found on *Acacia spirorbis* at Petit Gouli and Tiebaghi and the adult was reared. The larva (Plate 55) is pale green with paler stripes dorsolaterally and dark reddish longitudinal dashes dorsally. It is a yellower green than the *Acacia* leaves and, when holding itself erect, it resembles the flower stems after the flowers have dropped. Larvae that spun up to pupate in leaves on 15th July emerged on 8th August. Larval and pupal material will be deposited in the BMNH.

*Prasinocyma* near *ocyptera* (Meyrick) (Plate 54: 9)

*Iodis ocyptera* Meyrick, 1888, *Proc. Linn. Soc. N. S. Wales* (2), 887, type examined.

A single female taken at Mt. Aoupinié is similar in appearance to the Australian *ocyptera* but the genitalia, though on the same general pattern, are slightly different with a shorter bursa and a more complex lamella vaginalis. Further material is needed before this taxon can be described.

*Pyrrhorachis pyrrhogona augustata* Prout (Plate 54: 6)

*Pyrrhorachis pyrrhogona augustata* Prout; ROBINSON (1975: 255).

This subspecies was described from Lifu and occurs also in New Caledonia, the New Hebrides and Fiji, though specimens from the last locality have a much broader diffuse cream-coloured zone interior to the red margins of the wings and perhaps merit subspecific status. The species occurs in the Indian Region, the Himalaya, Sundaland, Sulawesi, Australia (Queensland, Northern Territory) and Norfolk I. It is recorded from New Guinea but not from the Bismarcks or Solomons.

In New Caledonia it was recorded as follows: Grand Lac site 2 (3), site 3 (1),

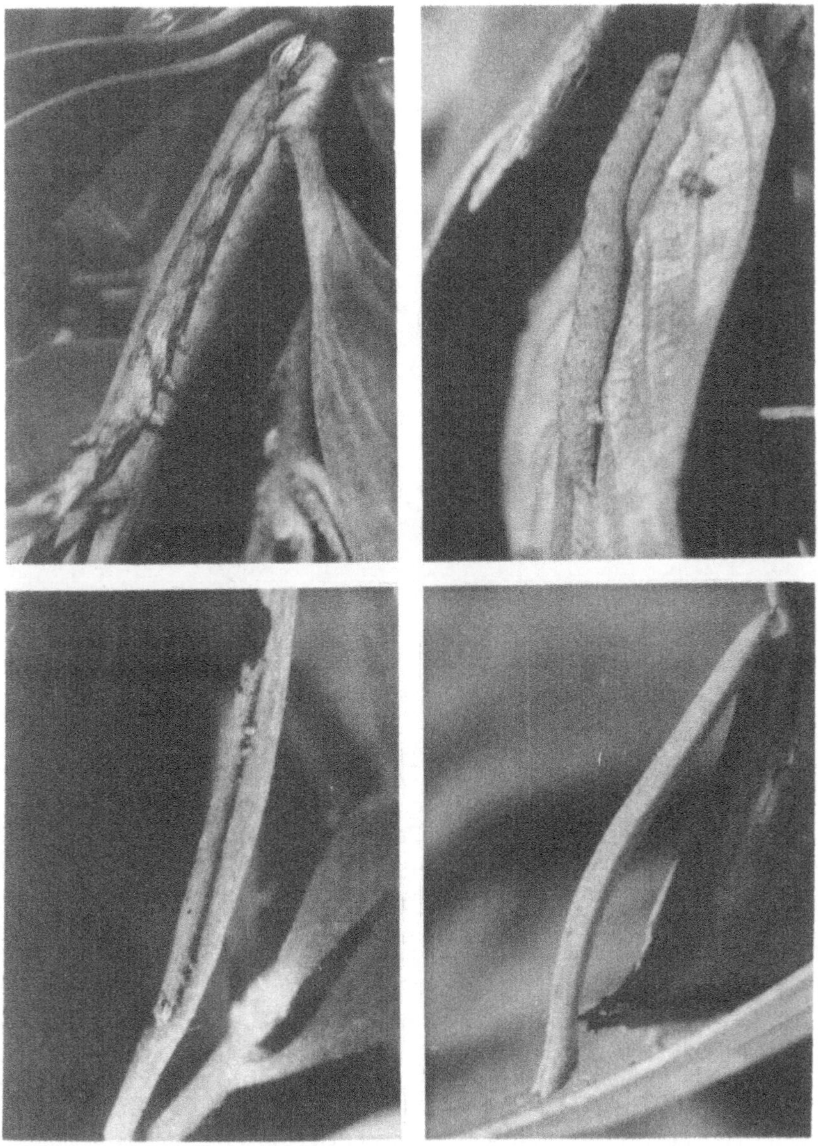

*Plate 55.* Top. Early and penultimate instars of *Ophiusa discriminans intacta*. Bottom. Final instars of *Prasinocyma albicosta* (left) and *Casbia rectaria* (right). The plate should be viewed with the left hand margin at the top.

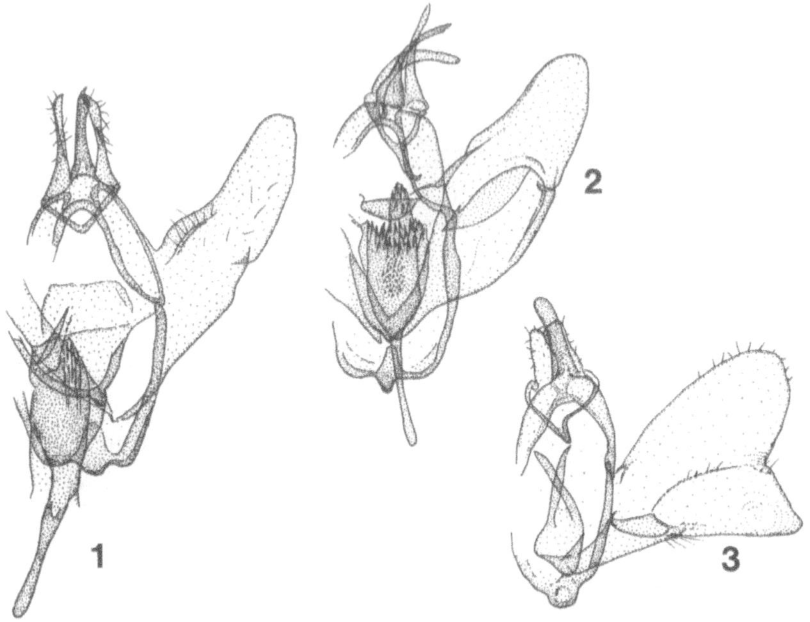


*Fig. 47.* Male genitalia of *Prasinocyma albicosta rufitincta* (1), *Thalassodes baladensis* (2), and *Chlorocoma octoplagiata* (3).

site 4 (1); R. Bleue site 8 (1), site 10 (2); Dothio site 18 (1), site 19 (1); Ouatou R. (3); Bréhoa R. (4); Ouambéo R. (2); Col d'Amieu (3); Col Toma (3); Sarraméa (1); Mt. Mou site 33 (1); Probokonoro (2); Mt. Krapé site 55 (1); Houailou (1); Mt. Panié site 61 (1); R. Koué (1); Les Dalmates site 77 (1). It is a widespread species but perhaps favours rainforest.

### *Pyrrhorachis viridula* Warren

*Pyrrhorachis viridula* Warren *sensu* Prout, 1913, *Lepidopterorum Catalogus*, 14: 162, and VIETTE (1949: 327).

This record from Lifu requires further investigation. The specimen has not been located but might possibly have originated from the ROTHSCHILD collection where there was considerable Lifu material, now at the BMNH. *P. viridula* is bright green compared with the pale bluish green of the previous species and was described from New Guinea. There are specimens in the BMNH of a bright green species from Santo and Aneityum in the New Hebrides. This is definitely not *viridula* but could be that recorded from Lifu, considering the biogeographical relationship between the New Hebrides and the New Caledonian group. A further bright green species, *rhodoselas* Prout, is found in Samoa.

### *Thalassodes baladensis* Holloway (Figure 47: 2)

*Thalassodes baladensis* Holloway, 1977: 120.

This is a taxonomically rather isolated species found in New Caledonia and on Norfolk I. It was recorded in New Caledonia as follows: Grand Lac site 2 (1); R. Bleue site 8 (1), site 9 (2); Mt. des Sources site 12 (1), site 14 (1); Dothio site 18 (1), site 19 (4); Ouatou R. (1); Bréhoa R. (2); Sarraméa (1); Kopéto (1); Probokonoro (6); Mt. Krapé site 55 (1); Mt. Panié site 61 (1); Les Dalmates site 77 (1). BMNH: Tinchialit (CHEESMAN). BPBM: Nouméa, Mt. Koghis, Le Crouen, Isle of Pines (Kanumera). The records are mainly from rainforest localities.

### **Thalassodes microchloropis sp. n.** (Plate 54: 3, Figure 48: 5)

♂♀ 14–16 mm. Facies as in *pseudochloropis* below except the fringes are pale green instead of yellowish and the insect is much smaller. The face is green as in *pseudochloropis* but the white scaling of the antennal shaft does not extend as far (one third) from the base. The species is externally similar to *grammonota* Prout (Bismarcks) and *dorsilinea* Warren (New Guinea, Queensland) but these species, together with *pseudochloropis*, have an entire dorsal stripe to the abdomen instead of a broken one. The male genitalia resemble those of *pseudochloropis* in having a diagonal fold of the valve that gives rise to a ventrally directed tooth at the ventral margin. However, the juxta is reduced to two lateral structures only narrowly joined instead of being a prominent v-shaped structure. The aedeagus of *microchloropis* is relatively much smaller. *T. dorsilinea* has a scobinate patch at the valve apex and a prominent scobinate spine at the valve base. *T. grammonota* has the valve markedly constricted centrally and scobinate dorsally, distal to the constriction.

HOLOTYPE ♂. New Caledonia: Grand Lac, 250 m Site 2 06975/75356 (J. D. HOLLOWAY) 16.5.1971, BM 1971–507. BM geometrid slide 8507.

PARATYPES. 8♂♂, 2♀♀ New Caledonia: Grand Lac site 2 (2), site 3 (1), Pic du Pin site 7 (6), Col d'Amoss site 46 (1); 1♂ La Coulée 4.12.1975 (J. GUTIERREZ) MNHN.

Though not common, the species would appear to be associated with the ultramafic maquis.

### *Thalassodes pilaria* Guenée (Figure 48: 4)

*Thalassodes pilaria* Guenée; ROBINSON (1975: 257).

A male and female of this species, bred from litchi at Nouméa, were presented to the author by M P. COCHEREAU. The species is distributed from Sulawesi and the Moluccas to Queensland, the New Hebrides, Fiji, the Society Is., Samoa and Pitcairn. In Fiji the species has been reared from a wide variety of host plants (ROBINSON, 1975).

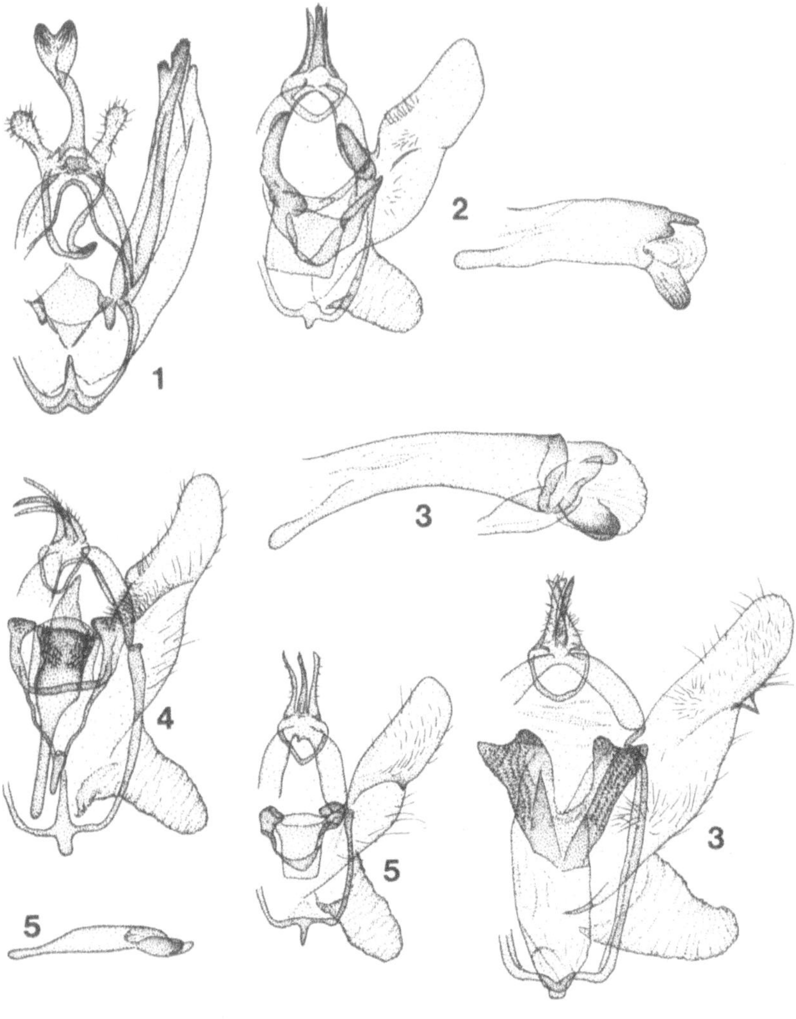


*Fig. 48.* Male genitalia of *Anisozyga caledonica* (1), *Thalassodes viridimargo* (2), *T. pseudochloropis* (3), *T. pilaria* (4) and *T. microchloropis* (5).

**Thalassodes pseudochloropis sp. n.** (Plate 54: 4, Figure 48: 3)

♂♀ 18–21 mm. The ground colour is generally pale bluish green marked in white as illustrated; the antennae, legs and forewing costa are white, tinged yellowish, and the fringes are also yellowish. The general appearance is as in *chloropis* Meyrick (New Hebrides, Fiji, Samoa, Tonga) and *hypoxantha* Fletcher (Bismarcks, Solomons) but the species is distinguished by the male genitalia which fall well outside the variation in morphology (primarily of the juxta) discussed for *chloropis* by ROBINSON (1975: 256). The valves are narrower apically, straighter, with a diagnostic angular projection subapically, ventrally. The subbasal valve ornamentation of *chloropis* (a flap) and *hypoxantha* (scobination, illustrated by FLETCHER (1957)) is absent. The aedeagus is very much broader than in the other species, the vesica containing a prominent, rounded cornutus.

HOLOTYPE ♂. New Caledonia: R. Kuenthio, 120 m Site 17 06153/75982 (J. D. HOLLOWAY) 6.6.1971, BM 1971–507. BM geometrid slide 8513.

PARATYPES. 50♂♂, 5♀♀ New Caledonia, various localities.

The species was recorded as follows: R. Bleue site 8 (2), site 9 (1), site 10 (1); Mt. des Sources site 14 (1); R. Kuenthio (1); Dothio site 18 (1), site 19 (1); Poupa (3); Ouatou R. (3); Bréhoa R. (6); Ouambéo R. (3); Col d'Amieu (3); Col Toma (13); Petit Gouli (1); Sarraméa (17); Kopéto (1); Mt. Mou site 33 (1); Kavatch (4); Ste. Therése (5); Probokonoro (6); Col des Roussettes (2); Mt. Krapé site 55 (4), site 56 (2); near Bâ (1); Mt. Aoupinié (16); Port Boisé site 74 (2); Mt. Koghis site 79 (10), site 80 (3); Ouen Toro site 81 (3). It is a widespread species, perhaps commoner in rainforest areas.

### **Thalassodes viridimargo sp. n.** (Plate 54: 7, Figure 48: 2)

♂♀ 18–20 mm. The ground colour is generally pale, slightly bluish green as in other New Caledonian *Thalassodes*. The fringes are very slightly darker green and the wings are striated and fasciated paler as illustrated. The striations are less abundant than in other species and the fasciae are broader, more diffuse and irregular. The hindwing postmedial is less angled, as is the hindwing margin. In the male genitalia the valves have a diagonal crease as in the *chloropis* complex but the ventral spur is absent from the distal extremity of the crease. The juxta consists of two lateral boomerang-shaped structures as illustrated. The aedeagus is short and broad, the apex unequally bilobed dorsally, and the vesica has a large semicircular cornutus as in *pseudochloropis*.

HOLOTYPE ♂. New Caledonia: Sarraméa, 160 m Site 29 05888/76065 (J. D. HOLLOWAY) 28.6.1971, BM 1971–507. BM geometrid slide 8529.

PARATYPES. 21♂♂, 3♀♀ New Caledonia, various localities.

The species was recorded as follows: R. Bleue site 8 (2); Ouatou R. (4); Bréhoa R. (5); Ouambéo R. (1); Col d'Amieu (2); Col Toma (2); Sarraméa (5); Mt. Mou site 33 (1); Kavatch (2); Probokonoro (5); Mt. Aoupinié (1); Mt. Koghis site 79 (1). It was found to be less abundant than the previous species and more strictly confined to rainforest.

### **Uliocnemis biplagiata reducta ssp. n.** (Plate 54: 12, Figure 45: 1)

*Comibaena biplagiata* Moore, 1887, *Lepid. Ceylon*, 3: 435.

♂♀ 11–14 mm. Facies with white markings on pale green as illustrated, the forewing discal dot and marginal flecks being black, those at the forewing tornus and hindwing apex surrounded by grey. The marginal flecks are, in most specimens, reduced to those at the torni of both wings, distinguishing this from other races. It is most similar to the New Guinea race, *elegans* Warren (type examined, St. Aignan I.), but this has extensive grey patches at the forewing tornus and hindwing apex. The species is recorded from the Indian region, Taiwan, Sundaland, the Philippines, Sulawesi, the Moluccas, New Guinea and the Bismarck Is. It is replaced by *U. woodfordi* Warren in the Solomons and New Hebrides.

HOLOTYPE ♂. New Caledonia: Bréhoa R., 320 m Site 24 05853/76139 (J. D. HOLLOWAY) 15.6.1971, BM 1971-507. BM geometrid slide 8514.

PARATYPES. 22♂♂, 2♀♀ New Caledonia, various localities.

The species was recorded as follows: Poupa (1); Ouatou R. (4); Bréhoa R. (2); Col d'Amieu (6); Col Toma (4); Sarraméa (2); Cap Colnett (1); R. Amoa (1); Ste. Therèse (3); Probokonoro (2); Mt. Krapé site 55 (1); Mt. Aoupinié (2). It was most frequently encountered in areas of rainforest on sedimentary rock.

### *Urolitha bipunctifera* (Walker) (Plate 54: 10, Figure 46: 2)

*Iodis bipunctifera* Walker, 1861, *List Specimens lepid. Insects Colln Br. Mus.*, 22: 546, type examined.

This species, occurring in Queensland, New South Wales, Lord Howe I., the New Hebrides and Uvea, was recorded as follows in New Caledonia: Grand Lac site 2 (1); R. Bleue site 10 (1); Dothio site 19 (2); Ciu (1); Koh (5); Ouatou R. (1); Bréhoa R. (5); Col d'Amieu (6); Col Toma (2); Sarraméa (4); Kopéto (2); Kavatch (1); Probokonoro (4); Col des Roussettes (3); Mt. Krapé site 55 (1); Mt. Aoupinié (1); Yaté Gorge (1); Les Dalmates site 77 (1), site 78 (1). BPBM: Le Crouen. MNHN: La Coulée. Most of these localities are in areas of rainforest.

## GEOMETRIDAE: STERRHINAE

### **Anisodes dalmatensis sp. n.** (Plate 56: 1, 2, Figure 49: 1)

♂♀ 14–16 mm. Ground colour slightly ochreous fawn, variably suffused dull rufous and fasciated dark grey with black flecks. One of the females has many of the darker markings black instead of grey and the hindwing discal spot is centred white (Plate 56: 2). The species belongs to a group with long slender palps and, in the male, prominent tufts of scales on a short, broad hind femur; the male hind tibia has no spurs and, with the exception of *denticulata* Hampson, has the straw scales raised; there is a pair of coremata on the male abdomen between the eighth segment and the genitalia. The male genitalia in the group are all very similar with scobinate, angled harpes to the valves and a pair of slender, descending processes to the uncus. Included in the group are *A. denticulata* (type examined; Assam, Sundaland, Sulawesi, Ceram, New Guinea), *porphyropis* Meyrick (type examined; New South Wales, Queensland)

*Plate 56.* 1 (♂), 2 (♀). *Anisodes dalmatensis;* 3. *Anisodes lateritica;* 4. *Anisodes kohensis;* 5. *Anisodes samoana* (12 mm); 6. *Anisodes paragloria;* 7. *Idaea scintillans flavior;* 8. *Idaea trypherophа* (♂, 7 mm); 9. *Idaea fruticeti;* 10. *Idaea paniensis;* 11, 12. *Idaea lateritica.*

and *microsticta* Turner (Lord Howe I.). The New Caledonian species is closest to *porphyropis* but has the forewing apex less produced, the forewing medial less oblique, adjacent to the discal spot instead of well distal to it. The male genitalia of the two species are very similar.

HOLOTYPE ♂. New Caledonia: Les Dalmates, 220 m site 77 06713/75420 (J. D. HOLLOWAY) 18.8.1971, BM 1971–507. BM geometrid slide 8588.

PARATYPES. New Caledonia: 1♀ Mt. des Sources site 11, 1♀ Mt. Koghis, 1♂ Col Toma (slide 8540); 1♂, 1♀ Mt. Koghis 500 m i, ii.1963 (R. STRAATMAN) BPBM.

**Anisodes kohensis sp. n.** (Plate 56: 4, Figure 50: 1)

♀ 11 mm. The wings are marked pinkish and black on fawn as illustrated. The facies bear a very close resemblance to those of *turneri* Prout (Queensland, New Guinea, Solomons) except the black dots at the base of the forewing are fewer; the hindwing medial passes through the discal spot instead of just distad. In the genitalia the bursa is similar to that of *turneri*, lightly scobinate, slender, tapering to the base; basally in *kohensis* it broadens into a squarish, more heavily sclerotised sac, but in *turneri* it tapers away to terminate in a slender, laterally crescent-shaped structure.
HOLOTYPE ♀. New Caledonia: Koh, 140 m site 22 05866/76177 (J. D. HOLLOWAY) 12.6.1971, BM 1971–507. BM geometrid slide 8580.

**Anisodes lateritica sp. n.** (Plate 56: 3, Figure 49: 2)

♂♀ 12–13 mm. Ground colour dull orange-red, finely stippled and striated with darker red and blackish. The discal spots, postmedials and forewing antemedial are picked out faintly darker greyish. There is a row of black dots submarginally on the veins and another marginally in the spaces. The species is similar to *compacta* Warren (type examined; Kei Is., New Guinea, Queensland) but is distinguished in the male genitalia where the aedeagus lacks a prominent cornutus, there is symmetry elsewhere and the basal processes to the tegumen lack scobination, being curved apically and bearing a diagnostic lateral spine; the harpe on the valve is longer and more distal than in *compacta*. The genitalia of *compacta* are described in the diagnosis of the next species.
HOLOTYPE ♂. New Caledonia: Yate Gorge, 120 m Site 68 06955/75499 (J. D. HOLLOWAY) 6.8.1971, BM 1971–507. BM geometrid slide 8536.
PARATYPES. 2♂♂, 6♀♀ New Caledonia: Grand Lac site 2 (1), site 3 (2), Pic du Pin site 7 (2), Mt. des Sources site 11 (2), Mt. Panié site 60 (1); 1♂ Col des Rousettes BPBM; 1♂ Mt. Koghis 78CH 12.1.77 (ad.) (A. DELOBEL) ex *Codia discolor*, MNHN.

Almost all specimens were taken in ultramafic maquis localities. *Codia discolor* is in the family Cunoniaceae. No description of the larva is available.

*Anisodes lautokensis* Prout **stat. n.** (Figure 50: 2)

*Anisodes* (*Pisoraca*) *compacta lautokensis* Prout, 1929, *Ann. Mag. nat. Hist* (10), 3: 595, type examined. ROBINSON (1975: 259).

The external characters of this Fijian species have been described by ROBINSON. The male genitalia, illustrated here, were dissected while the status of the previous species was being determined. They are very different from those of typical *compacta*. In *compacta* the basal processes of the tegumen are slender, scobinate and tapering, the left process half as long again as the uncus, the right slightly thicker and shorter than the uncus; the harpes of both valves are very small and the right valve bears a triangular process subapically from the costa.

### *Anisodes niveopuncta* Warren (Figure 49: 5)

*Anisodes niveopuncta* Warren, 1897, *Novit. zool.*, 4: 48, type examined (Queensland). *Brachycola decolorata* Warren, 1897, *Novit, zool.*, 4: 215, type examined (Lifu), **syn. n.** ROBINSON (1975: 259).

The male genitalia of Queensland and New Caledonian specimens are indistinguishable; those of Fiji specimens have a slightly smaller triangular projection in the subapical angle of the valve and in the New Hebrides form this projection is short and obtuse. In *A. gloria* Robinson, a forest duplex of the coastal *niveopuncta* in Fiji, this projection is absent and the apical portion of the valve is broader. A similar duplex occurs in New Caledonia where the forest species, described below, is genitalically much more distinct than *gloria.*

*A. niveopuncta* has been recorded from Queensland, New Guinea, Lifu, the New Hebrides, Fiji, the Gilberts, Tuvalu Is., Society Is., and the Tuamotus. It was taken as follows in New Caledonia: R. Kuenthio (1); Koh (1); Tiebaghi site 39 (1); Col d'Amoss site 45 (1); R. Amoa (1); near Bâ (1); a female was swept by day on the Col d'Amoss. These localities are mostly characterised by open, secondary vegetation.

In the Oriental Region there are further species in this complex such as *clandestina* Prout (type examined, Khasia Hills; the BMNH series contains a complex of species as males from southern India and Singapore were found to have different genitalia), *absconditaria* Walker and *melantroches* Prout. The male genitalia of the last two have been illustrated by HOLLOWAY (1976), though not from type material.

### **Anisodes paragloria sp. n.** (Plate 56: 6, Figure 49: 4)

♂♀ 14–17 mm. This species is externally very similar to *niveopuncta* (13–16 mm) but is slightly larger and of a darker, redder, less pink tint; the forewing postmedial is finer, more crenulate, and the hindwing postmedial passes diffusely distal to the discal spot rather than through it. In the male genitalia the valves are squarish, more as in *absconditaria* than in *niveopuncta.* In the female the bursa is as long as the ductus (half as long in *niveopuncta*); the ductus is narrow throughout rather than broadening towards the bursa; the bursa is set symmetrically or barely asymmetrically on the ductus instead of markedly asymmetrically.

HOLOTYPE ♂. New Caledonia: Col Toma, 460 m Site 27 05813/76123 (J. D. HOLLOWAY) 19.6.1971, BM 1971-507. BM geometrid slide 8589.

PARATYPES. 3♂♂ (slides 8539, 8590), 4♀♀ (slides 8414, 8416, 8427) New Caledonia: Grand Lac site 3 (1); Mt. des Sources site 14 (2); Col d'Amieu (1); Tiebaghi site 41 (1); Mt. Krapé site 55 (1); Mt. Koghis 1♀ taken by day.

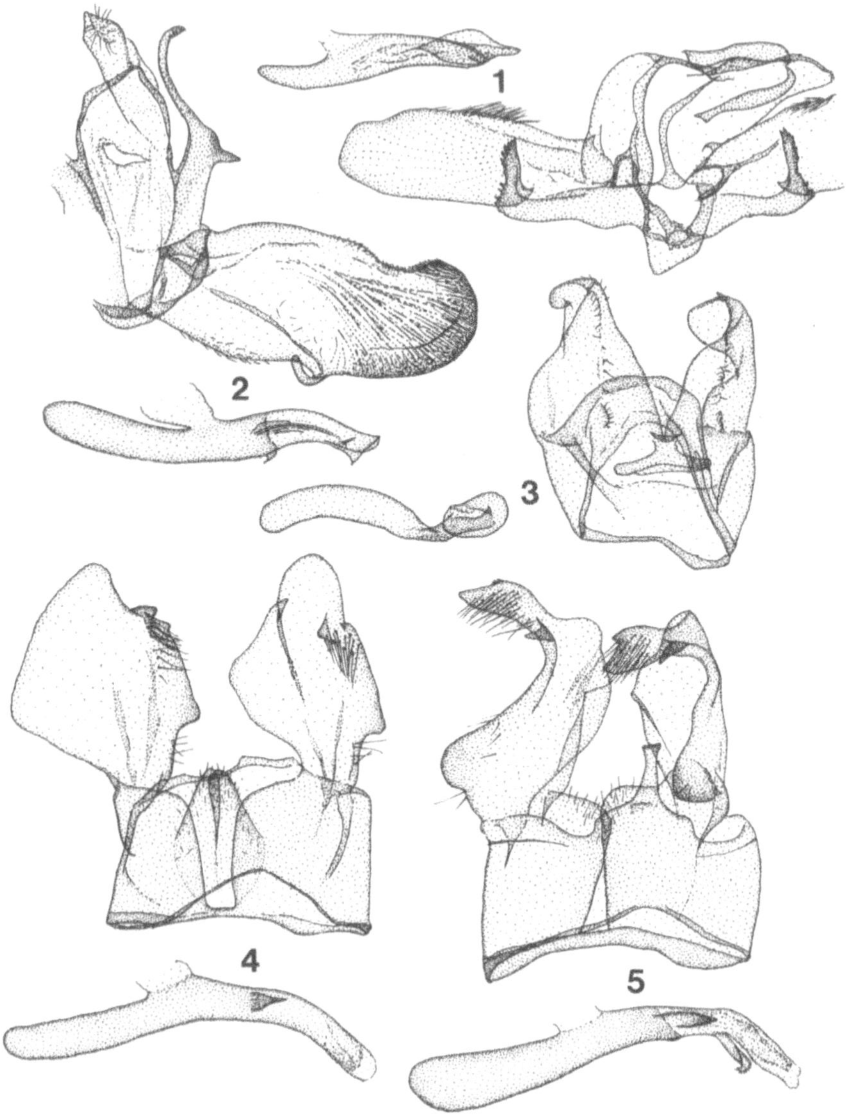


*Fig. 49.* Male genitalia of *Anisodes dalmatensis* (1), *A. lateritica* (2), *A. samoana* (3), *A. paragloria* (4), *A. niveopuncta* (5).

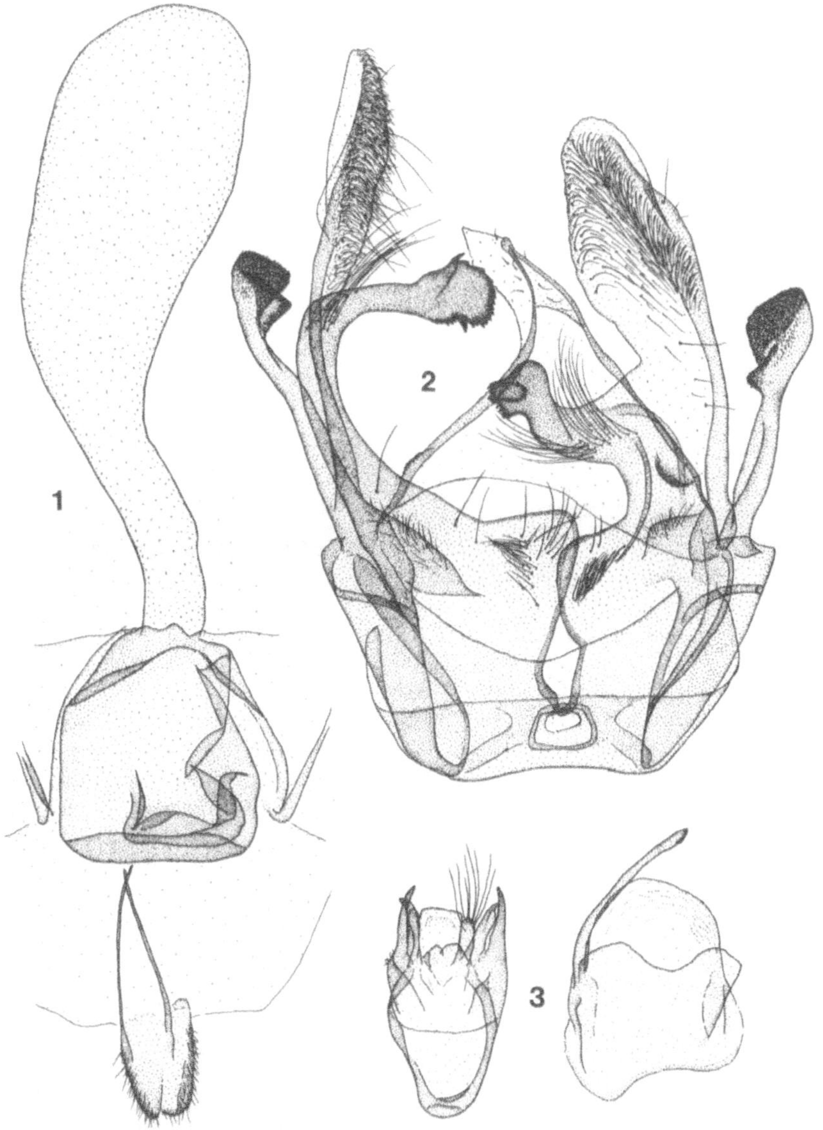


*Fig. 50.* Female genitalia of *Anisodes kohensis* (1), male genitalia of *A. lautokensis* (2) and male genitalia and eighth abdominal segment of *Scopula oppilata* (3).

### *Anisodes samoana* Warren (Plate 56: 5, Figure 49: 3)

*Anisodes samoana* Warren; VIETTE (1949: 327); ROBINSON (1975: 263).

This species occurs in the Loyalty Is., the New Hebrides, Fiji, Samoa, Tonga and the Society Is. Singletons were taken in New Caledonia from Grand Lac site 3, Kopéto and Probokonoro; the species has also been recorded at Le Crouen (BPBM).

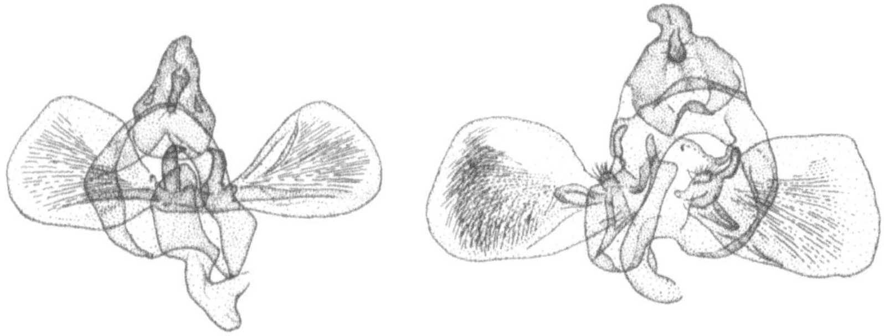

*Fig. 51.* Male genitalia of *Gnamptoloma rubra* (left) and *Traminda mundissima* (right).

### **Gnamptoloma rubra sp. n.** (Plate 57: 8, Figure 51: left)

♂♀ 10–14 mm. The ground colour of the wings is pale pinkish red with paler yellowish postmedials on both wings, usually edged basad with darker red. The species is very similar to *G. aventiaria* Guenée (type examined; Australia) which is widespread in the Oriental Region, extending to northern Australia and New Guinea. *G. aventiaria* is a pale dull green above with the darker line to the submarginal dully rufous. Below *rubra* is a similar colour to its upperside but *aventiaria* is greenish yellow. Australian specimens of *rubra* lack the leaden postmedial to the forewing underside that is found in most *aventiaria* specimens, though this postmedial is present in New Caledonian *rubra*. In the male genitalia the basal horn-like process to the valve is broader in *rubra*, less finely tapered or sinuous apically than in Australian *aventiaria*.

HOLOTYPE ♂. New Caledonia: Poupa, 200 m site 20 06120/76150 (J. D. HOLLOWAY) 10.6.1971, BM 1971-507. BM geometrid slide 8532.

PARATYPES. 40 ex. New Caledonia, various localities.

OTHER MATERIAL. There are numerous examples from Queensland, one from Northern Territory and three from Sumbawa in the BMNH. I. F. B. COMMON (*in litt.*) noted specimens also from New South Wales. A single male was taken on Uvea, Loyalty Is., by Mr. H. SUZUKI in May, 1974.

The species was recorded as follows in New Caledonia: Grand Lac site 1 (1), site 2 (6), site 3 (4), site 4 (1); Pic du Pin site 7 (4); R. Bleue site 10 (3); Mt. des Sources site 11 (3), site 12 (4), site 14 (3); Dothio site 18 (1), site 19 (1); Poupa (1); Koh (5); Ouatou R. (1); Bréhoa R. (5); Ouambéo R. (1); Col d'Amieu (1); Col Toma (4); Petit Gouli (1); Sarraméa (2); Kopéto (3); Anse

Longue site 32 (1); Mt. Mou site 33 (1); Tiebaghi site 41 (1); Mt. Krapé site 55 (2); Col Mouirange (2); Port Boisé site 73 (1); Ouen Toro site 81 (1). The species was assigned to the ultramafic association in the numerical analysis but this may be because it was commoner in the earlier part of the survey when mostly ultramafic localities were being sampled.

**Idaea fruticeti sp. n.** (Plate 56: 9, Figure 52: 3)

♂♀ 7–8 mm. This species, together with *I. lateritica* and *I. paniensis* below, has venation as in typical *Idaea* Treitschke but is otherwise isolated within the genus. The three species are distinguished by the male genitalia which have coremata on the saccus, broad lobes on the valve costa and a sclerotised strip in the vesica of the aedeagus that can give the impression of being a large cornutus. The wing pattern is reminiscent of that of *I. dicenea* (Prout) and *I. bathromyses* (Prout) from Fiji, the forewing fasciae being much darker at the costa. But the New Caledonian species do not have tufts of scales on the hind tibia of the males and are characterised by a broken dark streak at the apex of the forewing made up from darker dashes within the fasciae; the male genitalia are also diagnostic.

*I. fruticeti* is the smallest of the three. The ground colour of the wings is pale grey and is marked with medium grey and dark brown as illustrated. The hindwing discal spot is usually prominent and separate from the fasciae. The male genitalia are distinguished by the tapering of the saccus.

HOLOTYPE ♂. New Caledonia: Col Nassiboué, 80 m Site 63 06334/75497 (J. D. HOLLOWAY) 1.8.1971, BM 1971-507. BM geometrid slide 8548.

PARATYPES. 1♂ (slide 8549), 4♀♀ New Caledonia: Col Nassiboué (1), Ouen Toro site 82 (4); 3♂♂, 2♀♀ Nouméa, July, Sept 1940 (FXW) at light, BPBM.

All specimens so far have been recorded from west coast thicket areas.

**Idaea lateritica sp. n.** (Plate 56: 11, 12, Figure 52: 1)

♂♀ 9–11 mm. The ground colour of the wings is pale fawn, fasciated brownish grey and flecked blackish. Some specimens (from the Népoui valley particularly; Plate 56: 12) are less fasciated and more speckled with dark scales than in the type. The species is distinguished from the other two by the absence of a darker mark on the forewing where intersected by the medial. The hindwing discal spot is just basal to the medial but only faintly darker than the ground colour. The male genitalia are distinguished by the squarish saccus and the angular, somewhat separated costal lobes to the valve.

HOLOTYPE ♂. New Caledonia: Col Mouirange, 220 m Site 65 06701/75407 (J. D. HOLLOWAY) 3.8.1971, BM 1971-507. BM geometrid slide 8546.

PARATYPES. 2♀♀ New Caledonia: Pic du Pin site 7 (24.5., 7.8.1971); 2♂♂, 2♀♀ Col des Roussettes 1200 m 5-6.ii.1963 Malaise trap (C. YOSHIMOTO & N. KRAUSS) BPBM; 1♂, 2♀♀ Népoui Valley, July 1940, BPBM; 1♀ Nouméa, 1 July, in room, BPBM.

A further battered specimen has been taken on Mt. Koghis (BPBM). The majority of these localities are on or near ultramafic rock.

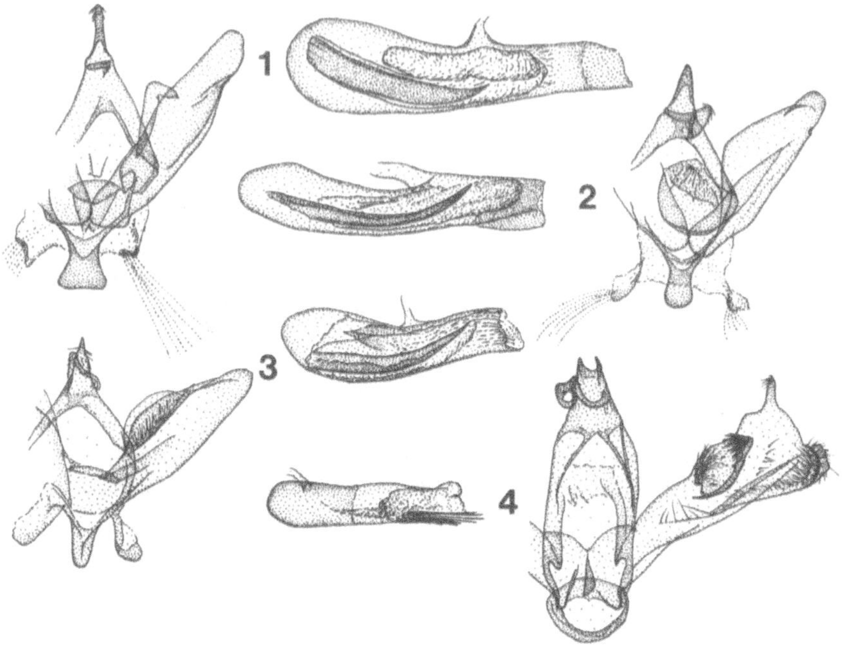


*Fig. 52.* Male genitalia of *Idaea lateritica* (1), *I. paniensis* (2), *I. fruticeti* (3) and *I. scintillans flavior* (4).

**Idaea paniensis sp. n.** (Plate 56: 10, Figure 52: 2)

♂ 11 mm. The wing ground colour is pale straw but is darker and greyer basal to the medials on both wings. The fasciae are greyish brown, the postmedial and more basal fasciae of the forewing tinged ochreous. The costa of the forewing is darker brownish and both wings are sparsely flecked with black scales. The discal spots are fused with the medials, though that of the forewing is set in a basally directed indentation and that of the hindwing is slightly darker with an area of ochreous suffusion distad. In the male genitalia the saccus is bulbous and the valve costa is angled rather than lobed. The aedeagus and vesica sclerotisation are slender compared to those of the other two species.
HOLOTYPE ♂. New Caledonia: Mt. Panié, 800 m Site 04764/77257 (J. D. HOLLOWAY) 27.7.1971, BM 1971-507. BM geometrid slide 8545.

**Idaea scintillans flavior ssp. n.** (Plate 56: 7, Figure 52: 4)

*Ptychopoda scintillans* Warren, 1898, *Novit. zool.*, 5: 243, type examined.
♂♀ 6–7 mm. Ground colour very pale yellow, banded pinkish; the basal half

of the forewing costa is tinged with red. The subspecies differs from typical *scintillans* (Queensland) in that the pink and yellow bands are of equal width instead of the pink predominating to restrict the yellow to narrow postmedial and submarginal bands. The male genitalia are as in the typical race except the uncus is bifid instead of single.

HOLOTYPE ♂. New Caledonia: Kaala-Gomen, 80 m Site 38 04351/77158 (J. D. HOLLOWAY) 8.7.1971, BM 1971-507. BM geometrid slide 8547.

PARATYPES: 3♂♂, 4♀♀ New Caledonia, various localities; 1♀ New Caledonia: Pouébo, coast-1,500 ft, 12-21.ix.1949 (L. E. CHEESMAN) BM 1950-1.

The species was recorded as follows in New Caledonia: Petit Gouli (3); Kopéto (1); Kaala-Gomen (1); Tiebaghi site 41 (1); Grottes de Koum (4); Bambou R. (1); Kavatch (1); Mt. Krapé site 55 (1), site 56 (1); Port Boisé site 73 (2). Two worn specimens have been taken at Le Crouen (BPBM), but the majority of records made during the survey were from the western side of the island.

The subspecies also occurs in the New Hebrides (Malekula, Efate, Erromango).

### *Idaea trypheropha* (Meyrick) (Plate 56: 8, Figure 60: 3)

*Mnesterodes trypheropha* Meyrick, 1889, *Trans. ent. Soc. Lond.*, 1889: 483.

A single female of this species was recorded at Ouen Toro site 82 during the survey and there are five males and two females from Nouméa in the BPBM. The species is distinguished within the genus by the prominent tuft of straw-coloured scales arising from the hindwing costa in the male. It occurs in the Kei Is., New Guinea, Queensland, the Bismarcks, Solomons, and Bellona I. in the Rennell group. The New Caledonian specimens lack the broad grey fascia adjacent to the postmedials of both wings that occurs in New Guinea specimens. Solomons specimens are generally intermediate but variable.

### **Scopula caledonica sp. n.** (Plate 57: 5, Figure 53: 4)

♂♀ 12-13 mm. The ground colour is pale straw, the forewing costa slightly darker. Both wings are fasciated pale grey and scattered sparsely with black scales. Only the forewing has an antemedial, angled at vein Rs and with three black dots on the veins. Four fasciae occur in the distal half of both wings. The medial is crenulate, on the hindwing basal to, and often encircling, the discal spot. In some specimens the medial is edged darker basad. There are two postmedial fasciae, the basal one narrow, crenulate, with black flecks on the veins; the more distal one is broader and more irregular. The submarginal is narrow, again with black flecks on the veins. There is a row of black dots marginally in the spaces of both wings. The male eighth sternite is without lateral processes, being broadly rounded distally. The processes of the valves are slightly broader and more curved than those of *quinquefasciata*, a related New Caledonian species described below.

*Plate 57.* 1. *Scopula oppilata* (9 mm); 2. *Scopula homodoxa* (8 mm); 3. *Scopula undilinea* (9 mm); 4. *Scopula* sp.; 5. *Scopula caledonica;* 6. *Scopula quinquefasciata;* 7. *Scopula desita* (12 mm); 8. *Gnamptoloma rubra;* 9. *Traminda mundissima* (11 mm); 10. *Eois tiebaghi;* 11. *Eoasthena sarramea.*

HOLOTYPE ♂. New Caledonia: Mt. Mou, 500 m Site 34 06386/75576 (J. D. HOLLOWAY) 1.7.1971, BM 1971-507. BM geometrid slide 8541.

PARATYPES. 7♂♂ (slide 8543), 4♀♀ New Caledonia: Grand Lac site 2 (1), site 72 (1); Pic du Pin site 7 (1); Koh (1); Ouambéo R. (1); Col Toma (1); Petit Gouli (1); Mt. Panié site 60 (1), 500 m in Malaise trap (1).

The species was taken mainly either in ultramafic maquis localities or those on sedimentary and metamorphic rocks where maquis elements are present.

*Scopula desita* (Walker) (Plate 57: 7, Figure 53: 1)

*Tephrosia desita* Walker, 1861, *List Specimens lepid. Insects Colln Br. Mus.*, 22: 41, type examined.

This is primarily an Australian species (Northern Territory, Queensland, New South Wales) but occurs also in the Lesser Sundas. It appeared to have an unusual distribution in New Caledonia, being recorded only from the central ultramafic massifs of the west coast as follows: Kopéto (1); Mt. Krapé site 55 (16), site 56 (16).

*Scopula homodoxa* (Meyrick) (Plate 57: 2, Figure 53: 3)

*Scopula homodoxa* (Meyrick); ROBINSON (1975: 266).

*S. homodoxa* was described from Tonga but has also been recorded from New Guinea, Australia (Northern Territory, Queensland), Lifu, Fiji, Rotuma and Samoa. It was recorded as follows in New Caledonia: Anse Longue site 15 (1), site 16 (5), site 32 (2); Poupa (1); Koh (1); Kopéto (1); Kaala-Gomen (2); Bambou R. (1); Probokonoro (1); Houailou (1); Ouen Toro site 64 (2), site 82 (1); Mt. Koghis site 79 (1). BPBM: Nouméa, Le Crouen, Népoui Valley; one specimen reared from a brown larva on *Casuarina collina*.

It is predominantly a species of coastal thicket and *Acacia spirorbis* associations; in Fiji it is a species of dry secondary vegetation (ROBINSON).

*Scopula oppilata* (Walker (Plate 57: 1, Figure 50: 3)

*Acidalia oppilata* Walker, 1861, *List Specimens lepid. Insects Colln Br. Mus.*, 23: 776, type examined. VIETTE (1949: 327).

There are three specimens from Lifu in the BMNH. The species has not been recorded from New Caledonia but occurs in the Lesser Sundas, Moluccas, New Guinea, Queensland (type. loc.), the Bismarcks and Solomons.

**Scopula quinquefasciata sp. n.** (Plate 57: 6, Figure 53: 5)

♂♂ 10 mm. The wings are pale fawn, fasciated grey and sparsely scattered with black scales. The discal spots of both wings are black, that of the hindwing ringed a browner grey than the fasciae, the ring being fused with the first fascia. The distal half of both wings is occupied by five crenulate fasciae, evenly spaced and of approximately even intensity; the most basal is slightly heavier and the central one includes black dots on the veins. On the margins of both wings there are black dots in the spaces. There is a faint antemedial on the forewing with black dots on the veins. The eighth sternite in the male is rounded distally, narrow, with no lateral processes. The genitalia are as illustrated, the valve processes slender and only slightly curved.

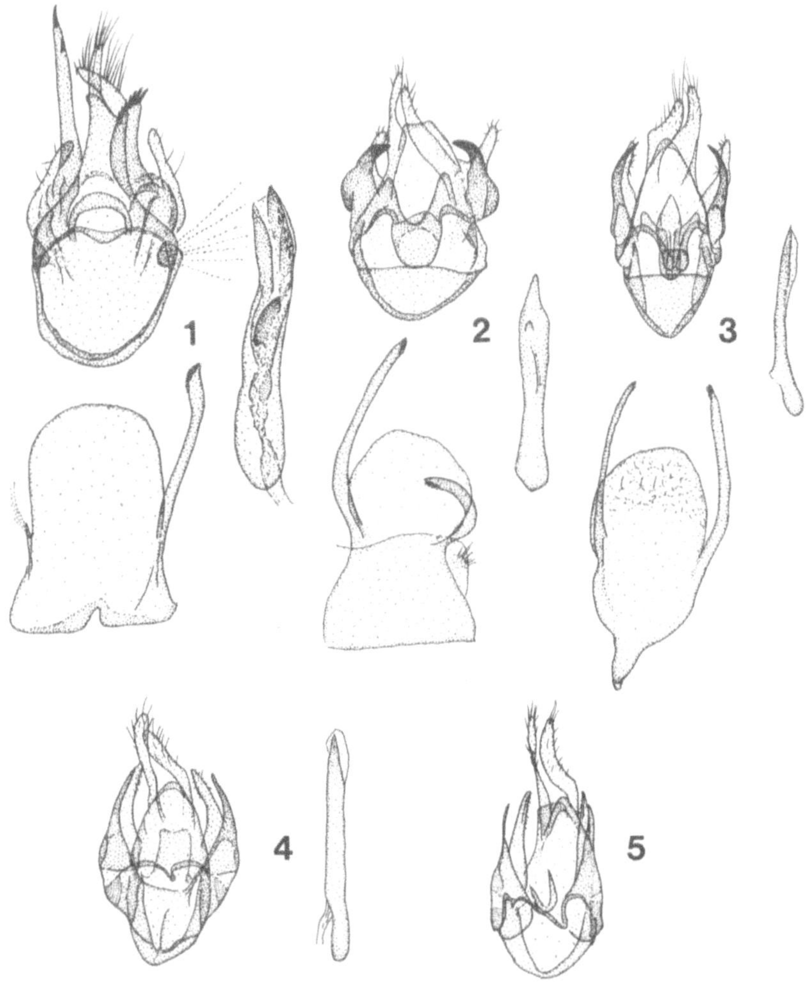


*Fig. 53.* Male genitalia of *Scopula* species, the first three species having the eighth abdominal segment featured also: *desita* (1), *undilinea* (2), *homodoxa* (3), *caledonica* (4) and *quinquefasciata* (5).

The species has no close relatives in the genus apart from *caledonica*. The five fasciae on each wing are diagnostic.

HOLOTYPE ♂. New Caledonia: R. Bleue, 150 m Site 8 06755/75530 (J. D. HOLLOWAY) 25.5.1971, BM 1971-507. BM geometrid slide 8544.

PARATYPES. 3♂♂ New Caledonia: Pic du Pin site 7, Col Toma, Port Boisé site 73.

It is possibly an ultramafic maquis species though most areas where it was recorded contained much rainforest.

*Scopula rubraria* (Doubleday)

*Scopula rubraria* (Doubleday); HOLLOWAY (1977: 124)

M. P. COCHEREAU sent the author a few specimens taken in Nouméa and two further examples from this locality (Anse Vata) have been seen (BPBM). The species occurs throughout Australia and has been recorded from New Zealand, Norfolk I. and the Kermadecs.

*Scopula undilinea* (Warren) (Plate 57: 3, Figure 53: 2)

*Craspedia undilinea* Warren, 1900, *Novit. zool.*, 7: 105, unique male type (Lifu) examined.

This species proved to be common in New Caledonia, the male genitalia resembling those of the type though the latter is much less heavily marked than most New Caledonian specimens. The species was recorded as follows: Anse Longue site 15 (3), site 16 (4), site 32 (17); Poupa (2); Koh (2); Col Toma (2); Kopéto (4); Grottes de Koum (2); Bambou R. (1); Col des Roussettes (1); Mt. Krapé site 55 (2); Col Nassiboué (3); Ouen Toro site 64 (1), site 81 (4), site 82 (1); Mt. Koghis site 79 (1). BPBM: Bourail, Le Crouen, Nouméa, Népoui Valley.

It is a west coast species in the main, being particularly abundant in the leguminous thicket at Anse Longue.

*Scopula* sp. (Plate 57: 4)

♀ 9 mm, from Anse Longue site 32. The ground colour of the wings is pale straw, marked and stippled with brownish grey as illustrated. The diffuse forewing medial is broader, more basal and slightly sigmoid compared with that of *undilinea*; the forewing postmedial is also more basal and both fasciae are more diffuse. This may be an aberration of *undilinea* but more material is needed to determine its status with confidence.

*Traminda mundissima* (Walker (Plate 57: 8, Figure 51: right)

*Acidalia mundissima* Walker, 1861, *List Specimens lepid. Insects Colln Br. Mus.*, 23: 795, type examined.

This is a widespread Indo-Australian species, recorded from the Indian region, S.E. Asia, Sundaland, Sulawesi, the Lesser Sundas, New Guinea and Australia (Northern Territory, Queensland). It was recorded as follows in New Caledonia: Grand Lac site 2 (1); Mt. des Sources site 14 (1); Anse Longue site 16 (6), site 32 (24); Sarraméa (1); Col Nassiboué (2); Ouen Toro site 81 (1), site 82 (1).

It was abundant at the Anse Longue sites with leguminous thicket but occurred only sparsely elsewhere. The larva has been recorded from *Acacia* in India (SINGH *et al.*, 1954–1960).

## CALEDASTHENA Gen. n.

The antennae are filiform in both sexes. The palps extend anteriorly beyond head to a distance the length of the head, the terminal segment very short (one sixth of the second); the second segment is scaled to give a rectangular appearance when viewed laterally. The legs are long, 1.5 × length of hindwing, the hind tibia with two unequal pairs of spurs.

The wing venation is typically larentiine (Figure 57: 5), the facies as described for the type species. As in *Gymnoscelis*, *Asthena*, *Poecilasthena*, *Eoasthena*, *Eois*, *Ziridava* and related genera the hindwing is marked similarly to the forewing. The extent to which this occurs in the Larentiinae is probably connected with the extent of the hindwing that is displayed in the resting posture; its value as a taxonomic criterion might be worth investigating. The white patch distal to the origin of Culb at the forewing cell is unusual, as is the small fan of straw-coloured androconia arising subdorsally on the male hindwing to cover the point where the anal vein meets the dorsum.

The male genitalia (Figure 61: 5) have an abnormally long vinculum suspending a broad saccus with lateral coremata. The transtillae are highly developed, possibly taking over the function of the uncus which is long and slender. The eighth abdominal segment is unmodified.

In the female genitalia (Figure 58: 1) the ovipositor lobes and apophyses are abnormally elongate for the subfamily. The bursa is lightly scobinate with a 'U' of greater sclerotisation, concave and thicker basad, extending almost the length of the bursa.

The single species appears to have no close relatives; the male androconia and extended vinculum, the elongate female genitalia with the U-shaped signum are highly distinctive within the subfamily. The type species is *montana*, described below.

**Caledasthena montana sp. n.** (Plate 58: 1, 2, Figures 57: 5, 58: 1, 61: 5)

♂♀ 11-14 mm. The general structure is described above. The thorax is banded narrowly black, broadly rufous fawn and pinkish grey. Each abdominal segment has the basal half pinkish grey and similarly the distal fringe, the rest of the distal half being broadly rufous fawn with a black dot dorsally, adjacent to a white dot at the margin. The ground colour of the wings is variably ochreous to pinkish fawn, fasciated darker pinkish grey with black at the veins, especially along the post- and antemedials. There is a faint black, crescent-shaped discal spot to the forewing with a streak of ground colour passing dorsally and basally to the angle of Culb with the cell that, in the majority of specimens, contains a white patch. Basal to the forewing antemedial and distal to the postmedials of both wings are narrow white fasciae. On both wings there is a thin, white. crenulate submarginal, that of the hindwing becoming broken and

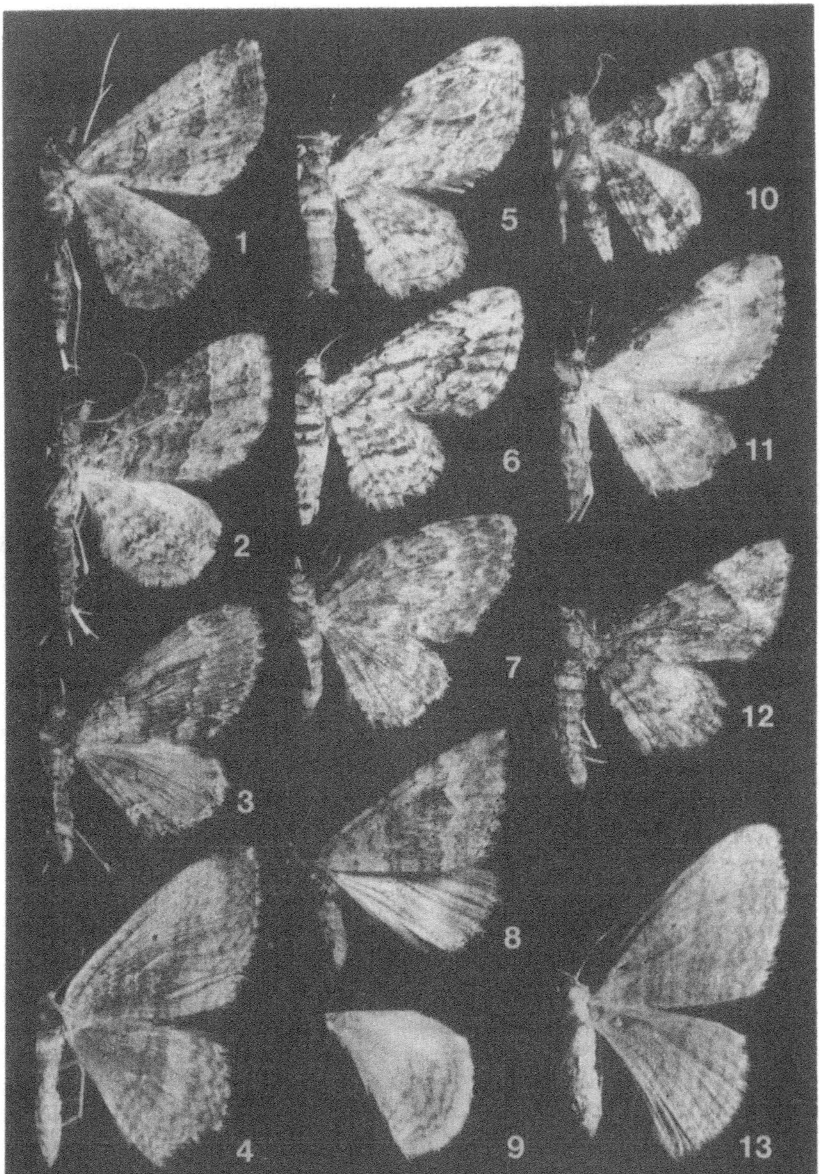


*Plate 58.* 1, 2. *Caledasthena montana;* 3. *Eupithecia rigida* (10 mm); 4. *Poecilasthena leucydra* (12 mm); 5 (♂), 6 (♀). *Chloroclystis lunifera;* 7. *Chloroclystis callinda;* 8. *Xanthorhoe finitima* (12 mm); 9. *Poecilasthena dimorpha,* ♂ hindwing; 10. *Chloroclystis scintillata* (7 mm); 11. *Chloroclystis pyrsodonta* (7 mm); 12. *Gymnoscelis sara hamata;* 13. *Poecilasthena dimorpha* ♀.

disappearing towards the dorsum. There is a thin, broken black line at the margin, strongest in the spaces, and the fringes are diffusely chequered pinkish grey and blackish.

HOLOTYPE ♂. New Caledonia: Mt. des Sources, 760 m Site 12a 06653/75513 (J. D. HOLLOWAY) 29.5.1971, BM 1971-507. BM geometrid slide 8523.

PARATYPES. 57 ex. New Caledonia, various localities. BM geometrid slides 8574 (♀), 9593 (venation).

The species was recorded as follows: Grand Lac site 1 (1), site 2 (10), site 3 (1); Pic du Pin site 7 (4); Mt. des Sources site 11 (39), site 12 (11), site 14 (12); Mt. Mou site 33 (1), site 34 (1), site 36 (11); Col Mouirange (1); Mt. Pouédihi (11); Mt. Koghis site 80 (1). It is restricted to open *Baeckea*-sedge maquis on the ultramafics and appears to be much commoner at altitudes above 500 m.

**Chloroclystis callinda sp. n.** (Plate 58: 7, Figure 54: 1)

♂ 7 mm. The wings are coloured deepish dull brown marked with pale fawn as illustrated. The species is close to *linda* Robinson (Fiji, New Hebrides) but the pale postmedial on both wings is simple, narrow (instead of double) with a diagnostic pale fleck subcostally, more reminiscent of the Oriental *modesta* Warren (type examined; Darjeeling) which resembles these two species in having the forewing costa prominently bowed. The insect is a darker, greyer brown than *linda*. In the male genitalia the valves are basally deeper and the sacculus is not separated apically from the main part of the valve by a cleft such as occurs in *linda*. The octavials of the abdomen are well separated, long, unequal in size (adjacent, equal and short in *linda*). *C. cuneativenis* Prout (unique male holotype examined; Sulawesi) also belongs to this section of the genus.

HOLOTYPE ♂. New Caledonia: Mt. Krapé, 460 m site 56 05274/76373 (J. D. HOLLOWAY) 22.7.1971, BM 1971-507. BM geometrid slide 8560.

**Chloroclystis lunifera sp. n.** (Plate 58: 5, 6, Figures 54: 2, 58: 5)

♂♀ 9–11 mm. Males with white ground, lightly marked in darkish brown as illustrated. The forewing costa is broad centrally, reflexed so that a marginal brush of dark grey scales partially (basally) covers an area of pale straw ones; similar pale straw scales occur in a slender crescent, anteriorly concave, extending from the discocellular area almost to the margin. The female also has a white ground but the dark brown markings are heavier, the postmedials of both wings, the antemedial and a submarginal row of dots on the forewing being black; the forewing has a marginal row of grey lunules, each bisected by a brown streak along the vein.

The species is very close to the Australian (vagrant to Norfolk I.) *insigillata* Walker (type examined) but in the Australian species the ground colour is grey with darker grey markings, giving a less contrasty effect; sexual dimorphism in colour is barely evident; in *insigillata* the forewing postmedial is more broken, less crenulate, and rounded rather than angled; the hindwing postmedial is less

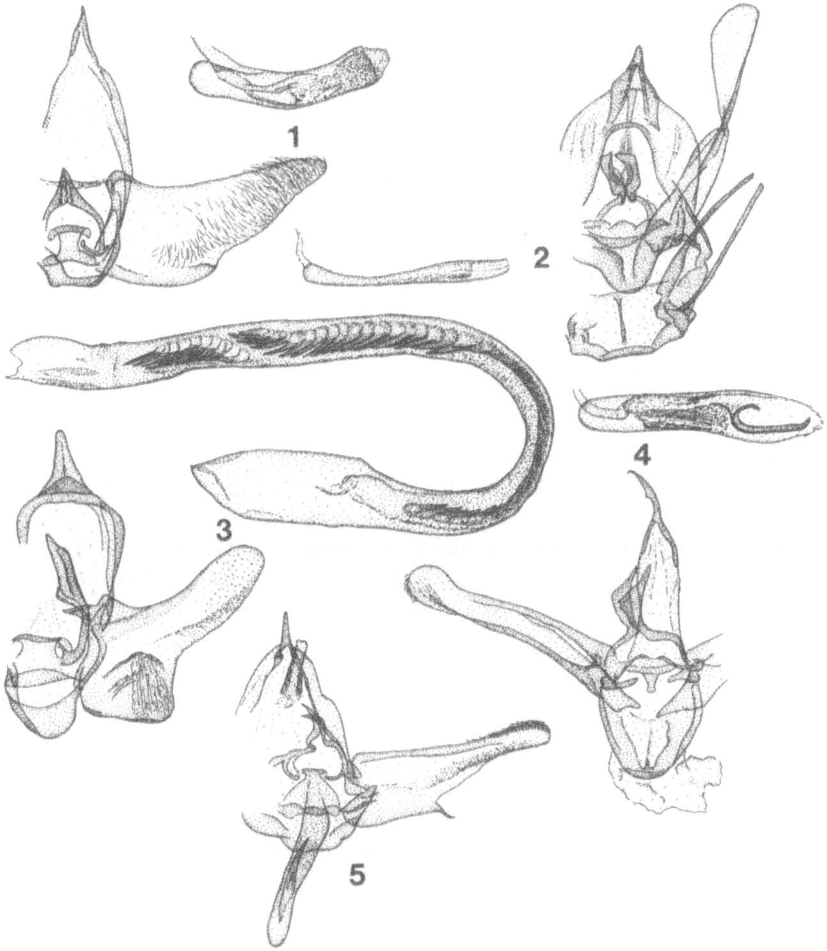


*Fig. 54.* Male genitalia of *Chloroclystis callinda* (1), *C. lunifera* (2), *C. macroaedeagus* (3), *Gymnoscelis chlorobapta* (4), *Chloroclystis scintillata* (5).

sigmoid. In the related, but smaller *encteta* Prout (New Hebrides, Fiji, Samoa, Cook Is.) the colour and dimorphism are similar but the male forewing crescent of straw-coloured scales is more elongate, straight centrally; the female postmedial is squarish at the angle and not crenulate. The male genitalia of the three species are indistinguishable.

HOLOTYPE ♂. New Caledonia: Mt. Dogny, 950 m Site 31 05902/76088 (J. D. HOLLOWAY) 27.6.1971, BM 1971-507. BM geometrid slide 8556.

PARATYPES: 18♂♂, 37♀♀ (slide 8575) New Caledonia, various localities; 1♂, 4♀♀ N. New Caledonia: Tinchialit, 2020 ft 22.viii-10.ix.1949 (L. E. CHEESMAN) BM 1950-1.

The species was recorded as follows: Grand Lac site 1 (3), site 2 (11), site 3 (10), site 4 (1); Pic du Pin site 7 (17); R. Bleue site 9 (3), site 10 (6); Mt. des Sources site 11 (34), site 12 (35), site 14 (30); Dothio site 19 (1); Col d'Amieu (1); Col Toma (1); Mt. Dogny (3); Mt. Mou site 33 (7), site 34 (1); Tiebaghi site 41 (1); Col d'Amoss site 47 (1); Col des Roussettes (1); Mt. Panié site 60 (15), site 61 (1); Mt. Aoupinié (2); Col Mouirange (1); Mt. Pouédihi (3); Anna-Madeleine (1); Les Dalmates site 77 (2); Mt Koghis site 80 (1). It is primarily an ultramafic maquis species, often common at high altitudes.

**Chloroclystis macroaedeagus sp. n.** (Plate 59: 1, Figure 54: 3)

♂ 9 mm. Markings brown on fawn as illustrated. The species is generally similar to *invisibilis* Warren (female type examined; New Guinea) but lacks the dark brown postmedials on both wings of that species, there being a paler double one, referable to the fasciae distal from the dark one in *invisibilis* but more distinct. On the hindwing the basal dark edging to the postmedial fuses with the next fascia basad at the dorsum; they are distinct in *invisibilis.* In the male genitalia both species have the distinctive, large, reflexed aedeagus, but it is relatively larger with larger cornuti in the New Caledonian species. In *macroaedeagus* the transtillae of the valves are longer and the valves themselves are narrower, straight instead of curved, and with a diagnostic angular ventral projection.

Though placed in *Chloroclystis* this species and its New Guinea relative show great similarities in facies and the packing of the aedeagus with cornuti to the two *Gymnoscelis* species described below, *acidna* and *pseudofluctuosa,* especially the latter. *G. pseudofluctuosa* is only known from females and it cannot be totally excluded that these are the females of *macroaedeagus.* However, they are significantly smaller, the pale fasciae are more diffuse and the forewing lacks the dark brown submarginal dashes of *macroaedeagus.*
HOLOTYPE ♂. New Caledonia: Grand Lac, 250 m Site 2L 06975/75356 (J. D. HOLLOWAY) 10.8.1971, BM 1971-507. BM geometrid slide 8569.

*Chloroclystis pyrsodonta* Turner (Plate 58: 11, Figure 58: 4)

*Chloroclystis pyrsodonta Turner;* ROBINSON (1975: 275).
A singleton of this Queensland species was taken at the Bréhoa R. site. ROBINSON recorded it from Fiji.

*Chloroclystis scintillata* Prout (Plate 58: 10, Figure 54: 5)

*Chloroclystis scintillata* Prout; ROBINSON (1975: 277).
Four specimens were taken, from Pic du Pin site 7, Kopéto, Mt. Mou site 33 and Mt. Krapé site 55. All are ultramafic localities but the last three are adjacent to dry west coast habitats. In Fiji and Rotuma the species inhabits dry secondary vegetation but was taken in more forested localities in the New Hebrides (ROBINSON).

*Collix ghosha subligata* Warren **stat. n.** (Plate 60: 8, Figure 55)

*Collix ghosha* Walker, 1862, *List Specimens lepid. Insects Colln Br. Mus.*, 24: 1294, type examined. *Collix subligata* Warren, 1896, *Novit. zool.*, 3: 385, type examined (Lifu).

The species is found in India, the Himalaya, Taiwan, Sundaland, the Lesser Sundas, Sulawesi, New Guinea (ssp. *dichobathra* Prout), Queensland (ssp. *puncticulata* Prout), the Solomons (ssp. *sticticata* Warren), the New Hebrides, Lifu and New Caledonia. The female type of *subligata* is larger than New Caledonian examples, the black markings of the underside somewhat more extensive and fused. In New Caledonia examples are variable but generally more contrasty than in other races, with the fawn postmedial band paler. The species was recorded as follows: R. Bleue site 8 (1); Petit Gouli (2); Mt. Mou site 33 (1); Col d'Amoss site 45 (1).

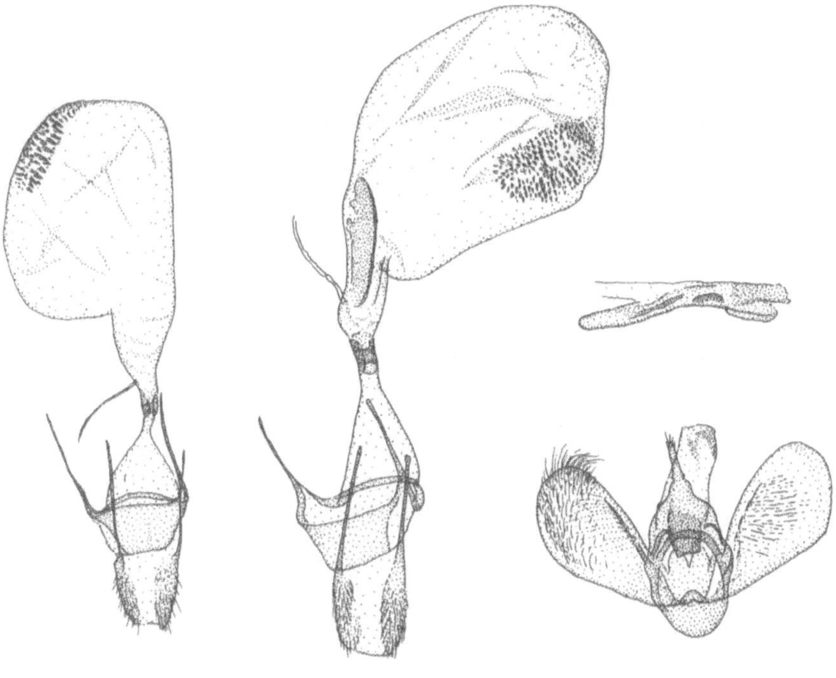

*Fig. 55.* Male and female genitalia of *Collix ghosha subligata*.

**Eoasthena sarramea sp. n.** (Plate 57: 11)

♀ 12 mm. The wings have numerous, slightly diffuse, crenulate red fasciae on a yellow ground as illustrated. The genus is centred on Fiji where there are six species (ROBINSON, 1975: 279); there are two in the New Hebrides as yet undescribed and one, *eurychora* (Prout), in Samoa. The New Caledonian

species is perhaps closest to the Fijian *stygna* Prout but the fascia immediately distal to the broader medial on both wings is relatively much more prominent (faint in *stygna*) though thinner than the next two distad; there are only these two fasciae basal to the narrow submarginal and distal to the faint one but there are three in *stygna*. The longitudinal streak dividing the forewing from margin to medial is more prominent as in one of the New Hebrides species, which has generally similar facies to the New Caledonian species. The female genitalia of *sarramea* have a more elongate, more finely scobinate bursa and a more triangular and distinctively umbonate base to the ductus bursae compared with the New Hebrides species.
HOLOTYPE ♀. New Caledonia: Sarraméa, 160 m Site 29 05888/76065 (J. D. HOLLOWAY) 21.6.1971, BM 1971-507. BM geometrid slide 8571.

**Eois tiebaghi sp. n.** (Plate 57: 10, Figure 59: 3)

♂♀ 8–10 mm. Ground colour yellow, mottled with orange, dull red, dark grey and black, the darker markings forming numerous, ill-defined, crenulate fasciae on the wings. On both wings there is a medial band where the ground colour is less obscured and that contains the dark discocellular dots; the fascia immediately distad is sharper than most and blackish. This medial band is not evident in *atrostrigata* Warren (Queensland; type examined), a more uniformly pale reddish insect, but is more prominent in *phaneroscia* Prout (type examined; Assam, Sundaland, and an undescribed subspecies with similar male genitalia in Queensland) which is generally more lightly marked and has a sharper angle to the distal border of the pale medial area of the forewing, this angle being nearer the costa. The male genitalia differ from those of *phaneroscia* in having a square, instead of rounded, saccus and in lacking a finely setose lobe on the valve sacculus. The tegumen in *phaneroscia* has a pair of lateral coremata. The aedeagus of *tiebaghi* (vesica not everted) contains an irregular mass of cornuti basally and a scattering of short ones apically, whereas in *phaneroscia* the basal group is more regular, heavily sclerotised, with a smaller group just distad; there is a large, broad cornutus in the latter species subapically, adjacent to a spined, sclerotised lobe.
HOLOTYPE ♂. New Caledonia: Tiebaghi, 440 m Site 39 04168/77382 (J. D. HOLLOWAY) 9.7.1971, BM 1971-507. BM geometrid slide 8565.
PARATYPES: 1♂, 1♀ Tiebaghi site 41; 1♀ Tiebaghi site 42; 1♂ Nouméa 13.ii.1963 Malaise trap (C. YOSHIMOTO & N. KRAUSS) BPBM.
The recorded occurrences of this species, though few, may be indicative of some association with *Acacia spirorbis*.

**Euphyia perseverata sp. n.** (Plate 60: 1, 2, Figures 56: 1, 58: 5)

♂♂ 14–16 mm, ♀♀ 15–17 mm. Both sexes are coloured various shades of grey as illustrated but the female is overall a much more uniform grey though fasciated as in the male. The hindwing over the basal two thirds is tinged brownish. The species is close to, and probably derived from, *E. severata* Guenée (E. and S.

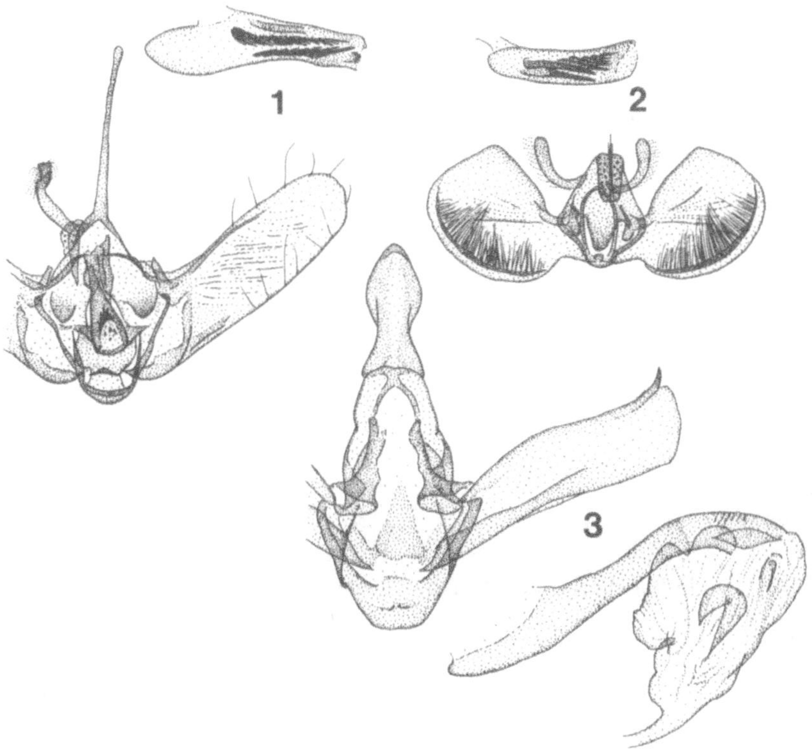


*Fig. 56.* Male genitalia of *Euphyia perseverata* (1), *Horisme grisearia* (2) and *Gonanticlea pulcherrima* (3).

Australia) but in both sexes the hindwing is much more weakly fasciated, the male forewing is much more contrastedly marked with the subbasal compound pale band splaying anteriorly instead of being of even thickness; the female forewing is less brown, more ashy grey. *E. severata* is smaller (12–14 mm). In the male genitalia the aedeagus vesica has two prominent bands of cornuti instead of one, and the valves and uncus are more elongate.

HOLOTYPE ♂. New Caledonia: Grand Lac 250 m Site 2 06975/75356 (J. D. HOLLOWAY) 15.5.1971, BM 1971-507. BM geometrid slide 8517.

PARATYPES: 65♂♂, 29♀♀ (slides 8523, 8593, 8593) New Caledonia, various localities; 1♂ N. New Caledonia: Tinchialit, 2020 ft 22.viii-10.ix.1949 (L. E. CHEESMAN) BM 1950-1; 1♂, 2♀♀ Mt. Koghis 500 m i.1963 (N. L. H. KRAUSS) BPBM.

The species was recorded in New Caledonia as follows: Grand Lac site 1 (7), site 2 (100), site 3 (17), site 4 (12), site 72 (10); Plaine des Lacs site 5 (5); Pic du Pine site 7 (53); R. Bleue site 8 (1), site 10 (2); Mt. des Sources site 11 (26), site 12 (41), site 14 (39); Poupa (1); Col Toma (2); Mt. Mou site 33 (5), site 34

(3), site 35 (2), site 36 (1); Tiebaghi site 41 (2); Col d'Amoss site 45 (2), site 46 (4), site 47 (21); R. Amoa (6); Mt. Krapé site 55 (2); Col de Boa (2); Houailou (2); near Bâ (38); Mt. Panié site 60 (1); Col Mouirange (39); Mt. Pouédihi (19); R. Fausse Yaté (3); Yate Gorgé (3); R. Koué (4); Anna-Madeleine (1); Les Dalmates site 77 (1), site 78 (1); Mt. Koghis site 80 (9). It is one of the most abundant ultramafic maquis species but can also be frequent in areas of maquis vegetation off the ultramafics.

*Eupithecia eupitheciata* (Walker) (Figure 58: 3)

*Eupithecia eupitheciata* (Walker); ROBINSON (1975: 283); HOLLOWAY (1976: Figure 476, ♂ genitalia).

This is a widespread Indo-Australian species, recorded from the Indian region, the Himalaya, Taiwan, S.E. Asia, Sundaland, the Philippines, Sulawesi, New Guinea, Australia (Queensland, New South Wales), the New Hebrides, Fiji, Samoa and Tonga. In New Caledonia it was recorded as follows: Bréhoa R. (2); Probokonoro (1); near Bâ (1). Miss CHEESMAN took a specimen at Pouébo.

*Eupithecia rigida* Swinhoe (Plate 58: 3, Figure 58: 6)

*Eupithecia rigida* Swinhoe, 1892, *Trans. ent. Soc. Lond.*, 1892: 2. HOLLOWAY (1976: Figure 472, ♂ genitalia).

This is a widespread but infrequently recorded Palaeotropical species, known to occur in Africa, India, the Himalaya, Borneo, Timor, Rossell I. and the Solomons. Single females were taken on the Mt. des Sources at site 12 and at Sarraméa and a further male has been taken on the Col des Roussettes in 1963 (BPBM).

**Gonanticlea pulcherrima sp. n.** (Plate 60: 5, 6, Figures 56: 3, 57: 1, 2, 58: 8)

♂♂ 16-18 mm, ♀♀ 18-21 mm. Though in general facies, sexual dimorphism and female genitalia structure (the signum of the bursa especially) this species is obviously a *Gonanticlea*, it is isolated within the genus and has several unique characters. Primarily it is distinguished by the marked modification of the male hindwing (Figure 57). The insect is much more robust than other members of the genus, except possibly *sublustris* Warren (Moluccas, New Guinea, Bismarcks, Solomons) which also stands somewhat isolated within the genus. The subapical marks of the forewing are typical but the angle of the postmedial is narrower, more dentate and central than in other congeners. The male is generally greyish, tinged brown on the forewing broadly submarginally and along the costa; there are broad markings of brownish black with a pale cream medial band which is suffused rufous anteriorly and posteriorly. The female is more uniform grey, more finely fasciated; the medial band is grey or yellowish, in the latter case the fasciae standing out strongly rufous on the yellow background. The male genitalia are very robust compared to those of other

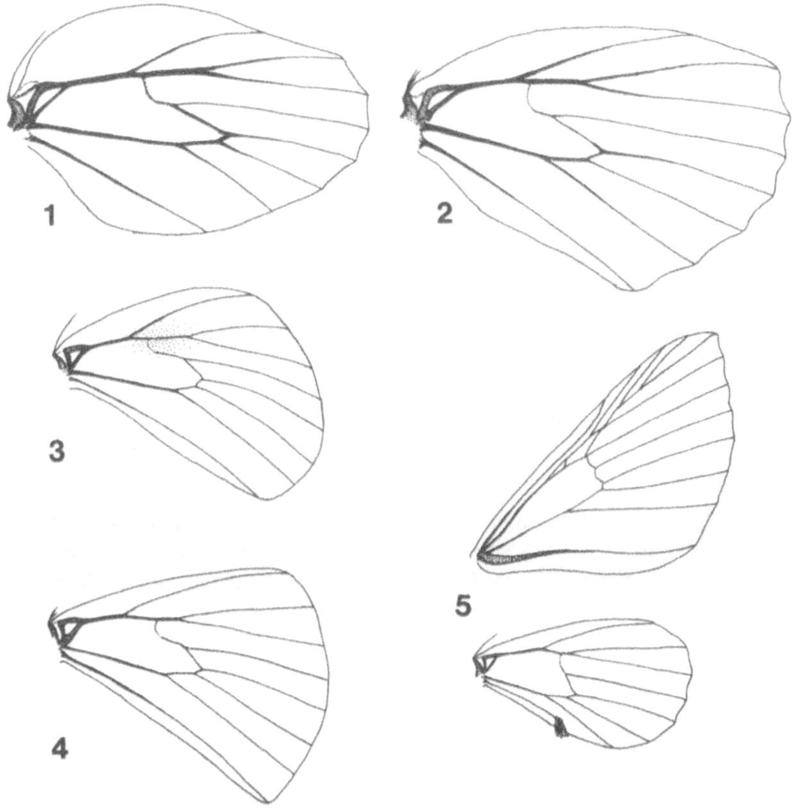


*Fig. 57.* Venation of the hindwings of (1) male and (2) female *Gonanticlea pulcherrima* and of (3) male and (4) female *Poecilasthena dimorpha,* and both wings of male *Caledasthena montana* (5). The stipple in (3) indicates the area of cream scales mentioned in the text.

species examined, the uncus being short and broad; the valves are relatively broad and their apical projection is distinctive; there are no slender projections from the juxta as are found in *sublustris* and in the type species *aversa* Swinhoe (Himalaya).

HOLOTYPE ♂. New Caledonia: Mt. Panié, 800 m Site 60 04764/77257 (J. D. HOLLOWAY) 27.7.1971, BM 1971-507. BM geometrid slide 8524.

PARATYPES: 15♂♂, 16♀♀ (slides 8596, 9591 (venation)) New Caledonia, various localities; 1♂ Mt. Pouédihi 4.8.1971, MNHN; 1♀ La Coulée 1.3.1976 (J. GUTIERREZ) MNHN; 1♀ La Coulée 29.6.1976 (J. GUTIERREZ) MNHN.

The species was recorded as follows: Grand Lac site 1 (1), site 2 (5), site 3 (3), site 71 (1); Pic du Pin site 7 (3); Mt. des Sources site 11 (2), site 12 (2), site 14 (1); Mt. Mou site 33 (1); Tiebaghi site 41 (2), site 42 (3); Mt. Krapé site 56 (2); Col de Boa (1); near Bâ (1); Mt. Panié site 60 (2); Col Mouirange (5); Mt.

Pouédihi (4); Yaté Gorge (1); R. Koué (1); Anna-Madeleine (1); Les Dalmates site 77 (1). It is restricted to ultramafic maquis localities apart from the records from Mt. Panié (where maquis elements are present).

*Gymnoscelis acidna* Turner (Plate 59: 3)

*Gymnoscelis acidna* Turner; ROBINSON (1975: 285), as subspecies of *minutissima* Swinhoe; HOLLOWAY (1977: 130)

The species has been recorded from Queensland, Norfolk I., Lifu, the New Hebrides and Fiji. In the last locality the larva has been recorded as feeding on the flowers of *Acacia farnesiana* and *Myrtus vitiensis*. *Acacia farnesiana* is an important component of the Anse Longue thicket and the moth is often common in west coast localities though it is also characteristic of ultramafic maquis localities. It was recorded as follows: Grand Lac site 1 (3), site 2 (14), site 3 (10), site 72 (3); Pic du Pin site 7 (1); R. Bleue site 8 (1), site 10 (1); Mt. des Sources site 11 (14), site 12 (6) site 14 (8); Anse Longue site 15 (2), site 16 (15), site 32 (9); Poupa (8); Koh (4); Ouatou R. (2); Bréhoa R. (4); Ouambéo R. (2); Col d'Amieu (1); Petit Gouli (3); Sarraméa (4); Mt. Mou site 33 (2); Tiebaghi site 41 (1), site 42 (1); Bambou R. (1); Kavatch (1); Mt. Panié site 60 (1); Col Nassiboué (4); Col Mouirange (2); Yaté Gorge (4); Port Boisé site 73 (7), site 74 (12); Les Dalmates site 77 (2), site 78 (1); Ouen Toro site 81 (1), site 82 (1). BMNH: Pouébo (CHEESMAN). BPBM: Mt. Koghis.

*Plate 59.* 1. *Chloroclystis macroaedeagus;* 2. *Gymnoscelis pseudofluctuosa;* 3. *Gymnoscelis acidna* (6 mm); 4. *Gymnoscelis tristrigosa* (8 mm); 5. *Micrulia pacifica;* 6. *Gymnoscelis chlorobapta* (9 mm).

*Gymnoscelis chlorobapta* Turner (Plate 59: 6, Figure 54: 4)

*Gymnoscelis chlorobapta* Turner, 1907, *Proc. Linn. Soc. N. S. Wales*, 31: 687.

New Caledonian specimens resemble a series from Queensland in the BMNH attributed to this species and fitting the original description. In New Caledonia singletons were taken at Koh, Ouatou R. and Ouambéo R.

**Gymnoscelis pseudofluctuosa sp. n.** (Plate 59: 2, Figure 58: 7)

♀♀ 6–7 mm. The wings are marked dark brown on a yellowish fawn ground as illustrated. This is a close relative of *G. acidna* (*q.v.*) and of *G. minutissima* Swinhoe (Bali, Sumbawa), but the postmedials of both wings are rounded rather than angled and that of the hindwing is slightly bilobed; the prominent spots of ground colour centrally at the margin of both wings are also a distinguishing feature, though somewhat poorly developed in some specimens. There is a superficial resemblance to *Chlorocystis lepta fluctuosa* Prout (Fiji) but the *Chloroclystis*, apart from having two pairs of spurs on the hind tibia, lacks the marginal pale spots, is more regularly fasciated, a paler, less rich brown, and does not have darker brown flecks basad to the forewing postmedial on the veins. The genitalia are intermediate between those of *acidna* and *minutissima*, the signum in the bursa of the former being a moderately simple plate, showing the beginnings of a spiral in some specimens and bearing scattered small spines, mainly peripheral. In *minutissima* the bursa is more elongate and slender with a delicate, strongly spiralled signum.

HOLOTYPE ♀. New Caledonia: Grand Lac, 260 m site 3 06965/75355 (J. D. HOLLOWAY) 18.5.1971, BM 1971-507. BM geometrid slide 8573.

PARATYPE ♀♀. New Caledonia: Grand Lac site 2 (1), site 3 (1); Bréhoa R. (1); Sarraméa (2); Kopéto (1); Mt. Mou site 33 (1); Mt. Panié site 60 (1); Yaté Gorge (1).

It is possible that there was some confusion between this species and *acidna* in counts made in the field for, though their distinction was suspected, discriminatory characters were not fully appreciated until subsequent examination. The new species is certainly rarer than *acidna* and would appear to favour more forested localities.

**Gymnoscelis sara hamata ssp. n.** (Plate 58: 12, Figure 59: 4)

*Gymnoscelis sara* Robinson, 1975: 286.

♂♂ 9 mm. The facies is as in the typical race from Fiji but the hindwing postmedial is not so markedly bilobed. The aedeagus vesica has a single hooked cornutus instead of two long, sinuous ones.

HOLOTYPE ♂. New Caledonia: Dothio, 400 m, Site 19 06133/76135 (J. D. HOLLOWAY) 9.6.1971, BM 1971-507. BM geometrid slide 8563.

PARATYPE ♂. New Caledonia: Mt. Koghis Site 79.

Two females taken by G. S. ROBINSON in the New Hebrides are almost certainly referable to *sara*.

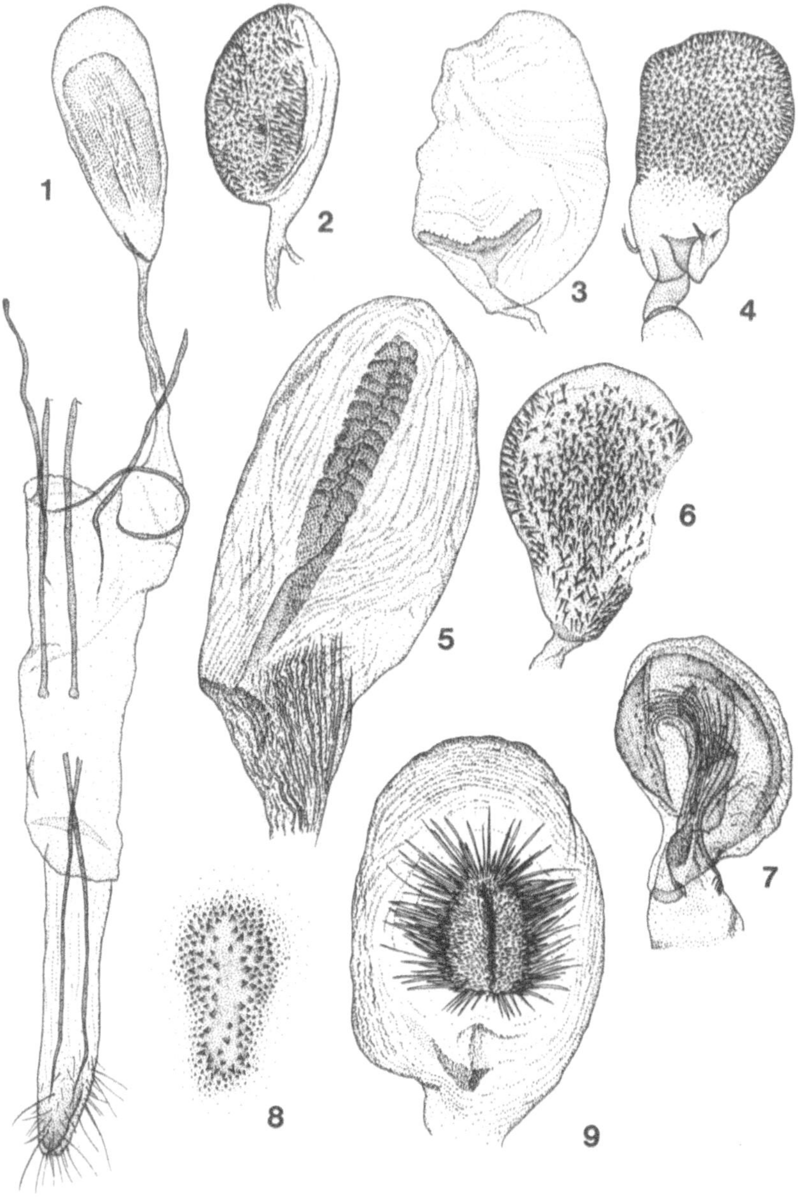


*Fig. 58.* Female genitalia of *Caledasthena montana* (1), bursae of *Chloroclystis lunifera* (2), *Eupithecia eupitheciata* (3), *Chloroclystis pyrsodonta* (4), *Euphyia perseverata* (5), *Eupithecia rigida* (6), *Gymnoscelis pseudofluctuosa* (7), *Poecilasthena dimorpha* (9), and signum of *Gonanticlea pulcherrima* (8). The posterior apophyses in (1) are broken.

*Fig. 59.* Male genitalia of *Papuanticlea ochroria* (1), *Sauris curvicosta* (2), *Eois tiebaghi* (3), *Gymnoscelis sara hamata* (4).

*Gymnoscelis tristrigosa* (Butler) (Plate 59: 4)

*Eupithecia tristrigosa* Butler, 1880, *Proc. zool. Soc. Lond.*, 1880: 688, type examined.

The female genitalia of New Caledonian specimens agree with those of typical *tristrigosa* from Sri Lanka. Also referable to this species is a single Fijian female, illustrated as *imparatalis* by ROBINSON (1975: Plate fig. 255) but later dissected by this author and noted as distinct from genuine Fijian *imparatalis* on an attached label. The species has been recorded from the Indian region, the Himalaya, Taiwan, the Philippines, Sulawesi, New Guinea, Queensland, Fiji and New Caledonia. Singletons were recorded from Koh and Col Toma and there is a specimen from Le Crouen in the BPBM.

## *Horisme suppressaria* (Walker)

*Phibalapteryx suppressaria* Walker, 1863, *List Specimens lepid. Insects* Colln Br. Mus., 26: 1721, type examined.
*Xanthorhoe suppressaria* (Walker); HUDSON (1928: 112).
*Xanthorhoe*? *suppressaria* (Walker); Prout, 1939, *Gross-Schmett. Erde*, 12: 251.

The New Caledonian species described below is most closely referable to the type of *suppressaria* and specimens associated with it in the BMNH collection. The female type is labelled as from Auckland, New Zealand. PROUT suggested that *suppressaria* was better placed in *Horisme* or *Euphyia* and that it was related to *severata* Guenée of the latter genus. HUDSON was not acquainted with the species and DUGDALE (1971 and *in litt.*) has not discovered more recent New Zealand material. The New Zealand locality may be erroneous as there is an identical female from Queensland in the BMNH collection. The two females referred to in the original descriptions lack abdomens but the Queensland specimen was dissected.

There are numerous '*Euphyia*' and '*Horisme*' species in Australia with facies of the *suppressaria*/*severata* type and the majority examined e.g. *Horisme mortuata* Guenée (type examined), *Euphyia arachnitis* Turner, *Horisme scotodes* Turner, have female genitalia similar to those illustrated for *perseverata* (Figure 58: 5), the New Caledonian derivative of *severata*. The Queensland *suppressaria* specimen and the related New Caledonian species have the bursa in the female almost round, with irregular scobination on one side. New Caledonian males have the valves very rounded as in *Euphyia squamulata* Warren (type examined; E. Australia) but *squamulata* lacks the prominent projections from the transtillae exhibited by the New Caledonian species.

Thus there would appear to be at least two distinct morphological groups amongst these finely fasciated, grey Australian larentiines, both only dubiously referable to *Euphyia* or *Horisme*. It is beyond the scope of this work to unravel the situation further and erect new genera. The New Caledonian species will be described as a *Horisme*, this being the genus to which *suppressaria* is assigned in the current BMNH arrangement.

## **Horisme grisearia sp. n.** (Plate 60: 3, 4, Figure 56: 2)

♂♀ 10–13 mm. The ground colour is grey, variable in shade, fasciated darker, again variably; some darker specimens have the medial area of the forewing paler grey, the paler specimens having this area often slightly darker or rufous tinged; a few female specimens have a broad black bar both at the antemedial and bisecting the apex as illustrated (Plate 60: 4). *H. suppressaria* is a browner insect and slightly larger; the fasciae in the medial area of the forewing are crenulate, dark extending along the veins, rather than gently sinuous as in *grisearia*. The two dark fasciae distal to the forewing postmedial are much more prominent in *grisearia* and the basal fasciation of the forewing and that over the dorsal half of the hindwing is better defined, especially the postmedial

*Plate 60.* 1 (♂), 2 (♀). *Euphyia perseverata;* 3, 4. *Horisme grisearia;* 5 (♂), 6 (♀). *Gonanticlea pulcherrima;* 7. *Papuanticlea ochroria;* 8. *Collix ghosha subligata* (13 mm); 9, 10. *Scotocyma legalis* (14–15 mm); 11. *Ziridava xylinaria* (15 mm); 12. *Horisme xylinata orientalis.*

of the latter, which is edged paler distad. The female genitalia differ from those of the Queensland *suppressaria* specimen in that the bursa is less scobinate generally and lacks scobination distally.

HOLOTYPE ♂. New Caledonia: Col Toma, 460 m Site 27 05813/76123 (J. D. HOLLOWAY) 19.6.1971, BM 1971-507. BM geometrid slide 8525.

PARATYPES. 79 ex. New Caledonia, various localities, BM geometrid slide 8594 (♀).

The species was recorded as follows: Grand Lac site 1 (3), site 2 (40), site 3 (8), site 4 (2), site 72 (3); Pic du Pin site 7 (37); Mt. des Sources site 11 (1); Dothio site 19 (1); Koh (4); Ouambéo R. (1); Col Toma (9); Mt. Dogny (19); Tiebaghi site 41 (3); Col d'Amoss site 47 (2); Ste. Therèse (2); Col de Boa (2); Houailou (1); near Bâ (5); Mt. Panié site 60 (3); Col Mouirange (1); Mt. Pouédihi (1); Les Dalmates site 77 (2); Mt. Koghis site 79 (1). BMNH: Tinchialit. The species is associated with areas of *Baeckea*-sedge maquis on and off the ultramafics.

**Horisme xylinata orientalis ssp. n.** (Plate 60: 12, Figure 61: 1)

*Coenocalpe xylinata* Warren, 1906, *Novit. zool.*, 13: 98, type examined.

♂♀ 13–15 mm. Wings with brown markings on pale brown as illustrated, not significantly different from typical *xylimata* (New Guinea, Queensland) and having similar pale grey blotching exterior to the postmedial in a few specimens. The male genitalia show the major differences: the sacculus of the valve has two slight angles interiorly instead of a distally directed ampulla; the aedeagus vesica has a cluster of long, slender cornuti but these are very short in the typical race.

HOLOTYPE ♂. New Caledonia: Col d'Amieu, 320 m Site 26 05828/76117 (J. D. HOLLOWAY) 18.6.1971, BM 1971-507. BM geometrid slide 8520.

PARATYPES: 13♂♂, 2♀♀ New Caledonia, various localities.

The species was recorded as follows: Dothio site 18 (1); Col d'Amieu (2); Petit Gouli (3); Sarramea (4); Mt. Dogny (1); Mt. Mou site 33 (1), site 36 (1); Probokonoro (1); Col des Roussettes (1); near Bâ (3); Col Nassiboué (1); Grand Lac site 72 (1); Port Boisé site 74 (1); Les Dalmates site 77 (1). It is probably a rainforest species. The subspecies also occurs in the New Hebrides (Santo).

**Micrulia pacifica sp. n.** (Plate 59: 5, Figure 60: 1)

♂♀ 8–9 mm. The species is slightly larger than *tenuilinea*, a richer brown in colour, having the discal spots of both wings more prominent. The male lacks the tuft of scales on the underside of the hindwing tornus that characterises *tenuilinea*. The female genitalia are as in *tenuilinea* (illustrated by HOLLOWAY, 1977) but those of the male differ considerably as can be seen in Figure 60. ROBINSON (1975) referred the type series listed below to *tenuilinea*.

HOLOTYPE ♂. Fiji: Rotuma I., Furoroa 22.iv.1971 (G. S. ROBINSON). BM geometrid slide 8343.

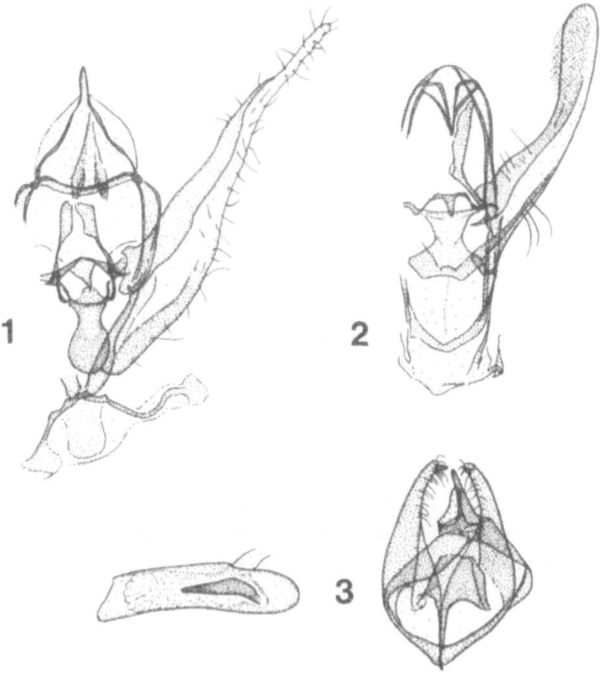


*Fig. 60.* Male genitalia of *Micrulia pacifica* (1), *M. tenuilinea* (2) and *Idaea trypherophа* (3).

PARATYPES. 3♀♀ Fiji: Rotuma (Noatau, Losa Rd., Binafa fdr. rd.) (G. S. ROBINSON). BM geometrid slide 8344.
OTHER MATERIAL. 1♂ New Caledonia: Tipindje 10.ii.1963 (YOSHIMOTO & KRAUSS) BPBM, dissected.

*Micrulia tenuilinea* Warren (Figure 60: 2)

*Micrulia tenuilinea* Warren; HOLLOWAY (1977: 131).
The species has been recorded from Sri Lanka, Assam, Singapore, Queensland, Norfolk I. and Samoa, though the Samoan species may prove to be *pacifica*. In the BPBM there is a male from Le Crouen and a female from La Foa.

## PAPUANTICLEA Gen. n.

The venation of the wings is as in typical *Gonanticlea* Swinhoe (type species *aversa* Swinhoe, examined) except the forewing cell is more elongate, less triangular, the more distal areole is much broader and the more basal one very much smaller; the forewing is generally more rounded apically and narrower.

The species do not show the sexual dimorphism of *Gonanticlea*. They have generally greenish forewings with a truncated paler band extending from the forewing costa to the discal spot and a postmedial with an irregular central angle.

The male genitalia differ from those of typical *Gonanticlea* in the great complexity of the valve structure (Figure 59: 1). The male of *G. aversa* has simple valves, and characteristic long slender projections from each side of the juxta with four setae towards the apex giving the appearance of a claw. This last structure is not found in all *Gonanticlea* species but a homolgous structure is found in the Papuan *sublustris* Warren.

The female bursa of *Papuanticlea* is lightly scobinate, the ductus sclerotised in a dorsal band; in *Gonanticlea* there is a definite signum (e.g. Figure 58: 8) and no widespread scobination.

*Papuanticlea* includes the following species previously placed in *Gonanticlea.*: *P. horia* (Prout) **comb. n.** (*Gonanticlea horia* Prout, 1939, *Gross-Schmett. Erde*, 12: 270, type examined, Queensland); *P. subcaesia* (Warren) **comb. n.** (*Gonanticlea subcaesia* Warren, 1903, *Novit. zool.*, 10: 373, type examined, New Guinea); *P. onaea* (Prout) **comb n.** (*Gonanticlea onaea* Prout, 1939, *Gross-Schmett. Erde*, 12: 269, type examined, New Guinea); *P. semiflava* (Warren) **comb. n.** (*Anticlea semiflava* Warren, 1906, *Novit. zool.*, 13: 93, type examined, New Guinea); a new species, *P. ochroria*, described below from New Caledonia and hereby designated as the type species of *Papuanticlea*.

The species *horia* and *ochroria* are closely related as discussed in the description of the latter below. *P. subcaesia* has similar male genitalia but the setae from the apex of the valve costa and submarginally on the valve are minutely spatulate, the spatulate tip being darker; the aedeagus is sharply angled with a long, irregular cornutus in the vesica. *P. semiflava* stands apart in the morphology of its male genitalia and its facies; the pattern of the forewing is more sharply defined and contrasted and the hindwing is bright yellow (*onaea* resembles *semiflava* in facies but was not dissected); there are two dorsal horns to the uncus, the ventral part of which is broadly three-lobed; the valve, though similarly rounded, has a central diagonal sclerotised flange and projections in positions different from the other species; the aedeagus vesica contains two large cornuti, one rather irregular in shape.

**Papuanticlea ochroria sp. n.** (Plate 60: 7, Figure 59: 1)

♂♀ 11–12 mm. The forewings are greenish, marked with brown as illustrated. The hindwings are yellow with a diffuse brown margin. The species is close to *horia* (Prout) but in that species the hindwings are entirely greyish brown, the aedeagus lacks a cornutus, the processes on the valves are invested with hair-like setae instead of thick ones, the central inwardly directed valve process is straight instead of angled, those from the back of the valves more slender and the processes from the tegumen generally broader.

HOLOTYPE ♂. New Caledonia: Kavatch, 40 m Site 50 04857/77068 (J. D. HOLLOWAY) 18.7.1971, BM 1971-507. BM geometrid slide 8564.
PARATYPES. 1♂, 3♀♀ New Caledonia: Koh (2); Ouatou R. (2). BM geometrid slide 8595 (♀).

*Phrissogonus laticostata* (Walker)

*Phrissogonus laticostata* (Walker); HOLLOWAY (1975: 132).

This species has been recorded from most of Australia (not the Northern Territory or Tasmania), New Zealand, Norfolk I., the Kermadecs and Lifu. It was recorded as follows in New Caledonia: Grand Lac site 2 (3); Mt. des Sources site 11 (1), site 12 (2), site 14 (2); Bréhoa R. (2); Col Nassiboué (4); Ouen Toro site 64 (3); Port Boisé site 73 (1); R. Koué (1). BPBM: Nouméa, Forêt de Thi.

**Poecilasthena dimorpha sp. n.** (Plate 58: 9, 13, Figures 57: 3, 4, 58: 9, 61: 4)

♂♂ 12–14 mm, ♀♀ 14–15 mm. Wings with emerald green fasciation on a satiny white ground, less translucent than in *leucydra* below. The male has the hindwing narrower than that of the female, more rounded at the apex and tornus (Figure 57); the anterior half is invested with less reflective, pale cream scales. There is a triple postmedial to both wings as in *leucydra* but four submarginals instead of two (in *leucydra* the first is faint and the fourth fused with the margin). The male genitalia lack the coremata of *leucydra* and there are differences also in the shape of the valves, the uncus and the transtillae. The signum of the female fursa is elongate instead of round. The eighth sternite of the male is unmodified (it tapers distally and there are two basal projections in *leucydra*).
HOLOTYPE ♂. New Caledonia: Pic du Pin, 240 m Site 7L 06879/75383 (J. D. HOLLOWAY) 8.8.1971, BM 1971-507. BM geometrid slide 8516.
PARATYPES. 26♂♂, 28♀♀ (slide 8577) New Caledonia: Grand Lac site 2 (15), site 3 (3), site 4 (3); Pic du Pin site 7 (10); R. Bleue site 10 (2); Mt. des Sources site 11 (1), site 12 (4); Col Toma (2); R. Amoa (1); Probokonoro (1); near Bâ (1); Col Mouirange (2); Les Dalmates site 77 (2); Mt. Koghis site 80 (1); 2♂♂ N. New Caledonia: Tinchialit, 2020 ft 22.viii-10.ix.1949 (L. E. CHEESMAN) BM 1950-1.

It was not recognised that there were two *Poecilasthena* species in New Caledonia when the field counts were made. Sufficient specimens were retained to indicate that *dimorpha* is primarily an ultramafic maquis species whilst *leucydra* is more associated with forest, being recorded frequently off the ultramafic rocks.

*Poecilasthena leucydra* Prout (Plate 58: 4, Figure 62)

*Poecilasthena leucydra* Prout; ROBINSON (1975: 290)

*P. leucydra* occurs also in Fiji and Samoa and was recorded in New

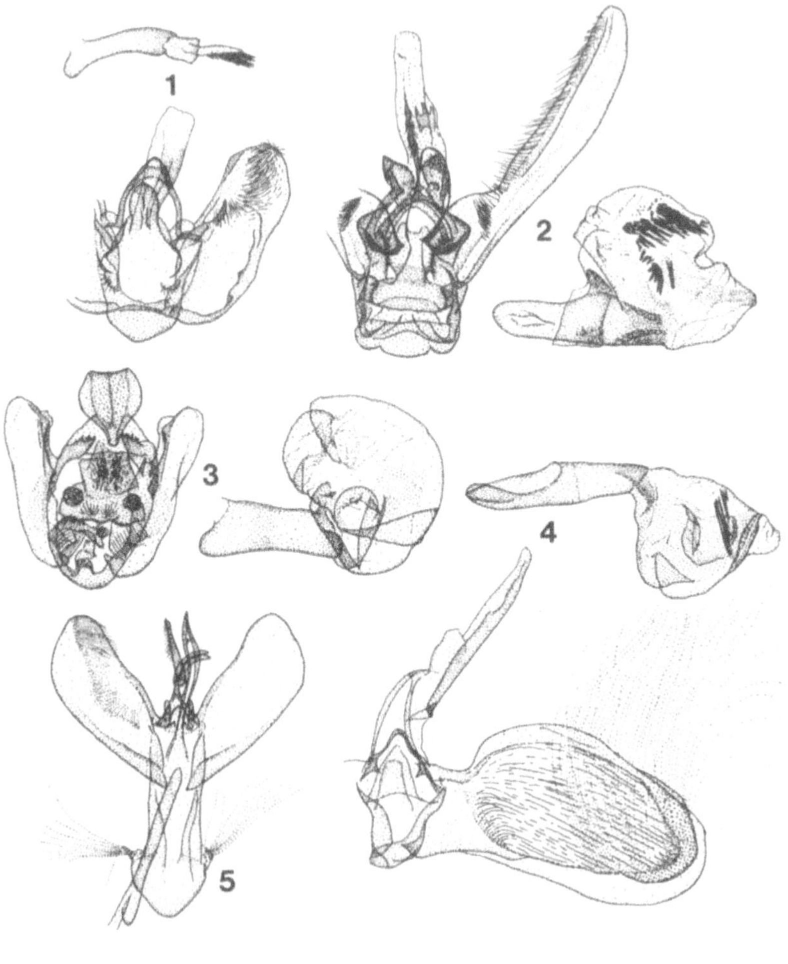


*Fig. 61.* Male genitalia of *Horisme xylinata orientalis* (1), *Ziridava xylinaria* (2), *Scotocyma legalis* (3), *Poecilasthena dimorpha* (4), and *Caledasthena montana* (5).

Caledonia as follows: Pic Du Pin site 7 (2); R. Bleue site 10 (2); Mt. des Sources site 11 (2), site 12 (1); R. Kuenthio (1); Dothio site 18 (1); Poupa (3); Koh (1); Ouatou R. (1); Col Toma (1); Sarraméa (2); Kavatch (1); Mt. Krapé site 55 (1); Mt. Panié site 60 (1); Mt. Aoupinié (2); Col Mouirange (1); Mt. Pouédihi (1). BPBM: Col des Roussettes.

### *Sauris curvicosta* Prout (Figure 59: 2)

*Sauris curvicosta* Prout; HOLLOWAY (1977: 133)

Described from the Loyalty Is., this species occurs in the forest element of

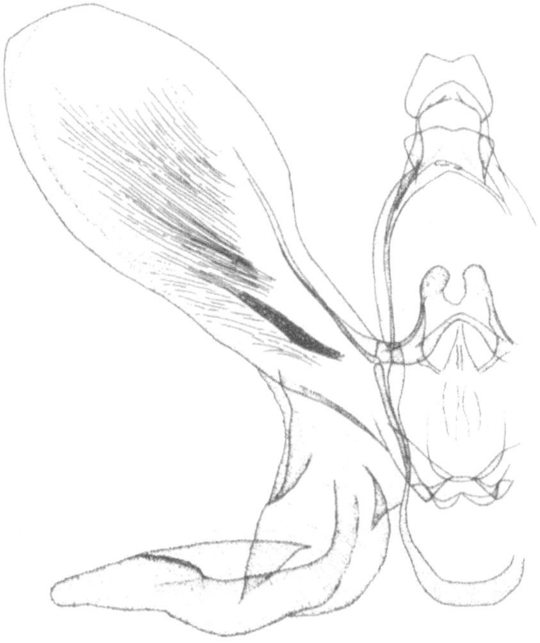

*Fig. 62.* Male genitalia of *Poecilasthena leucydra.*

Norfolk I. and was recorded as follows in New Caledonia: Pic du Pin site 7 (2); Koh (1); Col Toma (1); Tiebaghi site 41 (1); Probokonoro (2); Mt. Koghis site 79 (1). BMNH: Tinchialit.

### *Scotocyma legalis* (Warren) (Plate 60: 9, 10, Figure 61: 3)

*Coenocalpe? legalis* Warren, 1896, *Novit. zool.*, 3: 384, type examined, Lifu.

*S. legalis* has similar facies (including variation; some specimens have extensive white blotching on the wings) to *albinotata* (Walker) (type examined, Queensland) but no males of the latter were available to enable comparison of the male genitalia. *S. scotopepla* Prout (type examined, New Guinea) has a simple, narrow, tapering uncus, simple, unserrated socii (or transtillae) and a less elaborate valve costa. *S. miscix* (Fiji; Robinson (1975: 296)) has a highly scobinate vesica to the aedeagus. G. S. Robinson has taken a blackish female of the genus in Samoa. The species occurs in Lifu and was recorded from the following New Caledonian localities: Col des Roussettes (1); Mt. Krapé site 56 (1); Mt. Panié site 60 (1); Mt. Koghis site 79 (1). BPBM: Mt. Koghis (1).

### *Xanthorhoe finitima* (Walker) (Plate 58: 8)

*Larentia finitima* Walker, 1862, *List Specimens lepid. Insects Colln Br. Mus.*, 24: 1203, type examined.

This species was described from four females taken on the Isle of Pines in the middle of the last century by MACGILLIVRAY, naturalist on *H.M.S. Herald*, and presented to the BMNH by the Admiralty. It has not been recorded subsequently.

The genitalia are very similar to those of female *X. sodaliata* (*q. v.*) but the spherical bursa is set markedly asymmetrically on the ductus instead of only slightly so, and the ductus is relatively broader and square basally.

### *Xanthorhoe sodaliata* (Walker)

*Xanthorhoe sodaliata* (Walker); HOLLOWAY (1977: 133).

*X. sodaliata* occurs throughout mainland Australia except desert areas and is one of the more abundant species on Norfolk I. A single male was taken at Dothio site 18 and five females, very worn, at the Col d'Amieu site. The latter, following the arrival of a rain-bearing front from the west, could conceivably have been vagrants from Australia.

### *Ziridava xylinaria* Walker (Plate 60: 11, Figure 61: 2)

*Ziridava xylinaria* Walker, 1863, *List Specimens lepid. Insects Colln Br. Mus.*, 26: 1550, type examined.

The aedeagus vesica and apex of the New Caledonian male is more thickly scobinate than in typical *xylinaria* from Sarawak but otherwise there is little difference. The species is recorded from the Indian region, Taiwan, Sundaland, the Philippines, Sulawesi, Lesser Sundas, Moluccas, New Guinea (Rossell I.), Queensland (ssp. *leptomita* (Turner)) and the New Hebrides, and was taken in New Caledonia as follows: Col Toma (5); Mt. Koghis site 80 (2); Tinchialit (CHEESMAN).

## GEOMETRIDAE: ENNOMINAE

### *Aplochlora vivilaca* (Walker) (Plate 63: 8, Figure 66: 2)

*Aplochlora vivilaca* (Walker); FLETCHER (1957: 64); ROBINSON (1975: 299)

Both FLETCHER and ROBINSON included under this name forms from Sri Lanka (type examined), India, Taiwan, Malaya, Sulawesi, New Guinea (*subflava* Warren), Queensland (*pisochroa* Turner), the Bismarcks, Solomons (including Rennell I.) and Fiji. Singletons were taken in New Caledonia at Pic du Pin site 7 and Petit Gouli and there is a specimen from Nouméa in the BPBM.

Material from each locality is limited but generally invariable in external characters though Australian specimens are slightly larger and brighter green. The male genitalia examined (Sulawesi, Queensland, New Caledonia, Fiji) suggest that further material may facilitate the separation of distinct subspecies or even species.The primary difference is in the ornamentation of the aedeagus vesica though there are minor differences in valve shape (that of *pisochroa* being broader and rounder than in other forms) and the Fijian male dissected had a shorter, thicker uncus than the others. The Fijian vesica bears eleven slender cornuti in an even row, that from New Caledonia six larger ones in a very uneven row, the Sulawesi male had seven in an approximately even row. The Queensland male, referable to *pisochroa*, had a long row of approximately forty increasing in size towards the apex of the vesica.

*Casbia albinotata* Warren (Plate 61: 5, Figure 68: 3)

*Casbia albinotata* Warren, 1903, *Novit. zool.*, 10: 401, type examined.

The type of this species was from Queensland but it also occurs in New South Wales and New Guinea. A male was taken at Ouen Toro site 82 and a female from nearby at site 81. The genitalia indicate close relation within the genus to another Queensland species, *calliorma* Turner.

*Casbia rectaria* Walker

*Casbia rectaria* Walker, 1866, *List Specimens lepid. Insects Colln Br. Mus.*, 35: 1667, type examined (Queensland).

Specimens from New Caledonia and the New Hebrides were referred by ROBINSON (1975) to the Fijian *alphitoniae* Prout but compare in their male genitalia exactly with typical *rectaria*. *C. rectaria* has two harpes to the valve whereas *alphitoniae* has only one more needle-like one. *C. aedoea* Robinson, also from Fiji, resembles *alphitoniae* in the single harpe, which is more prominent. The two Fijian species are distinct externally and ecologically and no other species (the genus is primarily Australian with some extension into New Guinea) examined with approximately similar facies has a single harpe to the valve. It is possible therefore that this duplex evolved within Fiji rather than through a double invasion.

**Casbia rectaria lilacina ssp. n.** (Plate 61: 1, Figure 63: 2)

♂♀ 11–13 mm. This subspecies is generally smaller than typical *rectaria* (13 mm) and differs primarily in being pinkish grey overall rather than pale reddish brown. The darker fasciae on the wings tend to be more prominent.
HOLOTYPE ♂. New Caledonia: Mt. Pouédihi, 540 m Site 66 06728/75474 (J. D. HOLLOWAY) 4.8.1971, BM 1971-507. BM geometrid slide 8197.
PARATYPES: 54 ex. New Caledonia, various localities, including reared specimens from Col de Petchécara (1♂), Ouambéo R. (2♀♀) and Tiebaghi (1♂), all from *Alphitonia neocaledonica*, BM geometrid slide 8433 (♀); 1♂ La Coulée, 4.12.75 (J. GUTIERREZ) MNHN; 1♂ La Coulée 24.12.76 ex *Alphitonia neocaledonica* (A. DELOBEL) MNHN.

The species was recorded as follows: Grand Lac site 1 (5), site 2 (11), site 3 (9), site 4 (1), site 72 (2); Plaine des Lacs site 5 (1); Pic du Pin site 7 (18); R. Bleue site 8 (1), site 10 (2); Mt. des Sources site 11 (1), site 12 (1); Dothio site

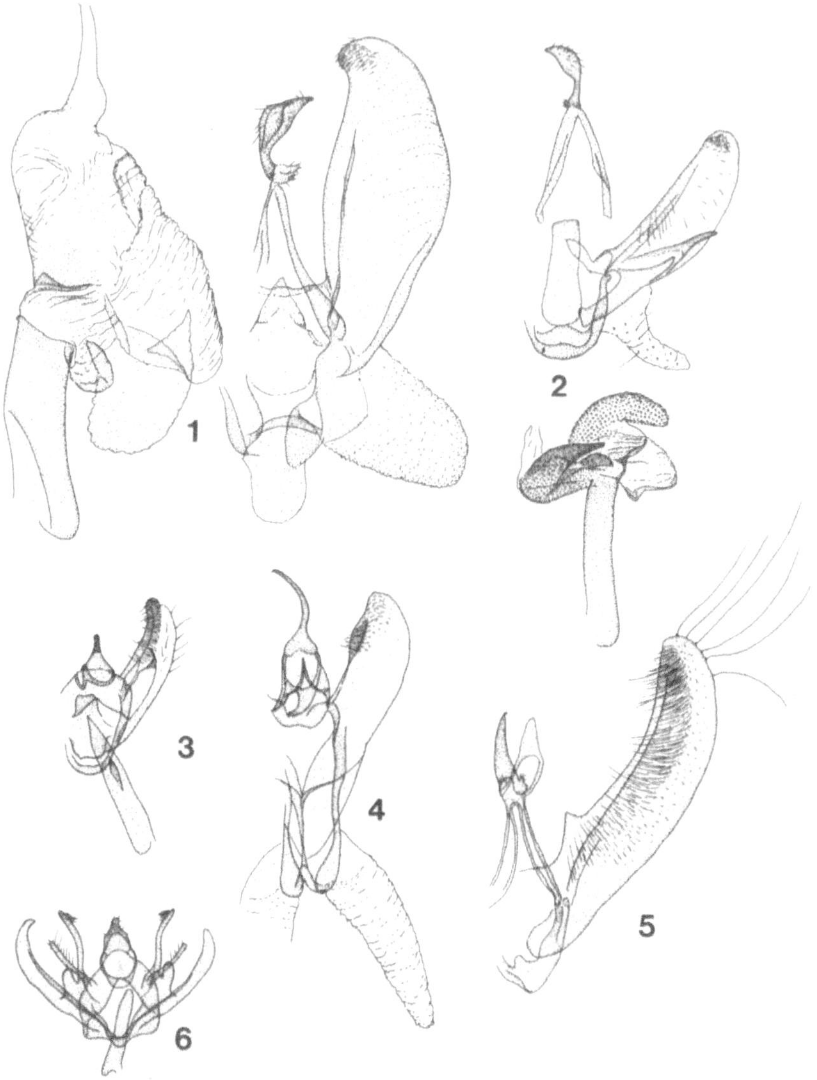


*Fig. 63.* Male genitalia and aedeagus of *Casbia serpentina* (1) and *C. rectaria lilacina* (2) and male genitalia of *Syneora strixoides* (3), *Scardamia eucampta* (4), *Eugnesia correspondens* (5) and *Symmetroctena variegata* (6).

18 (3); Poupa (1); Ouambéo R. (1); Petit Gouli (10); Kopéto (1); R. Amoa (1); Mt. Krapé site 55 (1), site 56 (4), Houailou (7); near Bâ (1); Ouen Toro site 64 (1), site 81 (3), site 82 (4); Mt. Pouédihi (1); Port Boisé site 74 (1); R. Koué (1); Les Dalmates site 77 (1), site 78 (1). BPBM: Isle of Pines.

The species is commonest in the ultramafic maquis but occurs also in drier

*Plate 61.* 1. *Casbia rectaria lilacina;* 2. *Casbia rectaria rosacea;* 3 (♂), 4 (♀). *Casbia serpentina;* 5. *Casbia albinotata* (10–12 mm); 6. *Scardamia eucampta* (13–14 mm).

semi-forested habitats off the ultramafics wherever the larval foodplant occurs. *Alphitonia neocaledonica* is a shrubby tree that occurs in the ultramafic maquis, particularly at the fringes of forests. It occurs in similar localities on the sedimentary rocks, often abundantly. The concentration of the moth in ultramafic habitats could conceivably be due to the greater competition it appears to experience as a larva off the ultramafics (p. 189). The reddish brown larva is illustrated in Plate 55.

**Casbia rectaria rosacea ssp. n.** (Plate 61: 2)

♂♀ 13 mm. This subspecies is a similar colour to the Australian race, perhaps slightly pinker, differing in the more prominent, darker post- and antemedial fasciae to both wings. The antemedial in each case is straighter, more basad than in the New Caledonian race, that of the forewing at right angles to the dorsum instead of meeting it obliquely, the acute angle distad.
HOLOTYPE ♂. New Hebrides: Santo, Mt. Tabwemasana, Nokowula 3700 ft 1.ix.1971 (G. S. ROBINSON) Roy. Soc. Expedition. BM geometrid slide 9502.
PARATYPES. New Hebrides: 1♂, 1♀ as holotype; 1♂, 1♀ Santo, Nokowula CK-3500 ft 3-4.ix.1971; 1♂ Aneityum I., Agathis Camp-1150 ft 19-21.vii.1971.

**Casbia serpentina sp. n.** (Plate 61: 3, 4, Figure 63: 1)

♂♀ 14–17 mm. The ground colour is a medium pinkish grey stippled darker and with darker fasciae as illustrated. On both wings there is a submarginal row of black-edged pale yellow dots, the central two displaced basad slightly; this row of dots is edged pale orange basad in some female specimens and the inset central dots are broadly surrounded by a black circle in others. Faded specimens appear more pale reddish. The species generally resembles *eccentritis* Meyrick (type examined; S.E. Australia and Tasmania) but is larger. *C. eccentritis* is redder with the central submarginal dots not inset, or only slightly so. In the male genitalia the valves of *serpentina* lack the harpe and basal constriction of *eccentritis*, the uncus is more bulbous and the aedeagus vesica is relatively very much larger.

HOLOTYPE ♂. New Caledonia: Grand Lac, 250 m Site 1 06980/75358 (J. D. HOLLOWAY) 13.5.1971, BM 1971-507. BM geometrid slide 8185.

PARATYPES. 69 ex. New Caledonia, various localities. BM geometrid slide 8432 (♀).

The species is closely associated with the ultramafic maquis and was recorded as follows: Grand Lac site 1 (12), site 2 (45), site 3 (6), site 4 (7), site 72 (2); Plaine des Lacs site 5 (1); Pic du Pin site 7 (18), site 69 (2), site 70 (2); R. Bleue site 9 (2), site 10 (7); Mt. des Sources site 11 (2); Ouambéo R. (1); Col d'Amieu (2); Mt. Dogny (1); Ste. Therèse (2); Mt. Krapé site 55 (3), site 56 (5); Houailou (1); near Bâ (9); Col Mouirange (2); Mt. Pouédihi (4); Les Dalmates site 78 (1); Mt. Koghis site 79 (1).

*Catoria camelaria carbonata* Warren (Plate 64: 1, Figure 67: 5)

*Boarmia camelaria* Guenée, 1858, *Hist. nat. Insectes* (*Lépid.*), 9: 256.
*Catoria camelaria carbonata* Warren, 1896, *Novit. zool.*, 3: 402, type examined (Lifu).

The male genitalia resemble those of the typical race from Queensland. There are also subspecies from New Guinea and Timor. Specimens from the New Hebrides of similar external appearance differ in having the aedeagus vesica less lobed and with a basal cluster of cornuti. These may belong to a distinct subspecies or even species.

The species was recorded from the following New Caledonian localities: Grand Lac site 3 (1); R. Bleue site 8 (2), site 9 (1), site 10 (2); Ouatou R. (1); Ouambéo R. (1); Col d'Amieu (13); Col Toma (1); Sarraméa (1); Mt. Mou site 36 (1); R. Amoa (3); Probokonoro (3); Col des Roussettes (1); Mt. Krapé site 55 (1); Mt. Aoupinié (2). BPBM: Tipindje. It is commonest in rainforest localities.

**Cleora acacicola sp. n.** (Plate 62: 10, Figure 64: 2)

♂♀ 13–16 mm. The ground colour is white, variably mottled and marked with blackish brown as illustrated. The species is smaller than other New Caledonian members of the genus and distinguished in most specimens by the prominent dark forewing medial which continues as a streak through the angle of the postmedial to join a dark line bisecting the apex. This pattern is unusual in the genus where the species appears to have no close relatives, but is often found in unrelated Australian geometrids and New Caledonian species of Australian affinity such as *Horisme grisearia*, *Scopula desita* and *Syneora strixoides*. The ecology of *acacicola* in New Caledonia suggests it may have relatives in Australia, especially Queensland, but possibly described in other genera; none such were located in the BMNH collection however. The male and female genitalia are typical of the genus but again give no indication as to its affinities; the aedeagus has a small group of slender cornuti at the apex of the vesica.

HOLOTYPE ♂. New Caledonia: Poupa, 200 m Site 20 06120/76150 (J. D. HOLLOWAY) 10.6.1971, BM 1971-507. BM geometrid slide 8176.

PARATYPES. 101 ex. New Caledonia, various localities.

The species was recorded as follows: Dothio site 18 (11), site 19 (1); Poupa (74); Koh (1); Ouatou R. (1); Bréhoa R. (1); Col d'Amieu (1); Petit Gouli (16); Sarraméa (1); Kaala-Gomen (1); Tiebaghi site 39 (1), site 40 (1), site 41 (3); Grottes de Koum (1); Bambou R. (3); Mt. Krapé site 55 (7); Col de Boa (2); Houailou (2); Col de Nassiboué (1); Ouen Toro site 64 (10), site 81 (5), site 82 (11). The species is commonest at sites with *Acacia spirorbis* abundant and the majority of records are from localities where this plant is present. In the numerical analyses it was associated with *Prasinocyma albicosta* which feeds on the *Acacia* as a larva.

### **Cleora dothionis sp. n.** (Plate 62: 11, Figure 64: 4)

♂♂ 15–17 mm. The species is marked and suffused with dark brown on a pale fawn ground as illustrated. The valves of the male genitalia are as in *idiocrossa* Turner (Norfolk I; HOLLOWAY, 1977: 135) but the bifurcations of the aedeagus vesica each bear two long cornuti, whereas in *idiocrossa* one bears two stout, short cornuti and the other bears five small ones. It is externally distinguished from *idiocrossa* by the more uniform, less mottled coloration and the less crenulate postmedials; the forewing postmedial is more sharply angled and is crossed at the angle by a diagnostic dark brown streak.

HOLOTYPE ♂. New Caledonia: Dothio, 400 m Site 19 06133/76135 (J. D. HOLLOWAY) 8.6.1971, BM 1971-507. BM geometrid slide 8183.

PARATYPES. 3♂♂ New Caledonia: Bréhoa R.; Sarraméa (2).

### **Cleora illustraria caledonica ssp. n.** (Plate 62: 6, 7, Figure 64: 1)

*Boarmia illustraria* Walker, 1863, *List Specimens lepid. Insects Colln Br. Mus.*, 26: 1539.

♂♂ 19–22 mm, ♀♀ 20–24 mm. Males with a bone-white ground, tinged yellowish in some specimens and variably marked and suffused dark brown. The females are greyer in tint, less contrastedly marked. The subspecies differs from typical *illustraria* (type examined, Queensland; other subspecies in New Guinea, the Bismarcks and Solomons) in the reduced or absent dark surround

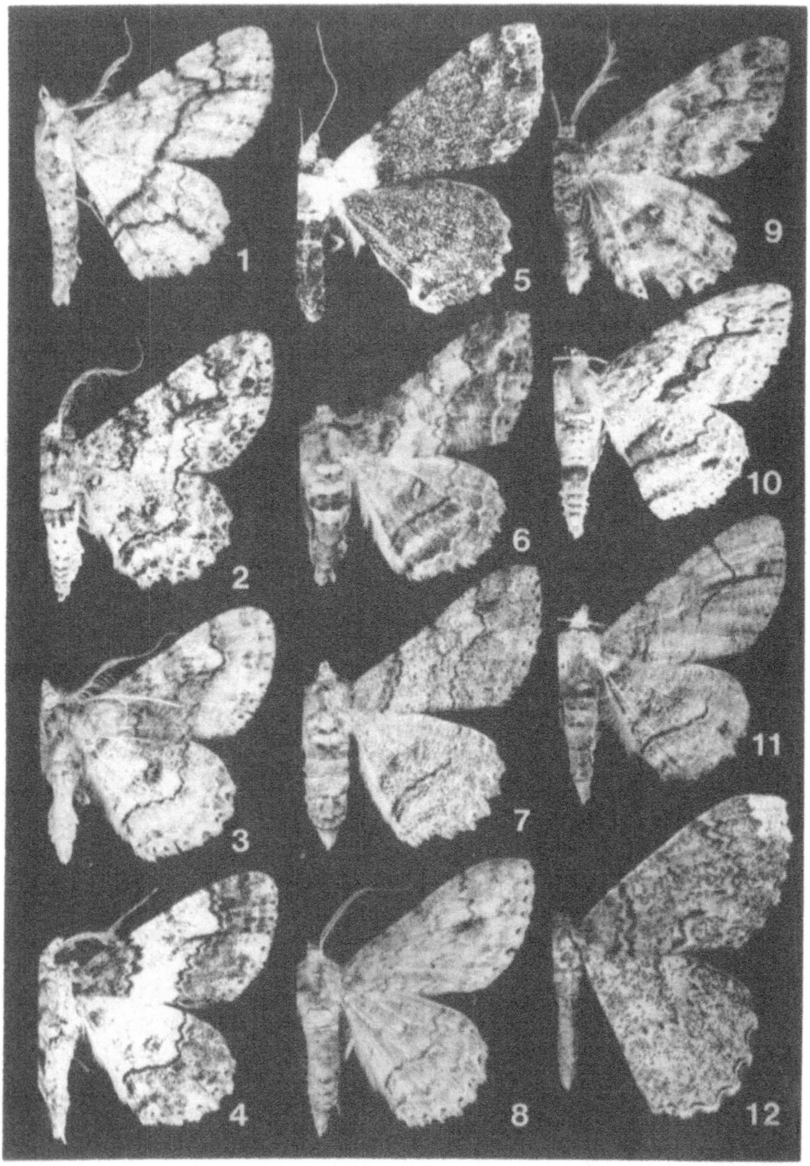


*Plate 62.* 1–5. *Cleora immemorata* (15–21 mm); 6 (♂), 7 (♀). *Cleora illustraria caledonica*; 8. *Cleora lacteata ecdees* (15–17 mm); 9. *Cleora injectaria* (16 mm); 10. *Cleora acacicola*; 11. *Cleora dothionis*; 12. *Elphos subrubida aoupiniensis.*

to the pale discal spot on both wings and in the less sinuous hindwing postmedial. In the male genitalia the uncus is shorter and the upper part of the harpe on the valve is very much broader and apically rounded instead of acute; the ornamentation of the aedeagus vesica is as in typical *illustraria*.

HOLOTYPE ♂. New Caledonia: Col Toma, 460 m Site 27 05813/76123 (J. D. HOLLOWAY) 19.6.1971, BM 1971-507. BM geometrid slide 8178.

PARATYPES: 6♂♂, 6♀♀ New Caledonia, various localities; 4♂♂, 2♀♀ N. New Caledonia, Tinchialit 2020 ft 22.viii-10.ix.1949 (L. E. CHEESMAN) BM 1950-1.

The species was recorded as follows: Grand Lac site 2 (3), site 72 (1); Pic du Pin site 7 (4); R. Bleue site 8 (1); Mt. des Sources site 11 (1); Dothio site 19 (7); Poupa (1); Ouatou R. (2); Bréhoa R. (14); Ouambéo R. (1); Col d'Amieu (8); Col Toma (8); Petit Gouli (3); Sarraméa (8); Mt. Dogny (4); Mt. Mou site 33 (2); Kavatch (1); R. Amoa (1); Ste. Therèse (1); Probokonoro (8); Col des Roussettes (1); Mt. Krapé site 55 (14), site 56 (2); near Bâ (1); Mt. Panié site 60 (14), site 61 (2); Mt. Aoupinié (14); Mt. Pouédihi (1); Port Boisé site 74 (1); Mt. Koghis site 79 (6), site 80 (6). It is a widespread rainforest species.

### *Cleora immemorata* (Walker) (Plate 62: 1-5, Figure 64: 5)

*Boarmia immemorata* Walker, 1863, *List Specimens lepid. Insects Colln Br. Mus.*, 26: 1540, type examined.

*Boarmia lichenina* Butler, 1877, *Ann. Mag. nat. Hist.* (4), 10: 358, type examined.

This is a highly variable species in all its localities (New Caledonia, Loyalty Is., New Hebrides) and several forms are illustrated here. Specimens from the drier west coast tend to be somewhat smaller than those from elsewhere.

It is the most abundant New Caledonian *Cleora* and was taken during the survey in a wide range of habitat types as follows: Grand Lac site 2 (13), site 3 (2), site 4 (3); Pic du Pin site 7 (4); R. Bleue site 8 (3), site 10 (1); Mt. des Sources site 11 (6), site 12 (2), site 14 (9); Anse Longue site 16 (13), site 32 (4); R. Kuenthio (1); Dothio site 18 (1), site 19 (1); Poupa (5); Ciu (1); Koh (4); Ouatou R. (3); Bréhoa R. (13); Ouambéo R. (3); Col d'Amieu (170); Col Toma (14); Petit Gouli (1); Sarraméa (9); Kopéto (10); Mt. Dogny (3); Mt. Mou site 33 (96), site 34 (8), site 36 (1); Col de Boguen (1); Tiebaghi site 41 (10); Grottes de Koum (2); Bambou R. (3); Col d'Amoss site 45 (2); Kavatch (1); R. Amoa (5); Ste. Therèse (3); Probokonoro (35); Col des Roussettes (5); Mt. Krapé site 55 (48), site 56 (3); near Bâ (8); Mt. Panié site 60 (1), site 61 (13); Mt. Aoupinié (7); Col de Nassiboué (1); Ouen Toro site 64 (4), site 81 (5), site 82 (5); Col Mouirange (10); Yaté Gorge (1); Port Boisé site 74 (8); R. Koué (2); Anna-Madeleine (2); Les Dalmates site 77 (8); Mt. Koghis site 79 (1). BPBM: Ile des Pins, Le Crouen, Mt. Koghis, La Foa, Nouméa. MNHN: La Coulée, Mt. Koghis.

### *Cleora injectaria* (Walker) (Plate 62: 9, Figure 64: 6)

*Cleora injectaria* (Walker); ROBINSON (1971b, 1975: 302)

Two worn males were taken at the Amoa R. site. More material is needed to

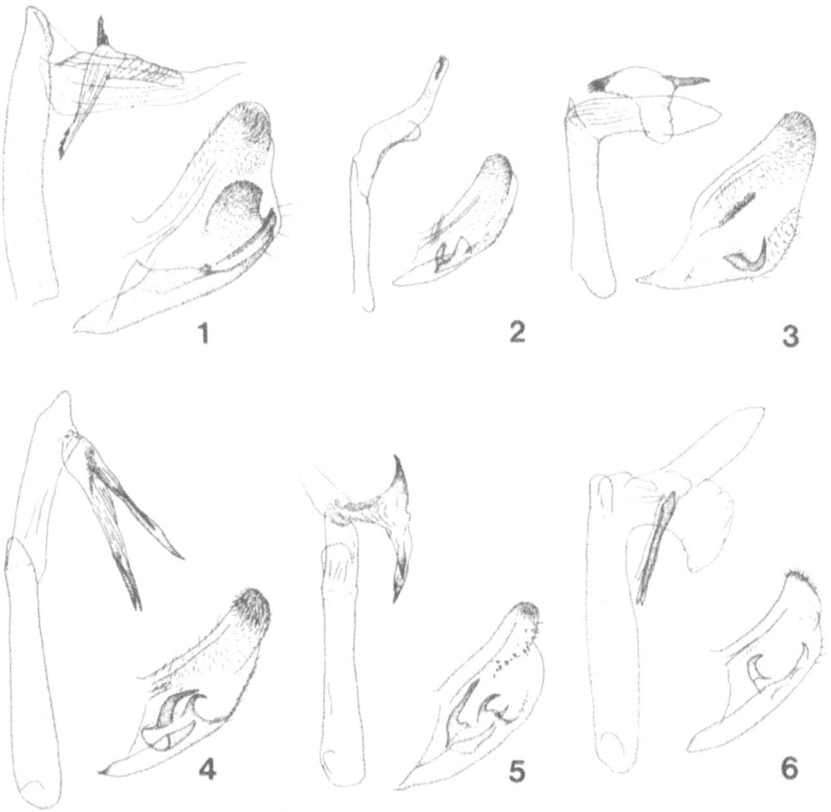


*Fig. 64.* Aedeagi and valves of *Cleora*: *illustraria caledonica* (1), *acacicola* (2), *lacteata ecdees* (3), *dothionis* (4), *immemorata* (5) and *injectaria* (6).

determine the subspecific status of New Caledonian *injectaria* but the aedeagus vesica lacks the additional curved cornutus of the Fijian race, *anidryta* Prout. The species occurs in the Indian region, S.E. Asia, Sundaland, the Philippines, New Guinea, Queensland, Micronesia and Fiji.

*Cleora lacteata ecdees* Prout **stat. n.** (Plate 62: 8, Figure 64: 3)

*Chogada lacteata* Warren, 1897, *Novit. zool.*, 4: 247, type examined.
*Cleora hemiopa ecdees* Prout, 1929, *Bull. Hill Mus. Witley*, 3: 207, type examined (Lifu).

*Cleora lacteata* is here taken to include as subspecies, apart from *ecdees* (Loyalties, New Caledonia) and the typical race (Bismarcks, Solomons), *hemiopa* Prout (1928, *Insects Samoa*, 3: 160, type examined, Samoa), *quirosi* Prout (1929, *Bull. Hill Mus. Witley*, 3: 207, type examined, New Hebrides) and *perbona* Prout (1937, *Novit. zool.*, 40: 193, type examined, Goodenough Is., also in mainland New Guinea). These races are combined under *lacteata* because of similarity of facies and male genitalia, and must be distinguished from the sympatric *callicrossa* Meyrick (type examined, New Guinea) and *nigristigma* Prout (type examined, Bismarcks, Solomons) which are best placed as subspecies of *decisaria* Walker (Sundaland to Solomons) as in the treatment by Prout (1929, *Bull. Hill Mus. Witley*, 3: 179–222). The *lacteata* group of races were also included with *decisaria* but may be distinguished as follows: externally the former has a broader, straighter (less crenulate) white fascia in the centre of the double dark submarginal; the harpe on the male valve of the former is generally a single, curved spine (a secondary spine, smaller, was present in the *quirosi* male dissected and the spine is longer in *ecdees*) with no basal ornamentation but in the latter the harpe is shorter, apically rounded and bearing a comb of spines extending basad to a basal lobe bearing further spines; the vesica of the aedeagus bifurcates in both species and on one lobe there is one, sometimes two cornuti, but on the other there are several in a comb in the former species as distinct from two in the latter.

*C. lacteata* was recorded in New Caledonia as follows: R. Bleue site 9 (1); R. Kuenthio (1); Dothio site 19 (8); Koh (1); Ouatou R. (7); Bréhoa R. (17); Ouambéo R. (1); Col d'Amieu (11); Col Toma (6); Petit Gouli (1); Sarraméa (8); Mt. Dogny (1); Mt. Mou site 33 (3); Grottes de Koum (8); Kavatch (7); R. Amoa (1); Ste. Therèse (3); Probokonoro (3); Col des Roussettes (2); Mt. Krapé site 55 (9), site 56 (1); Col de Boa (1); Mt. Aoupinié (17); Col Mouirange (1); Mt. Pouédihi (3); Port Boisé site 73 (2), site 74 (27); Mt. Koghis site 79 (11). The species appears to be strongly associated with rainforest.

### *Clepsimelia phryganeoides* Warren (Plate 63: 4, Figure 66: 1)

*Clepsimelia phryganeoides* Warren; Robinson (1975: 305).

This species was recorded in low numbers from areas of montane or ultramafic rainforest in New Caledonia: Mt. Mou site 33 (1); Mt. Krapé site 55 (2); Mt. Panié site 60 (2); Mt. Koghis site 79 (3). It has been recorded from Sulawesi, the Moluccas, New Guinea, the New Hebrides and Fiji.

### *Ctimene fulvimacula* (Warren) (Plate 63: 10)

*Bursada fulvimacula* Warren, 1895, *Novit. zool.*, 2: 121, type examined.
*Bursada fulvinaeata* (sic) Warren; Viette (1949: 327).

This species is brownish black, marked with orange. It occurs in the Loyalty Is. and New Caledonia but was not recorded during the survey. It resembles most closely an unnamed series in the BMNH taken in N. Australia.

### *Ctimene rubropicta* (Thierry-Mieg) (Plate 63: 11)

*Bursada rubropicta* Thierry-Mieg, 1915, *Miscnea ent.*, 22: 41, unique male type from New Caledonia examined (MNHN).

The species is dark brown, marked with dull reddish orange as illustrated. It has no obviously close relatives within the genus but probably falls within the same general group as the previous species.

**Ctimene nocturnignis sp. n.** (Plate 63: 12)

♀♀ 16 mm. The antennae are bipectinate, the pectinations depressed ventrally. The ground colour is very dark, slightly reflective blue, reflecting paler over the base of the hindwings. The forewings have a bright orange discal spot above and below; the fringes are white in two places as illustrated and the wing margin is slightly indented where these patches occur. In coloration the species is reminiscent of some *Milionia* Walker and *Craspedosis* Butler species, such as *Craspedosis aurigutta* Warren from New Guinea. These two genera have filiform antennae in the female whereas *Ctimene* Boisduval often has them bipectinate. *Ctimene hysginospila* Prout (type examined, New Guinea) is similar in general size and shape to the New Caledonian species but is very dark brown with an elongate white discal patch on the forewing and an orange postdiscal one on the hindwing; these patches recur on the underside, the hindwing postdiscal spot being joined by a subbasal orange patch. In several specimens the fringes are paler and the wing margin indented just subtornally as in *nocturnignis*.

HOLOTYPE ♂. New Caledonia: Mt. des Sources, 760 m Dusk 06653/75513 (J. D. HOLLOWAY) 29.5.1971, BM 1971-507. Hovering about and visiting a flowering shrub, possibly *Styphelia*.

PARATYPE ♀. New Caledonia: nr. Bâ, 190 m Dusk 05531/76586 (J. D. HOLLOWAY) 25.7.1971, BM 1971-507. BM geometrid slide 8195, bursa eaten by ants.

At the paratype locality several more specimens were observed in flight over the maquis at the same time. Both localities are in extensive ultramafic maquis.

**Elphos subrubida aoupiniensis ssp. n.** (Plate 62: 12, Figure 65, right)

*Elphos subrubida* Warren, 1896, *Novit. zool.*, 3: 300, type examined, Fergusson Is., New Guinea.

♂♂ 30 mm. Above, the wings have a pale grey ground, mottled with richer brown, stippled and fasciated dark brown; the forewing apical patch is white. Below, the hindwing is generally dark grey with small marginal white patches and, basally, a central ellipse of orange extending to the disc. The forewing below has the apical half dark grey, the basal border of this grey area running straight from the tornus to two thirds along the costa; there is an apical white patch and the basal half of the wing is also white, speckled grey, with a dark grey discal patch. The subspecies is smaller than the typical race (40-45 mm). Typical *subrubida* has cream areas where *aoupiniensis* is white; the orange area of the hindwing below usually extends broadly over the basal two thirds; the basal area of the forewing below is more heavily mottled grey in the typical

*Plate 63.* 1. *Neogyne elongata* (13 mm); 2 (♂), 6 (♀). *Syneora strixoides;* 3. *Eugnesia correspondens* (15 mm); 4. *Clepsimelia phryganeoides* (13 mm); 5. *Parosteodes grisea;* 7. *Petelia trifascia;* 8. *Aplochlora vivilaca* (13 mm); 9. *Symmetroctena variegata;* 10. *Ctimene fulvimacula* (11 mm); 11. *Ctimene rubropicta* (12 mm); 12. *Ctimene nocturnignis.*

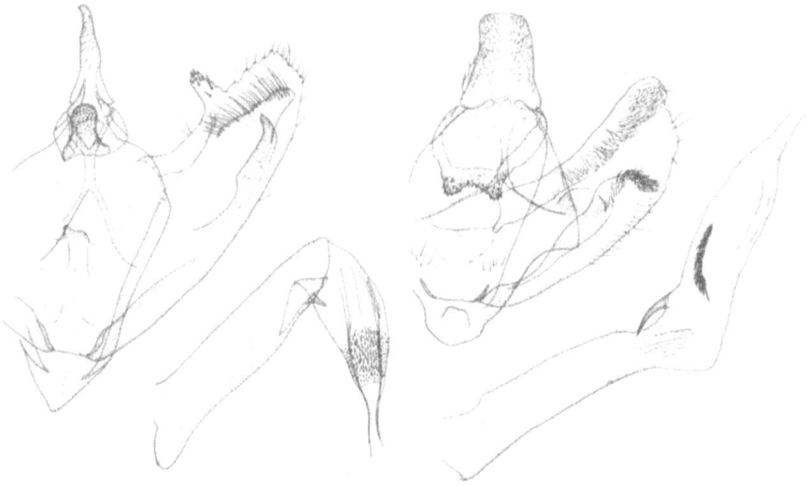

*Fig. 65.* Male genitalia and aedeagus of *Eunoumeana fletcheri* (left) and *Elphos subrubida aoupiniensis* (right).

race and the straight border to the apical grey area meets the dorsum well basally from the tornus. The genitalia of *aoupiniensis* are very similar to those of the typical race but the valve apex is broader, the fused gnathi less scobinate and the harpe less thickly setose.

HOLOTYPE ♂. New Caledonia: Mt. Aoupinié, 520 m Site 62 05298/76594 (J. D. HOLLOWAY) 31.7.1971, BM 1971-507. BM geometrid slide 8179.

PARATYPE ♂. Data as holotype.

### *Eugnesia correspondens* Warren (Plate 63: 3, Figure 63: 5)

***Eugnesia correspondens*** Warren, 1897, ***Novit. zool.***, 4: 76-77, type examined, Luzon, Philippines; type species of ***Eugnesia*** Warren (***Ibid.***).

As well as the type locality in the Philippines, *correspondens* has been recorded from the Moluccas and the New Hebrides. In the BMNH series there is a specimen from New Guinea only dubiously of this species, but it is obviously rare and might well be expected to turn up in intervening Papuan localities. Singletons were recorded from Mt. Mou site 33, Col Toma and Mt. Koghis during the survey.

### *Eunoumeana fletcheri* (Viette) (Plate 64; 11, 12, Figure 65, left)

***Noumea fletcheri*** Viette, 1949, ***Bull. Soc. zool. France,*** 74: 241, male type examined (MNHN).

*Eunoumeana fletcheri* (Viette); Viette, 1952, *Bull. Soc. ent. France.*, 1952: 95 (***Noumea*** praeocc.).

The second ever male of *fletcheri* was taken at Port Boisé site 74. It differs from the almost uniform grey-brown type in being slightly more variegated (basally paler, more rufous on the hindwing and paler along the forewing costa). In general shape, size and facies the species shows close affinities with the genus *Ruttellerona* Swinhoe, especially the Fijian *presbytica* Prout. However, the genitalia with their three subapical projections to the valve, narrow uncus, prominent fused gnathi and the band of scobination on the aedeagus vesica indicate a degree of taxonomic isolation for the species that justifies its retention in the monotypic genus *Eunoumeana*. In *presbytica* the valve has a simple, thickly setose costa and two well-defined central patches bearing stouter setae.

*Gonodontis clelia* (Cramer) (Plate 64: 2, Figure 67: 2)

*Gonodontis clelia* (Cramer); ROBINSON (1975: 305).

This species is widespread in the Indo-Australian tropics, being recorded from the Indian region, the Himalaya, China, Sundaland, New Guinea, Queensland, the Solomons, Fiji and Samoa. The genitalia are similar to those of a Borneo male dissected and illustrated by HOLLOWAY (1976: Figure 537), but the socii from the tegumen are very much smaller and without a lateral projection. The species was recorded from the following New Caledonian localities, mainly in rainforest: R. Bleue site 9 (1), site 10 (1); Dothio site 18 (1); Bréhoa R. (1); Ouambéo R. (1); Col d'Amieu (10); Sarraméa (1); Mt. Mou site 33 (3); R. Amoa (2).

**Nadagara serpentina sp. n.** (Plate 64: 9, 10, Figures 41: 2, 42: 5)

♂♀ 15–18 mm. The antennae of both sexes are filiform as in the rest of the genus. The ground colour of the wings is white, mainly obscured by pinkish grey in which dark brown scales are scattered. These dark scales are concentrated in the ante- and postmedial fasciae either in a narrow band or as points on the veins. The white ground colour is more evident adjacent to these fasciae on the side away from the medial area; it is also evident in a diffuse submarginal fascia. There are dark brown triangles marginally in the spaces.

The build is slender, particularly in the male. The male hind tibia is longer and broader than those of the other legs. It is invested interiorly with straw-coloured scales that cover a groove extending almost the length of the tibia and containing a hairpencil. The hind tibia bears two unequal pairs of spurs, one almost centrally and one distally.

The species is isolated within the genus. In facies and build it resembles *synocha* Prout but this Sundaland species is a straw colour and the male hind tibia is only barely modified, just slightly broader than the others. *N. irretracta* Warren (Queensland, New Guinea, Bismarcks, Solomons, New Hebrides, Fiji) has less elongate wings and legs, finer, more clearly defined fasciation and, with the next two species mentioned, the hind tibia is broadened and has a basal hairpencil though this is broadly based and not set in a groove. *N. xylotrema*

Lower (Queensland) and *N. extensipennis* Prout (New Guinea) are much more robust in body, more rufous in tint and the fasciae are also sharply defined.

The male genitalia of the type species *vigaia* Moore, of *synocha* and *irretracta* have the valves long, narrow and simple with basal coremata (slight only in *irretracta*) and a tuft of hairlike setae subbasally and interiorly. The aedeagus vesica contains two large cornuti, irregular, often serrate, a character shared with *xylotrema* and *serpentina*. In *xylotrema* the uncus is short and broad, the valves sinuous and not running adjacent to the tegumen as in the other species; the valve is corematous and the tuft of hairlike setae is replaced by a group of long sinuous blades, some of which are angled subapically. In *serpentina* the valves lack coremata and setal tufts and are rounder with a diagnostic projection from the costa; the bicornute juxta is also diagnostic.

The female genitalia of *serpentina* are as illustrated (Figure 41). The signum is more irregular and more scobinate than in *vigaia*, central rather than apical, the bursa more elongate.

HOLOTYPE ♂. New Caledonia: Tiebaghi, 330 m Site 41 04156/77366 (J. D. HOLLOWAY) 10.7.1971, BM 1971-507. BM geometrid slide 9550.

PARATYPES: 34♂♂, 24♀♀ New Caledonia, various localities, BM geometrid slides 8418, 9549 (♂♂), 8505 (♀), 9548 (venation); 1♀ La Coulée, 28.7.76 (J. GUTIERREZ) MNHN.

The species was only taken on the ultramafics except for four individuals on Ouen Toro. It was recorded as follows: Grand Lac site 2 (34), site 3 (3), site 72 (20); Pic du Pin site 7 (36); Mt. Mou site 33 (4), site 34 (3), site 36 (2); Tiebaghi site 41 (12), site 42 (1); near Bâ (1); Ouen Toro site 64 (4); Col Mouirange (5); R. Fausse Yaté (2); Yaté Gorge (2); Anna-Madeleine (4); Les Dalmates site 77 (5), site 78 (15); Mt. Koghis site 79 (1), site 80 (2).

### *Neogyne elongata* Warren (Plate 63: 1, Figure 66: 3)

*Neogyne elongata* Warren, 1898, *Novit. zool.*, 5: 256, type examined.

A single male of this Australian (Queensland, New South Wales) species, unique to its genus, was taken at Dothio site 18.

### *Oxymacaria insularis* (Warren) (Plate 64: 3, 4, Figure 67: 4)

*Thamnonoma insularis* Warren, 1896, *Novit. zool.*, 3: 413, type examined (Lifu).

The male is very variable with brown fasciae on a straw ground, variously stippled and mottled with brown. The female is pale greyish brown with darker fasciae and a broad, darker marginal border to the wings. The species is similar to *O. ochreata* (Warren) (Solomons, with the sympatric and genitalically similar *O. persimilis* (Warren) almost certainly conspecific) but in that species the female is more straw coloured. In the male genitalia the uncus of *insularis* is narrower, the upper arm of the valve downcurved instead of straight, the lower arm longer, more rounded, separated from the upper by a shallower

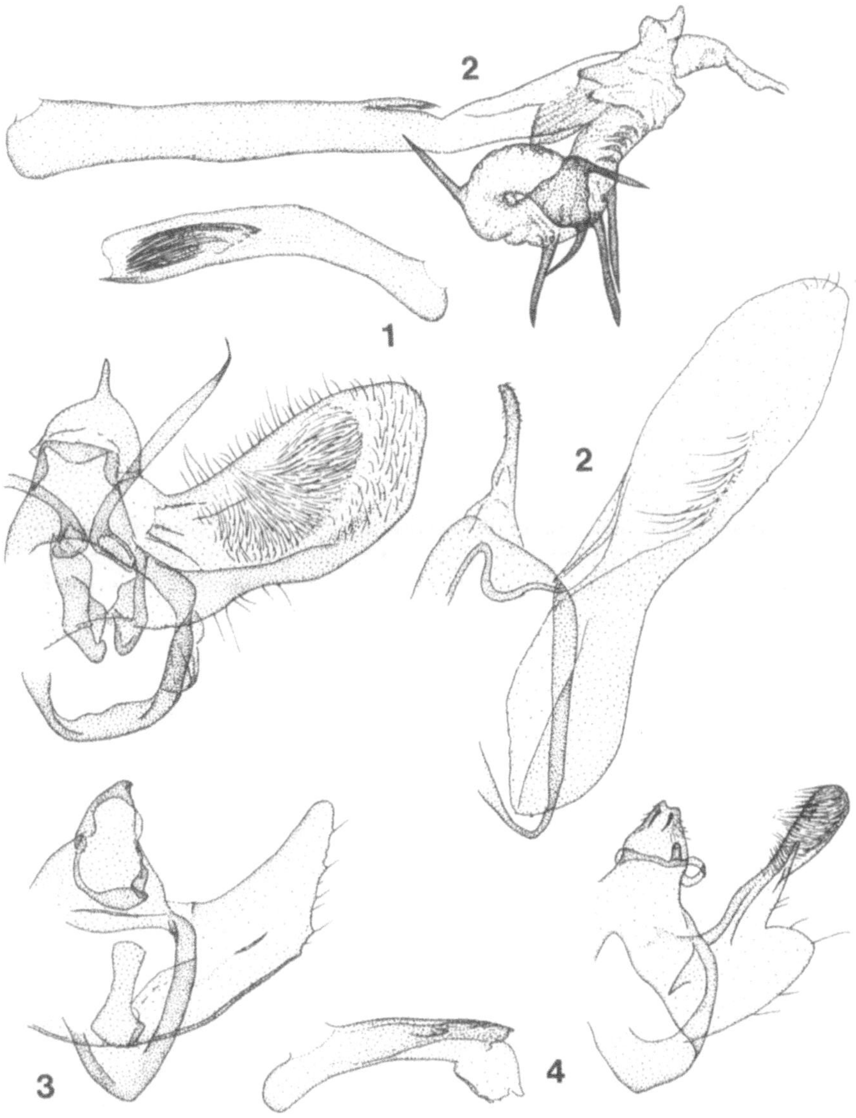


*Fig. 66.* Male genitalia of *Clepsimelia phryganeoides* (1), *Aplochlora vivilaca* (2), *Neogyne elongata* (3) and *Parosteodes grisea* (4).

cleft. The aedeagus vesica is less scobinate than in *ochreata* and lacks the spined sclerotisation at the end of a long lobe of that species. A species externally similar but stouter, with very different male genitalia (perhaps more characteristic of *Semiothisa* Hübner) occurs in the New Hebrides.

The species was recorded as follows: Grand Lac site 2 (1), site 3 (2); R.

Bleue site 8 (1), site 9 (3), site 10 (2); Dothio site 18 (1); Poupa (1); Ciu (1); Koh (4); Ouatou R. (6); Bréhoa R. (3); Ouambéo R. (4); Col d'Amieu (2); Petit Gouli (2); Sarraméa (4); Mt. Mou site 33 (4), site 34 (2); R. Amoa (1); Probokonoro (10); Mt. Krapé site 55 (5), site 56 (2); Mt. Pouédihi (1); Les Dalmates site 77 (1); Mt. Koghis site 79 (4). BPBM: Mt. Koghis, St. Louis Valley, Forêt de Thi. MNHN: La Coulée, Col d'Amieu.

**Parosteodes grisea sp. n.** (Plate 63: 5, Figure 66: 4)

♂♀ 9–12 mm. The ground colour is pinkish grey, finely stippled darker. The forewing has a dark grey antemedial, a black discal dot and a yellowish postmedial edged blackish basad. A similar but more diffuse and fainter postmedial occurs on the hindwing. The species differs from its only congener, *fictiliaria* (Guenée) (type examined (MNHN); Queensland, New South Wales, South Australia, Tasmania, Lord Howe I.) in the straight instead of sinuous forewing postmedial, the more oblique antemedial, concave distad instead of basad, and the entire hindwing postmedial (pronounced only at the dorsum in *fictiliaria*). The ground colour of *fictiliaria* is more rufous grey in the male and pinkish brown in the female. In the male genitalia the lower lobe of the valve is more rounded, the saccus slightly squarer and the cornutus of the aedeagus vesica has a less bulbous base.
HOLOTYPE ♂. New Caledonia: Grand Lac, 260 m Site 3 06965/75355 (J. D. HOLLOWAY) 17.5.1971, BM 1971-507. BM geometrid slide 8504.
PARATYPES: 5♂♂, 1♀ New Caledonia, Grand Lac site 2, site 3, Koh, Tiebaghi site 39, Houailou (2); 1♀ La Coulée 3.6.76 (J. GUTIERREZ) MNHN.
Apart from Koh all these are ultramafic localities.

**Petelia trifascia sp. n.** (Plate 63: 7, Figure 67: 1)

♂♀ 18–19 mm. The ground colour of the wings is pinkish grey, darker distal to the submarginal on both wings, irrorated with fine, dark brown striae and traversed by rufous fasciae. There is a black discal dot to the forewing and a pale yellow one to the hindwing. Below, the ground is fawn, marginally greyer, with a broad submarginal blackish band and a faint subbasal one on both wings; there are large, rounded, black discal spots on both wings.

The species belongs to the *medardaria* (Cramer) complex which needs revision (discussed by HOLLOWAY, 1975: 74). The specific status of *trifascia* has been determined by comparison with specimens from neighbouring land areas. Solomons, New Hebrides and Queensland specimens have the male forewing more produced; the forewing antemedial is narrower, nearer the discal dot than the postmedial (the reverse holds in *trifascia* where the antemedial and postmedial are straight, oblique and, with the submarginal, evenly spaced); below, the basal margin of the hindwing submarginal band is angled two thirds anteriorly in *trifascia* and its neighbours but the latter are distinguished by a

*Fig. 67.* Male genitalia of *Petelia trifascia* (1), *Gonodontis clelia* (2), *Polyacme dentata* (3), *Oxymacaria insularis* (4) and *Catoria camelaria carbonata* (4).

second angle subdorsally. In New Hebrides specimens the hindwing discal dot below is often reduced or centred pale yellow; the forewing submarginal band below extends more basally and has a sharp, sinuous boundary (similar in Solomons and Queensland specimens), but in *trifascia* this is diffuse and evenly concave. The hindwing margin of *trifascia* is more evenly curved than in neighbouring forms. *P. aesyla* Prout from Fiji differs in facies from *trifascia* in the same characters as the New Hebrides specimens but, below, the hindwing discal dot is usually absent, the submarginal bands are often very faint and the hindwing margin has two prominent angles.

In the male genitalia the valves of *trifascia* are intermediate in breadth between those of New Hebrides (broad) and Solomons (narrow) specimens. Both Solomons and New Hebrides specimens have a characteristic narrow uncus that is two-humped dorsally; that of *trifascia* has a shorter vertical section angling over (rather than smoothly curving) into a horizontal section distally with a narrow dorsal ridge and broadish lateral flanges. In *aesyla* the uncus is similar but the distal portion is longer and lacks the lateral flanges, the valves are apically acute instead of rounded and the aedeagus is centrally very much broader and the basal part is downturned.

HOLOTYPE ♂. New Caledonia: R. Bleue, 180 m Site 10 06704/75544 (J. D. HOLLOWAY) 28.5.1971, BM 1971-507. BM geometrid slide 8184.

PARATYPES: 12♂♂, 8♀♀ New Caledonia, various localities; 1♂ St. Louis Valley, Mar. 22, 1942 (H. E. MILLIRON) BPBM

The species was recorded as follows: Grand Lac site 3 (1); R. Bleue site 10 (2); Mt. des Sources site 11 (1), site 14 (6); Dothio site 19 (1); Poupa (1); Koh (1); Ouatou R. (1); Sarraméa (1); Bambou R. (1); R. Amoa (1); Probokonoro (1); Mt. Koghis site 79 (1); Ouen Toro site 81 (3), site 82 (2). The species is found in most habitats but the frequency at Ouen Toro might indicate a preference for drier ones.

### *Polyacme dentata* Warren (Plate 64: 5-8, Figures 67: 3, 68: 1)

*Polyacme dentata* Warren, 1896, *Novit. zool.*, 3: 418, type examined.

New Caledonian specimens are obviously referable to *dentata*, the type species of the genus, described from two females from Lifu. The female genitalia are identical and the bursa of the Lifu female dissected contains the deciduous cornuti characteristic of the aedeagus vesica of New Caledonian males. These are not found in the two closest related species from New Guinea. In *P. vagilinea* Warren (type examined) the aedeagus vesica has an apical lobe with three long, unequal spiral blades and in *P. straminea* Warren (type examined) there is a central group of short cornuti. The acute projection from the valve sacculus also distinguishes *dentata* from the New Guinea species.

The species is very variable, the ground colour ranging from grey to brown, rufous or ochreous fawn; the fasciation also varies in intensity. In some specimens there are bands and patches darker than the ground, giving a more variegated appearance. No specimen taken in New Caledonia is exactly comparable in facies with those from Lifu where the basal half of both wings is mottled ochreous and the distal half is greyish brown (Plate 64: 8).

The recorded distribution in New Caledonia was: Grand Lac site 2 (4), site 4 (1); R. Kuenthio (1); Ouambéo R. (1); Col d'Amieu (1); Col Toma (4); Sarraméa (1); R. Amoa (1); Probokonoro (2); Mt. Krapé site 55 (1); Port Boisé site 73 (1); Mt. Koghis site 79 (1). This may be a general rainforest species.

*Plate 64.* 1. *Catoria camelaria carbonata* (18–20 mm); 2. *Gonodontis clelia* (22 mm); 3, 4 (♂♂). *Oxymacaria insularis* (15–17 mm); 5–7, 8 (Lifu holotype). *Polyacme dentata* (16–19 mm); 9 (♂), 10 (♀). *Nadagara serpentina;* 11 (holotype), 12. *Eunnoumeana fletcheri* (20 mm).

### *Scardamia eucampta* Prout (Plate 61: 6, Figure 63: 4)

*Scardamia eucampta* Prout; ROBINSON (1975: 308).

Described from Fiji, this species also occurs in the New Hebrides, the Loyalty Is. (Maré) and New Caledonia (singletons from Grand Lac site 2, Mt. des Sources site 14, Port Boisé sites 73 and 74 and, in the BPBM, from Mt. Koghis and four specimens from Tipindje). In New Caledonia it has been recorded mainly from areas of ultramafic rainforest.

The genitalia of New Caledonian males differ from those of Fijian ones in the relatively smaller uncus and in possessing a prominent band of scobination in the aedeagus vesica. The closest relative is *chrysolina* Meyrick from Queensland and Lord Howe I., which differs externally as described by ROBINSON (1975). Together with an undescribed series in the BMNH from the Kei Is. these species are distinguished from their geographically close congeners by the curved (cf. straight) forewing postmedial. The genitalia of *chrysolina* are similar but much less elongate with the valves basally broader; the aedeagus vesica is not obviously scobinate.

### **Symmetroctena variegata sp. n.** (Plate 63: 9, Figures 63: 6, 68: 2)

♂♀ 11–14 mm. The species has brownish black markings on a white ground as illustrated, the forewing submarginal (immediately distal to the postmedial) being more rufous, especially over the posterior half. The forewings are variably suffused blackish, perhaps more so in the females where the hindwing ground is often pale grey. There is usually much dark suffusion between the straight antemedial and the subbasal on the forewing, producing a prominent dark band. This occurs also in *S. fumosa* Warren (Queensland, type examined) but in *fumosa* the forewing is more evenly suffused with brownish black scales and the forewing postmedial runs obliquely to the dorsum from distal to the discal spot instead of curving back distally to give a sigmoid as in *variegata*. The male genitalia differ markedly from those of *fumosa*, suggesting that the New Caledonian species is highly divergent or, alternatively, is not related to *Symmetroctena*, merely showing convergence in facies. No closer allies have been located however. The valve is quadrifid instead of simple and the aedeagus lacks the spine-like apex and large cornutus of *fumosa*. The genus is otherwise restricted to Australia.

HOLOTYPE ♂. New Caledonia: Yaté Gorge, 120 m Site 68 06955/75499 (J. D. HOLLOWAY) 6.8.1971, BM 1971-507. BM geometrid slide 8201.

PARATYPES: 53 ex. New Caledonia, various localities, BM geometrid slide 8201 (♂); 1♀ N. New Caledonia, Tinchialit 2020 ft 22.viii-10.ix.1949 (L. E. CHEESMAN) BM 1950-1. BM geometrid slide 8434.

The species was recorded in New Caledonia as follows: Grand Lac site 2 (3); Plaine des Lacs site 5 (5); Pic du Pin site 7 (25); R. Bleue site 10 (7); Poupa (2); Ouambéo R. (1); Petit Gouli (1); Kopéto (1); R. Amoa (1); Mt. Krapé site 55 (1), site 56 (1); near Bâ (3); Mt. Panié site 60 (7); Col Mouirange (1); Mt.

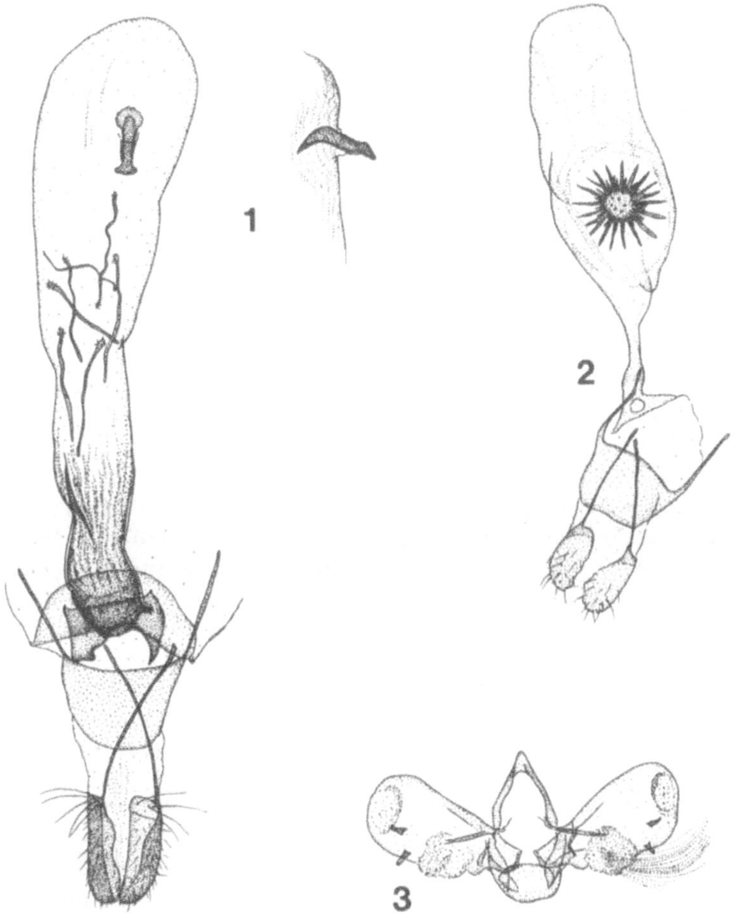


*Fig. 68.* Female genitalia of *Polyacme dentata* (1) and *Symmetroctena variegata* (2), and male genitalia of *Casbia albinotata* (3)

Pouédihi (3); Yaté Gorge (5); R. Koué (11); Mt. Koghis site 79 (1), site 80 (1). The species is almost entirely restricted to ultramafic localities, the main exception being the seven specimens taken in the montane forest of Mt. Panié.

**Syneora strixoides sp. n.** (Plate 63: 2, 6, Figure 63: 3)

♂♀ 11–14 mm. The ground colour is pale grey to fawn or greyish brown, the females being generally greyer. The wings are fasciated dark brown as illustrated. The fasciation is variable as can be seen from the two examples in the

Plate, variation being mainly in the intensity of the dark band from the forewing apex to the centre of the tornus and its counterpart on the hindwing. The species is very close to *strixata* Walker (♀ type examined, Queensland; the only specimen in the BMNH) but the forewing postmedial is less sinuous and its adjacent band distad is broader; the submarginal pale fascia of the forewing is straighter at the tornus and distinguished by the broader central portion; the hindwing has a basal dark band that is absent in *strixata*; the hindwing discal spot is more elongate and closer to the medial, or fused with it, instead of closer to the postmedial as in *strixata*; the hindwing postmedial extends to the costa instead of fading away just anterior to the discal spot.

HOLOTYPE ♂+pupa. New Caledonia: Grand Lac, 250 m 06965/75355 (J. D. HOLLOWAY) May 1971, BM 1971-507. Reared from larva on *Baeckea ericoides* (Myrtaceae). Emerged 12.6.71.

PARATYPES: 1♂, 2♀♀ as holotype; 79 ex. New Caledonia, various localities, BM geometrid slides 8198, 8199 (♂♂), 8430 (♀).

The species was recorded as follows: Grand Lac site 1 (2), site 2 (15), site 3 (17), site 4 (6), site 72 (1); Pic du Pin site 7 (29); R. Bleue site 10 (2); Mt. des Sources site 14 (1); R. Kuenthio (2); Dothio site 19 (1); Poupa (1); Koh (7); Col Toma (7); Petit Gouli (5); Kopeto (2); Mt. Dogny (10); Mt. Mou site 33 (5), site 34 (1), site 36 (2); Bambou R. (2); Col d'Amoss site 45 (1), site 47 (1); Cap Colnett (1); R. Amoa (3); Ste. Therèse (1); Col de Boa (1); Houailou (1); near Bâ (3); Ouen Toro site 64 (12); Col Mouirange (5); R. Koué (2); Mt. Koghis site 79 (1). Though primarily a species of *Baeckea*-sedge maquis on ultramafic rocks, *strixoides* occurs wherever *Baeckea* is frequent.

Several larvae were beaten from *Baeckea ericoides* at Grand Lac. They were in what proved to be their last instar, dorsally medium to dark grey-brown with darker flecks and paler streaks centrally, laterally blotched irregularly paler, and ventrally with a narrow central pale band containing dark brown streaks. Preserved material is to be deposited at the BMNH. During the day these larvae held themselves flattened against the *Baeckea* stems, which they closely resemble in coloration. They descended to the floor of the box to pupate, spinning a cocoon in any detritus there. The adults emerged after two or three weeks.

## URANIIDAE: MICRONIINAE

### **Micronia cynthiae sp. n.** (Plate 65: 1)

♀ 32 mm. The ground colour is pale cream (whitish in the most closely related species) with fine brown striae and pale grey-brown fasciae on both wings. At the hindwing tail are two black dots, one submarginally on the vein at the tail, the other marginally on the vein posterior to the tail; related species have more marginal spots. The striations are finer, more evenly distributed and more numerous than in any other species examined (e.g. *aculeata* Guenée from the Oriental Region or *pluvialis* Warren from Mt. Kinabalu (omitted from the lists of HOLLOWAY (1976))). The striations are longitudinal at the hindwing dorsum but transverse in other species. Further diagnostic characteristics are: the striae become darker marginally where they cross the forewing veins; the hindwing costa and some of the hindwing veins are broadly white, breaking the fasciae and the striae.

The nearest named relatives geographically are three closely related, possibly conspecific species in the Papuan Region. *M. fuscifimbria* Warren (Queensland) has much narrower, more ochreous fasciae and much lighter striation; *M. justaria* Walker from New Guinea and the Bismarcks is very similar and in *M. dilatistriga* Warren from the Solomons the striae are almost absent. In these three species the forewing fasciae tend to be parallel, but this is far from the case in *cynthiae*; the postmedial of the hindwing in *cynthiae* is much more adjacent to the margin posterior to the tail.

HOLOTYPE ♀. New Caledonia: Pic du Pin, 240 m Site 7 06879/75383 (J. D. HOLLOWAY) 22.5.1971, BM 1971-507.

## URANIIDAE: EPIPLEMINAE

### *Dirades latibrunnea* Warren (Plate 65: 5, 6, Figure 69: 2)

*Dirades latibrunnea* Warren, 1896, *Novit. zool.* 3: 345, unique female type from Lifu examined.

♂♀ 12–14 mm. The facies are shades of dull brown, the male having the hindwing dorsum somewhat modified, invested with a thick array of long, straw-coloured scales. It appears to have no close relatives amongst the species represented in the BMNH apart from the New Hebrides species described below. However, scattered amongst a variety of genera (the Epipleminae are in dire need of revision) are a number of generally similar species in the south-west Pacific area such as *Epiplema lypera* Tams (Samoa), *Epiplema lomalangi* Robinson, *Gathynia cythera* Swinhoe and *Gathynia yasawa* Robinson (all Fiji). The New Caledonian species could perhaps be a derivative of the widespread *Dirades mutans* (Butler), which extends from India to Queensland and the New Hebrides.

*D. latibrunnea* was recorded as follows: Pic du Pin site 7 (2); R. Amoa (1); Mt. Panié site 60 (3); Port Boisé site 74 (1). MNHN: La Coulée.

### **Dirades robinsoni sp. n.** (Plate 65: 9, 10)

♂♀ 14–18 mm. This species is much larger than *latibrunnea*, possibly the largest in the genus. It is generally the same colour as *latibrunnea* but the male forewing is almost uniform apart from a darker patch central on the dorsum. In the female the hindwing postmedial is more acutely angled, with a diagnostically even band distad that is slightly paler than the ground (in *latibrunnea* this is narrow anteriorly, broader and more diffuse posteriorly). In the male genitalia the uncus is relatively larger, the aedeagus vesica slightly more

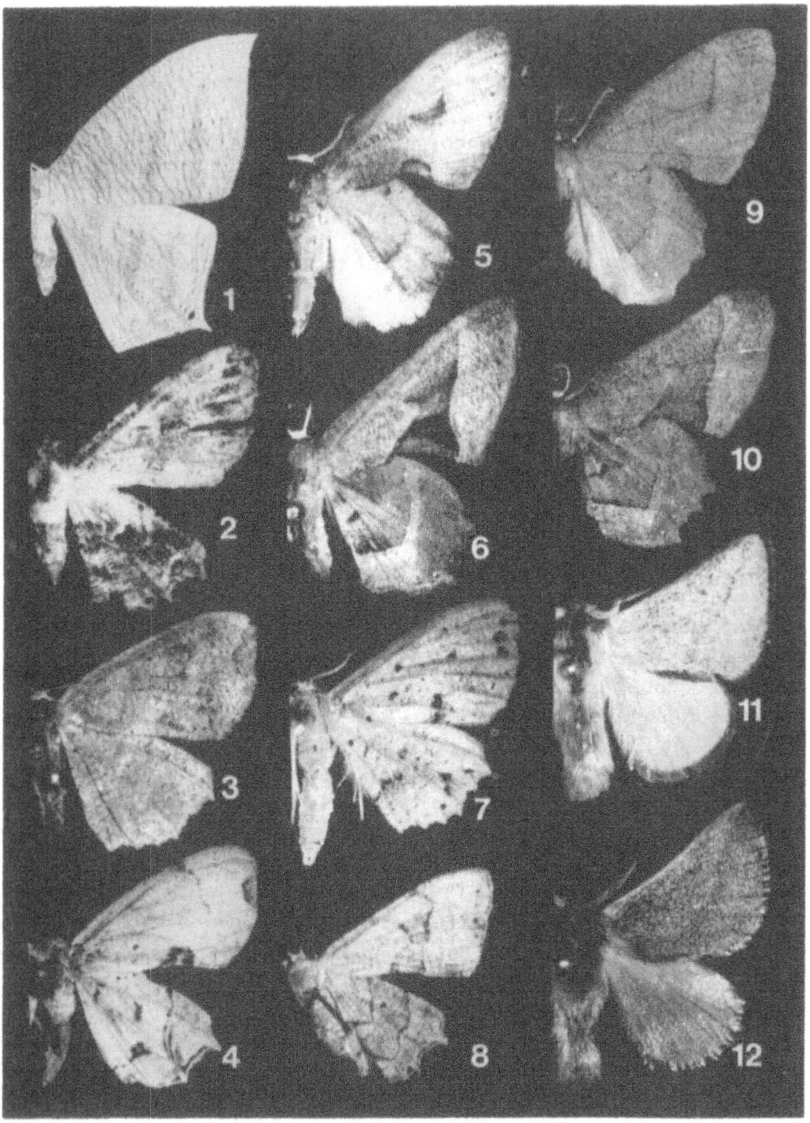


*Plate 65.* 1. *Micronia cynthiae;* 2. *Epiplema desistaria instabilata* (7 mm); 3 (♀), 7 (♂). *Epiplema lateritica;* 4. *Epiplema spatulata;* 5 (♂), 6 (♀). *Dirades latibrunnea* (12–13 mm); 8. *Epiplema yoshimotoi;* 9 (♂), 10 (♀). *Dirades robinsoni;* 11, 12. *Miresa obscurimargo.*

scobinate; the major difference is in the hair pencils on the valve where the hairs are filamentous instead of broad and flattened.

HOLOTYPE ♂. New Hebrides: Aneityum I., Agathis Camp-1150 ft, 19-21.vii.1971 (G. S. ROBINSON) Roy. Soc. Expedition. BM uraniid slide 85.

PARATYPES: 8♂♂ as holotype; 3♂♂, 2♀♀ New Hebrides: Santo, Mt. Tabwemasana, Nokowula CK. 3500 ft 3-4.ix.1971 (G. S. ROBINSON) Roy. Soc. Expedition. BM uraniid slide 86.

The Santo locality was in disturbed montane forest and that at Aneityum in rainforest with scattered *Agathis*.

### *Epiplema desistaria instabilata* (Walker) **stat. n.** (Plate 65: 2)

*Erosia* (?) *desistaria* Walker, 1861, *List Specimens lepid. Insects Colln Br. Mus.*, 23: 853.

*Erosia instabilata* Walker, 1886, *Ibid.*, 35: 1646.

*Epiplema desistaria* (Walker); Swinhoe, 1900, *Cat. east and Aust. Lepid. Heterocera*, 2: 220.

SWINHOE regarded *desistaria* as covering material and several names from Sri Lanka and the Himalaya to Queensland and Fiji. This range of material is remarkably constant as regards wing pattern. A single female was taken in New Caledonia at Tiebaghi site 41. It differs in facies from the above material in the much reduced white areas at the forewing dorsum and hindwing costa, and in having a white discal spot and submarginal line to the hindwing. More material is needed to determine its status.

*E. desistaria* was described from Borneo and *instabilata* from Mysol. No Mysol females were available in the BMNH for comparison with the New Caledonian female, so one from Buru was dissected, and also a female from Borneo and one from the New Hebrides. Apart from the Borneo specimen the genitalia were very similar, having two evenly spined signa in the bursa centrally, one slightly smaller than the other, more laterally elongate, and a helical ductus seminalis arising from a basal, lateral lobe on the bursa; the lobe is partially sclerotised, partially sparsely scobinate. The Bornean specimen has a single signum corresponding to the larger of the two above, slightly larger, more finely spined, squarish with two larger spines, one at each of the more apical corners of the square; the ductus seminalis arises more basally, less on a lobe, with the scobination much reduced.

Though more dissections are necessary, general biogeographic patterns in the area suggest that the name *instabilata* will refer to Papuan Region (Moluccas to Queensland, New Caledonia and Fiji) specimens of *desistaria*.

### **Epiplema yoshimotoi sp. n.** (Plate 65: 8)

♀♀ 8 mm. The wings are lightly marked with rich brown on a dull, creamy white ground as illustrated. The species is related to *polei* Hampson (Sri Lanka; type examined) and *transgrisea* Holloway (Borneo) in general colour and wing pattern: a centrally angled antemedial on both wings, that of the forewing passing through the discal spot, the dark discal spot of the hindwing being separate but close to the centre of the anterior half of the antmedial; a marginal dark dot on the hindwing just posterior to the posterior tail. It differs from *polei* and *transgrisea* in its smaller size and the anterior angle to the forewing postmedial, which is approximately straight in the other two species; the hindwing postmedial is angled nearer to the dorsum, significantly so compared with *polei*, slightly so compared with *transgrisea*; the submarginal over the anterior half of the forewing is much less prominent than in *polei*. Queensland specimens in the BMNH, possibly referable to *oxytype* Turner, have the forewing more blotched brown, especially submarginally, and the postmedials of both wings are angled almost centrally as with the antemedials. In the female genitalia the bursa of *yoshimotoi* is not ornamented. No males have been taken but the genitalia of *transgrisea* illustrated by HOLLOWAY (1976: Figure 680) indicate that this group of species may eventually be assigned to a distinct genus or subgenus.

HOLOTYPE ♀. New Caledonia: Grottes de Koum, 20 m Site 43 04314/77286 (J. D. HOLLOWAY) 11.7.1971, BM 1971-507. BM uraniid slide 57.

PARATYPE ♀. New Caledonia: La Crouen, 31.i.1963 (C. YOSHIMOTO & N. KRAUSS) light trap, BPBM.

**Epiplema lateritica sp. n.** (Plate 65: 3, 7, Figure 69: 3)

♂♀ 10–14 mm. The ground colour is variable warm greyish fawn with more ochreous, paler bands faintly evident in most specimens distad from the postmedial and basad from the antemedial on the forewing. The ground is variably finely striated and speckled with darker brown (darker in females which often tend to be greyer with darker forewing medial and submarginal areas). The males generally have three dark spots submarginally on the anterior half of the forewing, this being more a thin crenulate line separating off a darker area in the female. There are variable dark spots on the forewing and hindwing postmedials and antemedials in the males, one having a dark discal spot to the hindwing (the Plate shows one of the more heavily spotted individuals).

The species is generally similar to three Papuan Region species, two from New Guinea and one from the Solomons, but these differ as follows. *E. sparsipunctata* Warren (type examined; Solomons) has a similar ground colour to the wings but is smaller, the hindwing margin less angled and the forewing has two subapical spots only; the forewing postmedial is lightly marked with three evenly spaced dark spots; the species is not significantly variable. *E. casbiata* Warren (type examined; New Guinea) is more sparsely speckled on a paler, fawner ground, the fasciae being more regularly dotted dark brown. *E. despecta* Warren (type examined; New Guinea) is closest in appearance but smaller with the forewing submarginal dots in a more curved row so that the central ones are further from the margin.

The male genitalia show more pronounced differences. Those of *lateritica* differ from those of *casbiata* (which are identical with those of the unique type of *boarmiata* Rothschild from New Guinea, probably synonymous) in the more triangular upper portion of the valves (cf. elongate) with basal and apical groups of fine setae instead of a central group of short, broad, scale-like ones; the uncus is bifid instead of long and single, the gnathi short, spine-like instead

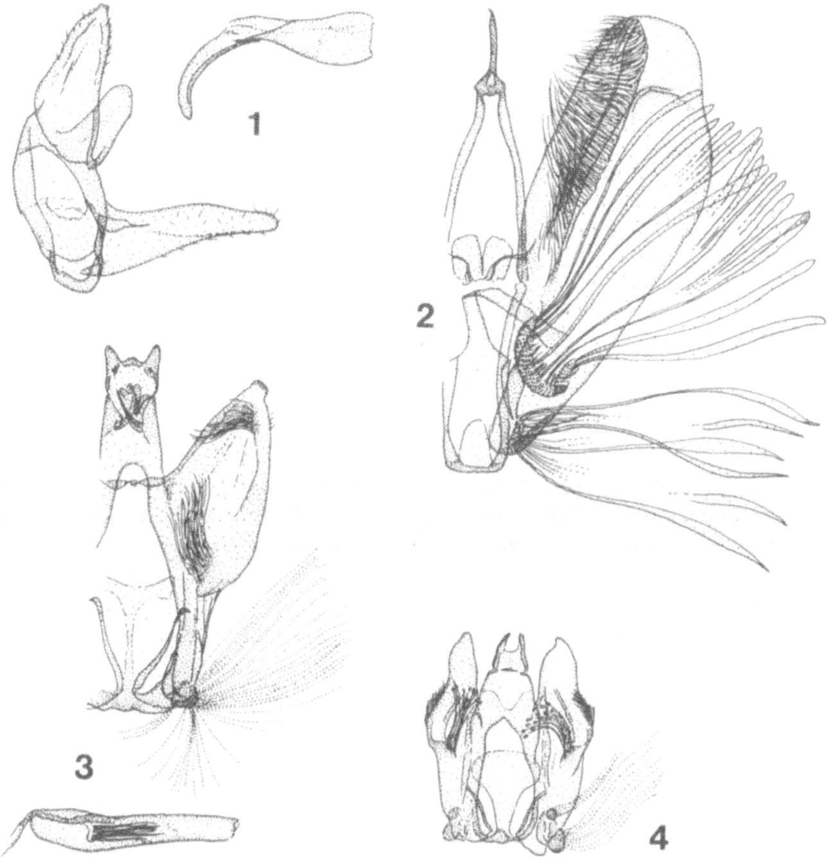


*Fig. 69.* Male genitalia of *Miresa obscurimargo* (1), *Dirades latibrunnea* (2), *Epiplema lateritica* (3) and *Epiplema spatulata* (4).

of long and spatulate; the aedeagus vesica of *lateritica* contains a group of long, slender cornuti instead of being finely scobinate. *E. sparsipunctata* has the valves almost simple, not divided as in *lateritica* into a broad upper part and a narrower lower part with a basal corema; the valve of *sparsipunctata* is triangular with a central patch of fine setae arising from a large oval area; the uncus and gnathi are similar to those of *lateritica* but the bifurcations of the uncus are directed laterally instead of dorsally and there is additionally a weak central lobe; the aedeagus vesica is scobinate but there is also a single slender cornutus. *E. despecta* has the uncus as in *sparsipunctata* and the valves the same shape but with a central, ventral row of short setae with an ill-defined group of finer ones just dorsad; the aedeagus vesica is finely scobinate with no

cornuti. Thus the two groups of fine setae on the valve and the group of cornuti in the vesica are diagnostic for *lateritica*.
HOLOTYPE ♂ New Caledonia: Grand Lac, 250 m Site 1 06980/75358 (J. D. HOLLOWAY) 13.5.1971, BM 1971-507. BM uraniid slide 60.
PARATYPES. 14♂♂, 8♀♀ New Caledonia, various localities, BM uraniid slide 61 (♂).

The species was recorded in New Caledonia as follows: Grand Lac site 1 (1), site 2 (3); Pic du Pine site 7 (5), site 69 (1); R. Bleue site 9 (1), site 10 (2); Mt. des Sources site 11 (1); Mt. Mou site 35 (1); Col d'Amoss site 47 (1); R. Amoa (1); Mt. Krapé site 55 (1), site 56 (3); Mt. Panié site 60 (2); Mt. Pouédihi (1); Port Boisé site 74 (1). The species is a member of the ultramafic association.

**Epiplema spatulata sp. n.** (Plate 65: 4, Figure 69: 4)

♂♂ 8–9 mm. The ground colour is creamy white, marked with dark, slightly rufous brown as illustrated. The species is part of the *conflictaria* Walker complex (female type examined; Ceram) that occurs throughout the Indo-Australian tropics. Compared with typical *conflictaria* the forewing markings are reduced and the hindwings are also less heavily marked, without the rufous striae over the posterior half exhibited by *conflictaria* and most other species (undescribed as yet) of the complex. The posterior section of the hindwing postmedial is straight or slightly sinuous, but angled in *conflictaria*.

A male from Amboina similar to the Ceram type was dissected (slide 88); the valve in this specimen is simple with a basal hair pencil but lacking the spatulate setae on the centre of the valve that characterise *spatulata*; the transtillae and juxta are highly modified to give two pairs of angled structures with a small plate, the juxta, between the upper pair (these structures are simple in *spatulata*); the uncus is broad, simple and round in the Amboina specimen but bifid in *spatulata*; the aedeagus is long, slender and curved instead of short and slender with a single cornutus.

The geographically nearest relative to *spatulata* occurs in the New Hebrides, Queensland and, with more tapering valves in the genitalia, the Solomons; material from the Bismarcks and New Guinea may also be referable. It is undescribed. It differs in having both wings lightly suffused with brown striae, more heavily over the posterior half of the hindwing. The valve is more even in width, not bulging centrally, rounded apically (except in Solomons material); the central setae on the valve are not spatulate; the uncus is trifid and the aedeagus is slightly stouter, the vesica immaculate.
HOLOTYPE ♂. New Caledonia: Grottes de Koum, 20 m Site 43 04314/77286 (J. D. HOLLOWAY) 11.7.1971, BM 1971-507. BM uraniid slide 59.
PARATYPES: 3♂♂ New Caledonia (Anse Longue site 16, site 32, Poupa); 1♂ Tipindjé, 10.ii.1963 (C. YOSHIMOTO & N. KRAUSS) Malaise trap, BPBM.

A female taken on Uvea, Loyalty Is., in May 1974 by Mr. H. SUZUKI may be referable to this species as it had similar markings. The ground colour, however, was pale grey.

### *Agrius convolvuli* (Linnaeus) (Plate 66: 3 (*ab.*), Figure 70: 3)

*Agrius convolvuli* (Linnaeus); VIETTE (1951b); ROBINSON (1975: 244); HOLLOWAY (1977: 116).

This species is widespread in the Old World and Pacific tropics and subtropics and is vagrant to more temperate latitudes. Three specimens were taken during the survey at Mt. Mou sites 33, 34 and at Probokonoro. VETTE recorded the species from Hienghène and COHIC (1956) gave *Ipomoea* species as New Caledonian host plants.

The site 34 specimen, a male, had the forewing darkened almost to black, reflecting deep blue slightly, except for a medial grey band. G. S. ROBINSON took a similar but larger specimen in the New Hebrides (S.W. Bay, Wintoua, Malekula). The genitalia of the New Caledonian specimen do not differ from those of typical *convolvuli*, so this is probably only an aberrant form.

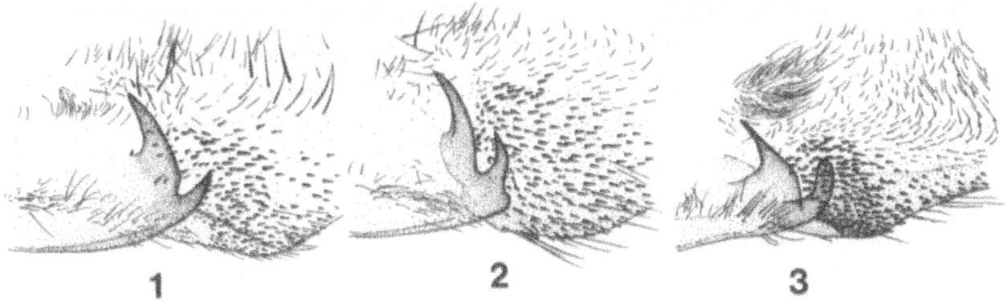


*Fig. 70.* The harpe of the male valve of *Agrius luctiferus* (1), *A. fasciatus* (2) and *A. convolvuli* (3).

### *Agrius fasciatus* (Rothschild) (Plate 66: 2, Figure 70: 2)

*Sphinx fasciatus* Rothschild, 1894, *Novit. zool.*, 1: 94. VIETTE (1949: 325)

This species is closely related to *luciferus* Walker from the S. Moluccas, New Guinea and the Bismarcks. It can be distinguished by the harpe of the male genitalia (Figure 70), and the lateral spots immediately posterior to the thorax are white instead of yellow.

Previously recorded only from the Loyalty Is., it was taken in New Caledonia as follows: Plaine des Lacs site 5 (6); Pic du Pin site 7 (1); Poupa (2); Koh (2); Bréhoa R. (1); Col d'Amieu (2); Sarraméa (1); Mt. Mou site 33 (3); Grottes de Koum (1); Col d'Amoss site 45 (1); Cap Colnett (1); Amoa R. (8); Probokonoro (1); Mt. Krapé site 55 (2); Col de Boa (1); near Bâ (2); Port Boisé site 74 (1); R. Koué (1).

## COMPSULYX Gen. n.

VIETTE (1971) described a new sphingid from New Caledonia in the genus *Compsogene* Rothschild & Jordan. He had only two males on which to base his diagnosis but drew attention to the fact that the new species did not fall readily into any genus. It bore similarity also to *Oxyambulyx* Rothschild & Jordan. During the 1971 survey it was taken frequently throughout the island and a series of 55 specimens was retained, including females. This permitted a more detailed analysis of its position in relation to *Compsogene* and *Oxyambulyx*.

The species resembles *Oxyambulyx* in the following characters: there is a dark bar between the antennae (the head is uniform in *Compsogene*); the thorax is uniform with the abdomen and forewing (darker in *Compsogene*, the tegular bands darker in *Oxyambulyx*); the abdomen above is uniform (the apical half is darker and patterned in *Compsogene*, and some *Oxyambulyx* have a thin, dark line dorsally); all the external surface of the fore tibia and the base of the mid and hind tibiae are white (rufous in *Compsogene*); the forewing subbasal fascia is broken into sharply defined costal and subdorsal spots (entire in *Compsogene*); the hindwing ground is yellow with three principal brown or black fasciae (in *Compsogene* the ground is basally pink, distally fawn, the fasciation is complex and the veins are distally picked out in dark brown); in *Oxyambulyx* there is a small dark brown spot submarginally on Culb and this is presumed to be homologous with the large black circle ringed with pale blue (also ringed slightly paler in *Oxyambulyx*) that forms the hindwing ocellus in the New Caledonian species (in *Compsogene* there is no such spot, though there are pale blue scales on the distal edge of the submarginal near the anal angle); the gnathi of the male genitalia are separate as in *O. subocellata* Felder, compared with their fused state in *O. substrigilis* (Westwood), the type species, and *Compsogene*.

The species resembles *Compsogene* in the following characters: the forewing postmedial is oblique and straight (curved and irregular in *Oxyambulyx*); the male genitalia have only a single lobe from the valve sacculus (this is accompanied by a prominent harpe in *Oxyambulyx*); the ductus bursae of the female lacks an appendix (a small one occurs in *O. substrigilis*); the signum is finely scobinate and basally notched (coarsely scobinate and rounded in *Oxyambulyx*), but much more elongate in *C. panopus* (Cramer).

In the following characters the New Caledonian species bears no close resemblance to either of the other two genera: the abdomen below is the same colour as the upperside instead of yellow as in the other taxa; the legs are generally greyish instead of rufous yellow, except the forelegs, palps below and the intervening area of the thorax are reddish in dark specimens and rufous orange in paler ones; the forewing apex is rounded as in *Compsogene* (acute in *Oxyambulyx*) but much more falcate than in either; similarly with the forewing anal angle; the submarginal above and below is diffuse only on the forewing, whereas it is sharply defined in the other genera; the forewing below is basally bright yellow, diffusing into the apical half which is the same colour as the upperside (in the other two genera it is yellow throughout); the aedeagus is unspined apically (spined to a greater or lesser extent in the other genera); the gnathus apex is uniquely scobinate; the bursa and ductus of the female extend almost the length of the abdomen as compared with two thirds in *C. panopus* and a half in *O. substrigilis*; the bursa is not corrugate (moderately so in *C. panopus* and strongly so in *O. substrigilis*).

There is great uniformity of morphology within *Compsogene* (two species, excluding *cochereaui*) and *Oxyambulyx* (many species). The former is restricted to the Oriental Region and the latter is primarily Oriental with some radiation in New Guinea and a species in the Solomons. The New Caledonian species bears closer resemblance to *Oxyambulyx* but certainly represents an offshoot of ambulicine stock prior to the main radiation within the other two genera. It therefore seems logical to place it in a new genus, *Compsulyx*, with type species *cochereaui* Viette.

### *Compsulyx cochereaui* (Viette) **comb. n.** (Plate 66: 5-9)

*Compsogene cochereaui* Viette, 1971, *Entomops*, 23: 237-240.

The species shows much variability. The forewings range from pale ochreous or rufous fawn or grey to deeper rufous or purplish grey. The intensity of fasciation varies. In the females the discocellular dot and patches in the spaces of the postmedial and double submarginal are often greyish white. The thorax and abdomen are always the same colour as the forewings.

*C. cochereaui* is endemic to New Caledonia. The type locality is Mt. Koghis and it was recorded as follows during the survey: Grand Lac site 2 (8), site 3 (6); Plaine des Lacs site 5 (4); Pic du Pin site 7 (16); R. Bleue site 8 (2), site 9 (3), site 10 (2); Mt. des Sources site 11 (1), site 13 (2), site 14 (2); Dothio site 19 (2); Poupa (1); Ciu (1); Ouatou R. (1); Bréhoa R. (6); Ouambéo R. (1); Col d'Amieu (15); Col Toma (17); Petit Gouli (1); Sarraméa (1); Kopéto (3); Mt. Mou site 33 (7), site 34 (1); Col d'Amoss site 45 (1); Kavatch (1); R. Amoa (1); Ste. Therèse (3); Probokonoro (3); Col des Roussettes (1); Mt. Krapé site 55 (10), site 56 (2); Col Mouirange (3); Mt. Pouédihi (2); Yaté Gorge (1); Port Boisé site 73 (1), site 74 (3); R. Koué (1); Les Dalmates site 77 (6), site 78 (2); Mt. Koghis site 80 (1). The species is frequent in areas of rainforest, perhaps more particularly on ultramafic rock.

### *Psilogramma menephron lifuense* Rothschild & Jordan (Plate 66: 1)

*Psilogramma menephron lifuense* Rothschild & Jordan, 1903, *Novit. zool.*, 9, Suppl: 43.

*P. menephron* (Cramer) is found throughout the Indo-Australian tropics to Australia (Northern Territory to New South Wales), New Caledonia and the Loyalty Is. A distinct species occurs in Fiji. The New Caledonian and Loyalty subspecies is a pale, ashy grey on the forewings, thorax and abdomen; races from the Oriental Region are very much darker, reminiscent of *Agrius luctiferus*. The species was recorded as follows in New Caledonia: Dothio site 19 (1); Poupa (3); Ouatou R. (1); Col d'Amieu (1); Col Toma (2); Petit Gouli

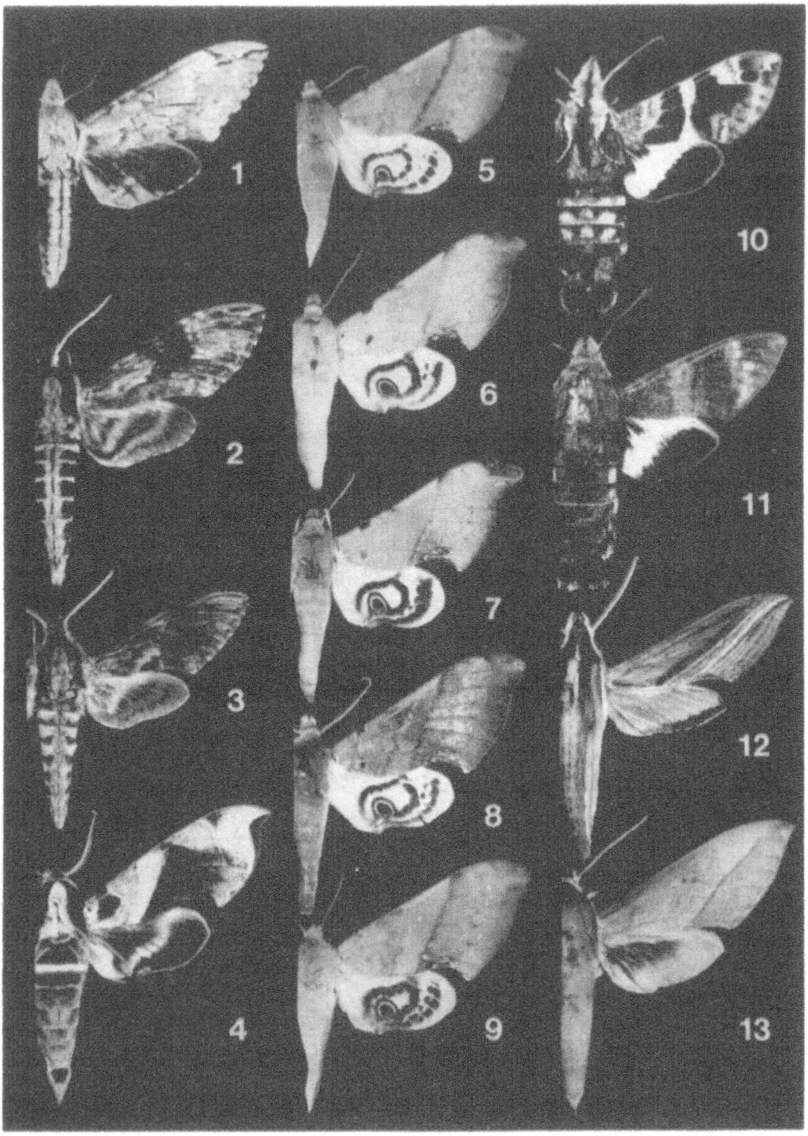


*Plate 66. Psilogramma menephron lifuense* (52 mm); 2. *Agrius fasciatus* (42 mm); 3. *Agrius convolvuli* dark aberration (38 mm); 4. *Daphnis placida* (40 mm); 5–9. *Compsulyx cochereaui* (48–54 mm); 10. *Macroglossum hirundo lifuensis* (26 mm); 11. *Macroglossum corythus fuscicauda* (30 mm); 12. *Theretra boerhaviae* (28 mm); 13. *Theretra clotho* (37 mm, Lifu).

(1); Sarraméa (3); Mt. Mou site 33 (2). site 34 (1); Grottes de Koum (1); Cap Colnett (1); R. Amoa (8); Probokonoro (3); Mt. Krapé site 55 (7); Col de Boa (1); Houailou (1); Port Boisé site 73 (2), site 74 (4). BPBM: Bourail, Nouméa.

## SPHINGIDAE: MACROGLOSSINAE

### *Cephonodes janus simplex* Rothschild (Plate 67: right)

*Cephonodes janus* Miskin, 1891, *Proc. Roy. Soc. Queensland*, 8: 6.
*Cephonodes janus simplex* Rothschild, 1894, *Novit. zool.*, 1: 66, type examined (illustrated).

This subspecies was described from Lifu and belongs to a species recorded from Queensland, the Lesser Sundas and Toekan Besi I. It may be distinguished from the next species in the field by the absence of a transverse dark band on the abdomen, the uniform straw colour of the abdomen below and the much narrower forewing margin.

In the male genitalia the valves extend well beyond the uncus and are asymmetric, the right one being longer, the left having a lobe basally on the ventral margin, distinguishing it from the *picus* group where this is merely angled. The triangular juxta is narrower, apically more acute than in the *picus* group. The gnathus is deeply bifid, the two arms splayed, and the double uncus is roughly three times as broad as in the *picus* group.

*Plate 67.* Left. *Cephonodes lifuensis* (26 mm); Right. *Cephonodes janus simplex* (27 mm).

### *Cephonodes lifuensis* Rothschild (Plate 67: left)

*Cephonodes lifuensis* Rothschild, 1894, *Novit. zool.*, 1: 66, type examined (illustrated).

This species, recorded from New Caledonia (Nouméa), Lifu and the New Hebrides, belongs to a group of three species distinguished within the genus by an apical 'thorn' to the fore tibia (ROTHSCHILD & JORDAN, 1903, *Novit. zool.*, 9 (Suppl.): 460–471). It is distinguished from the other two species by the broader margin, especially apically, to the forewing. The other two species are *picus* (Cramer), from the Oriental Region (including islands in the Indian Ocean such as the Maldive and Chagos groups), New Guinea and Queensland, and *armatus* Rothschild & Jordan, from Fiji, Samoa, Tonga and the Marianas. The former is distinguished from the other two in the group by the black (instead of rufous) markings of the ventral surface of the abdomen, and the latter is distinguished by the absence or reduction of the transverse band on the dorsal surface of the abdomen. Otherwise the species are very similar.

All three specimens of *lifuensis* in the BMNH are females and a male was therefore not dissected though there is one with another female in the BPBM. The male genitalia of *picus* and *armatus* are so similar as to suggest conspecificity. An Indian region specimen of the former and a Fijian specimen of the latter were dissected. The valves were asymmetric as in *janus* but not extending significantly beyond the uncus. The gnathus is single, though more serrate and unequally divided apically in the *picus* specimen. The *picus* specimen dissected (with one other) was labelled as having been taken at sea 500 miles from Colombo (Sri Lanka), indicating that the species is probably highly mobile.

### *Daphnis placida placida* (Walker) (Plate 66: 4)

*Darapsa placida* Walker, 1856, *List Specimens lepid. Insects Colln Br. Mus.*, 8: 186, type examined.

The typical race occurs throughout the Oriental Region and extends to New Guinea and Queensland, thence to New Caledonia, the Loyalties and the New Hebrides. A distinct race, *salomonis* Rothschild & Jordan, occurs in the Solomons and another, *torenia* Druce, in Fiji and Samoa (the Samoan form has been described as a separate race, *steffanyi* Clark, but ROBINSON could find nothing to distinguish it from *torenia*).

In New Caledonia the species was recorded as follows: Pic du Pin site 7 (1); R. Bleue site 10 (1); Dothio site 18 (1), site 19 (1); Ouatou R. (1); Mt. Dogny (1); Anse Longue site 32 (1); Tiebaghi site 41 (1); Grottes de Koum (1); Ste. Therèse (1); Mt. Krapé site 55 (9); Les Dalmates site 78 (1).

### *Gnathothlibus erotus eras* (Boisduval)

*Gnathothlibus erotus eras* (Boisduval); ROBINSON (1975: 246).

The species is recorded widely in the Oriental Region and extends east to New Guinea and Australia (Northern Territory to New South Wales); but it is perhaps most characteristic of the Pacific area from the Bismarcks to the Marquesas and Rapa. It is the commonest sphingid in New Caledonia and was recorded as follows: Grand Lac site 1 (4), site 2 (10), site 3 (14), site 4 (3); Plaine des Lacs site 5 (12); Pic du Pin site 7 (14); Mt. des Sources site 11 (1), site 12 (3), site 13 (1), site 14 (5); Anse Longue site 16 (2); R. Kuenthio (3); Dothio site 18 (1), site 19 (2); Poupa (1); Koh (43); Ouatou R. (2); Bréhoa R. (11); Col d'Amieu (39), Col Toma (6); Sarraméa (7); Mt. Dogny (2); Mt. Mou site 33 (10); Grottes de Koum (6); Bambou R. (6); Col d'Amoss site 45 (1), site 46 (1), site 47 (2); Mt. Ignambi (1); Cap Colnett (12); R. Amoa (39); Ste. Therèse (1); Probokonoro (32); Col des Roussettes (1); Mt. Krapé site 55 (21), site 56 (2); Col de Boa (2); near Bâ (15); Mt. Aoupinié (5); Port Boisé site 73 (1), site 74 (12), R. Koué (1); Les Dalmates site 77 (3). The species is widespread throughout the island, perhaps commonest in areas of rich secondary vegetation on sedimentary rock. It comes abundantly to light on warm, humid nights. COHIC (1956) recorded the larva as feeding on *Ochrosia oppositifolia* (Apocynaceae) and *Morinda citrifolia* (Rubiaceae) in New Caledonia. The latter is a foodplant in Fiji (ROBINSON).

*Hippotion boerhaviae* (Fabricius) (Plate 66: 12)

*Sphinx boerhaviae* Fabricius, 1775, *Syst. Ent.*: 542.

*H. boerhaviae* is found throughout the Oriental and Papuan Regions, extending to the Northern Territory, northern Queensland and the Solomons. A single specimen was taken on Mt. Mou site 33 in New Caledonia.

*Hippotion celerio* (Linnaeus)

*Hippotion celerio* (Linnaeus); ROBINSON (1975: 246); HOLLOWAY (1977: 116).

*H. celerio*, widespread in the Old World except at high latitudes, was recorded as follows: Cap Colnett (1); Probokonoro (1). VIETTE (1949): Nouméa, Paita. BPBM: Bourail, Hienghène, Népoui, Le Crouen. COHIC (1956) recorded as host plants: *Alocasia macrorrhiza, Arum, Caladium bicolor, Colocasia antiquorum, Ipomoea batatas*, and *Xanthostoma sagittifolium*.

*Hippotion scrofa* (Boisduval)

*Hippotion scrofa* (Boisduval); HOLLOWAY (1977: 116).

Widespread in Australia and vagrant to Norfolk I., *scrofa* occurs also in the New Hebrides and was recorded as follows in New Caledonia: Anse Longue site 16 (1); Dothio site 19 (1); Col d'Amieu (1); Mt. Mou site 33 (5); Bambou R. (1); Col d'Amoss site 45 (1); Mt. Krapé site 55 (1); near Bâ (1); Col Nassiboué (2); Ouen Toro site 64 (1); Port Boisé site 74 (1). BPBM: Nouméa. MNHN: La Coulée. BMNH: Pouébo. The majority of these localities are on the drier western side of the island. COHIC (956) recorded the larva as feeding on *Ipomoea batatas*.

*Hippotion velox* (Fabricius)

*Hippotion velox* (Fabricius); VIETTE (1949: 326); ROBINSON (1975: 247); HOLLOWAY (1977: 117).

*Panacra lifuensis* Rothschild, 1894, *Novit. zool.*, 1: 79, type examined.

In the Loyalty Is. and New Caledonia there is an abnormally high proportion of individuals with the forewings almost uniformly fawn. The species is very variable in New Caledonia. It is found throughout the Indo-Australian tropics as far east as Fiji and Rotuma, and is an occasional vagrant to Norfolk I. It was recorded as follows in New Caledonia: Plaine des Lacs site 5 (1); R. Bleue site 10 (1); Mt. des Sources site 14 (1); Dothio site 19 (2); Ouatou R. (1); Bréhoa R. (3); Ouambéo R. (1); Col d'Amieu (3); Col Toma (1); Sarraméa (3); Mt. Mou site 33 (12), site 34 (2); Grottes de Koum (1); Cap Colnett (1); Kavatch (2); R. Amoa (1); Probokonoro (23); Mt. Krapé site 55 (5); near Bâ (1); Mt. Aoupinié (2). BPBM: Isle of Pines.

### *Macroglossum corythus fuscicauda* Rothschild & Jordan (Plate 66: 11)

*Macroglossa corythus* Walker, 1856, *List Specimens lepid. Insects Colln Br. Mus.*, 8: 92.
*Macroglossum corythus fuscicauda* Rothschild & Jordan, 1903, *Novit. zool.*, 9 (Suppl.): 663, type examined.

*M. corythus* is recorded throughout the Indo-Australian tropics as far as Queensland and New Caledonia. Subspecies *fuscicauda* is restricted to the Loyalty Is. and New Caledonia. VIETTE (1949) recorded *M. moluccensis* for the Loyalties but suggested this species was a race of *corythus* and presumably the record refers to *fuscicauda*.

Three specimens were taken during the survey, visiting flowers at dusk. Two were visiting the sweetly scented flowers on the trunk of a species of *Dysoxylum* at Dothio site 19 and the third came to lantana at Col d'Amoss site 45. In the BPBM there are specimens from Mt. Koghis, Forêt de Thi, Bourail and Wé (Lifu). Specimens taken by Miss CHEESMAN in Lifu are labelled as having fed at *Datura* and *Frangipani* blooms.

### *Macroglossum hirundo lifuensis* Rothschild & Jordan (Plate 66: 10)

*Macroglossum hirundo lifuensis* Rothschild & Jordan, 1903, *Novit. zool.*, 9 (Suppl.): 43, type examined. ROBINSON (1975: 248). HOLLOWAY (1977: 117).

The species ranges from the Lesser Sundas, New Guinea and northern Australia eastwards into the Pacific as far as Pitcairn I. where the typical race flies. The forewing of the New Caledonian and Loyalty subspecies *lifuensis* is variable, overlapping to a considerable extent with the range of variation in the Australian race, *errans* Walker. The latter is probably the race vagrant to Norfolk I. Unlike the previous species, *hirundo* was not seen during the day but was taken frequently at light as follows: Grand Lac site 2 (1); Dothio site 19 (1); Bréhoa R. (1); Petit Gouli (1); Mt. Mou site 33 (3); Cap Colnett (1); Amoa R. (1); Probokonoro (4); Mt. Krapé site 55 (6); near Bâ (9); Mt. Aoupinié (1); Col Nassiboué (2); Yaté Gorge (2); Port Boisé site 73 (2), site 74 (4); Les Dalmates site 77 (3).

*Theretra clotho celata* (Butler) (Plate 66: 13)

*Sphinx clotho* Drury, 1773, *Illustr. exot. Ins.*, 2: 48.
*Chaerocampa celata* Butler, 1877, *Proc. zool. Soc. Lond.*, 1877: 472.
*Theretra lifuensis* Rothschild, 1894, *Novit. zool.*, 1: 78, type examined (illustrated).

*T. clotho* is widespread in the Indo-Australian tropics, extending to Queensland, New South Wales, the New Hebrides and Lifu, but has not yet been recorded from New Caledonia; *celata* is the Papuan Region subspecies.

*Theretra nessus* (Drury)

*Sphinx nessus* Drury, 1773, *Illustr. exot. Ins.*, 2: 46.

This species is widespread in the Indo-Australian tropics, having its eastern limits in Queensland, New South Wales, the New Hebrides, Loyalty Is. and New Caledonia, though no Solomons specimens have been noted. Singletons were recorded at Grand Lac sites 2 and 3 early in the survey but there are other records as follows. VIETTE (1949): Nouméa. BPBM: Bourail, Le Crouen, Col d'Amieu, St. Louis Valley. The species may be more abundant during the summer months.

*Theretra pinastrina intersecta* (Butler)

*Theretra pinastrina intersecta* (Butler); ROBINSON (1975: 250).

There are two specimens of *intersecta* from Le Crouen in the BPBM. The species is widespread in the Indo-Australian tropics with its eastern limits in northern Australia, Fiji and Samoa.

THAUMETOPOEIDAE

*Teara circumfumata* (Felder) (Plate 68: 1-3, 5, 6, 9-11, Figures 71: 3, 72)

*Ochrogaster circumfumata* Felder, 1874, *Reise Novara, Lep.* 14: t. 94, type examined.
*Miresa sobrina* Druce, 1898, *Ann. Mag. nat. Hist* (7), 1: 215, type examined, **syn. n.**
*Miresa alma* Druce, 1898, *Ibid.*, type examined, **syn. n.**
*Teara circumfumata* (Felder); Kiriakoff, 1970, *Genera Insectorum*, 219e: 37.

The species exists in New Caledonia in a variety of forms that are largely correlated with specific vegetation types (Figure 72). The genus is otherwise restricted to Australia where the species concerned exhibit all the facies types noted for the New Caledonian species. Unfortunately there is no information on the ecological distribution of these facies types. MCFARLAND (*in litt.*) has reared the larvae of Australian species on *Acacia*, *Cassia* and *Eucalyptus*. The male genitalia indicate that the New Caledonian forms are probably all conspecific, perhaps closest to the Australian *contraria* (Walker) but differing from that species in the more angular margin of the valve and the smaller size

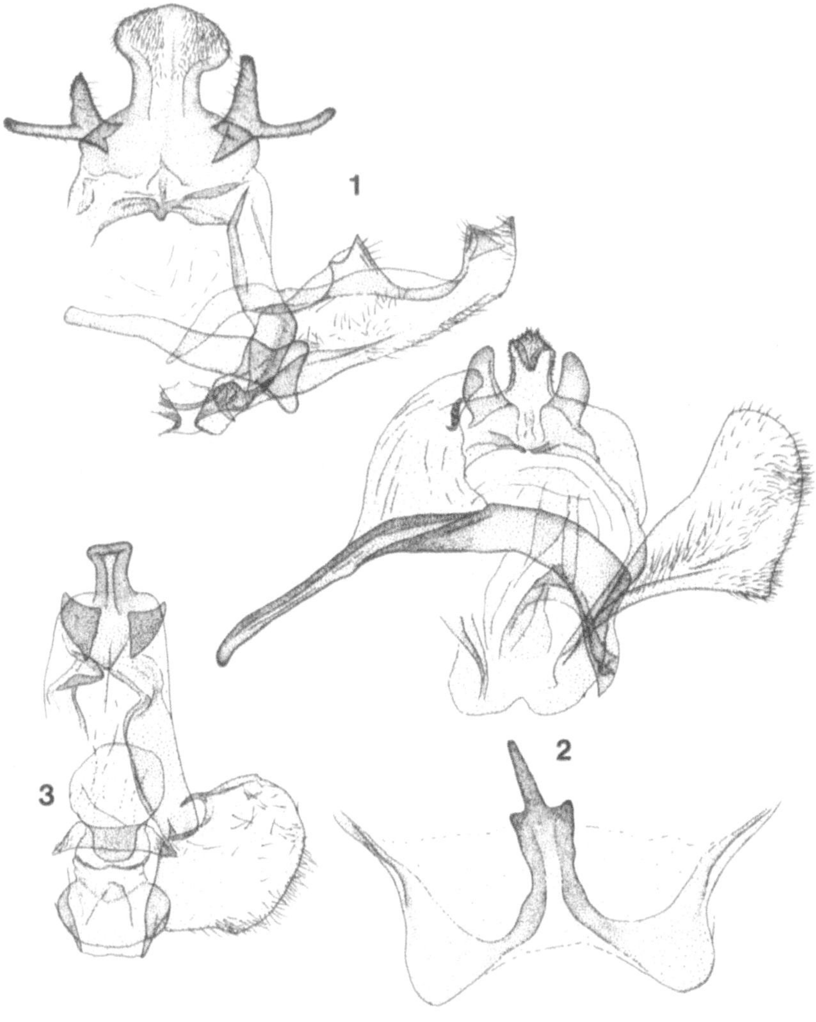


*Fig. 71.* Male genitalia of *Neola octofera* (1), *Cerura amoa* (with modified eighth sternite, 2) and *Teara circumfumata* (3).

relative to the tegumen of the valves uncus and gnathi. The various New Caledonian forms are as follows:

F. **alma**. The forewings are pale rufous orange to pale rufous brown, diffusely streaked darker, and the hindwings are pale yellowish orange (Pl. 68: 1, 2). DRUCE's type is smaller than average and faded. This form is characteristic of ultramafic *Baeckea*-sedge maquis (solid circles in Figure 72), and was recorded as follows: Grand Lac site 2 (5); Plaine des Lacs site 5 (2), site 6 (2); Pic du Pin site 7 (2); Dothio site 18 (1); Mt. Mou site 33 (3); near Bâ (1); Col

Mouirange (7); Mt. Pouédihi (1), Yaté Gorge (2); R. Koué (7); Anna-Madeleine (2).

F. **sobrina**. The wings and thorax are entirely medium to dark grey, only the abdomen being banded rufous. The forewing antemedial and postmedial are darker (Pl. 68: 5, 6). The form is more or less associated with dry forest, secondary growth and *Acacia spirorbis* woodland (open squares in Figure 72). It was recorded as follows: Poupa (4); Col d'Amieu (1); Col Toma (3); Petit Gouli (5); Sarraméa (1); Kopéto (5); Bambou R. (1); Kavatch (2); Mt. Krapé site 55 (14).

F. **silvicola** n. The wings are very dark grey, almost black, sometimes with slightly darker fasciae; the forewing costa and the thorax are suffused rufous (Pl. 68: 3, 11). The form is commonest in rainforest though it flies in several localities with *sobrina* and has been taken in more open localities where forest is absent (solid triangles in Figure 72). It was recorded as follows: Dothio site 19 (6); Poupa (8); Ciu (3); Koh (9); Ouatou R. (7); Bréhoa R. (11); Ouambéo R. (3); Col d'Amieu (5); Col Toma (3); Petit Gouli (4); Sarraméa (7); Grottes de Koum (1); Col d'Amoss site 47 (3); Kavatch (1); Col des Roussettes (1); Col de Boa (1); Mt. Aoupinié (5).

F. **striata** n. This form is basically a modification of *sobrina* with which it usually flies. The hindwings are tinged yellow and the forewing is streaked white in the spaces as illustrated in Pl. 68: 9. The form is restricted to the western ultramafic massifs (including the small area at Petit Gouli) and was recorded as follows: Petit Gouli (3); Kopéto (1); Mt. Krapé site 56 (3). It is indicated by stars in Figure 72.

The type of *circumfumata* is atypical, a similar specimen from Ouen Toro being shown in Pl. 68: 10 It is approximately intermediate between the *sobrina* and *alma* forms, having the forewing of the former except for an orange medial area and the hindwing of the latter. Two such specimens were recorded from Ouen Toro and there is one from Mt. Koghis in the BPBM. Another possibly intermediate form with fawnish grey uniform wings flies in the same localities and both were recorded as follows: Dothio site 18 (1); Ouen Toro site 64 (1), site 81 (2), site 82 (7); Mt. Koghis site 79 (1); Nouméa (1-MNHN). These forms are indicated by open circles in Figure 72. They occur more or less on the boundary between the ultramafics and other habitat types in the south.

## NOTODONTIDAE

### **Cerura amoa sp. n.** (Plate 68: 7, Figure 71: 2)

♂♂ 28–29 mm, ♀ 32 mm. The wings and body are white, marked with black as illustrated. The hindwings are often suffused greyish, more intensely in the postmedial and submarginal bands. The species is closest to *multipunctata* Bethune-Baker and undescribed relatives from New Guinea, but these species have the forewing fasciation made up of black dots rather than thin lines or

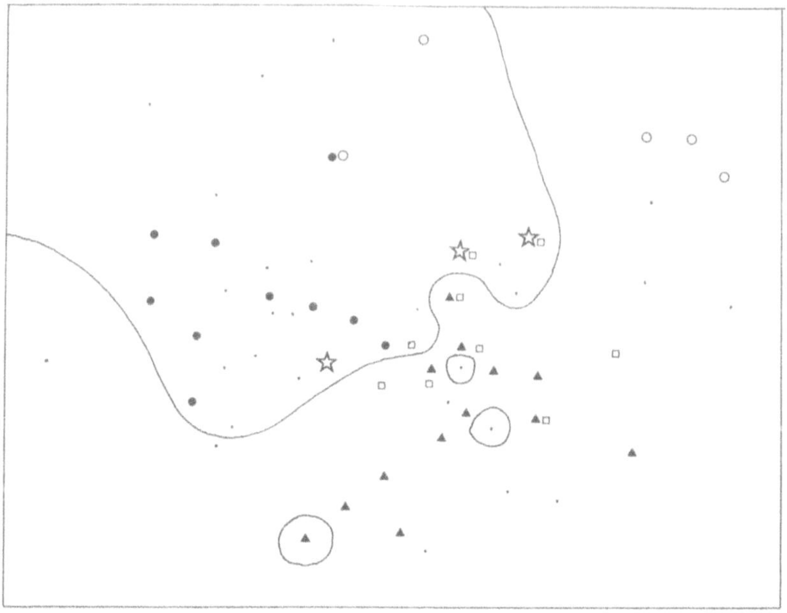

*Fig. 72.* Scaling plot of samples showing the distribution of forms of *Teara circumfumata*: *alma* (solid circles), *silvicola* (solid triangles), *sobrina* (open squares), *striata* (stars) and 'typical' and intermediate (open circles).

lunules as in *amoa*; their hindwings are generally only lightly marked and the abdomen is more heavily black banded, especially dorsally.

In the male genitalia *amoa* is distinguished by the greater basal constriction of the valve and by the aedeagus which is twice as produced and tapering beyond the exit of the vesica.

HOLOTYPE ♂. New Caledonia: Port Boisé, 20 m Site 74 07039/75276 (J. D. HOLLOWAY) 13.8.1971, BM 1971-507. BM notodontid slide 327.

PARATYPES: 12♂♂ New Caledonia, various localities; 1♀ Nouméa, ii, iii.1957 MNHN.

The species was recorded as follows: Bréhoa R. (1); Bambou R. (1); R. Amoa (1); Ste. Therése, Amoa valley (5); Probokonoro (2); Mt. Krapé site 55 (2); Port Boisé site 74 (2); R. Koué (1). It is probably a rainforest species and appears frequent on the ultramafic rocks.

## *LASIOCEROS* Bethune-Baker

*Lasioceros* was transferred by KIRIAKOFF (1968, *Genera Insectorum* 217c: 2) from Notodontidae to Hypsidae but WATSON (*pers. comm.* and MS) has

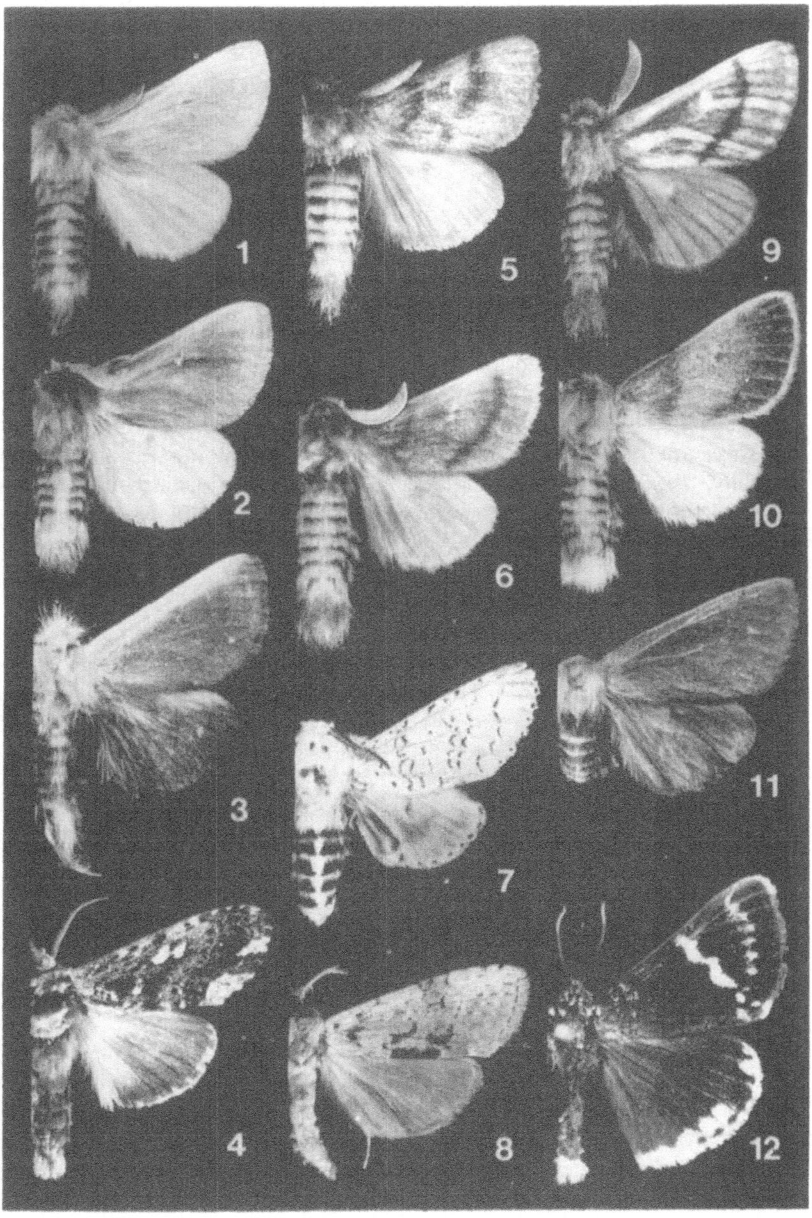


*Plate 68.* Forms of *Teara circumfumata*, *alma* (1, 2), *silvicola* (3, 11 (♀)), *sobrina* (5, 6), *striata* (9) and 'typical' (10), ♂♂ 20–25 mm, ♀ 32 mm; 4. *Neola octofera;* 7. *Cerura amoa;* 8. *Lasioceros aroa*; 12. *Eutrichopidia macchia.*

concluded from a study of the tympanal organs and venation that it is a genuine notodontid. As well as *aroa* Bethune-Baker, discussed below, the genus contains *dentilinea* Bethune-Baker from New Guinea, and *euteles* (West, 1932, *Novit. zool.*, 37: 213, *Gargetta*) **comb. n.** from Luzon and Sulawesi. The latter has facies almost identical to *aroa* and is therefore here combined with that species in the same genus; indeed it may prove conspecific.

### *Lasioceros aroa* Bethune-Baker (Plate 68: 8, Figure 87: 3)

*Lasioceros aroa* Bethune-Baker, 1904, *Novit. zool.*, 11: 380. ROBINSON (1975: 106).

The single male taken during the survey measured 15 mm as compared with 19 mm for typical New Guinea specimens. The male genitalia are as in typical *aroa* but the hindwings are more uniformly grey and the forewing dark markings are less heavy discally and more prominent towards the dorsum medially. The species occurs in New Guinea, there is a distinct subspecies in Fiji, and the New Caledonian specimen was taken at the R. Amoa locality.

### **Neola octofera sp. n.** (Plate 68: 4, Figure 71: 1)

♂♂ 23–25 mm, ♀ 32 mm. The forewing is black, marked and fasciated grey and white, slightly suffused rufous. The hindwing is fawnish yellow, heavily suffused brown except along the dorsum and basally. The abdomen is similar in colour to the hindwing, though more ochreous in some specimens. The species is close to *N. semiaurata* Walker (Queensland, New South Wales) but differs in being much smaller with more elongate forewings. The reniform and orbicular, usually fused, are whiter as are the medial fasciae posteriorly and the marginal pale patches. The hindwing of *semiaurata* has the basal two thirds yellow and there is a marginal brown band.

HOLOTYPE ♂. New Caledonia: R. Bleue, 150 m Site 8 06755/75530 (J. D. HOLLOWAY) 25.5.1971, BM 1971-507. BM notodontid slide 323.

PARATYPES: 34♂♂, 2♀♀ New Caledonia, various localities; 1♂ La Coulée 28.10.76 (J. GUTIERREZ) MNHN.

The species was recorded as follows: Grand Lac site 2 (5); Pic du Pin site 7 (2); R. Bleue site 8 (2), site 10 (3); Anse Longue site 16 (1), site 32 (2); Bréhoa R. (9); Ouambéo R. (2); Col d'Amieu (3); Col Toma (4); Sarraméa (1); Mt. Mou site 33 (5), site 34 (1); Cap Colnett (1); R. Amoa (1); Ste. Therèse (1); Probokonoro (1); Col des Roussettes (2); Mt. Krapé site 55 (8); Houailou (1); Mt. Aoupinié (2); Col Mouirange (2); Mt. Pouédihi (1); Port Boisé site 73 (1), site 74 (4); R. Koué (1); Anna-Madeleine (1); Les Dalmates site 78 (4); Mt. Koghis site 79 (1); Ouen Toro site 81 (1), site 82 (3). It is perhaps primarily a rainforest species, particularly on ultramafic rock. It also occurs in west coast localities. COMMON (1963) noted that *N. semiaurata* larvae feed on *Acacia* in Australia.

*Euproctis aliena* Butler (Figure 73: 1)

*Euproctis aliena* Butler, 1886, *Trans. ent. Soc. Lond.*, 1886: 386, type examined.

One male of this Australian (N. Territory, Queensland) species was taken at Mt. des Sources site 14. Its capture coincided with the arrival of a warm frontal system from the west, so it could conceivably have been a vagrant. A closely related species, *dolichocera* Collenette, occurs in the New Hebrides but this has the valve arms of the male genitalia more slender and the ventral projection of the juxta shorter and broader (Tanna specimen dissected). Several other species of similar appearance occur in Australia, such as *paradoxa* Butler.

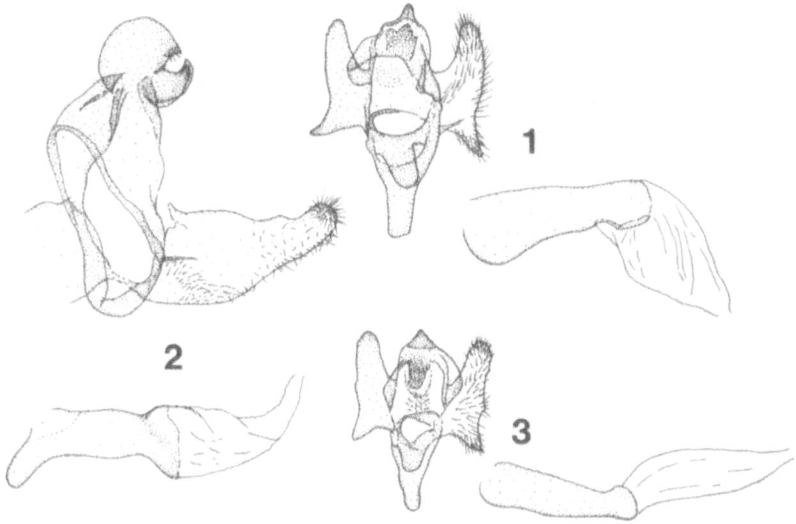


*Fig. 73.* Male genitalia of *Euproctis aliena* (1), *Orgyia sarramea* (2) and *Euproctis dimorphissima* (3).

**Euproctis dimorphissima sp. n.** (Plate 69: 2, 3, Figure 73: 3)

♂♂ 9–10 mm. The insect is entirely dark brown apart from the forewing, which is slightly paler and more rufous, and the legs and the thorax below, which are fawn. The genitalia resemble those of *aliena* and *dolichocera* in general pattern but the lower lobe of the valve is blunter and there is a bulge centrally between the lobes.

♀♀ 12–14 mm. The forewings are more produced apically than in the male and the insect is completely white apart from the brownish pectination of the antennae and the cream apex to the abdomen.

No other species of the group exhibits such extreme dimorphism.

HOLOTYPE ♂. New Caledonia: Anse Longue, 5 m Site 32 06218/75543 (J. D. HOLLOWAY) 30.6.1971, BM 1971-507. BM lymantriid slide 1038.

PARATYPES: 8♂♂, 19♀♀, New Caledonia, various localities; 1♀ Nouméa, Aug. BPBM; 4♂♂, 5♀♀ Nouméa Oct. 10 1940 ex white cocoon on *Poinciana*, BPBM; 1♂ Anse Vata, Nouméa xi.1956, MNHN; 1♂ La Coulée 25.10.76 (A. DELOBEL) MNHN; 1♂ La Coulée, *Phaseolus* phyllophage, ad. 9.5.1977 (A. DELOBEL) MNHN.

The species was recorded as follows: Grand Lac site 2 (3), site 3 (3); Pic du Pin site 7 (3); Mt. des Sources site 14 (1); Poupa (1); Brêhoa R. (2); Anse Longue site 32 (6); Grottes de Koum (4); R. Amoa (1); Mt. Aoupinié (6); Col Nassiboué (3); Ouen Toro site 64 (5), site 81 (1), site 82 (4); Yaté Gorge (1); Port Boisé site 73 (1), site 74 (2); Les Dalmates site 77 (1), site 78 (1). Though taken in a variety of habitats the species was perhaps commonest at west coast localities such as Ouen Toro.

The *Poinciana* host plant record refers to *Delonix regia* (Leguminosae), the Flamboyant. It may feed on Leguminosae generally, as during the survey several lymantriid larvae were beaten from *Acacia spirorbis* on the Dome de Tiebaghi. They were not reared out.

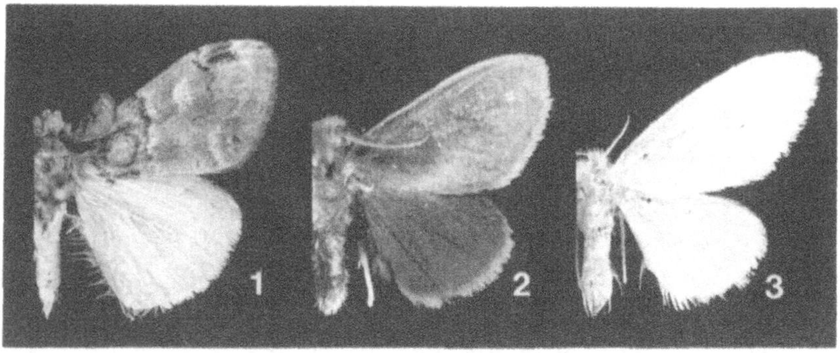


*Plate 69.* 1. *Orgyia sarramea;* 2 (♂), 3 (♀). *Euproctis dimorphissima.*

**Orgyia sarramea sp. n.** (Plate 69: 1, Figure 73: 2)

♂♂ 18–19 mm. The forewings and body are medium greyish brown, marked with darker brown as illustrated. The whitish edging distad to the almost straight, but irregular antemedial is diagnostic, though otherwise the species is similar in size and general pattern and coloration to *O. athlophora* Turner. The male genitalia are as illustrated. J. C. E. RIOTTE is shortly to resurrect the genus *Teia* Walker to include *athlophora* amongst other species and therefore *sarramea* should then be transferred to that genus.

HOLOTYPE ♂. New Caledonia: Petit Gouli, 30 m Site 28 05852/76047 (J. D. HOLLOWAY) 20.6.1971, BM 1971-507. BM lymantriid slide 1039.
PARATYPES. 14♂♂ New Caledonia, various localities.

The species was recorded as follows: Poupa (1); Ouambéo R. (2); Col d'Amieu (4); Col Toma (1); Petit Gouli (2); Sarraméa (9); Mt. Mou site 33 (12); Grottes de Koum (1); Kavatch (1); Ste. Therèse (1); Probokonoro (1); Mt. Aoupinié (4). It is probably a rainforest species. In the BPBM there is a single battered male from Wé, Lifu, Loyalty Is.

## ARCTIIDAE: LITHOSIINAE

### *EILEMA* Hübner

The type species of *Eilema* is the western Palaearctic *caniola* Hübner. In New Caledonia there occur two species best placed in *Eilema* until such a time as this large genus is revised. They belong to a well defined section of the genus characterised by small size and, in the male genitalia, a bifid saccus and a prominent, rod-like juxta. There is a roughly symmetrical branching arrangement of the radial veins and $M_1$ from the anterior angle of the forewing cell (Figure 75, left), whereas in *caniola* $R_4$ and $R_5$ arise from a straight $R_3$ and run almost parallel to $M_1$, which arises from the base of $R_3$ on the cell.

Three species are discussed below. A fourth, *E. amaurus* Rothschild (1916, *Novit. zool.*, 23: 325, unique male type examined, Manus, Admiralty Is.; Figure 74: 1, 2), has been examined and belongs to the section and others at present included under *Tigrioides* in the BMNH arrangement may also prove referable when dissected.

#### *Eilema nitens* (Walker) **comb. n.** (Figure 74: 3, 4, 10)

*Lithosia nitens* Walker, 1864, *List Specimens lepid. Insects Colln Br. Mus.*, 34: 321, type examined.

♂♀ 9–11 mm. The males have the forewings ochreous fawn, those of the females being darker and browner. The hindwings of both sexes are generally uniform straw but some Australian specimens have the hindwing coasta greyish. The species occurs in New Guinea (see below), Australia from the Northern Territory to New South Wales, and New Caledonia, where it was recorded as follows: Anse Longue site 16 (2); R. Kuenthio (7); Petit Gouli (5); Bambou R. (1); R. Amoa (1); Mt. Krapé site 55 (1); Ouen Toro site 64 (5), site 81 (3), site 82 (4). MNHN: La Coulée. The species is a member of the west coast association.

#### *Eilema nitens minima* (Hampson) **comb. & stat. n.**

*Lexis minima* Hampson, 1903, *Ann. Mag. nat. Hist.* (7), 11: 342, type examined.

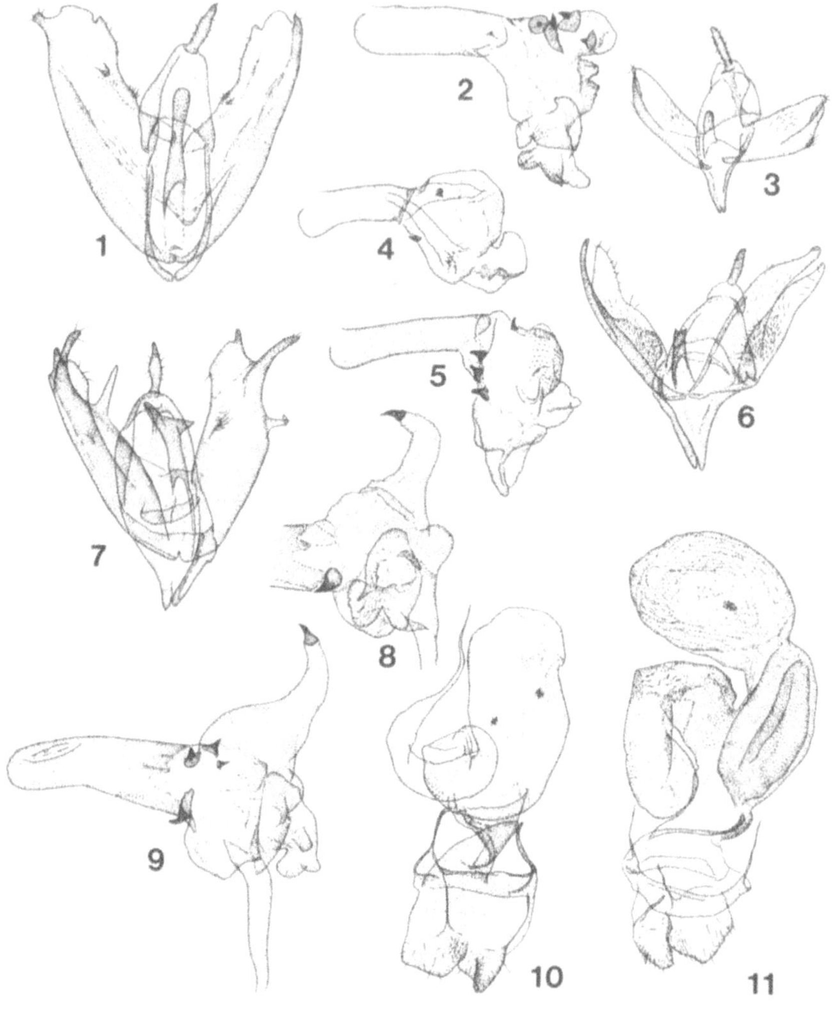


*Fig. 74.* Male genitalia, aedeagi and female genitalia of *Eilema* species: *amaurus* (1, 2), *nitens* (3, 4, 10), *diliensis* (5, 6) and *caledonica* (7, 8, 9, 11).

Dissection of a male with identical data and appearance to the male type (Port Moresby) revealed genitalia as in typical *nitens*. Though similar in appearance, *minima* is only two thirds the size of *nitens*, though it is possible the two specimens in the BMNH are starvelings.

**Eilema diliensis sp. n.** (Figure 74: 5, 6)

♂♀ 10 mm. Male antennae are filiform as in *nitens*. The forewings are greyish buff except for the costa which is ochreous; the hindwings are straw-buff, grading darker slightly towards the margin, especially in the female where they are greyer in tone. The male genitalia have the valve bifid as in *nitens* but relatively longer and with the ventral part rather than the distal part larger. The juxta is apically square instead of rounded and the aedeagus vesica contains larger cornuti.

HOLOTYPE ♂. (Timor) Dili, May 1892 (W. DOHERTY), Rothschild Bequest, BM 1939-1. BM arctiid slide 1945.

PARATYPES. 1♂, 1♀ as holotype.

OTHER MATERIAL. 1♂, 1♀ Bali: Low Country (W. DOHERTY) 96–252. BM arctiid slide (♂) 2127.

**Eilema caledonica sp. n.** (Plate 71: 8, Figure 74: 7–9, 11)

♂♀ 10–14 mm. The antennae are somewhat biserrate as in *amaurus*, the setation being more robust than in *nitens* or *diliensis*. The forewings are medium grey-brown apart from the straw-coloured costa, though they are often faded or worn to straw-buff. The hindwings are straw colour. The male genitalia have the valve undivided as in *amaurus* (Figure 74: 1, 2) but the apex has three prominent projections and the juxta broadens to give two acute angles instead of being rounded. The aedeagus vesica has a variable number of cornuti but the prominent lobe with an apical cornutus is diagnostic. The female genitalia have a prominent appendix bursae.

HOLOTYPE ♂. New Caledonia: Mt. Koghis, 600 m Site 80 06556/75456 (J. D. HOLLOWAY) 21.8.1971, BM 1971-507. BM arctiid slide 1894.

PARATYPES. 95 ex. New Caledonia, various localities. BM arctiid slides 1893, 1897, 1899, 1900, 1901, 1929, 1932, 1933 (♂♂), 1949 (♀), 1955 (venation).

The species was recorded as follows: Grand Lac site 1 (2), site 2 (10), site 4 (1), site 72 (9); Pic du Pin site 7 (8); R. Bleue site 8 (1), site 10 (1); Mt. des Sources site 11 (1), site 12 (2), site 14 (2); R. Kuenthio (2); Dothio site 18 (1); Bréhoa R. (2); Col d'Amieu (1); Col Toma (17); Sarramea (1); Mt. Dogny (5); Mt. Mou site 36 (1); Tiebaghi site 40 (1); Grottes de Koum (10); Kavatch (19); Ste. Therèse (3); Probokonoro (16); Mt. Krapé site 56 (6); Houailou (1); Mt. Panié site 60 (9); Mt. Aoupinié (3); Col Nassiboué (8); Col Mouirange (3); Yaté Gorge (4); Port Boisé site 73 (1), site 74 (2); R. Koué (6); Anna-Madeleine (2); Les Dalmates site 77 (4); Mt. Koghis site 79 (17), site 80 (14). BMNH: Tinchialit (CHEESMAN). The species is widespread in the island, perhaps commoner in wetter, more forested localities and on the ultramafics.

**FABRESEMA Gen. n.**

The male antennae are filiform with two large setae and several smaller ones per segment as in *Eilema nitens*. The mid tibiae have a single pair of spurs and the hind tibiae two as in *Eilema*.

The venation is generally similar to that of the *Eilema* section discussed above (Figure 75) with $R_1$ fused to Sc on the forewing and $Sc+R_1$ fused

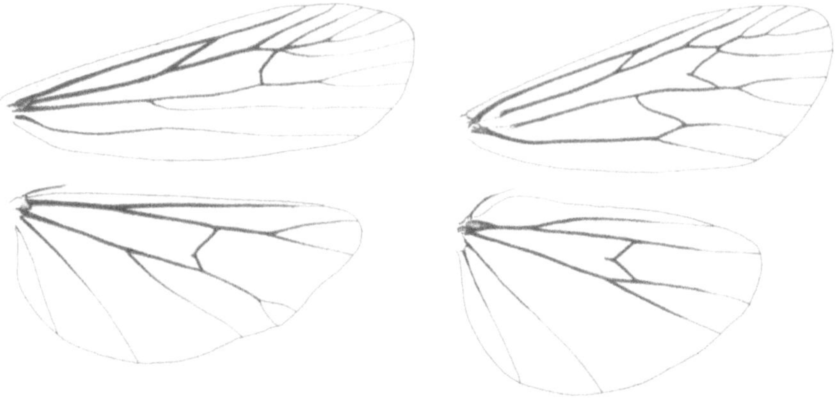

*Fig. 75.* Wing venation of *Eilema caledonica* (left) and *Fabresema elisabethae* (right).

basally on the hindwing. The most significant difference is the unusual cross-vein between 1A and Culb on the forewing. The crossveins on both cells are more deeply angled and $R_3$ is fused with $R_{4+5}$ over a greater length on the forewing. On the hindwing $R_3$ and $M_1$, and $M_3$ and Cula are usually entirely fused, whereas they are apically bifurcate in *Eilema*.

In most species the ground colour of the wings is straw, with dark brown markings on the forewing, the zig-zag medial following the venation, particularly the crossveins.

The male genitalia are characterised by the uncus being ridged apically and on each side, and by the valve having a harpe from the sacculus, usually a setose projection from the dorsal margin and often a central membranous flap. The aedeagus vesica is usually immaculate or with a single cornutus, and the aedeagus apex in several species bears ventral ridge with spines.

The female bursa contains a single signum consisting of small, obtuse spines (there are often two signa in *Eilema*) and the lamella vaginalis is often broad and folded.

The genus contains at least 14 species, described below. As several are represented by very few specimens or by singletons it is likely that more remain to be discovered. The genus is restricted to New Caledonia apart from one species on Lifu. It is the only genus to show major radiation in New Caledonia though there are duplexes and triplexes in several other genera. The species are found mainly in forest localities, perhaps with greater diversity on the ultramafic rocks (Figure 76).

It has not proved possible to relate *Fabresema* closely to any other lithosiine genus. It may therefore represent one of the oldest elements in the New Caledonian Lepidoptera fauna. Its species grade from distinctive, rather isolated forms to poorly differentiated duplexes and triplexes.

Because of the similarity of the various species externally, it has not proved possible in some cases to assign females to males; some of these females may

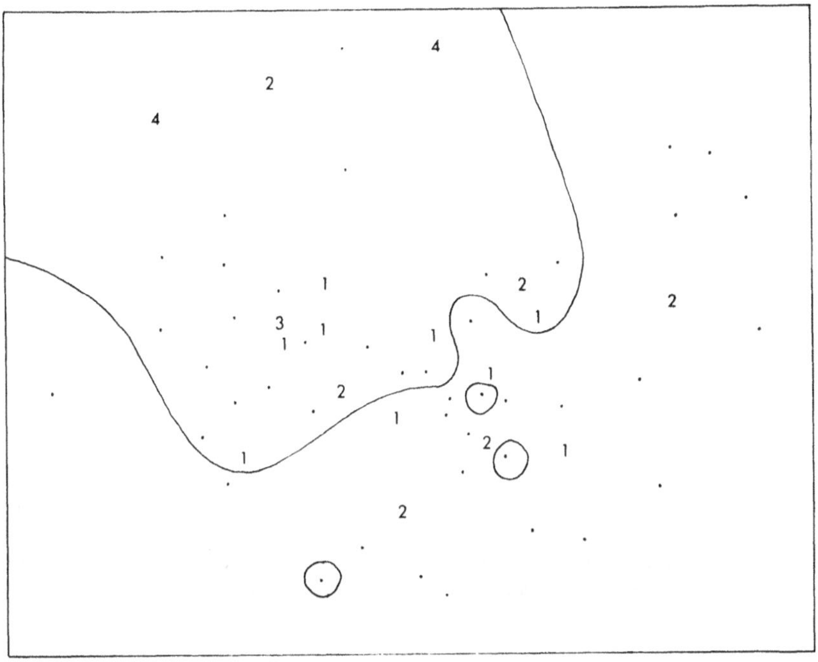


*Fig. 76.* Scaling plot showing the distribution of *Fabresema* species amongst the samples. For this purpose female forms have been treated as distinct species and summed with the species based on males.

indeed represent separate species and the figures in Figure 76 represent the sum of both male and female forms. Several females have been assigned tentatively and the rest have been described, but not named, separately.

The genus and several of the species have been named for M GÉRARD FABRES and his family who showed the author much kindness during the survey.

The type species is hereby designated as *gerardi.*

**Fabresema gerardi sp. n.** (Plate 70: 4, Figures 77: 9, 79: 2)

♂ 10 mm, ♀ 11 mm. The male forewing is dull straw with a basal suffusion of dark brown, an approximately transverse zig-zag medial, and the postmedial area with the veins picked out brown and crossed by a diffuse submarginal brown band. The hindwings are a yellower straw with the fringes basally brownish. The genitalia are typical. The valves have the harpe short, hooked; the setose processes are adpressed to the costa; the membranous flap is prominent, broad, rounded. The aedeagus has the apex acute, downturned as in the next species, the vesica with no cornutus.

The female (from the same locality, tentatively assigned) has similar darkening along the veins postmedially and the hindwing fringes are also dark. But the ground colour of both wings is a duller beige with the forewing medial fascia broader, more diffuse. The genitalia have a round signum near the base of the bursa; the ductus bursae is corrugated, broadly based on the lamella vaginalis, arising asymmetrically as illustrated, the ductus seminalis arising from the small, lateral lobe.

HOLOTYPE ♂. New Caledonia: Les Dalmates, 220 m Site 78 06714/75424 (J. D. HOLLOWAY) 18.8.1971, BM 1971-507. BM arctiid slide 1908.

PUTATIVE ♀. New Caledonia as holotype, 19.8.1971. BM arctiid slide 1948.

**Fabresema elisabethae sp. n.** (Plate 70: 1, Figures 77: 2, 3, 79: 1)

♂♀ 10–12 mm. The ground colour is straw, the forewings with a prominent oblique medial, a large dark brown patch basally from the costa to the anal vein, extending distad in the cell, and a broad submarginal suffusion of brown, more intense on the veins. The hindwing fringes are basally darker than the ground.

The male genitalia have the valve lacking the setose process and the membranous flap; the uncus is narrow, unridged; the saccus is prominently bilobed. The aedeagus is as in *gerardi* but with a long cornutus in the vesica. The female bursa has the signum one third from the apex, dorsal and lateral; the ductus bursae is corrugated, expanded subbasally at the origin of the ductus seminalis, constricted basally; the lamella vaginalis is broad, unfolded.

HOLOTYPE ♂. New Caledonia: Mt. Koghis, 600 m Site 80 06556/75456 (J. D. HOLLOWAY) 21.8.1971, BM 1971-507. BM arctiid slide 1895.

PARATYPES: 5♂♂, 2♀♀ New Caledonia (Col Nassiboué (1), Port Boisé site 74 (1), Les Dalmates site 77 (1); Mt. Koghis site 79 (3), site 80 (1)), BM arctiid slides 1903, 1925 (♂♂), 1930 (♀), 1956 (venation, Figure 75); 1♀ N. New Caledonia, Tinchialit 2020 ft 22.viii-10.ix.1949 (L. E. CHEESMAN), BM 1950-1.

**Fabresema catherinae sp. n.** (Plate 70: 8, Figure 77: 11)

♂ 11 mm. The forewing is uniform buff with only a faint trace of a darker medial; the wing apex is generally more rounded than in other species. The hindwings are more straw in tint, the fringes not darker. The male genitalia have the valves broadening apically, constricted just above the setose projection which is moderate in size; the membranous flap is prominent, shaped as in *gerardi* but the harpe is longer and narrower than in that species and directed distally rather than ventrally. The aedeagus is shorter than in *gerardi* with a small cornutus; the apex of the shaft has a dorsal scobinate bulge and two larger spines ventrally.

HOLOTYPE ♂. New Caledonia: Mt. Aoupinié, 520 m Site 62 05298/76594 J. D. HOLLOWAY) 31.7.1971, BM 1971-507. BM arctiid slide 1922.

**Fabresema sylviae sp. n.** (Plate 70: 7, Figure 77: 1)

♂♂ 8–10 mm. The forewings are medium brown-grey, uniform but for a faintly darker, diffuse medial. The hindwings are pale buff, grading slightly darker

*Plate 70.* The genus *Fabresema*. 1. *elisabethae*; 2. *costiplaga*; 3. *grisea*; 4. *gerardi*; 5. *valeriae*; 6. *flavibasalis*; 7. *sylviae*; 8. *catherinae*; 9, 17. *toma*; 10. *cheesmanae*; 11. *bicornuta*; 12. *suffusa*; 13. female (5); 14. female (2); 15. female (1); 16. female (3); 18. female (4).

over the margin. The male genitalia have an elongate, narrow harpe on the valve with the setose process displaced behind it. The latter is long, half the length of the valve, apically sinuous and the setae are small and adpressed. The membranous flap is absent and the aedeagus is short and unadorned.
HOLOTYPE ♂. New Caledonia: Les Dalmates, 220 m Site 78 06714/75424 (J. D. HOLLOWAY) 19.8.1971, BM 1971-507. BM arctiid slide 1923.
PARATYPE ♂. As holotype but 18.8.1971, BM arctiid slide 1924.

**Fabresema valeriae sp. n.** (Plate 70: 5, Figure 78: 3)

♂ 9 mm. This species is similar in pattern and colour to *gerardi* but the medial is at right angles to the costa instead of oblique, the submarginal band of brown is absent and the hindwings do not have darker fringes. The genitalia have a broad uncus, the setose process to the valve is broad and curved, the membranous flap is very short (obscured in the figure) and the aedeagus is of moderate length and unadorned.
HOLOTYPE ♂. New Caledonia: Tiebaghi, 330 m Site 41 04156/77366 (J. D. HOLLOWAY) 10.7.1971, BM 1971-507. BM arctiid slide 1907.

**Fabresema flavibasalis sp. n.** (Plate 70: 6, Figure 77: 10)

♂♂ 8–11 mm. In appearance this species is like a smaller *gerardi*, variable in the degree of brown suffusion on the forewing. It is distinguished by the marked yellow tint basally on the hindwings, which also have slightly darker fringes. The genitalia have the valve apex roundedly rhomboidal, but not so constricted above the setose projection as in *catherinae* which they resemble; the setose projection is broader than in *catherinae*, differently oriented. The aedeagus is short as in *catherinae*, apically bearing a ventral ridge with five short spines.
HOLOTYPE ♂. New Caledonia: Grand Lac, 250 m Site 2 06975/75356 (J. D. HOLLOWAY) 16.5.1971, BM 1971-507. BM arctiid slide 1904.
PARATYPES. 9♂♂ New Caledonia: Grand Lac site 2 (1), site 3 (1); Pic du Pin site 7 (1); Mt. Krapé site 56 (1); Les Dalmates site 78 (2); Mt. Koghis site 79 (3). BM arctiid slides 1902, 1909, 1921.

**Fabresema costiplaga sp. n.** (Plate 70: 2, Figures 78: 4, 80: 6)

♂♂ 10 mm. Both wings have a fawnish straw ground. The pattern of the forewing is generally as in *gerardi* but less heavily marked except for an almost square dark patch at the costa at the anterior of the medial and a dark streak along the costa basally. The hindwings do not have darker fringes. The genitalia have the harpe straight, directed ventrally from the valve; there is no membranous flap evident; the valve broadens apically to an angular, upturned, squarish apex, marginally serrated especially ventrally; the aedeagus is slender, relatively long, with a diagnostic curved cornutus.

Two females with an elongate signum basal and lateral in the bursa may be those of *costiplaga* as their genitalia resemble those of females assigned to the

next species where the male has similar genitalia to *costiplaga* also. They have more uniform patterning than the males and lack the costal darker markings; therefore this association is very tentative.
HOLOTYPE ♂. New Caledonia: Pic du Pin, 300 m Site 69 06880/75385 (J. D. HOLLOWAY) 7.8.1971. BM 1971-507. BM arctiid slide 1911.
PARATYPE ♂. New Caledonia: Pic du Pin, 350 m Site 70 06880/75386 (J. D. HOLLOWAY) 8.8.1971, BM 1971-507. BM arctiid slide 1912.
PUTATIVE ♀♀. Mt. Aoupinié (slide 2021), Mt. Koghis site 79 (slide 2024).

**Fabresema cheesmanae sp. n.** (Plate 70: 10, Figure 80: 5, 7)

♂ 9 mm, ♀♀ 8–9 mm. The ground colour is pale straw with fasciation as in *costiplaga*, but lacking the darker marks on the costa. The insect is smaller. The male genitalia are similar to those of *costiplaga* but the harpe is dorsally directed even on a splayed valve; the setose process arises from an angle rather than a curve; the valve apex is less serrated marginally and broadly three-pointed; the aedeagus is sinuous, upturned at the apex and lacking a cornutus.

The females tentatively assigned through similarity of size and facies and through proximity in collection to the male above have the signum of the bursa elongate, basal and lateral as in *costiplaga* but shorter; the ductus bursae is also shorter and not partially sclerotised as in *costiplaga*.
HOLOTYPE ♂. New Caledonia: Tinchialit, 2020 ft 22.vii-10.ix.1949 (L. E. CHEESMAN) BM 1950-1. BM arctiid slide 2016.
PUTATIVE ♀♀: 6 N. New Caledonia, Pouébo, coast-1500 ft x.1949 (L. E. CHEESMAN), BM arctiid slide 2017; 1 Loyalty Is., E. Lifu I., Cap des Pins 18.xi.1949-18.i.1950 (L. E. CHEESMAN), BM arctiid slide 2018.

**Fabresema grisea sp. n.** (Plate 70: 3, Figure 77: 5, 7)

♂ 8 mm. The ground colour is pale to medium greyish fawn. The forewing medial is transverse (displaced distad in cell to crossveins), terminating at the costa in a prominent square spot; the basal dark suffusion is largely restricted to the costa; the postmedial area has the veins picked out dark, though not intensely; the submarginal band is slight and diffuse. This and the next two species have similar fasciation and genitalia. There are slight but definite differences. In view of the diversity of the genus and the constancy of genitalia characters in species represented by longer series, all three are described as distinct despite only being represented by single males.

The male genitalia of *grisea* are of a similar pattern to those of the previous two species except the aedeagus is short as in *catherinae* and *flavibasalis* with a ventral ridge bearing three pointed but obtuse projections directed posteriorly. The valve apex is not so produced as in *costiplaga* and *cheesmanae*, the sides curving gently in to a point at the apex.
HOLOTYPE ♂. New Caledonia: Col des Roussettes, 340 m Site 54 05467/76304 (J. D. HOLLOWAY) 21.7.1971, BM 1971-507. BM arctiid slide 1916.

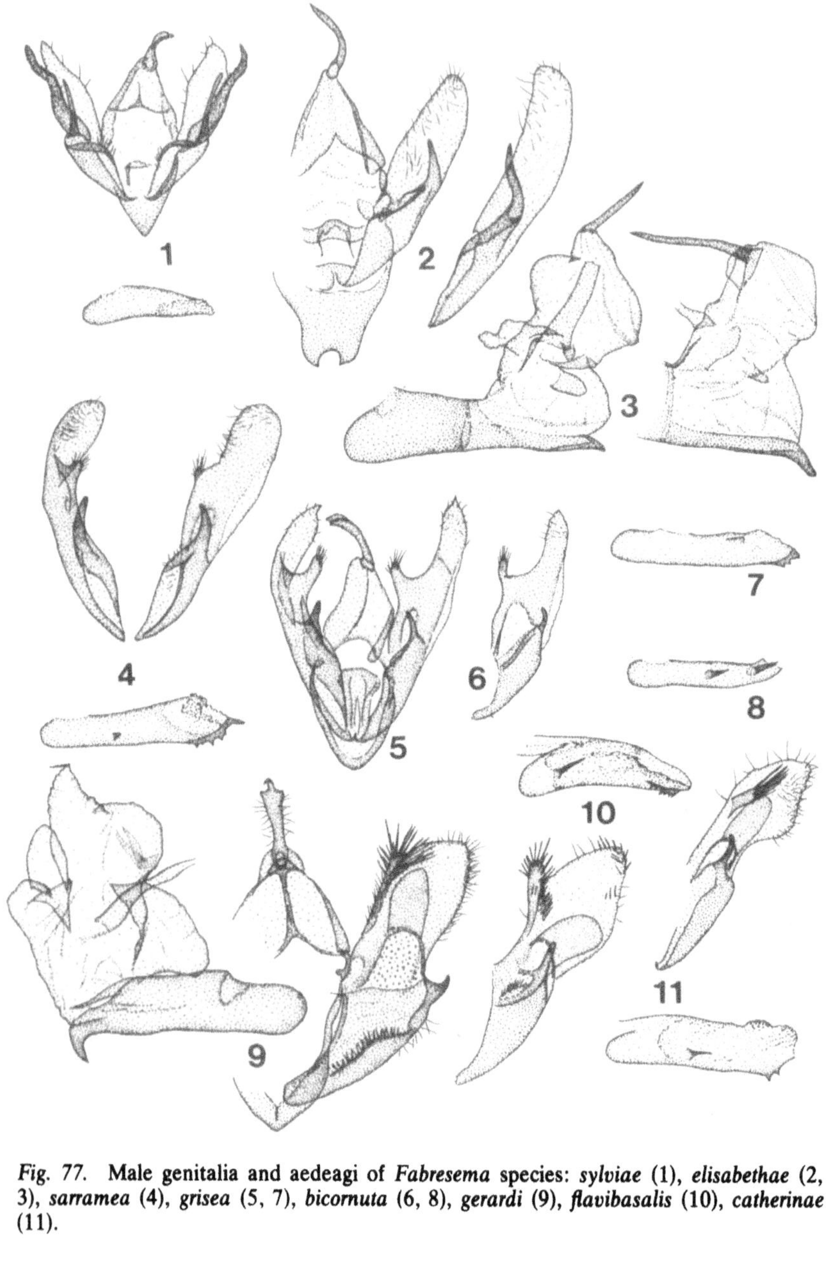


*Fig. 77.* Male genitalia and aedeagi of *Fabresema* species: *sylviae* (1), *elisabethae* (2, 3), *sarramea* (4), *grisea* (5, 7), *bicornuta* (6, 8), *gerardi* (9), *flavibasalis* (10), *catherinae* (11).

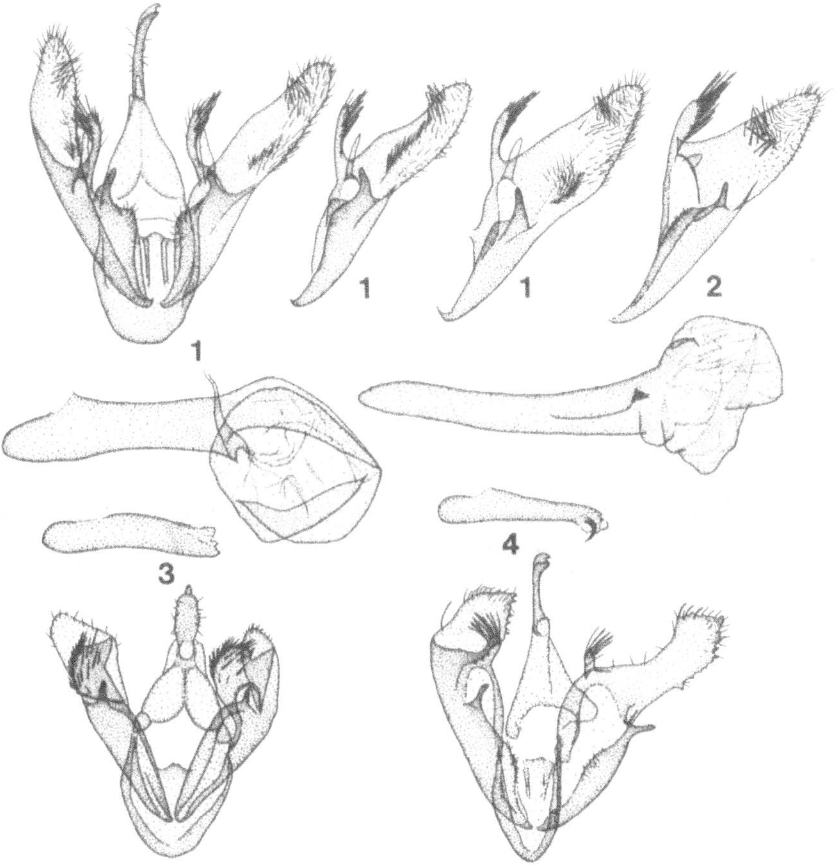


*Fig. 78.* Male genitalia and aedeagi of *Fabresema* species: *obliqua* (1), *suffusa* (2), *valeriae* (3), *costiplaga* (4).

**Fabresema sarramea sp. n.** (Figure 77: 4)

♂ 8 mm. The forewing pattern is as in *grisea* but on a straw ground. The genitalia have the valve apically rounded and the setose process much shorter than in *grisea*. The aedeagus is slightly longer with a small cornutus; the ventral, apical ridge has a posteriorly directed spine apically and five ventral spines of various sizes.

HOLOTYPE ♂. New Caledonia: Sarraméa, 160 m Site 29 05888/76065 (J. D. HOLLOWAY) 28.6.1971, BM 1971-507. BM arctiid slide 1896.

**Fabresema bicornuta sp. n.** (Plate 70: 11, Figure 77: 6, 8)

♂ 8 mm. This species is externally similar to *sarramea* but the most posterior section of the medial is longer, extending further basad along the dorsum. The genitalia have the valve as in *grisea* but with a larger cornutus in the aedeagus than that of *sarramea* and a second one of similar size apically; there is no spined ventral ridge.
HOLOTYPE ♂. New Caledonia: Kavatch, 40 m Site 50 04857/77068 (J. D. HOLLOWAY) 18.7.1971, BM 1971-507. BM arctiid slide 1906.

**Fabresema suffusa sp. n.** (Plate 70: 12, Figure 78: 2)

♂♂ 10 mm. In external appearance this species is very like *costiplaga*. The general course of the medial fascia of the forewing is slightly concave basad, meeting the dorsum obliquely; the forewings are more heavily suffused brownish than in *costiplaga*. The genitalia have the valve apex rounded and the distal half generally triangular or tapering; the setose process is narrow but well developed, set on an angle of the valve dorsal margin; the membranous flap is narrow. The aedeagus is relatively large and elongate within the genus, with a diagnostic group of slender spines at the apex, directed posteriorly, and a weak band of sclerotisation extending into the vesica which is globular when everted.
HOLOTYPE ♂. New Caledonia: Bambou R., 50 m Site 44 04304/77369 (J. D. HOLLOWAY) 12.7.1971, BM 1971-507. BM arctiid slide 1913.
PARATYPE ♂. New Caledonia: Sarraméa, 160 m Site 29 05888/76065 (J. D. HOLLOWAY) 21.6.1971, BM 1971-507. BM arctiid slide 1905.

**Fabresema toma sp. n.** (Plate 70: 9, 17)

♂ 11 mm. The species is very similar to *suffusa* but the forewing is less suffused with brown and the hindwing is a brighter, less greyish straw colour. The genitalia (not illustrated) are as in *suffusa* but relatively larger compared with general size, with the valve apex more rounded, the aedeagus thicker, lacking the apical spines, and the sclerotised band in the vesica more prominent.
HOLOTYPE ♂. New Caledonia: Col Toma, 460 m Site 27 05813/76123 (J. D. HOLLOWAY) 19.6.1971, BM 1971-507. BM arctiid slide 1910.

**Fabresema obliqua sp. n.** (Figure 78: 1)

♂♂ 8–11 mm. Externally this species is also similar to *suffusa* but the forewing medial meets the dorsum slightly more basally so that it is more oblique over the posterior two thirds. The genitalia have the valve narrower with the setose process not set on an angle of the dorsal margin of the valve, differing from the previous two species; the aedeagus is narrower and shorter relative to the size of the everted globular vesica; there is no sclerotised band in the vesica nor a group of spines at the apex of the aedeagus.

All specimens were taken in rainforest on the ultramafics, whereas those of the previous two species were taken in localities on sedimentary rock.
HOLOTYPE ♂. New Caledonia: Pic du Pin, 300 m Site 69 06880/75385 (J. D. HOLLOWAY) 7.8.1971, BM 1971–507. BM arctiid slide 1917.
PARATYPES. 5♂♂ New Caledonia: Mt. des Sources site 14; Pic du Pin site 69; Port Boisé site 73, site 74; Mt. Koghis site 79. BM arctiid slides 1914, 1915, 1918, 1919, 1920.

Indeterminate females

**Type 1.** Plate 70: 15, Figure 79: 3. 10 mm. The facies resemble *obliqua* but the medial is less zig-zag. The genitalia are distinguished by a broad, straight, parallel-sided ductus bursae; the ductus seminalis arises from the base of the bursa; the rounded signum is central and ventral. One specimen, Mt. Krapé site 56, slide 1947.
**Type 2.** Plate 70: 14, Figure 80: 1. 9 mm. Facies as in putative female *costiplaga*. The ductus bursae arises from a peripherally scobinate vaginal lamella; the basal third is triangular, sclerotised, and the distal two thirds is straight, membranous, with the ductus seminalis arising basally from it; the signum is ventral, one third from the base of the bursa. One specimen, Col Nassiboué, slide 2022.
**Type 3.** Plate 70: 16, Figure 80: 2. 9 mm. Externally this form resembles *obliqua* but the forewing medial is much finer, the hindwing greyish fawn. The ductus bursae is very long with the ductus seminalis arising from a lateral bulge subbasally; the lamella vaginalis is very broad but short; the signum is ventral, oval, just basad of central. One specimen, Tiebaghi site 41, slide 2019.
**Type 4.** Plate 70: 18, Figure 80: 3. 10-11 mm. The forewing is suffused brown uniformly except for the angled, oblique medial. The hindwing is buff. The genitalia are similar to type 3 but the ductus is shorter, broader, more asymmetrical basally at the origin of the ductus seminalis; the lamella vaginalis is less broad and, with the surrounding membranes, more sclerotised, scobinate and folded; the signum is central, dorsal, elongate, tapering at each end. Two specimens, Col Toma, Tinchialit, slides 2020, 2025.
**Type 5.** Plate 70: 13, Figure 80: 4. 10 mm. The forewings are suffused greyish brown with the medial obscure, oblique; the hindwing is greyish fawn. The genitalia are on the same pattern as type 4 but the ductus is basally more symmetric and the signum is shorter, more oval. One specimen, Les Dalmates site 78, slide 2023.

*Macaduma atriplaga* (Hampson) **comb. n.** (Plate 71: 1)

*Heliosia atriplaga* Hampson, 1914, *Cat. Lepid. Phalaenae Br. Mus. Suppl.* 1: 586, type examined, Lifu.

This species bears no relationship to *jucunda* Walker, the type species of *Heliosia*, but is generally similar in form, facies and genitalia to other *Macaduma* species from New Caledonia and those from Fiji and Samoa. The

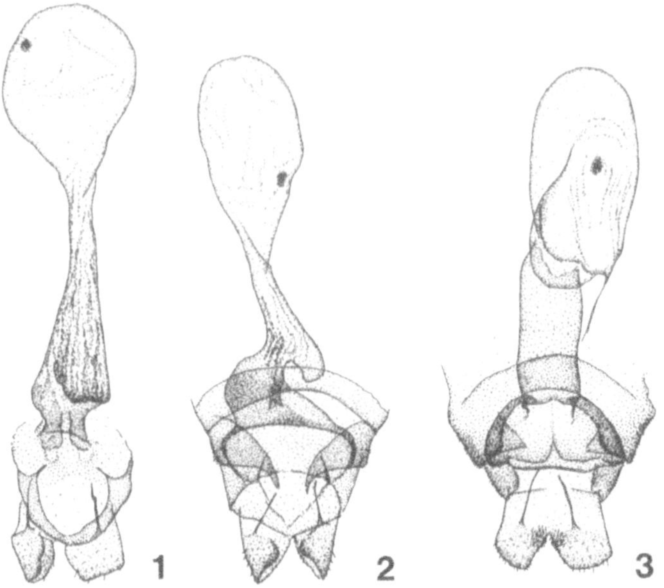


*Fig. 79.* Female genitalia of *Fabresema* species: *elisabethae* (1), *gerardi* (2), female type 1 (3).

abdomen bears three pairs of coremata as in these other species and the male genitalia are asymmetrical. The uncus of the holotype male was found to be missing and the only two other specimens lacked abdomens altogether. The left valve lacks the prominent lateral lobe of the Fijian *montana* Robinson and *biangulata*, described below. The right valve, longer, is relatively narrow. It shows affinities with *biangulata* in its bifurcate juxta but the two horns are narrower and bear diagnostic short spines exteriorly.

It is generally smaller (9 mm) than the other species and lacks transverse fasciae. The pattern of the forewing consists of brown patches, variable in position in the three specimens examined, on a straw ground. The holotype is illustrated. All three specimens are from Lifu, Loyalty Is.

**Macaduma biangulata sp. n.** (Plate 71: 2, Figure 81: 4)

♂♂ 9–10 mm. The wings have brown markings on a fawn ground as illustrated. The species is very variable, some specimens with the forewings almost entirely suffused brown; one specimen differs from that illustrated in having the medial area pale. *M. montana* Robinson (Fiji) is similarly variable with generally similar fasciation though the postmedial of that species is more regular and angled centrally; in *biangulata* there is a double angle one third from the costa. In the genitalia the dorsal lobe of the left valve is less prominent than in *montana*, the uncus is deeper at the apex with a less prominent ventral curve; the juxta is diagnostically bifid.

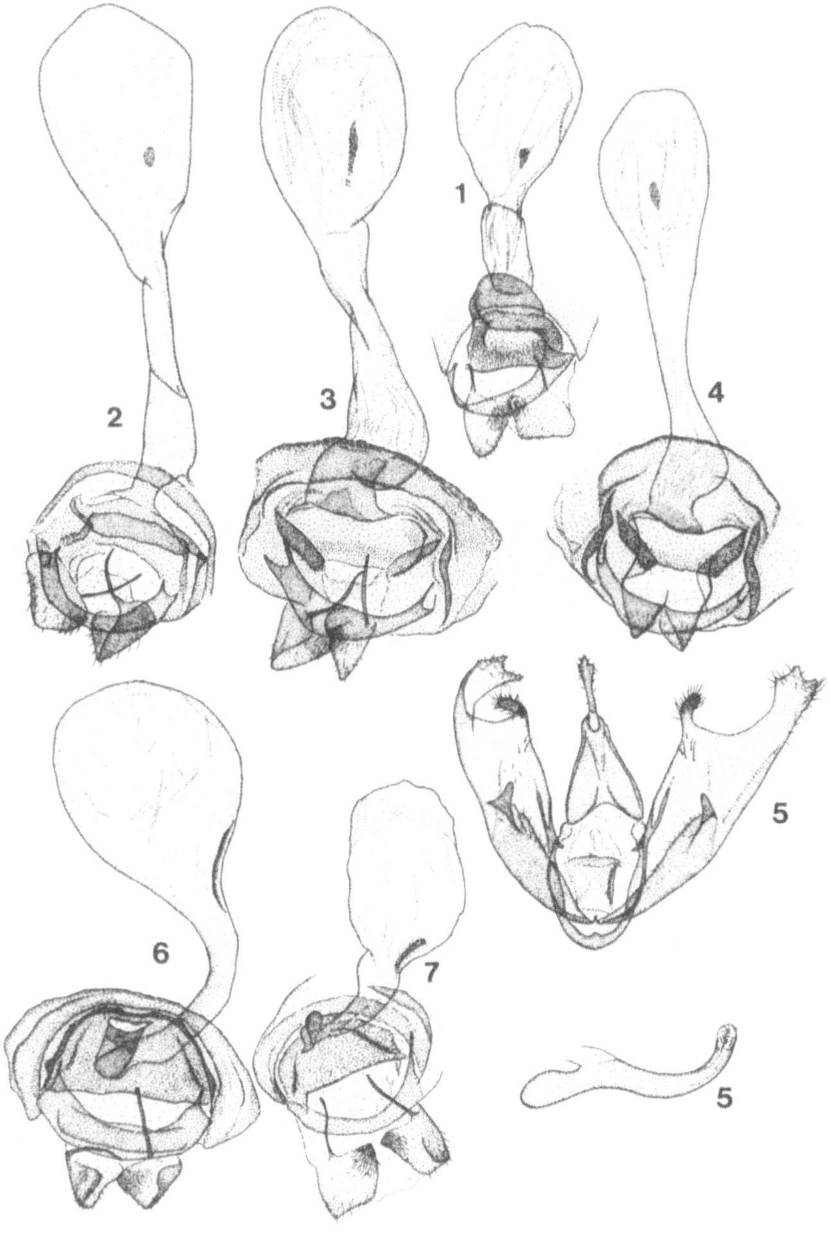


*Fig. 80.* Female genitalia of *Fabresema* type 2 (1), type 3 (2), type 4 (3), type 5 (4) and *costiplaga* (6), and male and female genitalia of *F. cheesmanae* (5, 7).

*Plate 71.* 1. *Macaduma atriplaga;* 2. *Macaduma biangulata;* 3. *Macaduma cristata;* 4 (♀), 5 (♂). *Thallarcha multifasciata;* 6. *Schistophleps subtilis;* 7. *Thumathoides fumosa;* 8. *Eilema caledonica.*

HOLOTYPE ♂. New Caledonia: Mt. Koghis, 500 m Site 79 06550/75463 (J. D. HOLLOWAY) 20.8.1971, BM 1971-507. BM arctiid slide 1927.
PARATYPE ♂♂. New Caledonia: Col des Roussettes (2); Ouatou R. (1).

**Macaduma cristata sp. n.** (Plate 71: 3)

♀ 10 mm. The forewings and thorax are a dark, dull, slightly vinous brown, with the hindwings medium brownish grey (reflecting paler in the illustration). The forewings bear a prominent tuft of long scales in the centre of the costa, distinguishing the species from other dark species such as *corvina* Felder (Fiji; much darker and forewings with a bluish sheen) or *fuliginosa* Rothschild (New Guinea; reddish in tint, with a faint, curved postmedial evident). The female genitalia are similar to those of *corvina.*
HOLOTYPE ♀. New Caledonia: Mt. Mou, 1100 m, Site 35, 06385/75594 (J. D. HOLLOWAY) 3.7.1971, BM 1971-507. BM arctiid slide 2015.

*Oeonistis entella delia* (Fabricius)

*Oeonistis entella delia* (Fabricius); ROBINSON (1975: 102).
The species is widespread in the Indo-Australian tropics, having its eastern

limits in Queensland, Fiji and Samoa. A single specimen was taken at Ciu during the survey and there is a specimen from Houailou in the BPBM.

*Oeonistis lifuensis* Rothschild (Plate 73: 3)

*Oeonistis lifuensis* Rothschild, 1912, *Novit. zool.*, 19: 225, type examined. VIETTE (1950b: 83).

This species differs from the previous in that the bluish black area near the anal angle of the forewing is entirely surrounded by pale ground colour, separated both from the margin and the dark along the costa. The differences in the genitalia were illustrated by VIETTE (1950b).

The species occurs on Lifu and Maré in the Loyalty Is. and was recorded as follows in New Caledonia: Grand Lac site 2 (2), site 3 (1), site 4 (1); R. Bleue site 8 (1), site 9 (1), site 10 (11); Mt. des Sources site 11 (3), site 14 (10); Dothio site 19 (5); Poupa (2); Ciu (2); Koh (3); Ouatou R. (3); Bréhoa R. (8); Ouambéo R. (1); Col d'Amieu (51); Col Toma (13); Petit Gouli (2); Sarraméa (3); Kopéto (1); Mt. Dogny (2); Mt. Mou site 33 (1); Col d'Amoss site 47 (1); Kavatch (1); R. Amoa (7); Ste. Therèse (5); Probokonoro (45); Mt. Krapé site 55 (27), site 56 (1); Col de Boa (5); near Bâ (1); Mt. Panié site 61 (4); Mt. Aoupinié (27); Col Mouirange (1); Mt. Pouédihi (1); Pic du Pin site 69 (2); Mt. Koghis site 79 (1). BPBM: Le Crouen, Mt. Koghis, Houailou, Sarramea. VIETTE: Canala.

**Schistophleps subtilis sp. n.** (Plate 71: 6, Figure 81: 3)

♂♀ 8–10 mm. The wings are translucent white, delicately marked in black on the forewing as illustrated. The intensity of this black marking is variable. The fasciation places this species close to *obducta* Lucas (Queensland; a female determined by TURNER in the BMNH was examined) but *obducta* is more brownish generally, the forewings suffused with brown medially and basally. The bursa of the female genitalia is rounder than in *obducta* and lacks the prominent scobination of that species.

HOLOTYPE ♂. New Caledonia: R. Bleue, 180 m Site 10 06704/75544 (J. D. HOLLOWAY) 27.5.1971, BM 1971-507. BM arctiid slide 1890.

PARATYPES. 2♂♂, 7♀♀ New Caledonia, various localities. BM arctiid slide 1946 (♀).

The species was recorded as follows: R. Bleue site 10 (1); Col Toma (2); Petit Gouli (1); Sarraméa (1); Ste. Therèse (1); near Bâ (1); Mt. Aoupinié (1); Mt. Koghis site 79 (5). It is probably a rainforest species.

**Thallarcha multifasciata sp. n.** (Plate 71: 4, 5, Figure 81: 1)

♂ 8 mm, ♀ 7 mm. The ground of the forewing and thorax is pale grey, marked with dark blackish brown. The hindwings are pale yellow, grading greyish at the margins and with a diffuse greyish discal spot. The insect is generally more finely fasciated on the forewing, the fasciae more numerous, than other

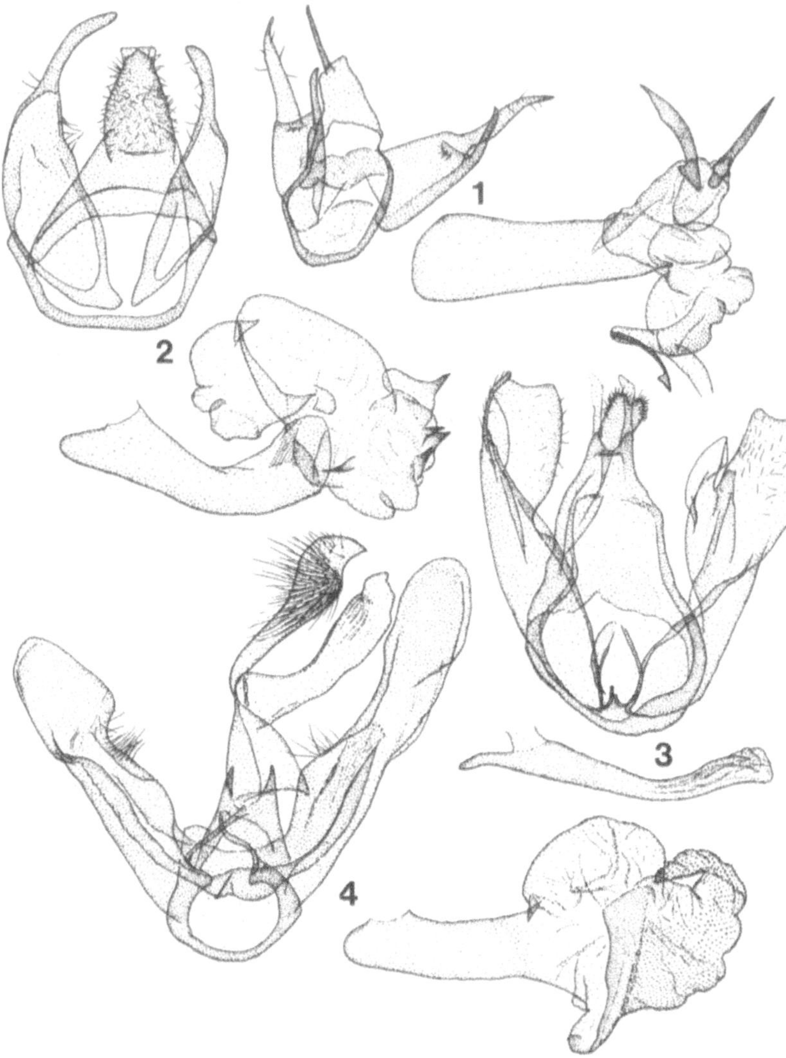


*Fig. 81.* Male genitalia and aedeagi of *Thallarcha multifasciata* (1), *Thumathoides fumosa* (2), *Schistophleps subtilis* (3) and *Macaduma biangulata* (4).

members of the genus represented in the BMNH collection such as *fuscogrisea* Rothschild. The species is similar to *isophragma* Meyrick (Tasmania) which has several narrow fasciae, the submarginal broadening at the costa to give a dark triangle as in the New Caledonian species, but *isophragma* is much larger (11 mm) with the fasciae somewhat more broken. The female specimen of *multifasciata* has the area of the forewing from the antemedial to the submarginal suffused dark grey.

The genus is otherwise restricted to Australia. A number of species described by Australian authors are not represented in the BMNH and *multifasciata* may prove to have closer relatives amongst these. There is an unnamed series in the BMNH from N. Queensland very similar to *multifasciata* but with the forewing fasciae less definite, less angled, the forewing being generally more uniform grey in tone.

HOLOTYPE ♂. New Caledonia: Grand Lac, 250 m Site 21 06975/75356 (J. D. HOLLOWAY) 10.8.1971, BM 1971-507. BM arctiid slide 1934.

PARATYPE ♀. New Caledonia: Mt. Koghis, 500 m Site 79 06550/75463 (J. D. HOLLOWAY) 21.8.1971, BM 1971-507.

## THUMATHOIDES Gen. n.

The singleton male described in this genus did not fit readily into any described genus and had several very distinctive characters, serving to separate it from species of *Thumatha* Walker, which it somewhat resembles. Examination was made of the type species *senex* Hübner (Palaearctic) and the Indo-Australian *fuscescens* Walker (type examined; Australia).

The antennae are filiform, slightly serrate ventrally, with a pair of long setae and several small ones per segment. This is as in *senex* but *fuscescens* has markedly bipectinate antennae in the male. The genus resembles *Thumatha* in having a single pair of spurs on the mid tibia and two pairs on the hind tibia.

The forewings are somewhat rounded as in *Thumatha*, with a similarly curved costa. Sc and $R_1$ are separate as in *Thumatha* but roughly parallel instead of being adjacent centrally and separate at each end. $R_4$ runs straight from the cell, $R_3$ branching anteriorly and $R_5$ more distally, posteriorly; in *Thumatha* $R_3$ is straight with $R_4$ and $R_5$ arising posteriorly. $M_2$, $M_3$ and Cula arise separately from the cell but in *Thumatha* the first two have their origins fused at the cell.

In the hindwing *Thumathoides* has Sc separate, but in *Thumatha* it arises from the cell not far from the origin of $Rs+M_1$. *Thumathoides* resembles *fuscescens* in having $M_3$ and Cula fused, separating two thirds from the cell to the margin; in *senex* they are entirely separate.

The forewing pattern is vaguely similar to that in *Thumatha* with a discocellular dot and faint antemedial, medial and postmedial fasciae, but the hindwing of *Thumathoides* lacks the discal spot of *Thumatha*. The ground colour is dark grey instead of pale buff.

The male eighth sternite has two anteriorly directed apophysis-like projections not present in *Thumatha*. The uncus of the male genitalia is large, broadly triangular, lobed to give a square apex; that in *Thumatha* is small, narrow and simple, more as in *Eilema*. The saccus is broad, not projecting much below the base of the valves, but it is narrower in *Thumatha* and projects considerably. The valves are unequal, simple, narrow over the apical third, whereas in *Thumatha* the valve is broad at the apex with prominent central and ventral-marginal spine-like projections. The aedeagus vesica has three stout cornuti, while in *Thumatha* the cornuti are narrower with six moderate ones in *fuscescens* and numerous small ones in *senex*.

The type species is hereby designated as *fumosa*, described below.

**Thumathoides fumosa sp. n.** (Plate 71: 7, Figure 81: 2)

♂ 10 mm. Diagnosis and description as above.

HOLOTYPE ♂. New Caledonia: Grand Lac, 270 m Site 72 06965/75354 (J. D. HOLLOWAY) 11.8.1971, BM 1971-507. BM arctiid slide 1931.

ARCTIIDAE: ARCTIINAE

*Rhodogastria albivitrea* Hampson (Plate 72: 4, Figure 82)

*Rhodogastria albivitrea* Hampson, 1901, *Cat. Lepid. Phalaenae Br. Mus.* 3: 506, type examined. VIETTE (1950b: 87).

This species differs from the majority of Indo-Australian *Rhodogastria* and resembles *rubripes* Walker and *alberti* Rothschild from northern Australia in having both wings pale and uniform cream. It produces the characteristic acrid smelling froth from the thorax when disturbed that is common to other members of the genus.

It occurs in the Loyalty Is. and was recorded in New Caledonia as follows: Grand Lac site 2 (15), site 4 (2); Pic du Pin site 7 (3); R. Bleue site 8 (2), site 9 (1); Mt. des Sources site 11 (2), site 12 (6), site 14 (31); Anse Longue site 15 (1), site 32 (11); R. Kuenthio (2); Dothio site 19 (1); Poupa (11); Koh (36), Ouatou R. (3); Bréhoa R. (43); Col d'Amieu (142); Col Toma (9); Petit Gouli (9); Sarraméa (6); Kopéto (3); Mt. Mou site 33 (50), site 34 (4); Tiebaghi site 41 (1); Grottes de Koum (2); Col d'Amoss site 45 (1); Kavatch (6); Probokonoro (31); Col des Roussettes (1); Mt. Krapé site 55 (7); near Bâ (4); Mt. Panié site 61 (1); Mt. Aoupinié (3); Ouen Toro site 64 (1); Col Mouirange (2); Mt. Pouédihi (2); Anna-Madeleine (2); Les Dalmates site 77 (1). VIETTE: Nouméa. BPBM: Népoui, Isle of Pines. The species is very widely distributed with no indications of strong habitat preference.

**Spilosoma lateritica sp. n.** (Plate 72: 7-9, Figure 83: 3-7)

♂♂ 20–25 mm, ♀♀ 29–32 mm. The forewings are cream with dark brown markings as illustrated. The markings are variable, ranging from specimens with almost entire fasciae to those with fasciae absent. The forewing is occasionally lightly to heavily suffused uniformly with brown scales. The hindwings are orange-pink, creamer towards the costa, usually with only a faint, diffuse discal brown mark, but occasionally with small submarginal dark markings. The abdomen is red, banded dorsally with dark brown. The thorax is cream, occasionally with a longitudinal brown streak.

The species is closest to *lifuensis* below but differs in the cream coloration and the entirely reddish hindwings, the presence of a discal spot on the

*Fig. 82.* Male and female genitalia of *Rhodogastria albivitrea.*

forewing and in the generally much heavier fasciation. In the male genitalia there is general similarity with *lifuensis* but the valve, though variable, is much broader. Similar, as yet undescribed forms taken by ROBINSON in the New Hebrides (possibly a subspecies) have longer valves and a less heavily spined apex to the aedeagus. VIETTE's (1950b) reference to *Ardices glatignyi* (Le Guillou) in the New Hebrides could conceivably relate to this form but his description and illustration of the juxta and valve are not comparable exactly with ROBINSON's material. *A. glatignyi* is otherwise Australian and is externally similar to *lateritica,* though the valve is longer, markedly bilobed.

Externally similar cream and red species from New Guinea, e.g. *artichroa* Druce, are more heavily and extensively marked black on both wings and have more elaborate valves and greater ornamentation of the vesica in their male genitalia. *S. erythrastis* (Meyrick) from Queensland has similar vesica ornamentation and lobing, but the valves are longer, curved and tapering.

HOLOTYPE ♂. New Caledonia: Grand Lac, 250 m Site 2L 06975/75356 (J. D. HOLLOWAY) 9.8.1971, BM 1971-507. BM arctiid slide 1882.

PARATYPES: 65♂♂, 2♀♀ New Caledonia, various localities; 1♂ New Caledonia, Col des Roussettes, 450–500 m 4-6.ii.1963 (N. L. H. KRAUSS) BPBM; 1♂ New Caledonia 10 km S. of Koh 31.i.63 (C. M. YOSHIMOTO) BPBM; 1♀ New

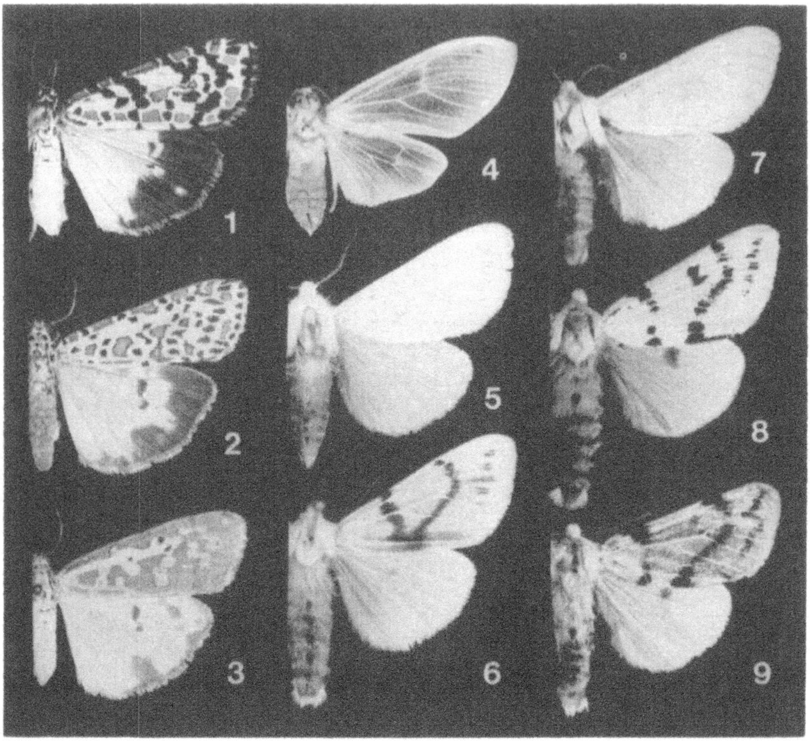


*Plate 72.* 1. *Utetheisa salomonis* (16 mm); 2, 3 (f. *rubra*). *Utetheisa lotrix stigmata* (18 mm); 4. *Rhodogastria albivitrea* (22 mm); 5, 6. *Spilosoma lifuensis* (20–24 mm); 7–9. *Spilosoma lateritica.*

Caledonia, Hienghène Oct. 8, 1940 (FXW) BPBM; 1♂ La Coulée 28.10.76 (J. GUTIERREZ) MNHN.

The species was recorded as follows: Grand Lac site 2 (7), site 3 (10), site 4 (10), site 71 (1); Plaine des Lacs site 5 (1); Pic du Pin site 7 (25); R. Bleue site 8 (2), site 9 (2), site 10 (1); Mt. des Sources site 11 (11); R. Kuenthio (1); Dothio site 18 (5); Ouatou R. (2); Bréhoa R. (3); Col d'Amieu (7); Col Toma (3); Petit Gouli (3); Sarraméa (19); Kopeto (3); Mt. Mou site 33 (6), site 34 (1); Tiebaghi site 40 (1), site 41 (7), site 42 (1); Cap Colnett (3); R. Amoa (3); Ste. Therèse (6); Probokonoro (3); Col des Roussettes (2); Mt. Krapé site 55 (13), site 56 (3); Col de Boa (1); Houailou (1); near Bâ (3); Mt. Aoupinié (5); Mt. Pouédihi (2); Yaté Gorge (1); Port Boisé site 73 (1), site 74 (1); R. Koué (4); Anna-Madeleine (2); Les Dalmates site 77 (3), site 78 (1).

The species falls in the overlap between the ultramafic and forest and secondary associations in the numerical analyses. It is usually the only species taken on the ultramafics, overlaps with *lifuensis* in localities with rainforest on sedimentary rocks, and is entirely replaced by *lifuensis* in drier, open habitats on the latter (Figure 84). These two species and the New Hebrides form mentioned form a closely related group within the genus, rather isolated taxonomically and geographically from other species.

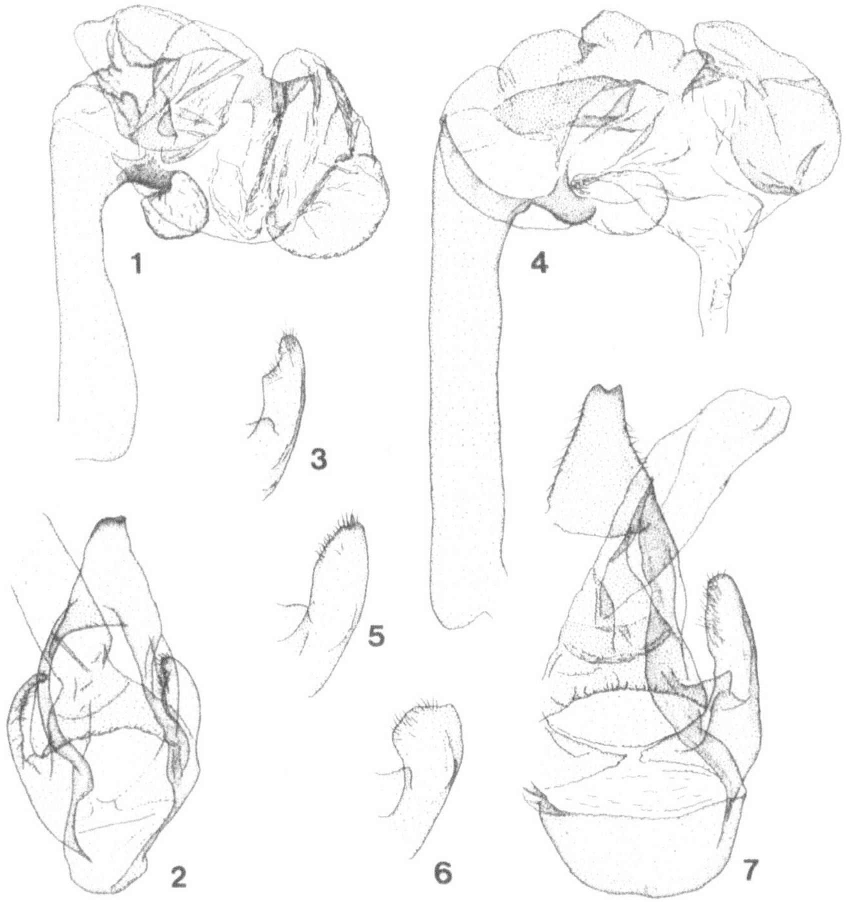


*Fig. 83.* Aedeagi and male genitalia of *Spilosoma lifuensis* (1, 2) and *S. lateritica* (3–7), illustrating variability in the valves of the latter.

### *Spilosoma lifuensis* (Rothschild) (Plate 72: 5, 6, Figure 83: 1, 2)

*Diacrisia lifuensis* Rothschild, 1910, *Novit. zool.*, 17: 122, type examined. VIETTE (1950b: 85).

Like the last, this species is also variable, smaller, usually with immaculate white forewings in coastal localities, larger, with brown-grey broken fasciation from inland localities with rainforest. A series from Mt. Koghis and a single specimen from Pic du Pin have almost complete fasciae on the forewing, two extreme specimens having a suffusion of grey-brown along the dorsum (illustrated). The insect is white with a pinkish abdomen and pinkish suffusion to the hindwing dorsum.

The species was recorded in New Caledonia as follows: Pic du Pin site 7 (1); Anse Longue site 16 (2), site 32 (1); Dothio site 19 (2); Koh (2); Bréhoa R. (6); Ouambéo R. (1); Col Toma (1); Petit Gouli (5); Sarraméa (20); Kopéto (1); Mt. Mou site 33 (2); Grottes de Koum (10); Kavatch (1); R. Amoa (9); Ste. Therèse (8); Probokonoro (17); Col des Roussettes (3); Mt. Krapé site 55 (3); Mt. Aoupinié (18); Ouen Toro site 64 (1), site 82 (2); Col Mouirange (1); Port Boisé site 73 (5), site 74 (8), Mt. Koghis site 79 (11), site 80 (1). BPBM: Hienghène. CAS: Isle of Pines. VIETTE: Nouméa. MNHN: La Coulée. The species occurs also in the Loyalty Is., the type locality.

There are records at ORSTOM, Nouméa, of the larva feeding on the introduced genus *Erythrina* (Leguminosae). In common with *lateritica*, the adult comes to light in the later part of the night, after the peak influx of most other species.

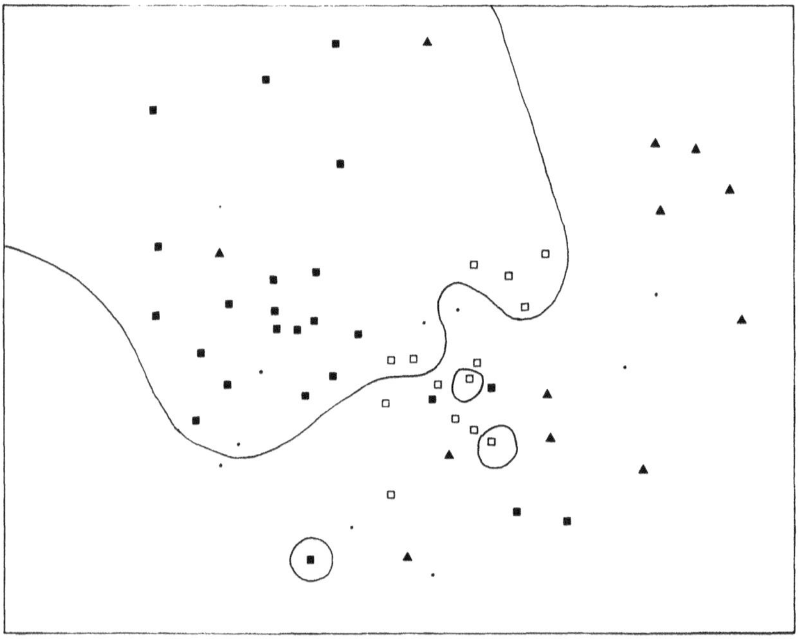

*Fig. 84.* Scaling plot showing the distributions of the two *Spilosoma* species: *lateritica* (solid squares), *lifuensis* (solid triangles) and both together (open squares).

*Utetheisa lotrix stigmata* Rothschild (Plate 72: 2, 3)

*Utetheisa lotrix stigmata* Rothschild; VIETTE (1950b: 86); ROBINSON (1971a, 1975: 104).

This is a widespread palaeotropical species with its eastern limit in Fiji. It is frequently encountered in open secondary vegetation off the ultramafics in New Caledonia by day but rarely comes to light. The larvae feed on Leguminosae (ROBINSON). It was recorded at light as follows: Poupa (1); Koh (2); Petit Gouli (1); Col de Boguen (1); Probokonoro (2); Houailou (1). Daytime records: Col d'Amieu, Poupa, Anse Longue, Fonwhary, Oubatche, R. Kuenthio, Col de Petchécara. VIETTE: Nouméa, Paita, Mt. Mou. BPBM: Yambé, Humboldt R. BMNH: Pouébo.

ROTHSCHILD (1914, *Gross-Schmett. Erde,* 10: 260) gave the aberration name *rubra* to specimens from 'New Caledonia and the Friendship Is.' but there is only one specimen from Lifu (Plate 72: 3) in the BMNH that may be referable. The black markings of the wings are reduced and the red patches of the forewings are correspondingly expanded and fused.

### *Utetheisa salomonis* Rothschild (Plate 72: 1)

*Utetheisa salomonis* Rothschild, 1910, *Novit. zool.,* 17: 181. VIETTE (1950b: 85), placed as ssp. of *pulchelloides* Hampson. ROBINSON (1971a); as *bon. sp.*

The species is recorded from the Bismarcks, Solomons, New Hebrides, Loyalties and New Caledonia. Specimens from the last three localities have the hindwing dark band more extensive than in the typical Solomons form, completely enclosing the white indentation in space 3 of the latter as a white spot. The subspecific name *ruberrima* Rothschild (1910, *Novit. zool.,* 17: 182, type examined, Lifu) is available for these forms.

The species was recorded from Pic du Pin site 7, and specimens were noted in the ORSTOM collection, Nouméa, one reared from *Messerschmidia* (Boraginaceae), the usual foodplant of species of the *pulchelloides* group of species discussed by ROBINSON (1971a).

## CTENUCHIDAE

### *Euchromia cincta* (Montrouzier) (Plate 73: 1, Figure 85)

*Glaucopis cincta* Montrouzier, 1864, *Ann. Soc. Linn. Lyon,* 11: 252, type not examined but identity not in doubt.

This *Euchromia* stands apart from other Papuan Region species in having markedly pectinate antennae in the male and yellow and white patches on a dark brown ground as distinct from the blue-black, white and red colour scheme of other species. It is the only yellow form in eastern Melanesia and appears to fly only by night instead of being taken visiting flowers by day. Males predominate markedly in light-trap catches.

The species was recorded as follows: Grand Lac site 4 (1); Plaine des Lacs site 5 (4); R. Bleue site 8 (1); Ouatou R. (1); Bréhoa R. (2); Col d'Amieu (2); Sarraméa (2); Mt. Mou site 33 (3); Tiebaghi site 41 (1); Grottes de Koum (1); Port Boisé site 74 (2); R. Koué (3); Les Dalmates site 77 (1); Mt. Koghis site 79 (1). BMNH: Isle of Pines.

*Fig. 85.* Male genitalia of *Euchromia cincta.*

### *Euchromia creusa* (Linnaeus)

*Euchromia creusa* (Linnaeus); Viette (1949: 327); Robinson (1975: 99).

A single example was taken at Grottes de Koum visiting lantana. The species is recorded from Amboina, Kei, New Guinea, Queensland, the New Hebrides and Fiji as well as New Caledonia.

### *Euchromia rubricollis* Walker

*Euchromia rubricollis* Walker, 1864, *List Specimens lepid. Insects Colln Br. Mus.*, 31: 99, type examined.

Viette (1949) recorded *rubricollis* from Lifu. It occurs in the New Hebrides and could remain to be discovered in New Caledonia.

## Hypsidae

### *AGAPE* Felder

In order to determine the correct status of the New Caledonian form *lifuensis* (Rothschild) it was necessary to dissect representatives of other species of the genus. All named forms were examined and dissected. In addition there is a large species, as yet undescribed, from Sulawesi and two specimens from the Moluccas (Amboina, Halmahera) which, if not aberrations of *chloropyga* (Walker), may represent a further species. These unnamed forms were not dissected. The species are all predominantly bright yellow with dark rufous-brown markings on the forewings, black dots at the base of the forewing and on the thorax, and blue-black bands and patches on the abdomen.

*Agape chloropyga* (Walker), generic type species. *Hypsa chloropyga* Walker, 1854, *List Specimens lepid. Insects Colln. Br. Mus.*, 2: 455, type examined, Queensland. *Agape cyanopyga* Felder, 1874, *Reise Novara, Lep.*, 2: t. 106, f. 4. The species is recorded from Malaya, Borneo, the Lesser Sundas, Buru, Amboina, New Guinea and Queensland. It is like a larger version of *lifuensis* (described in detail below) but has several black spots on the thorax; the terminal segments of the abdomen are entirely blue-black. The genitalia of the male have the socii narrower than in *lifuensis*, the sides concave rather than convex; the valves are long, extending just beyond the uncus when flattened on a slide, narrower centrally; the saccus is two thirds the length of the valves and heavily sclerotised; the juxta is long and rectangular. In the aedeagus the left lobe of the vesica (as arrayed in Figure 86) is longer, more simple than in *lifuensis*, but not quite as long as the right lobe.

*Agape leonina* Butler, 1879, *Proc. zool. Soc. Lond.*, 1879: 161, type examined, New Ireland. The species was described from the Bismarck Is. but also flies with the previous species in New Guinea; otherwise it might merely be regarded as a subspecies of it. Externally it is very similar though the spots representing the forewing fasciae are somewhat larger and more irregular in shape and the abdomen is banded blue-black throughout, rather than having a solid area of this colour apically. The genitalia are as in *chloropyga* but the left lobe of the vesica is as long as the right lobe. It is genitalically distinct from *lifuensis*.

*Agape arctioides* Butler, 1887, *Ann. Mag. nat. Hist* (5), 20: 123, type examined, Solomon Is. This species is found throughout the Solomon group. The forewings have entire but separate postmedials and antemedials. In some specimens they are much reduced, almost to spots as in *leonina*. The antemedial is generally straight, slightly oblique. The postmedial extends in a squarish loop towards the margin distal to the reniform (larger than in the previous species), is broadest at the extremity of this loop and irregularly bordered distad there. The bands of the abdomen are reduced relative to those of the previous species. In the genitalia the socii are absent; the valves are long, extending just beyond the uncus, and are much broader and rounded apically; the saccus is two-thirds the length of the valve and highly sclerotised; the juxta is broad, irregularly rhomboidal. The aedeagus has a simple, single lobe corresponding to the right lobe of *lifuensis*; this lobe has the basal half much broader than the distal half.

*Agape stapialis* Gaede, 1914, *Ent. Rundsch.*, 31: 74, type examined. *Agape novapommeranicola* Strand, 1917, *Societas Entom.*, 32: 48, **syn. n.,** type not examined but description unambiguous. This species is known only from islands in the Bismarck group, namely New Britain, New Ireland and Rook I. The forewing is distinctively marked. The antemedial is sometimes entire, sometimes broken basal to the orbicular; it fuses broadly with the postmedial at the dorsum. The postmedial is narrowest distal to the enlarged reniform where it forms a rounded loop that fuses with the reniform to form a ring. The margins of these fasciae curve smoothly and are never irregular. In the genitalia the socii are very small, very acute; the valves are very long, extending well beyond the uncus and are slightly broader apically; the saccus is one third of the length of the valves and lightly sclerotised; the juxta is narrow, elongate, semi-tubular. The aedeagus vesica is similar to that of *arctioides* but the distal half of the right lobe is broader, less tapered, and the broad left lobe is present, though much reduced.

The last two species appear to be related. *A. lifuensis* is closer in appearance to *chloropyga* and *leonina* but is distinguished by the genitalia characters discussed.

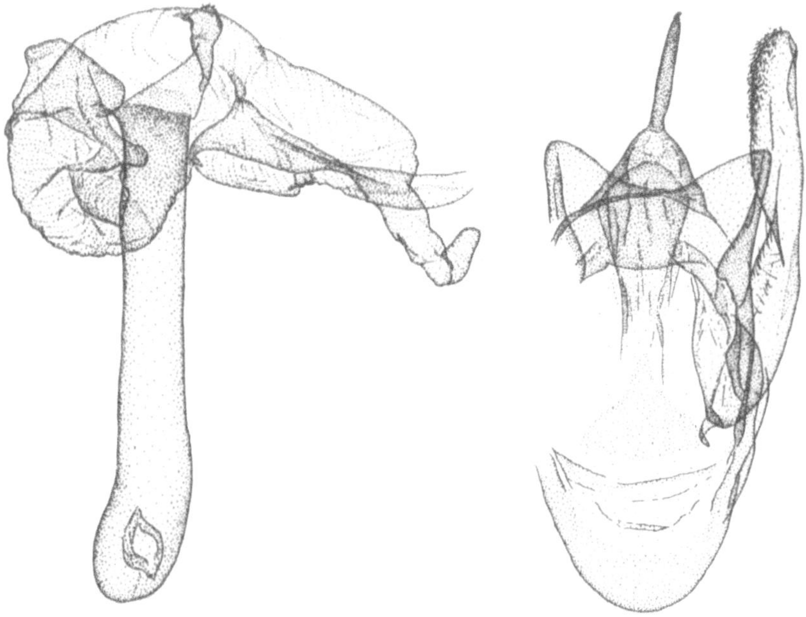

*Fig. 86.* Male genitalia of *Agape lifuensis lifuensis.*

*Agape lifuensis* (Rothschild) **stat. n.** (Plate 73: 2, Figure 86)

*Spilobotys leonina lifuensis* Rothschild, 1896, *Novit. zool.*, 3: 93, type examined. VIETTE (1950b: 91).

There are five brown spots on the forewing, these being the orbicular, the posterior part of the reniform, one posterior and slightly basal to the orbicular (the remnant of the antemedial), and two more (the remnants of the postmedial), one distal to, one posterior and slightly basal to the reniform. The abdominal segments are basally banded with blue-black and there is a dorsal blue-black triangle over the terminal two segments as illustrated. There is a black spot at the base of the forewing but none on the thorax, a distinctive feature. The male genitalia are as in Figure 86, differing from those of *chloropyga* and *leonina* in the proportions of various features, particularly the shorter valves and saccus, and in the shape of the juxta.

The species was described from the Loyalty Is., a distinct subspecies occurs in the New Hebrides (described below), and records from New Caledonia were as follows: Grand Lac site 1 (2), site 2 (1), site 4 (1); Plaine de Lacs site 5 (1); Pic du Pin site 7 (2); R. Bleue site 8 (1), site 10 (1); Mt. des Sources site 14 (1); R. Kuenthio (1); Dothio site 18 (3), site 19 (10); Koh (1); Ouatou R. (1); Bréhoa R. (7); Ouambéo R. (3); Col d'Amieu (21); Col Toma (5); Petit Gouli (2); Sarraméa (12); Kopéto (1); Anse Longue site 32 (4); Mt. Mou site 33 (17); Col de Boguen (1); Col d'Amoss site 45 (1); Cap Colnett (2); Kavatch (1); Amoa R. (9); Ste. Therèse (2); Probokonoro (19); Col des Roussettes (2); Mt. Krapé site 55 (3), site 56 (1); near Bâ (1); Mt. Aoupinié (11); Port Boisé site 73 (2), site 74 (3); Anna-Madeleine (1); Les Dalmates site 77 (1). MNHN: La Coulée. It is widely distributed but commoner in areas of rainforest.

**Agape lifuensis hebridesia ssp. n.**

♂♀ 25–27 mm. This race differs from typical *lifuensis* in having a black spot on each patagium and in lacking the blue-black triangle apically on the abdomen. The male genitalia are similar.

HOLOTYPE ♂. New Hebrides: Efate I., Vila, 26.vi.1971 (G. S. ROBINSON) Roy. Soc. Expedition. BM arctiid slide 2129.

PARATYPES: 6 ex. New Hebrides: Efate I., Narabut Camp 30.vi-7.vii 1971 (G. S. ROBINSON) Roy. Soc. Expedition; 4 ex. Port Vila, 1975 (J. A. BURGESS).

There are also specimens from Erromanga, Malekula and Aneityum in the BMNH

*Argina cribraria* (Clerck)

*Argina cribraria* (Clerck); VIETTE (1950b: 88); ROBINSON (1975: 105).

This species is widespread in the Old World tropics. The larva feeds on *Crotalaria* (Leguminosae) in New Caledonia as elsewhere; its biology in Fiji was described by ROBINSON. The species occurs in Lifu and was recorded as follows in New Caledonia: Dothio site 19 (1); Koh (5); Ouatou R. (3); Bréhoa R. (7); Col d'Amieu (13); Petit Gouli (2); Sarraméa (4); Kopéto (1); Mt. Mou site 33 (2); Cap Colnett (12); Kavatch (2); Amoa R. (3); Probokonoro (1). BPBM: Prony Bay. In Fiji the species is associated with secondary vegetation and wasteland. It probably has a similar ecology in New Caledonia, associated with such habitats in areas of primary rainforest on sedimentary rock.

*Asota alienata* (Walker) (Plate 73: 6)

*Agarista alienata* Walker, 1864, *List Specimens lepid. Insects Colln. Br. Mus.*, 31: 46.

*Asota alienata* (Walker); VIETTE (1950b: 92), genitalia illustrated.

*Plate 73.* 1. *Euchromia cincta* (20 mm); 2. *Agape lifuensis* (23 mm); 3. *Oeonistis lifuensis* (19 mm); 4. *Asota caledonica;* 5. *Digama marmorea clinchorum;* 6. *Asota alienata* (32 mm).

VIETTE recorded this species from the New Hebrides only. There is a single specimen known from Nouméa in New Caledonia (illustrated). The species is closely related to *diana* Butler, a Solomons species.

**Asota caledonica sp. n.** (Plate 73: 4, Figure 87: 1)

♂♂ 20 mm. The head, thorax, abdomen and base of the forewings are orange, marked with dark brown and white. The forewings are dark brown distally, (though not as dark as the patches on the creamy white hindwing) white with orange and brown markings basally. The species is similar to the Fijian *woodfordi* Druce but differs in the forewing where the marginal white circle of that species is lacking, the medial white patch on the costa is acutely angled distad, (cf. right angled) the dark brown marking in the basal orange is more streaky and the subbasal spot in space 1 of *woodfordi* is absent; the hindwing anterior marginal brown patch extends further towards the dorsum and the posterior patch is reduced. The genitalia differ markedly (Figure 87).
HOLOTYPE ♂. (New Caledonia) *caledonica* Bd. N. Caled (in Boisduval handwriting), Ex Musaeo Dris. Boisduval, très voisin de Hypsa Woodfordi Moore P.Z.S. 1888 pl xiii fig 2 de Fiji (in Oberthur handwriting), Ex Oberthur Coll. Brit. Mus. 1927-3.
PARATYPE ♂. New Caledonia, capturée Station Forestière du Col d'Amieu, Mars 64. Ex coll. ORSTOM, Nouméa. BM arctiid slide 1878.

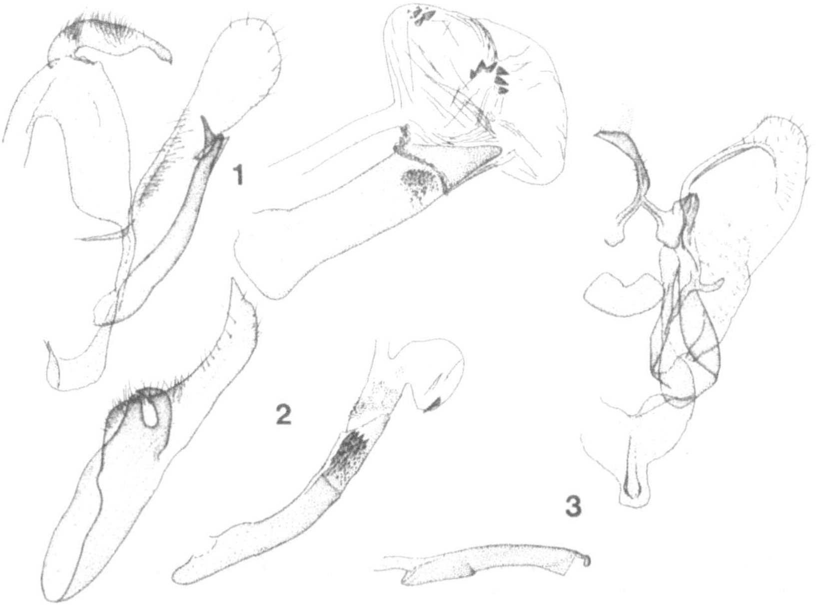


*Fig. 87.* Male genitalia and aedeagi of *Asota caledonica* (1), *A. woodfordi* (valve and aedeagus, 2) and *Lasioceros aroa* (3).

**Digama marmorea clinchorum ssp. n.** (Plate 73: 5, Figure 88)

♂ 12 mm. ♀ 14 mm. The forewing fasciation is as in typical *marmorea* Butler (1877, *Trans. ent. Soc. Lond.*, 1877: 363) from Australia (Northern Territory to New South Wales, one record from south-west Australia), New Guinea, Kisser I. and the New Hebrides. However, it is suffused black uniformly over all except the base of the medial area and a small discal spot where it is pale grey; in *marmorea* the pale grey traverses the medial area and extends distally along the costa and basally along the dorsum. The hindwing is yellow with a brownish black apex and discal spot, differing from *marmorea* in that the inner margin of the apical marking is narrower and tapers away posteriorly instead of being abruptly truncated. The male genitalia are similar apart from the aedeagus vesica which has a single basal sclerotised area with broad, short spines compared with two more weakly sclerotised and spined areas in typical *marmorea*.

The subspecies is named for STAN and JOY CLINCH of Anse Longue who showed the author much hospitality during his travels.

HOLOTYPE ♂. New Caledonia: Anse Longue, 5 m site 32 06218/75543 (J. D. HOLLOWAY) 30.6.1971, BM 1971-507. BM arctiid slide 1935.

PARATYPE ♀. New Caledonia: Kopéto, 160 m Site 30 05102/76542 (J. D. HOLLOWAY) 24.6.1971, BM 1971-507.

In Australia the larva feeds on *Carissa ovata* (COMMON, 1970), a member of

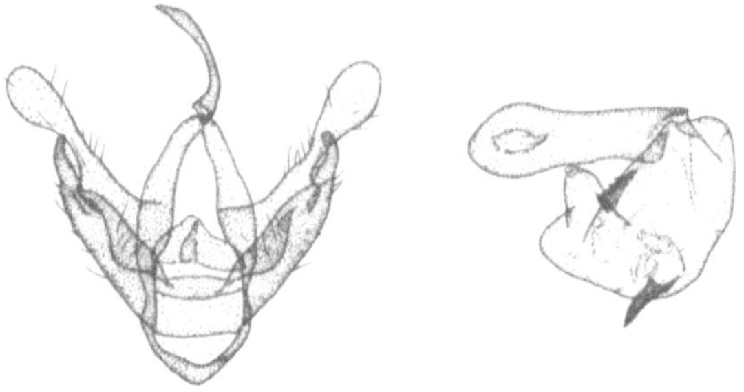

*Fig. 88.* Male genitalia of *Digama marmorea clinchorum.*

the Apocynaceae. The genus is not represented in New Caledonia but the family is.

### *Nyctemera baulus fasciata* Walker

*Nyctemera baulus fasciata* Walker; VIETTE (1950b: 90); ROBINSON (1975: 107).

This species is widespread in the Indo-Australian tropics, reaching its eastern limits in Fiji, Samoa and Tonga. In New Caledonia it came to light as follows: Grand Lac site 3 (1); Pic du Pin site 7 (1); Mt. des Sources site 12 (1); Anse Longue site 16 (1); Dothio site 19 (1); Poupa (1); Koh (1); Col d'Amieu (2); Col Toma (1); Sarraméa (3); Kopéto (3); Mt. Mou site 33 (1); Bambou R. (1); Col d'Amoss site 46 (1); Cap Colnett (1); Amoa R. (1); Mt. Krapé site 55 (1); Ouen Toro site 81 (1), site 82 (5). Daytime records: Port Boisé, Poupa, Dothio, R. Kuenthio, Bambou R., Col de Petchécara, Ouatou R., Col d'Amieu. BMNH: Pouébo. BPBM: Le Crouen. VIETTE: Paita, Nouméa. CAS: Isle of Pines.

ROBINSON (1975) has described the biology of the species in Fiji, where it is commonest in areas of secondary vegetation. It would appear to favour similar habitats in New Caledonia.

## NOLIDAE

### **Nola bitransversata sp. n.** (Plate 74: 2, Figure 89: 3)

♂♀ 6 mm. The ground colour of the wings is pale grey and the forewing is fasciated with darker grey and black as illustrated. The antmedial and postmedial fasciae are prominent, whereas in *transversata* Robinson it is the medial that is the most developed. The species appears to have no close relatives in the south-west Pacific. The male genitalia are similarly distinctive with the valve apex bilobed and setose; there is a central harpe to each valve, that on the left valve being twice as large as that illustrated on the right valve.
HOLOTYPE ♂. New Caledonia: Grand Lac, 260 m Site 3 06965/75356 (J. D. HOLLOWAY) 18.5.1971, BM 1971-507. BM arctiid slide 1939.

PARATYPES. 3♂♂, 1♀ New Caledonia: Grand Lac site 2; Bambou R., Mt. Aoupinié; Tiebaghi site 41.

*Plate 74.* 1. *Nola major* (12 mm); 2. *Nola bitransversata;* 3. *Nola scabralis* (8 mm); 4. *Nola lichenosa* (5 mm); 5, 6. *Nola samoana* (7–9 mm).

*Nola insularum* (Collenette)

*Nola insularum* (Collenette); ROBINSON (1975: 108); HOLLOWAY (1977: 56).

This is a widespread Pacific species that may prove to be conspecific with species further west in Indonesia (HOLLOWAY, 1977). It was recorded as follows: Sarraméa (2); Col des Roussettes (1); Mt. Panié site 60 (1); Ouen Toro site 64 (1); Port Boisé site 74 (1). BPBM: Tao, Bourail, Le Crouen, Yahoué, Col des Roussettes.

*Nola lichenosa* Robinson (Plate 74: 4, Figure 89: 2)

*Nola lichenosa* Robinson, 1975: 109.

This species occurs in Fiji and Rotuma. Forms with similar facies from the New Hebrides may also be referable. It was recorded in New Caledonia as follows: Grand Lac site 2 (2), site 4 (1), Tiebaghi site 41 (1); Mt. Panié site 60 (3). BPBM: Col des Roussettes.

*Nola major* Hampson (Plate 74: 1, Figure 94: 1)

*Nola major* Hampson, 1891, *Illust. typical Specimens Lepid. Heterocera Colln Br. Mus.* 8: 48, type examined.

A single worn male from the Isle of Pines is in the BPBM collection. G. S. ROBINSON took a series on Santo I. in the New Hebrides. Otherwise the species is known from Sri Lanka, India, the Himalaya, Burma and Singapore. In Taiwan there flies a greyer race, *N. m. formosana* Wileman & West (1929, *Ann. Mag. nat. Hist.* (10), 3: 190) **stat. n.**

*Nola samoana* (Hampson) (Plate 74: 5, 6, Figure 89: 4, 5)

*Nola samoana* (Hampson); ROBINSON (1975: 109).

The patterning and genitalia of the male are as in typical *samoana* though the former is extremely variable and the projections of the eighth abdominal segment are much broader in the latter. Apart from Samoa the species occurs in Rotuma and the New Hebrides but is replaced by *fijiensis* in Fiji. Both these species are probably derivatives of *insularum*. Specimens were brought back from the following New Caledonian localities: Bréhoa R., Ouambéo R, Grottes de Koum, Col d'Amoss site 46, Kavatch, Col des Roussettes, Port Boisé site 74. BPBM: Tao, Le Crouen, Yahoué, Tipindjé, Nouméa.

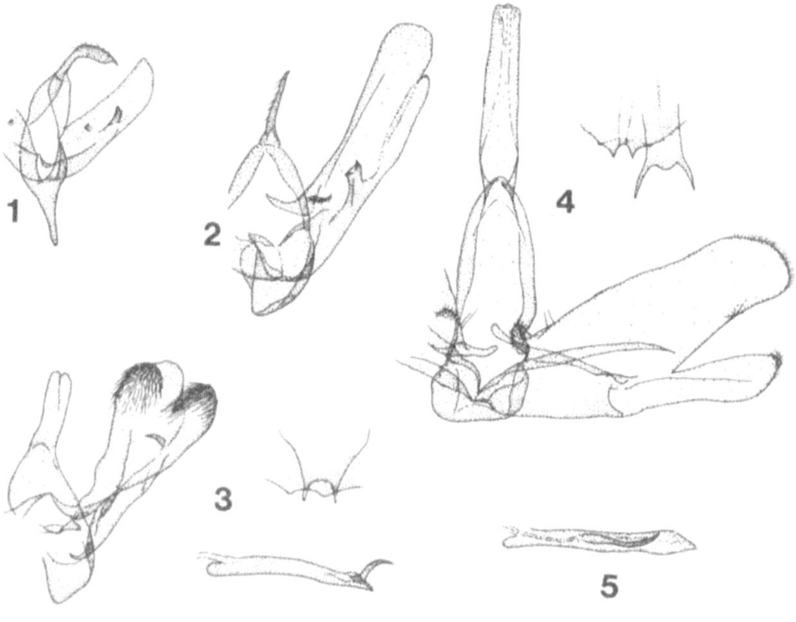


*Fig. 89.* Male genitalia of *Nola* species: *scabralis* (1), *lichenosa* (2), *bitransversata* (3) and *samoana* (4, 5), with the modification of the eighth abdominal segment shown for the last two.

*Nola scabralis* (Walker) (Plate 74: 3, Figure 89: 1)

*Tribunta scabralis* Walker, 1865, *List Specimens lepid. Insects Colln. Br. Mus.*, 34: 1507, type examined.

New Caledonian specimens are slightly larger and paler than typical ones from New South Wales and the projections from the male eighth segment are slightly longer. Specimens brought back were all from southern ultramafic localities: Grand Lac site 2 and site 3, Pic du Pin site 7.

## NOCTUIDAE: NOCTUINAE

### *Agrotis ipsilon aneituma* (Walker)

*Agrotis ipsilon aneituma* (Walker); ROBINSON (1975: 112); HOLLOWAY (1977: 60).

This subspecies of a widespread Old World tropical and subtropical species extends from Australia east into the Pacific as far as Easter I. It was recorded as follows during the survey: Mt. des Sources site 12 (1); Col Mouirange (1). VIETTE (1949): Voh. BPBM: Nouméa, Forêt de Thi. Records at ORSTOM show the larva to be a pest of lettuce in New Caledonia.

The records from the Mt. des Sources of this and the next species occurred prior to the passage over the island of a rain-bearing front from the west. These specimens, taken in an inhospitable locality, may have been vagrants from Australia; vagrant arrivals have been recorded in similar condition on Norfolk I. (HOLLOWAY, 1977). The Col Mouirange record and that of the next species with the pyrale *Nomophila australica* Munroe at Les Dalmates were again in unhospitable localities and were within a fortnight of each other. Though the two species undoubtedly breed on the island at least from time to time, neither was recorded from more favourable localities of secondary growth and cultivation on sedimentary rock at any time during the survey. However, *A. munda* larvae are found commonly on various garden crops (e.g. *Brassica, Phaseolus*) in the northern part of the island (A. DELOBEL, *in litt.*).

### *Agrotis munda* Walker

*Agrotis munda* Walker; ROBINSON (1975: 112); HOLLOWAY (1977: 60). Records of *Euxoa radians* Guenée and *Agrotis compta* Walker (in VIETTE, 1949) from New Caledonia probably are of *munda*.

The species occurs throughout Australia and has been recorded from Fiji and Tonga. It is vagrant to Norfolk I. and may also occur in the Loyalties. It was recorded during the survey as follows: Mt. des Sources site 11 (1); Les Dalmates site 77 (1). ORSTOM: Nouméa. BPBM: Nouméa, Isle of Pines.

### *Agrotis fumipennis* (Warren) **comb. n.** (Plate 75: 5, holotype)

*Euxoa fumipennis* Warren, 1912, *Novit. zool.*, 19: 5, type examined.

This is a very similar species to *munda* and may prove merely to be a subspecies if it can be shown that the two forms do not fly together in the Loyalties. Capture and dissection of males of *fumipennis* would also be valuable as the species was described from four females from Lifu and no further material is available. The forewings of these are deeper, darker and slightly browner than in *munda*. The hindwing is more extensively suffused brown from the margin. The genitalia are similar to those of female *munda* as illustrated by COMMON (1958) and do not resemble those of *A. radians* Guenée, the Australian species with which WARREN compared *fumipennis* in his description.

*Plate 75.* 1. *Leucania loreyi* (18 mm); 2. *Leucania loreyimima* (19 mm); 3. *Leucania pseudoformosana* (17 mm); 4. *Leucania latericia;* 5. *Agrotis fumipennis* (20 mm); 6. *Mythimna consanguis* (16 mm); 7. *Leucogonia amarginata;* 8. *Tiracola* ♀, Efate; 9. *Tiracola tabwemasana;* 10. *Tiracola circularis*, showing forewing variability.

## NOCTUIDAE: HELIOTHINAE

### *Helicoverpa armigera conferta* (Walker)

*Helicoverpa armigera conferta* (Walker); ROBINSON (1975: 114); HOLLOWAY (1977: 62).

This widespread Old World tropical and subtropical species was only recorded as a singleton at Anse Longue site 16 during the survey, but HARDWICK (1965, *Mem. ent. Soc. Can.*, 40) recorded it from Nouméa, Népoui and Hienghène. COHIC (1956) and more recent ORSTOM records indicate it (possibly not discriminated from the next species) to be a pest of *Passiflora*, tomato, *Brassica*, *Dahlia*, *Gossypium*, *Ipomoea*, *Lycopersicum*, *Nicotiana*, *Phaseolus*, *Pisum* and *Zea* in New Caledonia, many of which were also recorded as host plants in Fiji by ROBINSON (1975).

### *Helicoverpa assulta assulta* (Guenée)

*Helicoverpa assulta assulta* (Guenée); ROBINSON (1975: 114); HOLLOWAY (1977: 63).

Like the previous species, *assulta* is widespread in the Old World tropics. It was recorded during the survey as follows: Mt. Mou site 33 (1); Col Nassiboué (1).

## NOCTUIDAE: HADENINAE

### **Leucania latericia sp. n.** (Plate 75: 4, Figures 90: 5, 91: 3)

♂♀ 13–16 mm. The forewing, like the thorax, is pale brick red, the veins picked out with fawn and edged darker brown; the postmedial is a curved row of dark brown dots on the veins, and the antemedial is similar, acutely angled in space 1b; the discal spot is pale fawn. The abdomen and hindwings are fawn, the latter suffused reddish towards the margin and along the veins.

The male genitalia place this species in a group that includes the grey Oriental species *celebensis* Tams, *simillima* Walker and *nabalua* Holloway. These species are characterised by the reduction of the cucullus of the valve to a setose rod and the relative enlargement of the sacculus and development of the triple harpe, especially the most ventral arm of the latter. SUGI (1970) described the new subgenus *Xipholeucania* to include these species, referring to the characteristics mentioned. Unfortunately he designated as type species *roseilinea* Walker to which TAMS had related his *celebensis* in error. The type of *roseilinea* (Oxford; examined and dissected by W. W. BRANDT) is not related to this group; the genitalia were illustrated by HOLLOWAY (1976) and by SUGI (1970), the latter placing the species under the name *compta* (Moore). The Oriental *roseilinea* also has rather unusual genitalia within the *Mythimna*/*Leucania* complex; it is related to *L. pinna* (Sallmüller) from Africa. Thus the name *Xipholeucania* must relate to the latter group.

The New Caledonian species is perhaps closest to the Queensland *dasycnema* Turner (1912, *Mem. Mus. Melbourne*, 4: 21; the name proposed for material referred to *adjuncta* Walker by HAMPSON, 1905, *Cat. Lepid. Phalaenae Br. Mus.*, 5: 489, some of which is in the BMNH; the ab. names *adjunctoides* Strand and *obscura* Warren were based on this material and must therefore refer to *dasycnema* rather than *adjuncta*). *L. dasycnema* differs from the New Caledonian species in having a uniformly pale pink forewing, the veins and discal spot barely distinguished, and there are immense fuscous purple tufts of scales on the male fore tibia that are not present in *latericia*. *L. latericia* has a more rounded sacculus than *dasycnema*, a broader, shorter ventral arm to the harpe and a longer, more curved central arm; the cucullus is much narrower.

HOLOTYPE ♂. New Caledonia: Mt. des Sources, 760 m Site 11 06650/75512 (J. D. HOLLOWAY) 29.5.1971, BM 1971-507. BM noctuid slide 7403.

PARATYPES: 28 ex. New Caledonia, various localities; 17 ex. N. New Caledonia, Tinchialit 2020 ft 22.viii-10.ix. 1949 (L. E. CHEESMAN) BM 1950-1: 1♂ Mt. Koghis 500 m 26-30.i.1963 (C. M. YOSHIMOTO) BPBM; 1♂ La Coulée 5.1.76 (J. GUTIERREZ) MNHN; 1♀ Mt. Koghis 20.v.1971 (P. COCHEREAU) MNHN.

The species was recorded as follows during the survey; Grand Lac site 2 (1); Mt des Sources site 11 (1), site 12 (2), site 14 (3); R. Kuenthio (5); Dothio site 18 (1), site 19 (9); Poupa (3); Ciu (2); Koh (38); Ouatou R. (8); Bréhoa R. (24); Ouambéo R. (1); Col d'Amieu (4); Col Toma (1); Sarraméa (3); Kopéto (4); Mt. Dogny (1); Mt. Mou site 33 (1); Col de Boguen (1); Kaala-Gomen (1); Tiebaghi site 40 (1); Grottes de Koum (6); Bambou R. (12); Col d'Amoss site 45 (1), site 47 (2); Cap Colnett (2); Kavatch (1); R. Amoa (20); Ste. Therèse (3); Probokonoro (62); Col des Roussettes (5); Mt. Krapé site 55 (60), site 56 (2); Col de Boa (2); near Bâ (6); Mt. Aoupinié (2); Col Nassiboué (4); Ouen Toro site 64 (1), site 81 (5), site 82 (1); Port Boisé site 73 (2), site 74 (6); Les Dalmates site 77 (1); Mt. Koghis site 79 (3), site 80 (4). Localities on sedimentary rock with rainforest and rich secondary vegetation appear to be preferred.

### *Leucania loreyi* (Duponchel) (Plate 75: 1, Figures 90: 1, 91: 1)

*Leucania loreyi* (Duponchel); ROBINSON (1975: 116).

This species is widespread in the Indo-Australian tropics and has its eastern limits in Fiji and Tonga apart from an outlying population on Rapa I. It was recorded as follows in New Caledonia: Anse Longue site 16 (1); Kopéto (1); Mt. Mou site 34 (1); Col Nassiboué (1). These are all localities on the drier western side of the island. The species also flies in Lifu.

New Caledonian specimens are slightly smaller with darker forewings than those from Fiji.

### *Leucania loreyimima* Rungs (Plate 75: 2, Figures 90: 2, 91: 2)

*Leucania loreyimima* Rungs; HOLLOWAY (1977: 64).

This species, though recorded from scattered localities in the Oriental Region (Indochina, Sumatra; it is the only one of the related pair on the Ryukyu Is. (SUGI, 1970)), is commoner in northern Australia and New Guinea. It occurs on Norfolk I. but is only really abundant there in periods of abnormally dry weather (HOLLOWAY, 1977, and subsequent observations). In New Caledonia it was recorded mainly from the drier western side of the island as follows: Grand Lac site 2 (1); Anse Longue site 16 (2); Mt. Mou site 33 (3); R. Amoa (1); Ouen Toro site 82 (3). BMNH: Pouébo. MNHN; La Tamoa. There are records of the larva feeding on sorghum (ORSTOM).

*Leucania pseudoformosana* Robinson (Plate 75: 3, Figures 90: 4, 91: 5)

*Leucania pseudoformosana* Robinson (1975: 116).

New Caledonian specimens differ from typical Fijian ones in having slightly narrower forewings and more extensive brown suffusion on the hindwings. The species has been noted from Singapore, Borneo, New Guinea, the Solomons, New Hebrides and Fiji. It is part of a group of related Indo-Australian species that includes *formosana* (Butler) and *semicana* Pagenstecher, the former widespread in the Oriental Region and the latter known only from the islands east of New Guinea (ROBINSON). To these species must be added *uruma* (Sugi) from the Ryukyu Is.

The species was recorded as follows in New Caledonia: Mt. des Sources site 11 (1); R. Kuenthio (3); Dothio site 19 (9); Koh (20); Ouatou R. (4); Ouambéo R. (1); Col d'Amieu (8); Col Toma (10); Petit Gouli (6); Sarraméa (10); Kopéto (1); Mt. Mou site 33 (5); Col d'Amoss site 45 (3). site 47 (1); Cap Colnett (10); Kavatch (3); Amoa R. (12); Ste. Therèse (7); Probokonoro (53); Col des Roussettes (9); Mt. Krapé site 55 (3); Col de Boa (2); Houailou (1); near Bâ (16); Mt. Aoupinié (1); Col Nassiboué (1). BMNH: Pouébo, Tinchialit. MNHN: La Coulée. The species would appear from its distribution to have similar ecological requirements to its congener, *latericia*. The larva has been reared from sorghum (ORSTOM records).

*Leucania scottii* Butler (Figure 90: 6)

*Leucania scottii* Butler; ROBINSON (1975: 117).

The smallest New Caledonian *Leucania, scottii* has been recorded from northern Australia, New Guinea, the New Hebrides, Lifu and Fiji. It is a species of the west coast association in New Caledonia and was recorded as follows: Grand Lac site 1 (1), site 2 (1); Mt. des Sources site 11 (2); Anse Longue site 15 (7); site 16 (22), site 32 (50); R. Kuenthio (14); Poupa (6); Koh (8); Ouatou R. (1); Col d'Amieu (3); Col Toma (3); Petit Gouli (1); Sarraméa (4); Kopéto (1); Mt. Mou site 33 (3), site 34 (1); Kaala-Gomen (2); Tiebaghi site 41 (1); Bambou R. (1); Cap Colnett (1); Kavatch (2); R. Amoa (40); Ste. Therèse (10); Col des Roussettes (4); Mt. Krapé site 55 (9), site 56 (1); Houailou (1); near Bâ (2); Col Nassiboué (4); Ouen Toro site 64 (16), site 81 (4), site 82 (21); Port Boisé site 74 (3). BMNH: Pouébo, Tinchialit. MNHN: Mt. Koghis. BPBM: Nouméa, La Foa, Forêt de Thi, Le Crouen, Bourail, Népoui valley.

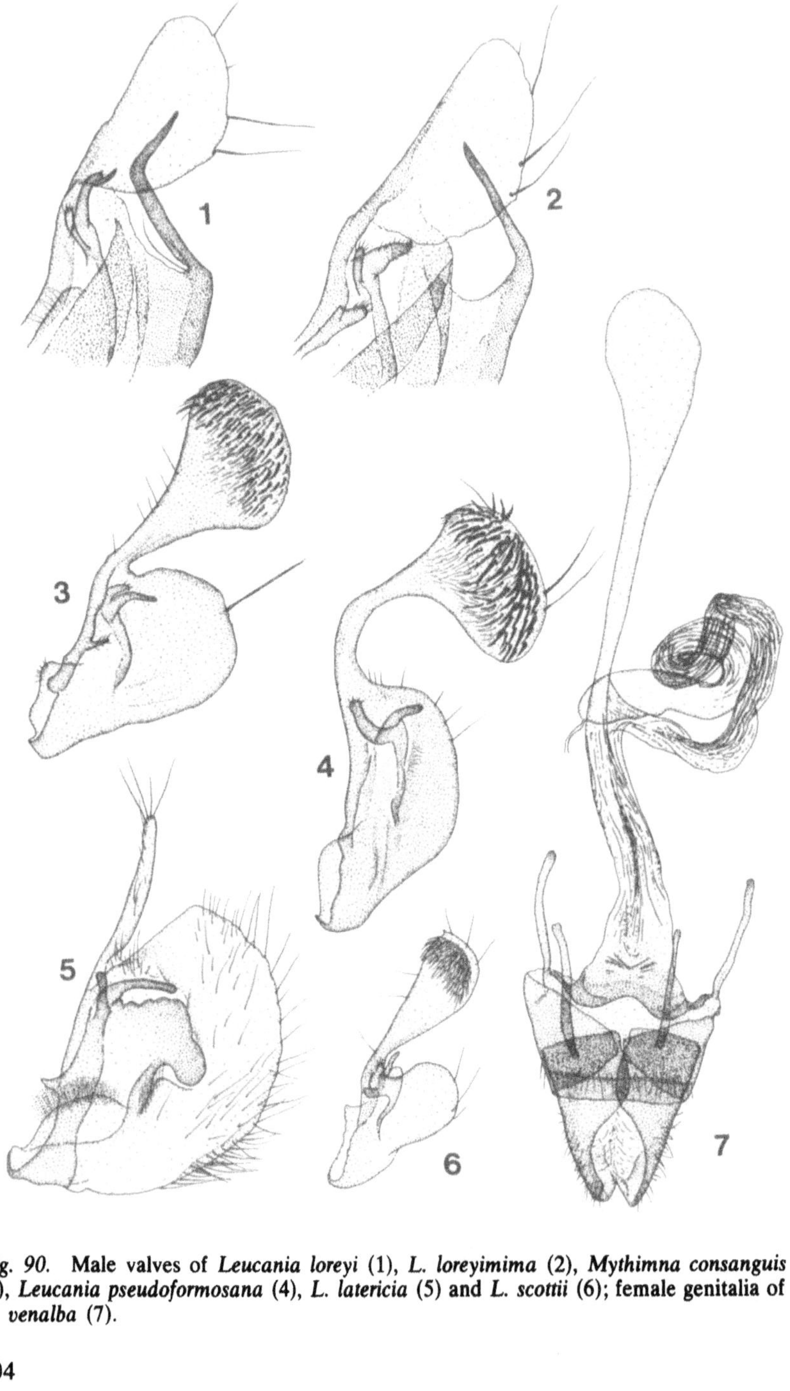


*Fig. 90.* Male valves of *Leucania loreyi* (1), *L. loreyimima* (2), *Mythimna consanguis* (3), *Leucania pseudoformosana* (4), *L. latericia* (5) and *L. scottii* (6); female genitalia of *L. venalba* (7).

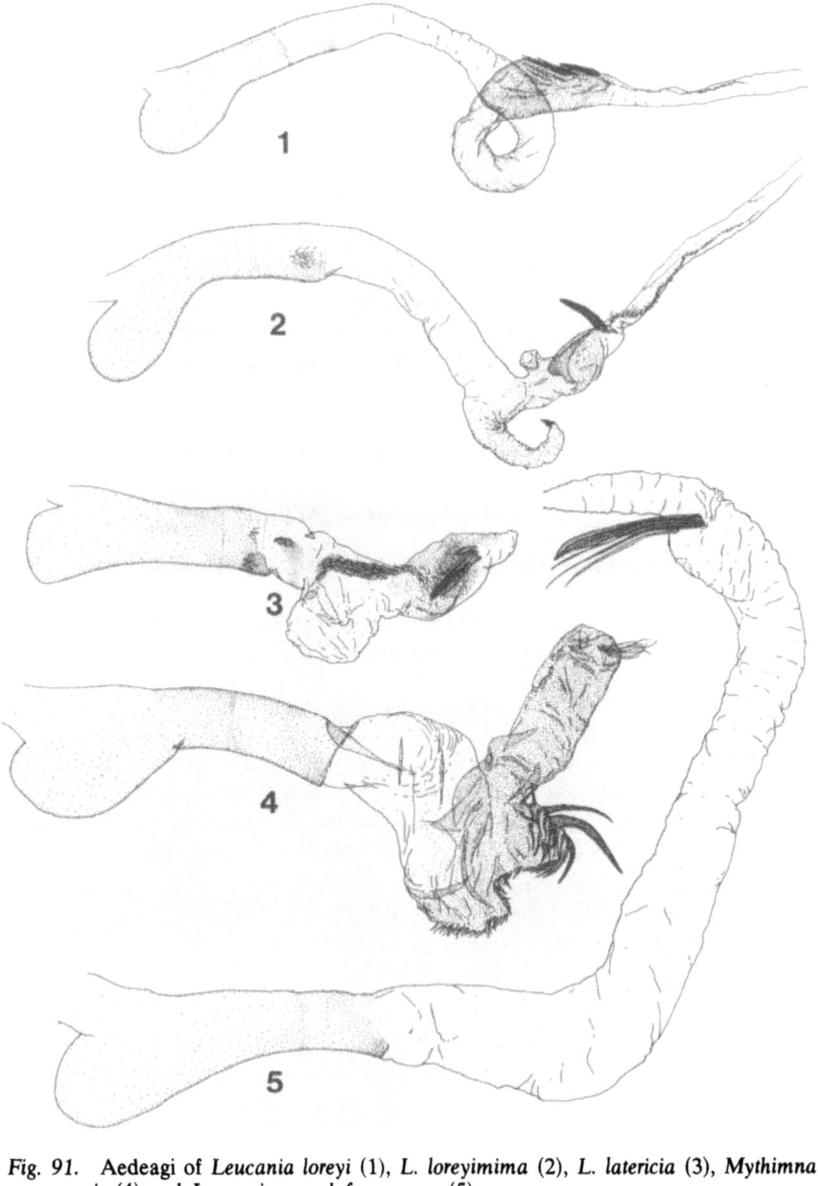


*Fig. 91.* Aedeagi of *Leucania loreyi* (1), *L. loreyimima* (2), *L. latericia* (3), *Mythimna consanguis* (4) and *Leucania pseudoformosana* (5).

*Leucania venalba* Moore (Figure 90: 7)

*Leucania venalba* Moore; ROBINSON (1975: 118).

This species is mainly Oriental but occurs in Queensland, Fiji and Tonga. Only two specimens were taken during the survey, at Koh and the R. Amoa site, but it has been recorded from Bourail (BPBM) and La Coulée (MNHN).

*Mythimna consanguis* (Guenée) (Plate 75: 6, Figures 90: 3, 91: 4)

*Hadena consanguis* Guenée, 1852, *Hist. nat. Insectes* (*Lépid.*), 6: 97.

This species is widespread in the Oriental Region to Sundaland but has only been recorded from the Kei Is., Australia (Northern Territory to New South Wales) and New Caledonia in Australasia. It was recorded as follows: Anse Longue site 32 (2); Mt. Mou site 36 (1); Ouen Toro site 64 (1). BPBM: Nouméa, Isle of Pines.

In the male genitalia the cucullus of the valve is slightly broader and the triple harpe slightly more developed than in an Indian male examined.

*Pseudaletia separata* (Walker)

*Pseudaletia separata* (Walker); ROBINSON (1975: 118); HOLLOWAY (1977: 65).

One female of this widespread Indo-Australian tropical species was taken on Mt. Koghis site 79. There is also a specimen in the MNHN from La Tamoa.

*TIRACOLA* Moore

Species of *Tiracola* show an interesting distribution pattern through the Papuan Region and south-west Pacific. *T. plagiata* (Walker) is widespread through the Indo-Australian tropics and Pacific, extending to Norfolk I., Henderson I. and the Marquesas. *T. rufimargo* Warren, the largest species, is uniformly pale reddish brown with a darker margin to the forewing. The typical race flies in New Guinea, with ssp. *concolor* Prout **stat. n.** (the genitalia are as in the typical race unlike those of the sympatric *versicolor* Prout which was originally ascribed to *rufimargo*) from Ceram and ssp. *samoensis* Tams from Samoa. In the male genitalia of *rufimargo* the cucullus has a ventral, subapical projection (much reduced in *plagiata*; Figure 92) and does not project much above the sacculus. The aedeagus vesica has two basal patches of fine cornuti, one on a lateral lobe, much as in *plagiata* though less well-developed. The subspecies are generally similar, *concolor* having the sacculus projection more pronounced and *samoensis* having a broader cucullus that is produced dorsally and has its coronal setae divided into a dorsal group of moderate-sized ones and a smaller, ventral group of stouter ones.

The second Ceram species, *T. versicolor* Prout **stat. n.** is slightly smaller than *rufimargo* with a more variegated forewing. It is distinguished mainly in the genitalia where the aedeagus vesica has no lateral lobe and the more basal group of cornuti is reduced compared with *rufimargo* and *plagiata*.

Distinct species occur in the New Hebrides (possibly two) and New Caledonia and are described below; further work needs to be done on the genus in the Bismarcks and Solomons.

**Tiracola circularis sp. n.** (Plate 75: 10, Figure 92: 2, 4)

♂♀ 22–28 mm. The species is very variable but the forewings and thorax are generally some shade of reddish brown or ochre (as distinct from the greyish brown of *plagiata*), extremes being fawn and almost black, the paler forms often being suffused dark brown broadly over the forewing disc and darker forms sometimes having the reniform and orbicular picked out in ochre or white. In a few examples the forewing is mottled pale grey. The most reliable character to distinguish this species from *plagiata* is the postmedial on the underside of both wings which is curved markedly concave basad, parallel to the margin on the hindwing; in *plagiata* the postmedial is straight or slightly sinuous on the forewing and only slightly curved on the hindwing. Other differences are that the forewing lacks paler prominence of the veins distally, the reniform is not so often dark and fused with the medial to produce a dark triangular area on the costa, the postmedial is straight and oblique at the costa, often prominently pale (usually indistinct in *plagiata*), and the hindwing is a warmer almost reddish grey, lacking the marginal black triangles in the spaces of *plagiata*. The abdomen is generally a much more ochreous or fawn grey.

The male genitalia are relatively much smaller than in *plagiata* and are distinguished by a well-developed ventral projection to the cucullus and an aedeagus vesica lacking a lateral lobe and having one much larger cornutus amid a cluster of finer ones in the only group of cornuti.

HOLOTYPE ♂. New Caledonia: Mt. Dogny, 950 m Site 31 05902/76088 (J. D. HOLLOWAY) 27.6.1971, BM 1971-507. BM noctuid slide 7402.

PARATYPES: 39 ex. of both sexes, New Caledonia, localities as follows: Grand Lac site 2 (2), site 3 (1); Pic du Pin Site 7 (3); R. Bleue site 10 (1); Mt. des Sources site 11 (1); Koh (1); Bréhoa R. (3); Mt. Dogny (8); Mt. Mou site 34 (1), site 35 (2), site 36 (1); Col d'Amoss site 47 (1); Kavatch (1); near Bâ (1); Les Dalmates site 77 (3); Mt. Koghis site 79 (1), site 80 (1); 1♀ Mt. Koghis, Forêt de Thi 530 m 8.iii.61 (J. SEDLACEK) BPBM; 1♀ La Coulée 5.1.76 (J. GUTIERREZ) MNHN.

The species was not distinguished from *plagiata* until after field counts had been made but it appears to be more characteristic of ultramafic maquis localities.

*Tiracola plagiata* (Walker) (Figure 92: 1, 3)

*Tiracola plagiata* (Walker); ROBINSON (1975: 119); HOLLOWAY (1977: 66).

This species was taken abundantly during the survey but because of confusion with the previous species its precise distribution was not ascertained. It is certainly widespread and probably associated with areas of rainforest and secondary growth on sedimentary rock. A New Caledonian host plant is *Erythrina* (Leguminosae) according to ORSTOM records.

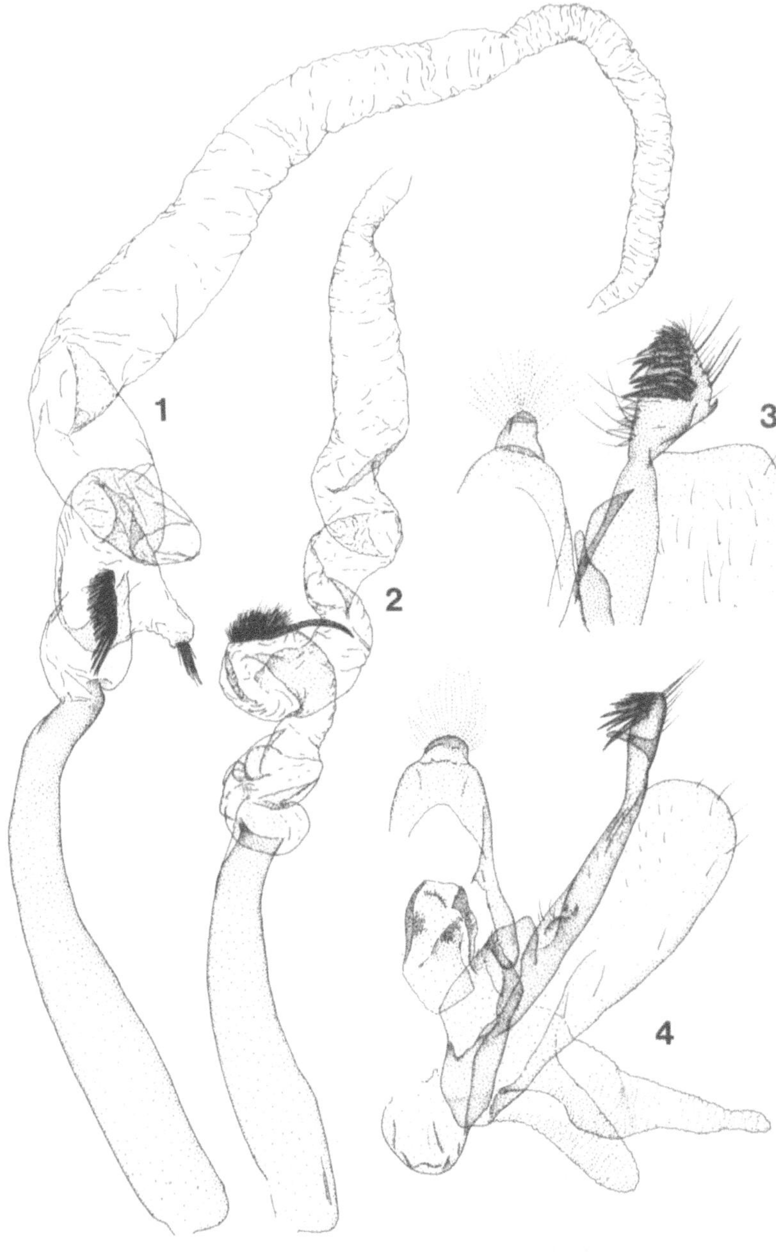


*Fig. 92.* Aedeagi and male genitalia of *Tiracola plagiata* (1, 3) and *T. circularis* (2, 4).

**Tiracola tabwemasana sp. n.** (Plate 75: 9)

♂♂ 27–28 mm. The forewings and most of the thorax are uniformly deep reddish brown apart from the patagia, the forewing costa broadly over the basal third and, slightly, the reniform and a thin submarginal which are paler, more ochreous. The hindwing is a uniform medium grey with no black triangles in the spaces. The male genitalia have the uncus very reduced, the cucullus very broad, not deeply divided from the sacculus, lacking a ventral projection and having two groups of stout setae well separated. The vesica of the aedeagus has a basal lobe that is unspined, a basal group of cornuti and another group distad; the latter is distinctive, the cornuti being long, slender and forming a fan when the vesica is everted.

HOLOTYPE ♂. New Hebrides: Santo, Mt. Tabwemasana, Ridge-4500 ft 2.ix.1971 (G. S. ROBINSON) Roy. Soc. Expedition.

PARATYPES. 4♂♂ New Hebrides: data as holotype. BM noctuid slide 7595.

A female specimen (New Hebrides: Efate I., Marabut Camp, 30.vi–7.vii.1971 (G. S. ROBINSON)), 29 mm, with the forewings and thorax very dark brown, the former with some crisp black fasciation and mottled pale ochre basally (Plate 75: 8), may be of a distinct species.

NOCTUIDAE: ACRONICTINAE

*Acrapex exsanguis* Lower (Plate 76: 8, Figure 93: 2)

*Acrapex exsanguis* Lower, 1902, *Proc. Linn. Soc. N. S. Wales*, 26: 655.

The New Caledonian specimens listed below agree in appearance and male genitalia with the series from Queensland in the BMNH attributed to *exsanguis*. The species also occurs in New South Wales. New Caledonian records were as follows: Koh (1); Kopéto (3).

**Acrapex pacifica sp. n.** (Plate 76 9, Figure 94: 3)

♂ 10 mm. The species is smaller than *exsanguis*, less fawn in colour, the forewings with a bone white ground marked in pinkish grey as illustrated. The posterior margin of the central pale streak curves anteriorly to the apex, distinguishing the species from *exsanguis* and indicating its relation to the darker *leucophlebia* Hampson (India) and the brown, more ochreous *azumai* Sugi (1970; Japan, Ryukyus). There is a postmedial row of elongate black flecks on the veins extending from the dorsum to two thirds anteriorly. The hindwings are pale grey, grading paler basad.

The male genitalia are as illustrated, the valve and its processes being much narrower than that of *leucophloebia*. The aedeagus vesica has a single long cornutus whereas that of *leucophloebia* has two short ones. The genitalia of *azumai* are very similar to those of *pacifica* but in the former there is a small projection from the ventral curvature of the downturned distal third of the valve.

HOLOTYPE ♂. New Caledonia: Col des Roussettes, 340 m Site 54 05467/76304 (J. D. HOLLOWAY) 21.7.1971, BM 1971-507. Bm noctuid slide 9519.

A distinct subspecies flies in the New Hebrides.

*Fig. 93.* Male genitalia of *Leucogonia amarginata* (1) and *Acrapex exsanguis* (2).

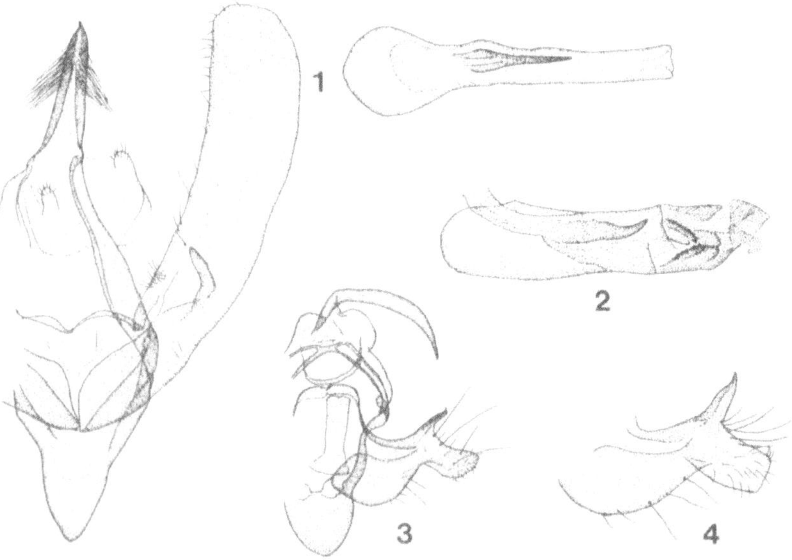


*Fig. 94.* Male genitalia of *Nola major* (1), *Acrapex pacifica pacifica* (3) and *A. p. santonis* (2, 4)

**Acrapex pacifica santonis ssp. n.** (Figure 94: 2, 4)

♂♂ 12–13 mm. This subspecies is slightly larger than the typical race though this may not hold when more material is available. The wings are generally a creamer tint. The genitalia are one third as large again with both the valve apex and the costal spine somewhat broader; the cornutus of the aedeagus is relatively larger.

HOLOTYPE ♂. New Hebrides: Santo, Mt. Tabwemasana, Nokowula-3700 ft 1.ix.1971 (G. S. ROBINSON) Roy. Soc. Expedition. BM noctuid slide 8714.

PARATYPE ♂. Data as holotype.

**Athetis postpuncta sp. n.** (Plate 76: 6, Figure 95; 4)

♂♀ 14–16 mm. The forewings, thorax and abdomen are variable, medium pinkish brown-grey as illustrated; the hindwings are brownish white with darker costal area, veins and discal spot, and a postmedial row of dark brown spots on the veins parallel to the margin, distinguishing this species from other Indo-Australian *Athetis*.

The closest resemblance of the facies is to the female type of ab. *notanda* Warren (1913, *Gross-Schmett. Erde*, 11), referred to the Papuan *insignifica* Bethune-Baker. The type locality is Bowen, Queensland. A similar specimen with similar female genitalia to *notanda* and also from Queensland is referred to *maculatra* (Lower) (1902, *Proc. Linn. Soc. N. S. Wales*, 26: 651). The resemblance between *postpuncta* and these two specimens is close in the forewing but the latter lack the hindwing postmedial. Their genitalia differ from those of *postpuncta* in having a long, narrow bursa instead of a rounded one, and a small, basal, posteriorly directed appendix bursae instead of a broad and prominent one closely adpressed and directed anteriorly. The male valves of *postpuncta* are of typical *Athetis* type but unusually narrow.

HOLOTYPE ♂. New Caledonia: Anse Longue, 5 m Site 32 06218/75543 (J. D. HOLLOWAY) 30.6.1971, BM 1971-507. BM noctuid slide 7480.

PARATYPES: 58 ex. New Caledonia, various localities; 13 ex. N. New Caledonia, Tinchialit 2020 ft 22.viii–10.ix.1949 (L. E. CHEESMAN) BM 1950–1; 1♀ La Coulée 4.12.75 (J. GUTIERREZ) MNHN.

The species was recorded as follows during the survey: Grand Lac site 1 (1), site 2 (1); Mt. des Sources site 11 (3), site 12 (3), site 14 (5); R. Kuenthio (1); Dothio site 18 (1), site 19 (2); Poupa (1); Koh (2); Ouatou R. (2); Bréhoa R. (2); Sarraméa (2); Mt. Dogny (2); Anse Longue site 32 (3); Mt. Mou site 33 (11), site 34 (2), site 36 (1); Tiebaghi site 40 (1), site 41 (2); Grottes de Koum (2); Bambou R. (3); Col d'Amoss site 45 (3), site 46 (2), site 47 (3); Kavatch (2); R. Amoa (2); Ste. Therèse (3); Probokonoro (5); Mt. Krapé site 55 (5), site 56 (1); Mt. Panié site 60 (2); Les Dalmates site 77 (4); Mt. Koghis site 80 (1). No particular habitat preference is indicated.

*Athetis reclusa* (Walker) (Figure 95: 3)

*Athetis reclusa* (Walker); ROBINSON (1975: 121).

Four specimens of this widespread species (Sundaland to Queensland, New South Wales and the New Hebrides) were taken in New Caledonia as follows: Anse Longue; Probokonoro; Ouen Toro site 64; Mt. Koghis site 80. Other material from New Caledonia consists of four specimens from Tipindjé (BPBM) and one reared from carrot at La Coulée (MNHN).

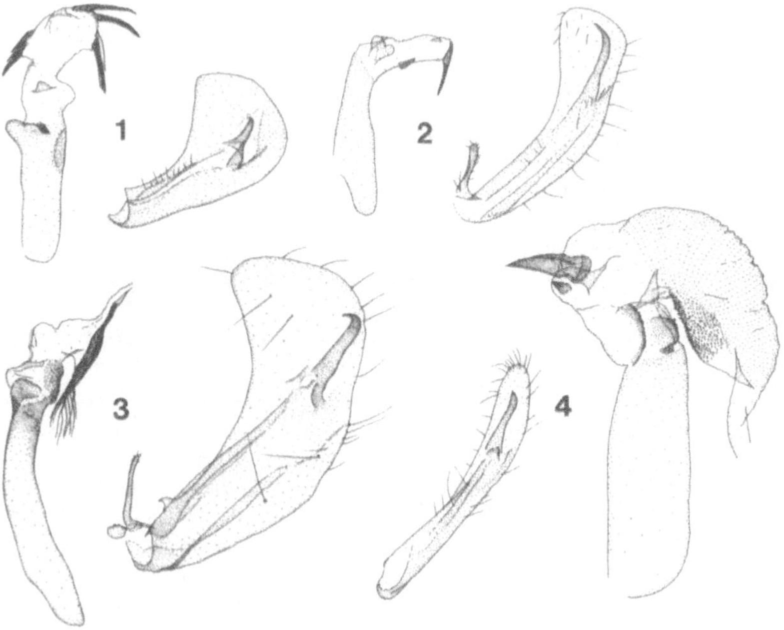


*Fig. 95.* Aedeagi and male genitalia of *Athetis* species: *tenuis* (1), *striolata* (2), *reclusa* (3) and *postpuncta* (4).

### *Athetis striolata* (Butler) (Figure 95: 2)

*Athetis striolata* (Butler); Robinson (1975: 121).

New Caledonian specimens differ slightly from typical Fijian *striolata* in the male genitalia: the harpe of the valve is less sinuous and the dorsal projection from the base of the valve (ampulla) is shorter and thicker. The species occurs in New Guinea, Queensland, Fiji, Samoa and Tonga. It is a member of the west coast/*Acacia spirorbis* association in New Caledonia and was recorded as follows: Grand Lac site 4 (1); Pic du Pin site 7 (1); Mt. des Sources site 11 (4); R. Kuenthio (13); Dothio site 18 (1); Koh (2); Bréhoa R. (1); Cap Colnett (1); Kavatch (3); Probokonoro (1); Col Nassiboué (5); Ouen Toro site 64 (2), site 82 (5); Les Dalmates site 78 (1).

### *Athetis tenuis* (Butler) (Figure 95: 1)

*Athetis tenuis* (Butler); Holloway (1977: 67).

*A. tenuis* is found throughout Australia and is resident on Norfolk I. where it is associated with secondary and cultivated habitats. It is vagrant to both Norfolk I. and New Zealand. A single specimen was taken in New Caledonia at Les Dalmates site 77. This could have been a vagrant individual as it was taken with two other potential vagrant species as discussed under *Agrotis ipsilon* (p. 399).

**Callopistria dimorpha sp. n.** (Plate 76: 10, 11, Figure 96: 2)

♂♀ 15–16 mm. The male is marked with rich brown on a fawn ground and the female is a more uniform dark brown with an ochreous tinge to the pale discal area of the forewing and in the band distad from the postmedial; the antemedial forms a whitish patch subdorsally in both sexes. The forewing fasciation is much more crenulate than in most *Callopistria*, for example *maillardi* (q.v.) to which the species appears related from the male genitalia. No other described species resembles it closely in facies, though G. S. ROBINSON has taken species in both the New Hebrides and Samoa that are similar, though darker and not exhibiting sexual dimorphism. This dimorphism is unusual in the genus though it is also exhibited by the other endemic New Caledonian species, *falcata*, described below.

The male genitalia are on a similar pattern to those of *maillardi* but the basal two thirds of the valve are very much broader and the aedeagus vesica lacks the ring of cornuti basal to one of the lateral lobes exhibited by *maillardi* but has larger cornuti apically.

HOLOTYPE ♂. New Caledonia: Kavatch, 40 m Site 50 04857/77068 (J. D. HOLLOWAY) 18.7.1971, BM 1971–507.

PARATYPES: 17♂♂, 17♀♀ New Caledonia, various localities, BM noctuid slide 7482 (♂); 1♀ Mt. Koghis 500 m i.1963 (N. L. H. KRAUSS) BPBM; 1♀ La Coulée 6.1.75 (J. GUTIERREZ) MNHN; 1♂, 1♀ Col d'Amieu 21.7.77 (A. DELOBEL) MNHN.

The species, perhaps more characteristic of rainforest than *maillardi*, was recorded as follows: Grand Lac site 3 (4); Pic du Pin site 7 (4); R. Bleue site 8 (1); Mt. des Sources site 14 (1); Dothio site 19 (1); Poupa (1); Koh (8); Ouatou R. (2); Col d'Amieu (2); Col Toma (9); Sarraméa (9); Kopéto (4); Mt. Mou site 33 (3); Bambou R. (3); Kavatch (4); Amoa R. (1); Ste. Therèse (1); Probokonoro (39); Col des Roussettes (2); Mt. Krapé site 55 (9), site 56 (1); Mt. Aoupinié (14); Port Boisé site 74 (1).

There is a female in the BPBM from Fayaoué, Ouvéa, Loyalty Is.

**Callopistria falcata sp. n.** (Plate 76: 12, 13, Figure 96: 3)

♂♀ 12–13 mm. The male ground colour is straw, marked with rich brown that is more rufous over the forewing disc; the hindwing markings are greyer brown, as is the dark mark at the centre of the forewing margin. The female is a more uniform pinkish brown, lacking the paler basal rhomboid of the male forewing; the hindwing is also uniform pale grey-brown, not fasciated as in the male. The

*Plate 76.* 1. *Platysenta praesecta* (21 mm); 2. *Eutamsia quadrispina;* 3. *Hypoperigia lunulata;* 4. *Chasmina basiflava* ♂; 5. *Molvena caledonica;* 6. *Athetis postpuncta;* 7. *Craniophora fasciata* (18 mm); 8. *Acrapex exsanguis* (11 mm); 9. *Acrapex pacifica pacifica;* 10 (♂), 11 (♀). *Callopistria dimorpha;* 12 (♂), 13 (♀). *Callopistria falcata.*

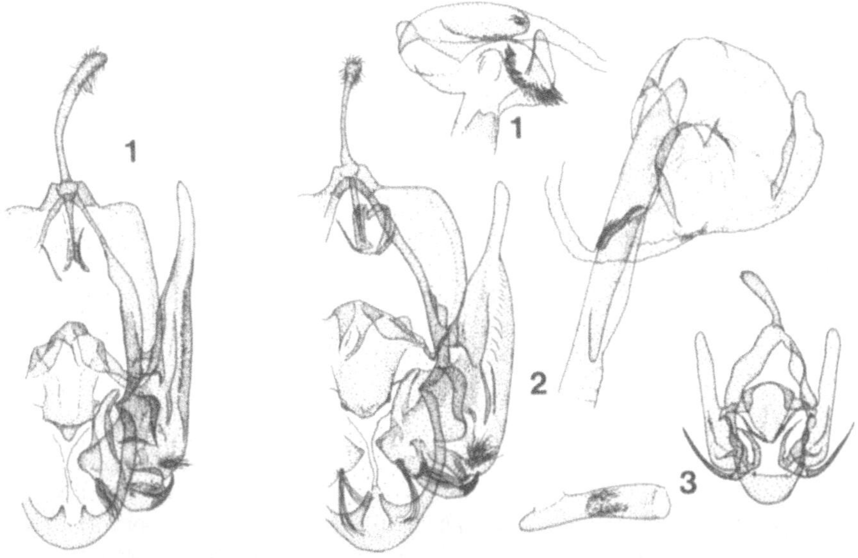


*Fig. 96.* Male genitalia and aedeagi of *Callopistria* species: *maillardi* (1), *dimorpha* (2) and *falcata* (3).

species is distinguished from its congeners by prominent falcation in both sexes submarginally on the dorsum, produced by elongation of the fringe scales. The legs of the male are straw colour, irregularly tufted, except the tarsi which are dark brown. The male genitalia are relatively simple compared with those of other species, except for a slender, curved sclerotised spine arising basally and ventrally from the valve. The elongation of this relative to that of *maillardi* is reminiscent of the Oriental species, *ventralis* Walker.
HOLOTYPE ♂. New Caledonia: Col Toma, 460 m Site 27 05813/76123 (J. D. HOLLOWAY) 19.6.1971, BM 1971-507. BM noctuid slide 7483.
PARATYPES: 1♂ Mt. Aoupinié; 1♀ Ouatou R.

*Callopistria maillardi* (Guenée) (Figure 96: 1)

*Eriopus maillardi* Guenée, 1862, *Maillard's Reunion Lep.*: 39.
*Callopistria recurvata* Moore, 1882, *Descr. new Indian lepid. Insects Colln Atkinson:* 144, type examined.
*Callopistria meridionalis nauticorum* Tams, 1935, *Ins. Samoa* 3: 199, type examined, **syn. n.** ROBINSON (1975: 122).

*C. maillardi* is widespread in the Indo-Australian tropics and the Pacific. It occurs on islands in the Indian Ocean and is represented by a distinct subspecies, *meridionalis* Collenette, in the Austral Is. and on Rapa. This subspecies is distinguished from the typical race by a marked subdorsal indentation basad to the forewing postmedial. Dissection of males throughout the range of *maillardi* and of those from the type locality and neighbouring areas of *nauticorum* revealed no differences to justify the segregation of the latter.

New Caledonian records were made as follows: Grand Lac site 2 (9), site 3 (11), site 4 (5); Pic du Pin site 7 (7); R. Bleue site 10 (1); Mt. des Sources site 11 (2), site 14 (5); Dothio site 19 (3); Poupa (3); Koh (57); Ouatou R. (3); Bréhoa R. (10); Col d'Amieu (9); Petit Gouli (2); Sarraméa (9); Mt. Mou site 33 (7); Bambou R. (6); Cap Colnett (8); Kavatch (1); R. Amoa (12); Ste. Therèse (2); Probokonoro (24); Col des Roussettes (1); Mt. Krapé site 55 (1); near Bâ (4). This species showed a marked decline in abundance through the survey period. Early records indicated it to prefer open habitats, particularly on sedimentary rock though not in west coast habitats.

*Calogramma picta* (Guérin-Méneville)

*Polia picta* Guérin-Méneville, 1838, *Voyage de Coquille* (Zool.) 2: 285.
*Calogramma festiva* (Donovan); ROBINSON (1975: 124). The former is the replacement name for the latter, which is a junior homonym (NYE, 1975: 98).

The species is distributed throughout the Indo-Australian tropics to New South Wales, Lifu, Fiji, Samoa and Tonga. In New Caledonia it was recorded from the following localities: Bréhoa R. (1); Ouambéo R. (1); Probokonoro (2). VIETTE (1949): Nouméa. COHIC (1956) listed *Crinum asiaticum* and *C. pedonculatum* as larval host plants in New Caledonia.

**Chasmina basiflava sp. n.** (Plate 76: 4, Figure 97: 1)

♂♀ 19–21 mm. The insect is mostly immaculate white but for brown eyes, antennae and a bar between the antennae and spots on the ochreous fore legs; the palps are also ochreous yellow. In the male there is a diagnostic thick tuft of long, pale yellow scales arising from the base of the hindwing above and passing forward and outward under the forewing. Some other species (e.g. *candida* (Walker)) have prominent coremata associated with the male genitalia (ROBINSON, 1975) but these do not occur in *basiflava*. A further distinctive characteristic is the angling, in the male, of the forewing costa two-thirds from the base; the forewings are characteristically rounded marginally.

The male genitalia are distinguished by the upturned tip to the uncus, the relative shortness of the valves and the absence of the row of cornuti in the vesica which are present in both of the other New Caledonian species.
HOLOTYPE ♂. New Caledonia: Pic du Pin, 240 m Site 7L 06879/75383 (J. D. HOLLOWAY) 8.8.1971, BM 1971-507. BM noctuid slide 7491.
PARATYPES. 15♂♂, 16♀♀ New Caledonia, various localities, BM noctuid slide 7504 (♂).

The species is a member of the ultramafic association and was recorded as follows: Grand Lac site 2 (15), site 3 (2); Pic du Pin site 7 (4); R. Bleue site 10 (1); Mt. des Sources site 11 (1), site 14 (1); Tiebaghi site 41 (1); Probokonoro (3); Mt. Krapé site 55 (12), site 56 (1); near Bâ (2); Grand Lac site 72 (1); R. Koué (1); Mt. Koghis site 80 (1).

*Chasmina candida* (Walker) (Figure 97: 2)

*Chasmina candida* (Walker); ROBINSON (1975: 125). VIETTE (1949: 329, as *sericea* Hampson).

The species is extensive in the Oriental Region and has been recorded from Sulawesi, New Guinea, the Solomons, Marianas, New Hebrides, Fiji, Rotuma, Samoa and Tonga. Apart from VIETTE's record the species has been taken at Thio (BPBM) and Nouméa (MNHN). It was not recorded during the survey.

*Chasmina tibialis* (Fabricius) (Figure 97: 3, 4)

*Chasmina tibialis* (Fabricius); VIETTE (1949: 329); ROBINSON (1975: 126).

*C. tibialis* is widespread in the Indo-Australian tropics, extending to Australia (Northern Territory to New South Wales), the Marquesas, Rapa and Pitcairn Is. In New Caledonia it was recorded as follows: Grand Lac site 2 (1); Koh (1); Anse Longue site 32 (1); Mt. Ignambi (1); R. Amoa (1). VIETTE: Nouméa, Canala. BPBM: Isle of Pines.

*Craniophora fasciata* (Moore) (Plate 76: 7, Figure 98: 2)

*Hyboma fasciata* Moore, 1884, *Lepid. Ceylon*, 3: 5.

The cornuti of the male vesica are perhaps one quarter of the length of those in a Bornean (Mt. Kinabalu) male dissected, but there are no other differences in the genitalia. The species is recorded from India, the Himalaya, Sundaland, the Lesser Sundas, Sulawesi, New Guinea, Queensland, New South Wales and the Bismarcks. In New Caledonia it is a member of the ultramafic association and was recorded as follows: Grand Lac site 2 (2), site 3 (7); Pic du Pin site 7 (10); Mt. des Sources site 14 (1); Dothio site 19 (1); Poupa (1); Bréhoa R. (1); Col d'Amieu (2); Col Toma (1); Kopéto (1); Tiebaghi site 40 (1), site 41 (3); Bambou R. (1); Col d'Amoss site 47 (1); Probokonoro (4); Mt. Krapé site 55 (7); Col de Boa (1); near Bâ (1); Col Mouirange (1); Port Boisé site 74 (1); Les Dalmates site 77 (2), site 78 (1).

**Eutamsia quadrispina sp. n.** (Plate 76: 2, Figure 98: 3)

♂♀ 18–19 mm. The wings and thorax are deep, rich brown as illustrated; two out of six specimens had the reniform picked out with white. This is one of a complex of Indo-Australian species that is reliably separated only by reference to the genitalia. As yet it has been poorly worked. The species are usually referred to *consummata* (Walker), described from South Africa and probably absent from the Oriental Region. *E. quadrispina* is distinguished by having only four stout setae to the valve corona (arranged in a row), a relatively narrow valve, and by lacking any significant spining on a sclerotised band extending from the aedeagus apex into the vesica; the apical half of the vesica is distinctively scobinate and has an apical area of much stouter spines. *E. consummata* ab. *fusca* (Strand) (Nilgiri Hills, type examined) also has four

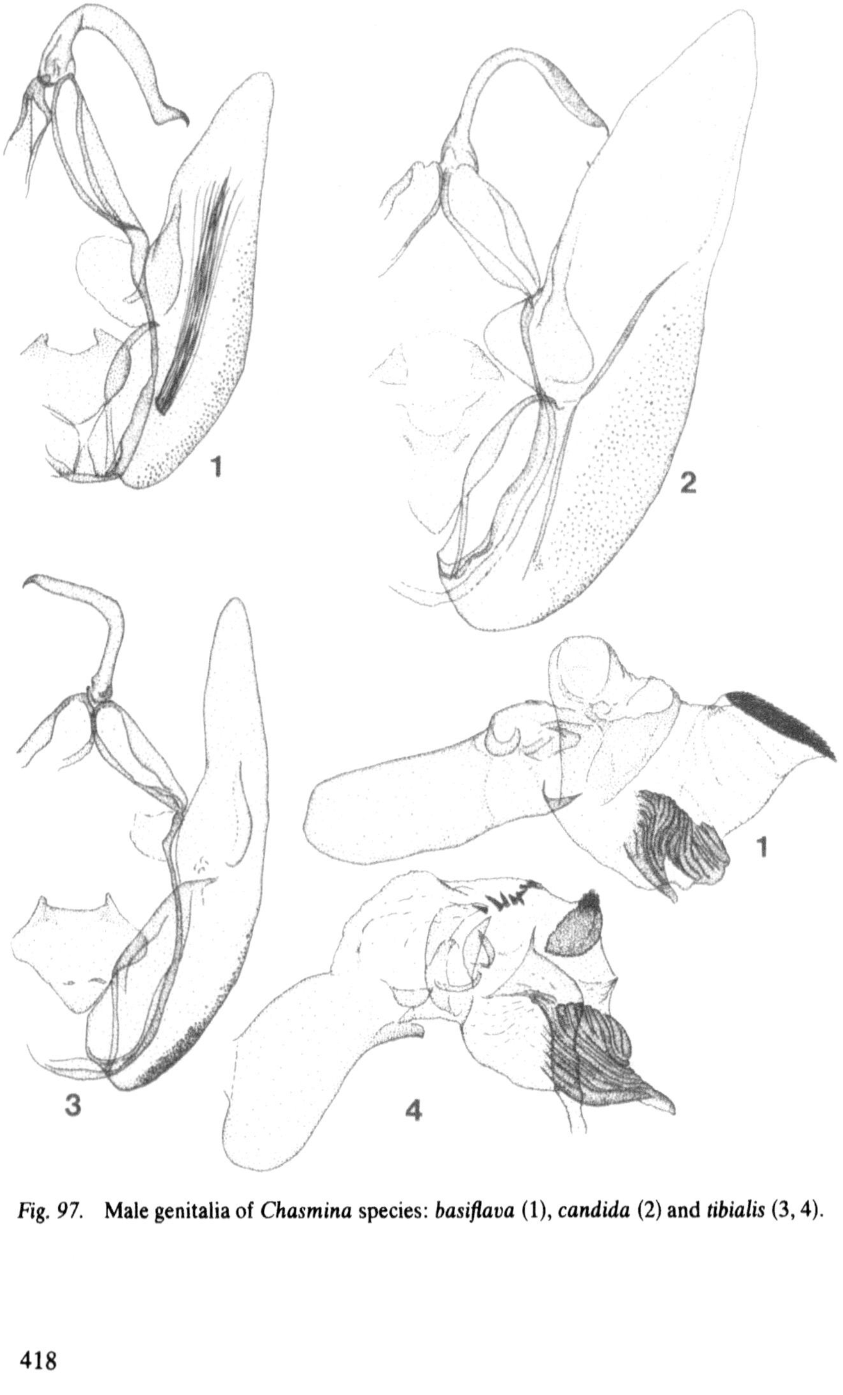


*Fig. 97.* Male genitalia of *Chasmina* species: *basiflava* (1), *candida* (2) and *tibialis* (3, 4).

*Fig. 98.* Aedeagus of ***Sasunaga tenebrosa*** (1), aedeagus and male genitalia of ***Craniophora fasciata*** (2), aedeagus and valve of ***Eutamsia quadrispina*** (3), aedeagus and male genitalia of ***Molvena caledonica*** (4).

setae to the corona but these are arranged in a rhomboidal pattern and the sclerotised band in the vesica is heavily spined. The female bursa copulatrix is elongate, finely scobinate, broadening to an asymmetrical apex where it is produced, corrugated and more thickly scobinate on one side only.

This species is probably endemic to New Caledonia. There are distinct species in Queensland and the New Hebrides (females from the Bismarcks and Solomons may be associated with the latter). They have prominent banding in the aedeagus vesica and there are numerous stout setae on the valve corona.

HOLOTYPE ♂. New Caledonia: Grottes de Koum, 20 m Site 43 04314/77286 (J. D. HOLLOWAY) 11.7.1971, BM 1971-507. BM noctuid slide 8745.

PARATYPES. 2♂♂, 3♀♀ New Caledonia: Bréhoa R. (2); Ouambéo R.; Col d'Amieu; Mt. Dogny. BM noctuid slides 7489 (♂), 8746, 8747 (♀♀).

**Hypoperigea lunulata sp. n.** (Plate 76: 3, Figure 99: 2)

♂♀ 13–15 mm. The species is varying shades of brownish grey as illustrated, some specimens having the pale areas of the forewing suffused rufous. It is very close to, and probably derived from, *tonsa* Guenée, an Australian species that is vagrant to Norfolk I. (HOLLOWAY, 1977: 67). It is greyer, less brown than *tonsa*, with more contrast in the forewing markings, the postmedial closer to the antemedial at the dorsum, both more markedly lunulate in spaces 1a and 1b.

In the male genitalia the valve is narrower apically and the harpe simple (lacking the obtuse lateral spine of *tonsa*); the aedeagus vesica resembles that of *tonsa* in having a basal collar of small cornuti on the left half and four large ones at the apex, but in the central portion of the vesica the cornuti are more symmetrically arranged and more even in size (in *tonsa* there is a central row of medium ones and approximately four rows of smaller ones on the left side only).

HOLOTYPE ♂. New Caledonia: Dothio, 90 m Site 18 06176/76128 (J. D. HOLLOWAY) 7.6.1971, BM 1971-507. BM noctuid slide 7484.

PARATYPES. 40 ex. New Caledonia, various localities.

The species was recorded as follows in New Caledonia: Grand Lac site 2 (8), site 3 (3), site 4 (3); Pic de Pin site 7 (7); Mt. des Sources site 14 (1); R. Kuenthio (1); Dothio site 18 (3), site 19 (1); Koh (2); Sarraméa (1); Kopéto (1); Mt. Mou site 33 (1); Kaala-Gomen (3); Tiebaghi site 41 (1); Grottes de Koum (2); Col d'Amoss site 47 (2); Cap Colnett (6); R. Amoa (3); Mt. Krapé site 56 (1); Col de Boa (1); Houailou (1); near Bâ (8); Col Mouirange (1); Anna-Madeleine (1); Les Dalmates site 77 (5). The species was assigned to the ultramafic association in the numerical analysis though it is by no means restricted to such localities. Some specimens may have been assessed as *Platysenta illecta* (Walker) in field counts.

*Leucocosmia nonagrica* (Walker)

*Leucocosmia nonagrica* (Walker); VIETTE (1949: 329); ROBINSON (1975: 128); HOLLOWAY (1977: 68).

*L. nonagrica* is widespread from India and S.E. Asia to Australia and has been recorded from the Pacific as far east as Rapa I. VIETTE recorded the species only from the Loyalty Is. New Caledonian records were as follows: Grand Lac site 1 (2), site 2 (12), site 3 (19), site 4 (1), site 72 (1); Plaine des Lacs site 5 (2), site 6 (2); Pic du Pin site 7 (4); Mt. des Sources site 11 (15), site 12 (2), site 13 (1), site 14 (74); Anse Longue site 15 (2), site 16 (6), site 32 (18); R. Kuenthio (3); Dothio site 18 (8), site 19 (24); Poupa (6); Koh (170); Ouatou R. (18); Bréhoa R. (52); Ouambéo R. (3); Col d'Amieu (39); Col Toma (7); Petit Gouli (3); Sarraméa (23); Kopéto (9); Mt. Dogny (4); Mt. Mou site 33 (16), site 34 (19); Kaala-Gomen (9); Tiebaghi site 40 (2), site 41 (3), site 42 (2); Grottes de Koum (14); Bambou R. (4); Col d'Amoss site 45 (19), site 46 (22), site 47 (26); Mt. Ignambi (5); Cap Colnett (4); Kavatch (11); Amoa R. (43); Ste. Therèse (17); Probokonoro (816); Col des Roussettes (11); Mt. Krapé site 55 (307), site 56 (19); Col de Boa (4); Houailou (1); near Bâ (41); Mt. Panié site 60 (1); Mt. Aoupinié (6); Col Nassiboué (12); Ouen Toro site 64 (1); Col Mouirange (1); R. Fausse Yaté (1); Yaté Gorge (1); Mt. Koghis site 80 (4). BMNH: Tinchialit. BPBM: Nouméa, La Foa, Le Crouen, Tipindjé, Mt. Koghis. It is an extremely widespread species, perhaps more abundant in areas of rainforest.

**Leucogonia amarginata sp. n.** (Plate 75: 7, Figure 93: 1)

♂♂ 15 mm, ♀♀ 16 mm. The head and the centre of the thorax are fawn, the tegulae ash grey and the abdomen pale, dull ochreous orange with brown dorsally and basally that tapers away distad. The forewings are dull ochreous orange with dorsal and submarginal markings that are dark brown suffused with bluish grey. The costa is finely edged white. The hindwings are dull yellow with a broad dark brown margin that extends up the dorsum and is marked pale grey in the anal angle.

The species differs from its two congeners, *ekeikei* Bethune-Baker (New Guinea, Queensland) and *kebeae* Bethune-Baker (New Guinea), in that the forewings have the dorsal markings more irregular, only extending to the postmedial, and the broad, uniform marginal band of the New Guinea species is absent. *Metaxanthiella cosmopis* (Lower), from Queensland, is also related, the genus monotypic.

The male genitalia of all four species are similar, with valves long, narrow, with a corona of setae and a simple, curved harpe one-third from the base. In *amarginata* the aedeagus is shorter and broader than in the New Guinea species, the valve apex less tapered and incurved, and the harpe slightly shorter and stouter. *M. cosmopis* has narrower valves that are not incurved, and the harpe is stout as in *amarginata* but angled two-thirds distad instead of smoothly curved.

HOLOTYPE ♀. New Caledonia: Port Boisé, 20 m Site 74 m 07039/75276 (J. D. HOLLOWAY) 13.8.1971, BM 1971-507. BM noctuid slide 7414.

PARATYPES. 3♂♂, 2♀♀ New Caledonia, various localities.

The species was recorded as follows: Petit Gouli (1); Grottes de Koum (1); Probokonoro (8); Port Boisé site 74 (1).

**Molvena caledonica sp. n.** (Plate 76: 5, Figure 98: 4)

♂♀ 15–17 mm. The head, thorax and forewings are white, finely marked with black as illustrated; rough, irregular, multiple medial and postmedial fasciae are evident. The hindwing is nacreous white with an irregular grey postmedial fascia and a grey marginal area at the apex, both variable in extent. The abdomen is dark grey, darkening further distad, with a white apical tuft and white lateral patches on each segment.

*M. caledonica* is one of a group of black and white species found in Australasia that are often atttributed to the genus *Thalatha* Walker (type species *sinens* Walker; male genitalia illustrated by HOLLOWAY (1976: fig. 83)), e.g. by TURNER (1936, *Proc. Roy. Soc. Qd*, 47: 31), with reference to *guttalis* Walker, the type species of *Molvena* Walker. *M. guttalis* (Queensland, type examined) and *M. melaleuca* (Hampson) **comb. n.** (New Guinea, type examined) are both black and white as in *caledonica* and both have slender valves with a setose corona and a ventrally based harpe as in that species. They are thus distinct from *Thalatha sinens* and from another group of Australian species generally included in *Thalatha* (*aridoxa* Turner (? = *trichroma* Meyrick), *artificiosa* Turner, *hippolopha* Turner and *melanophrica* Turner). The latter are generally patterned with greenish brown, dark brown and cream.

*M. melaleuca* has only five to six moderate black patches arranged marginally only on the forewing, the hindwing being uniformly grey except for a variably broad white area up the dorsum that diffuses into the grey. The abdomen in *melaleuca* is uniformly grey apart from an apical white tuft. The harpe is more basal on the valve, shorter, broadest apically and dorsally serrate compared with that of *caledonica*. The aedeagus of *caledonica* is more elongate with a more scobinate vesica.

*M. guttalis* is similar to *caledonica* in the grey and white hindwing and the general disposition of black on the white forewings, but the patches of black are more regularly spaced, fewer, slightly larger and less irregular. The harpe on the valve is broad over the basal half, thence continues narrowly and incurved.

HOLOTYPE ♂. New Caledonia: Mt. Krapé, 460 m Site 56 05274/76373 (J. D. HOLLOWAY) 22.7.1971, BM 1971-507. BM noctuid slide 7490.

PARATYPES. 37 ex. New Caledonia, various localities; 1♂ N. New Caledonia, Tinchialit 2020 ft 22.viii–10.ix.1949 (L. E. CHEESMAN) BM 1950–1, BM noctuid slide 6771; 1♂ La Coulée 5.1.76 (J. GUTIERREZ) MNHN.

The species was recorded as follows: Grand Lac site 2 (6), site 72 (1); Pic du Pin site 7 (1), site 70 (1); Dothio site 19 (2); Bréhoa R. (2); Col Toma (1); Mt. Mou site 33 (3); site 34 (1); Tiebaghi site 40 (1), site 41 (1); Probokonoro (6); Col des Roussettes (2); Mt. Krapé site 55 (36), site 56 (9); near Bâ (3); Mt. Panié site 60 (1); Mt. Aoupinié (1); Col Mouirange (1); Yaté Gorge (1); Port

Boisé site 74 (10); Mt. Koghis site 80 (2). This would appear to be a rainforest species, often common in ultramafic localities where rainforest has developed.

### *Neogalea esula* (Druce)

*Neogalea esula* (Druce); HOLLOWAY (1977: 68).

This is a tropical American species that is occasionally introduced for control of lantana. It is established along the eastern seaboard of Australia and on Norfolk I. A single specimen was taken by A. DELOBEL on the Col d'Amieu in February, 1977. There are no records of deliberate introduction to New Caledonia (DELOBEL, *in litt.*).

### *Platysenta conducta* (Walker)

*Platysenta conducta* (Walker); ROBINSON (1975: 129); HOLLOWAY (1977: 68)

The species is widely distributed in the Indo-Australian tropics, though never common and with a rather patchy distribution; it has been recorded in the Pacific from Fiji, the Society Is. and Marquesas. It is vagrant to Norfolk I. Two specimens were taken in Nouméa by P. COCHEREAU.

### *Platysenta dolorosa* (Walker)

*Platysenta dolorosa* (Walker); ROBINSON (1975: 129); HOLLOWAY (1977: 69).

This species is widespread in the Indo-Australian tropics, extending to Queensland, New South Wales, Norfolk I. (probably only vagrant), the Solomons, Lifu and Fiji. It was taken as follows in New Caledonia: Grand Lac site 2 (1); R. Kuenthio (1); Koh (1); Col d'Amieu (1); Mt. Mou site 33 (1); Probokonoro (8); Col des Roussettes (1); Col Nassiboué (1).

### *Platysenta illecta* (Walker)

*Platysenta illecta* (Walker); ROBINSON (1975: 130); HOLLOWAY (1977: 70).

This species is similarly widespread in the Indo-Australian area as *dolorosa*, but extends into the Pacific as far as Rapa I. It is resident on Norfolk I. and vagrant to New Zealand. As on Norfolk I. and Fiji, and in common with *L. nonagrica*, it is an abundant species of disturbed vegetation, secondary growth on sedimentary rocks, though perhaps not so abundant in areas of extensive rainforest. It was recorded in New Caledonia as follows: Grand Lac site 2 (20), site 3 (7); Pic du Pin site 7 (4); R. Bleue site 9 (1), site 10 (1); Mt. des Sources site 11 (4), site 12 (1), site 14 (90); Anse Longue site 15 (2), site 16 (5), site 32 (1); R. Kuenthio (17); Dothio site 19 (7); Poupa (2); Koh (80); Ouatou R. (14); Bréhoa R. (30); Ouambéo R. (1); Col d'Amieu (45); Col Toma (9); Petit Gouli (4); Sarraméa (31); Kopéto (2); Mt. Mou site 33 (28), site 34 (5); Bambou R. (90); Col d'Amoss site 45 (1); Kavatch (4); R. Amoa (9); Ste. Therèse (2); Probokonoro (77); Col des Roussettes (5); Mt. Krapé site 55 (29), site 56 (2); Col de Boa (2); near Bâ (7); Mt. Aoupinié (3); Col Nassiboué (4); Col Mouirange (1); Port Boisé site 73 (1), site 74 (3); Les Dalmates site 77 (2); Mt. Koghis site 80 (2). BMNH: Pouébo, Tinchialit. MNHN: La Coulée.

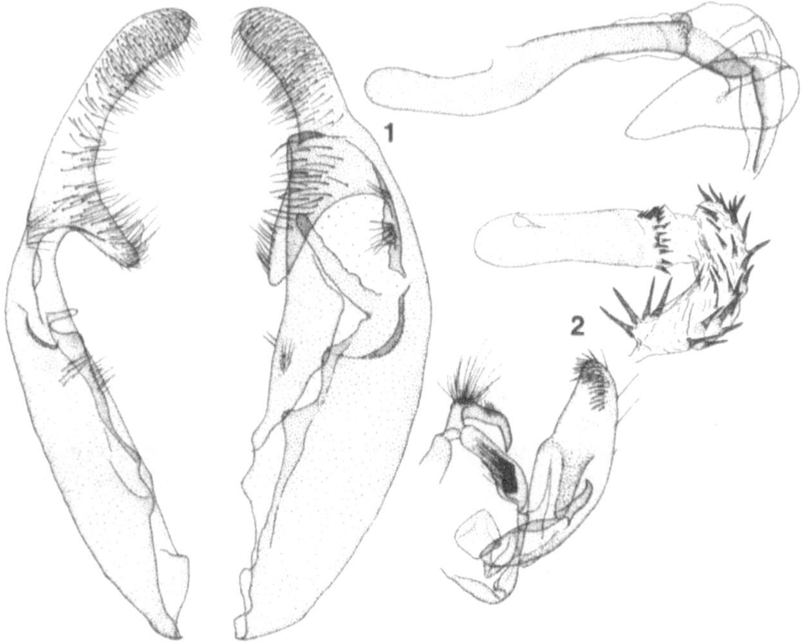


*Fig. 99.* Valves and aedeagus of *Platysenta praesecta* (1) and male genitalia and aedeagus of *Hypoperigea lunulata* (2).

*Platysenta praesecta* (Warren) (Plate 76: 1, Figure 99: 1)

*Prospalta praesecta* Warren, 1912, *Novit. zool.*, 19: 30, type examined.

*P. praesecta* has been recorded from Kei, the Moluccas, New Guinea, Queensland, the Solomons and New Hebrides. In New Caledonia three specimens were taken: Col d'Amieu (2); Mt. Mou site 33 (1).

*Sasunage tenebrosa* (Moore) (Figure 98: 1)

*Sasunaga tenebrosa* (Moore); ROBINSON (1975: 133).

Widespread in the Indo-Australian tropics to northern Australia, New South Wales, New Guinea, the Solomons, New Hebrides and Fiji, *tenebrosa* was recorded as follows in New Caledonia: Grand Lac site 2 (9), site 3 (6), site 4 (1), site 72 (1); Pic du Pin site 7 (14); R. Bleue site 10 (2); Mt. des Sources site 11 (5), site 12 (2), site 14 (3); R. Kuenthio (1); Dothio site 19 (2); Koh (8); Ouatou R. (1); Bréhoa R. (8); Ouambéo R. (1); Col d'Amieu (17); Sarraméa (1); Mt. Dogny (4); Mt. Mou site 33 (22), site 34 (7), site 36 (1); Bambou R. (1); Col d'Amoss site 45 (3), site 47 (1); Probokonoro (39); Col des Roussettes (1); Mt. Krapé site 55 (19), site 56 (2); near Bâ (3); Mt. Aoupinié (3); Col Mouirange (1); Port Boisé site 73 (2), site 74 (9). The species is found in a wide range of habitats. It is absent from the dry west coast areas sampled and was taken perhaps more commonly where there was rainforest.

*Spodoptera litura* (Fabricius)

*Spodoptera litura* (Fabricius); ROBINSON (1975: 134); HOLLOWAY (1977: 70).

This species is widespread in the Indo-Australian and Pacific tropics and subtropics where, with its congener below, it is often a serious agricultural pest. COHIC (1956) recorded the larva as feeding on *Alocasia, Brassica, Capsicum, Erythrina, Gossypium, Musa, Nicotiana, Rosa, Solanum, Xanthosoma* and *Zea*. ROBINSON (1975) gave many more host plants for Fiji. In New Caledonia the species is not as abundant as *mauritia* though widespread. It was recorded as follows: Grand Lac site 2 (2), site 3 (1), site 4 (1); Mt. des Sources site 11 (1), site 12 (1), site 14 (5); R. Kuenthio (2); Dothio site 19 (1); Poupa (1); Koh (2); Ouatou R. (2); Bréhoa R. (7); Col d'Amieu (9); Petit Gouli (1); Sarraméa (4); Kopéto (2); Anse Longue site 32 (3); Mt. Mou site 33 (18), site 34 (2); Tiebaghi site 40 (1); Grottes de Koum (9); Bambou R. (5); Cap Colnett (1); Kavatch (4); R. Amoa (2); Ste. Therèse (4); Probokonoro (9); Col des Roussettes (1); Mt. Krapé site 55 (9), site 56 (2); near Bâ (2); Mt. Aoupinié (1); Col Nassiboué (1); Port Boise site 73 (1), site 74 (2); Les Dalmates site 77 (1); Mt. Koghis site 80 (1); Ouen Toro site 81 (1), site 82 (2). VIETTE (1949): Baraoua, near Bourail. BPBM: Nouméa.

*Spodoptera mauritia acronyctoides* Guenée

*Spodoptera mauritia acronyctoides* Guenée; ROBINSON (1975: 134); HOLLOWAY (1977: 72).

The species has a distribution similar to the previous but also occurs in Africa. It is also a crop pest, though tending more towards the Gramineae. COHIC (1956) recorded the larva from *Oryza sativa* and *Saccharum officinarum* in New Caledonia. It was recorded as follows: Grand Lac site 2 (3), site 3 (4), site 4 (1); Mt. des Sources site 14 (6); Anse Longue site 16 (1), site 32 (4); R. Kuenthio (4); Dothio site 19 (1); Poupa (5); Koh (10); Ouatou R. (2); Bréhoa R. (18); Col d'Amieu (41); Col Toma (1); Petit Gouli (14); Sarraméa (89); Mt. Mou site 33 (15), site 34 (3); Grottes de Koum (4); Bambou R. (9); Col d'Amoss site 47 (1); Kavatch (12); Amoa R. (18); Ste. Therèse (9); Probokonoro (35); Col des Roussettes (1); Mt. Krapé site 55 (63), site 56 (2); near Bâ (7); Mt. Aoupinié (2); Mt. Koghis site 80 (1); Ouen Toro site 82 (1). BPBM: Nouméa, Tao, Le Crouen. MNHN: La Tamoa.

A note on the occurrence of *S. exempta* Walker in New Caledonia in 1978 can be found on p. 539.

*Syntheta nigerrima* (Guenée) (Plate 85: 3)

*Mamestra nigerrima* Guenée, 1852, *Hist. nat. Insectes (Lépid.)*, 5: 200.
*Hadena extima* Walker, 1857, *List Specimens lepid. Insects Colln Br. Mus.*, 11: 599.
*Eumichtis extima* (Walker); VIETTE (1949: 329).

There are two New Caledonian specimens of this species in the MNHN, from the THIERRY-MIEG collection, one of which is illustrated here. Otherwise the species only occurs in Australia, where it is found throughout.

## NOCTUIDAE: ACONTIINAE

### **Amyna flavirena sp. n.** (Plate 77: 15, Figure 106: 1)

Illatia aroa ab. *flavirena* Warren, 1913, *Gross-Schmett. Erde*, 11: 273.

♂♀ 11–13 mm. The species is very similar to *aroa* Bethune-Baker (New Guinea, Bismarcks, Solomons) but the forewing ground is black instead of brown, showing up the white flecks of the postmedial and the white patch at the apex; the reniform is ochre, less thickly ringed with white; the submarginal is pale brown, more prominent than in *aroa*. The male genitalia have the uncus apex bicornute rather than tricornute, and with two projections centrally instead of four; the dorsal arm of the valve is smoothly incurved with a basiventral spur, whereas it is sinuous, outcurved at the apex and unspurred in *aroa*; the harpe of the valve is broader, shorter, more flap-like. WARREN's male type of ab. *flavirena* (from Lifu) has similar genitalia but the uncus has only a short, single projection distal to the central ones, the harpe is shorter, the dorsal arms of the valves are more slender, especially the right one. More material is needed from both localities in order to test whether these differences are constant. Externally the Lifu specimen resembles those from New Caledonia.

HOLOTYPE ♂. New Caledonia: Koh, 140 m Site 22 05866/76177 (J. D. HOLLOWAY) 12.6.1971, BM 1971-507. BM noctuid slide 7523.

PARATYPES. 6♀♀ New Caledonia: Dothio site 19; Ouambéo R.; Kavatch (2); Mt. Krapé site 55 (2).

### *Amyna natalis* (Walker)

*Amyna natalis* (Walker); ROBINSON (1975: 137); HOLLOWAY (1977: 72).

This species is widespread in the Indo-Australian tropics, extending to the northern half of Australia, Norfolk I. (probably only vagrant), Fiji, Rotuma, Samoa and Tonga, with outlying records from Rapa I, the Gilbert Is. and the Tuvalu group. In New Caledonia it was recorded as follows: Anse Longue site 15 (1), site 16 (1), site 32 (2); Ouen Toro site 81 (4). BMNH: Pouébo. BPBM: La Foa, Nouméa, Isle of Pines. It has been taken mainly in dry west coast localities.

### *Amyna octo* (Guenée)

*Amyna octo* (Guenée); ROBINSON (1975: 137); HOLLOWAY (1977: 73).

The distribution is wider than that of *natalis*, embracing Africa and many more localities in Polynesia. In New Caledonia *octo* was recorded as follows: Anse Longue site 16 (1); Koh (1); Mt. Mou site 33 (1); Bambou R. (1);

Probokonoro (4); Port Boisé site 74 (1). BMNH: Pouébo. BPBM: Isle of Pines.

*Amyna punctum* (Fabricius)

*Amyna punctum* (Fabricius); ROBINSON (1975: 138).

This species is recorded from Africa and extends to Australia (Northern Territory to New South Wales) and as far east as Fiji. It occurs on Lifu and was recorded as follows in New Caledonia: Bréhoa R. (1); Mt. Mou site 33 (1); Ouen Toro site 81 (6), site 82 (1). A vagrant specimen was recorded on Norfolk I. in 1978.

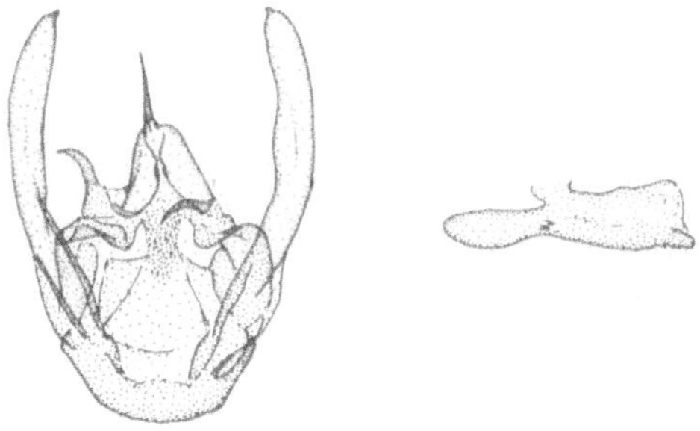

*Fig. 100.* Male genitalia of *Araeopteron canescens.*

*Araeopteron canescens* (Walker) (Plate 77: 1, Figure 100)

*Isopteryx canescens* Walker, 1865, *List Specimens lepid. Insects Colln Br. Mus.*, 34: 1318.

*Isopteryx favillalis* Walker, *Ibid*: 1319. Both types examined.

This Queensland species was probably overlooked during the survey and may be commoner than appears from the records, singletons from: Grand Lac site 3; R. Bleue site 10; Mt. Mou site 35; Col des Roussettes; Mt. Panié site 61. BPBM: La Foa.

## CODALITHIA Gen. n.

The male antennae are bipectinate, those of the female filiform. The palps are half as long again as the head, forwardly directed, moderately slender, the third segment half as long as the second. The legs have no distinctive modifications and the venation of the wings is illustrated in Figure 101.

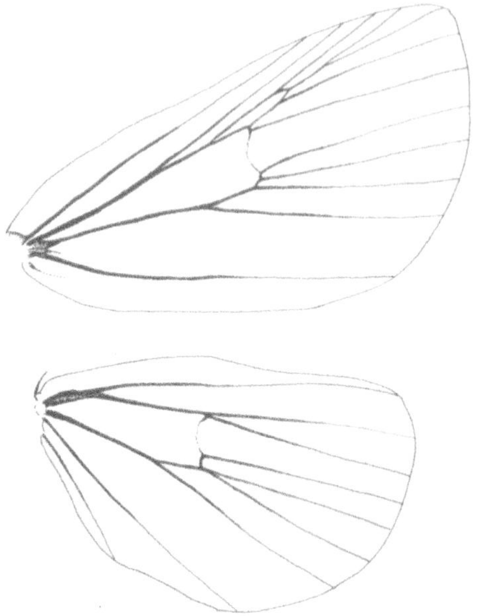

*Fig. 101.* Wing venation of *Codalithia quincuncialis.*

The structure of the genitalia is diagnostic. The males have a distinctive, almost divided sacculus to the valves and a single triangular cornutus in a short aedeagus. The female genitalia have the eighth segment modified, with lobed structures either side of the ostium bursae (Figure 103).

The genus is generally similar to *Lithacodia* Hübner but is distinguished by the genitalia characters and the chequered forewing pattern. It contains two species endemic to New Caledonia and largely restricted to ultramafic localities.

The type species is hereby designated as *quincuncialis*, a new species described below.

**Codalithia quincuncialis sp. n.** (Plate 78: 8, Figure 103: 1, 2)

♂♀ 13–17 mm. The forewings are boldly marked in black and dull green on a bone white ground as illustrated, the hindwings being pale grey, grading darker towards the margin. The genitalia are as illustrated.
HOLOTYPE ♂. New Caledonia: Pic du Pin, 240 m Site 7L 06879/75383 (J. D. HOLLOWAY) 8.8.1971, BM 1971-507. BM noctuid slide 8803.
PARATYPES: 12♂♂, 3♀♀ New Caledonia, Grand Lac site 2 (1), site 3 (2), site 72 (1), Pic du Pin site 7 (6), Col Mouirange (3), Mt. Pouédihi (1), Yaté Gorge (1), Les Dalmates site 78 (1), BM noctuid slides 7619 (venation), 8804 (♀); 1♀ La Coulée 26.8.76 (J. GUTIERREZ) MNHN.

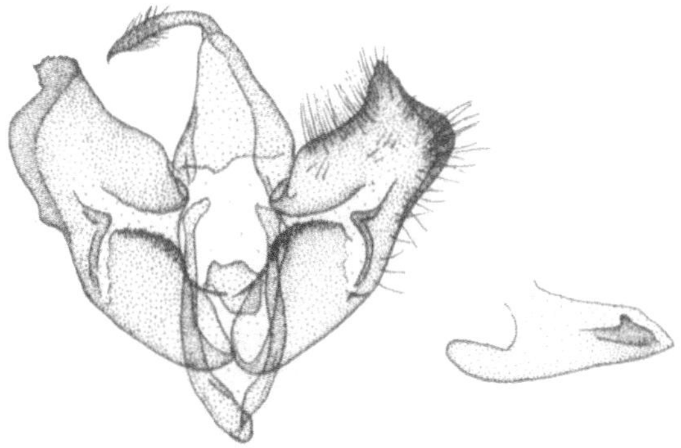

*Fig. 102.* Male genitalia of *Codalithia subtilis*.

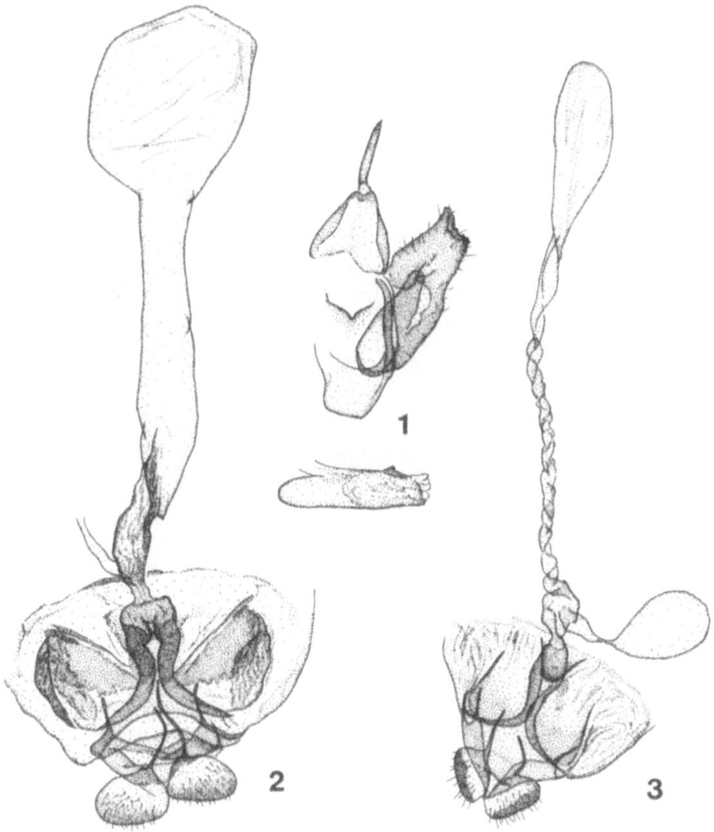


*Fig. 103.* Male genitalia of *Codalithia quincuncialis* (1), female genitalia of *C. quincuncialis* (2) and *C. subtilis* (3).

**Codalithia subtilis sp. n.** (Plate 78: 7, Figures 102, 103: 3)

♂♀ 12–15 mm. This species is very similar to its congener but is slightly smaller and more delicate in build. The forewings are similarly coloured though generally more uniformly suffused greenish and the black markings are more broken, giving a less chequered effect. The hindwings are slightly browner in tint and most specimens have a diagnostic darker postmedial, thin and sinuous. The genitalia differ as illustrated.

HOLOTYPE ♂. New Caledonia: Grand Lac, 270 m Site 72 06965/75354 (J. D. HOLLOWAY) 11.8.1971, BM 1971-507. BM noctuid slide 7410.

PARATYPES. 10♂♂, 1♀ New Caledonia: Pic du Pin site 7 (3); Poupa (1); Bréhoa R. (1); Ouambéo R. (1); Houailou (1); near Bâ (1); Mt. Panié site 60 (1), at 300 m in Malaise trap (1); R. Koué (1); BM noctuid slide 8807 (♀).

This species has a wider recorded distribution than its congener, occurring in northern localities, some of which are on sedimentary or metamorphic rocks. *C. quincuncialis* has not been taken away from the southern ultramafic massif.

*Corgatha concavata* (Walker) (Plate 78: 5, Figure 104: 3)

*Sophta concavata* Walker, 1862, *List Specimens lepid. Insects Colln Br. Mus.*, 26: 1759, type examined.

This is the type species of *Sophta* Walker (NYE, 1975), a genus that could be revived to include all bifalcate acontiines with similar facies such as *excisa* Hampson and *ruficeps* Walker. The type species of *Corgatha* is *zonalis* Walker, the male genitalia of which resemble those of the next species discussed. Further dissection of other species in these genera will be necessary before their separation can be confirmed.

The type locality of *concavata* is Queensland but it also flies in New Guinea and was recorded (singletons) from Koh and Mt. Koghis site 79 during the survey of New Caledonia.

**Corgatha figuralis paucifasciata ssp. n.** (Plate 78: 6, Figure 104: 2)

*Scopula figuralis* Walker, 1865, *List Specimens lepid. Insects Colln Br. Mus.*, 34: 1475, type examined.

♂♀ 8–10 mm. The insect is dark reddish brown, fading to pinkish beige in worn examples; the forewing costa and the fasciation of both wings is beige. The antemedial and postmedial fasciae are indistinct compared with typical Australian *figuralis* (Queensland, New South Wales) and the zig-zag submarginal of *figuralis* is absent in *paucifasciata*. The male genitalia are similar and also resemble those of the Oriental *zonalis* Walker; the discovery of similar forms in intermediate localities could indicate the conspecificity of the two species.

HOLOTYPE ♂. New Caledonia: R. Bleue, 180 m Site 10 06704/75544 (J. D. HOLLOWAY) 28.5.1971, BM 1971-507.

PARATYPES. 3♂♂, 6♀♀ New Caledonia, various localities, BM noctuid slide 7547 (♂).

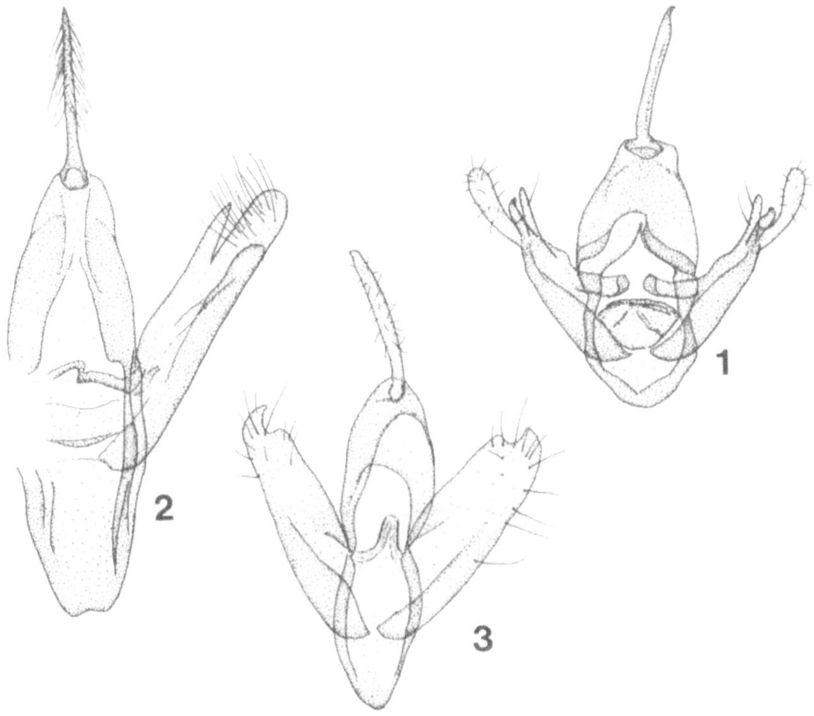


*Fig. 104.* Male genitalia of *Oruza kavatcha* (1), *Corgatha figuralis paucifasciata* (2) and *C. concavata* (3).

The species was recorded from the following mainly ultramafic localities: Grand Lac site 2 (2); Plaine des Lacs Site 5 (1); Pic du Pin site 7 (3); R. Bleue site 9 (1), site 10 (1); R. Bambou (1); Cap Colnett (1); Port Boisé site 74 (1).

A few specimens of *figuralis* have been taken in the New Hebrides (Malekula, Aneityum).

*Eublemma anachoresis* (Wallengren) (Plate 77: 4, Figure 105: 5)

*Eublemma anachoresis* (Wallengren); ROBINSON (1975: 138).

Two specimens were taken at Anse Longue site 16. The species has been recorded throughout the Oriental tropics, and from the Philippines, New Guinea, northern Australia, the Solomons, New Hebrides and Fiji.

*Eublemma baccalix* (Swinhoe) (Plate 77: 6, Figure 105: 4)

*Eublemma baccalix* (Swinhoe); ROBINSON (1975: 139)

Singletons were taken at R. Kuenthio and Port Boisé site 74. The species occurs in the Indian region, China, the New Hebrides and Fiji. The record of *quadrilineata* Moore from the New Hebrides noted by VIETTE (1949) may be referable to *baccalix*. The species are very similar and may prove synonymous though the type of *quadrilineata* has not been examined.

### *Eublemma cochylioides* (Guenée) (Plate 77: 3, Figure 105: 6)

*Eublemma cochylioides* (Guenée); ROBINSON (1975: 139).

The species is widespread, recorded throughout the African and Indo-Australian tropics, over most of Australia and in Fiji and Tonga, It was recorded as follows in New Caledonia: Anse Longue site 16 (1); R. Kuenthio (1); Cap Colnett (1); Port Boise site 74 (1); Ouen Toro site 82 (1). BPBM: Nouméa.

### *Eublemma glaucochroa* Hampson (Plate 77: 9, Figure 105: 9)

*Eublemma glaucochroa* Hampson, 1910, *Cat. Lepid. Phalaenae Br. Mus.*, 10: 128, type examined.

A singleton of this Australian species (Northern Territory, Queensland, S. Australia) was taken at the R. Kuenthio site. *E. marmorata* Wileman & West (Philippines) is a similar species, but probably not conspecific.

### *Eublemma miasma* (Hampson) (Plate 77: 14, Figure 105: 7)

*Erastia miasma* Hampson, 1891, *Illust. typical Specimens Lepid. Heterocera Colln Br. Mus.*, 8: 73, type examined.

*Trilophota tegulata* Wileman & West, 1929, *Novit. zool.*, 35: 14, **syn. n.,** type examined.

A single male was taken at Anna-Madeleine. The species occurs in the Indian region, China, the Philippines, Sulawesi, Queensland, St. Matthias I. and the New Hebrides (Aneityum).

### *Eublemma pudica* (Snellen) (Plate 77: 11, Figure 105: 10)

*Eublemma pudica* (Snellen); ROBINSON (1975: 140).

Singletons were taken at Petit Gouli and R. Amoa. The species occurs widely in the Oriental tropics, the Philippines, Queensland, Fiji and Samoa. In the BMNH collection there is a note by W. W. BRANDT that *E. stereoscia* Turner (1935, *Proc. Roy. Soc. Qd.*, p. 33) is synonymous with *pudica.*

### *Eublemma ragusana* (Freyer)

*Eublemma ragusana* (Freyer); ROBINSON (1975: 140).

This species occurs widely in the African and Oriental tropics and has been recorded from the Philippines, New Guinea, Queensland, the New Hebrides and Fiji. There is only one record from New Caledonia, a male in the BPBM from the Forêt de Thi at 530 m.

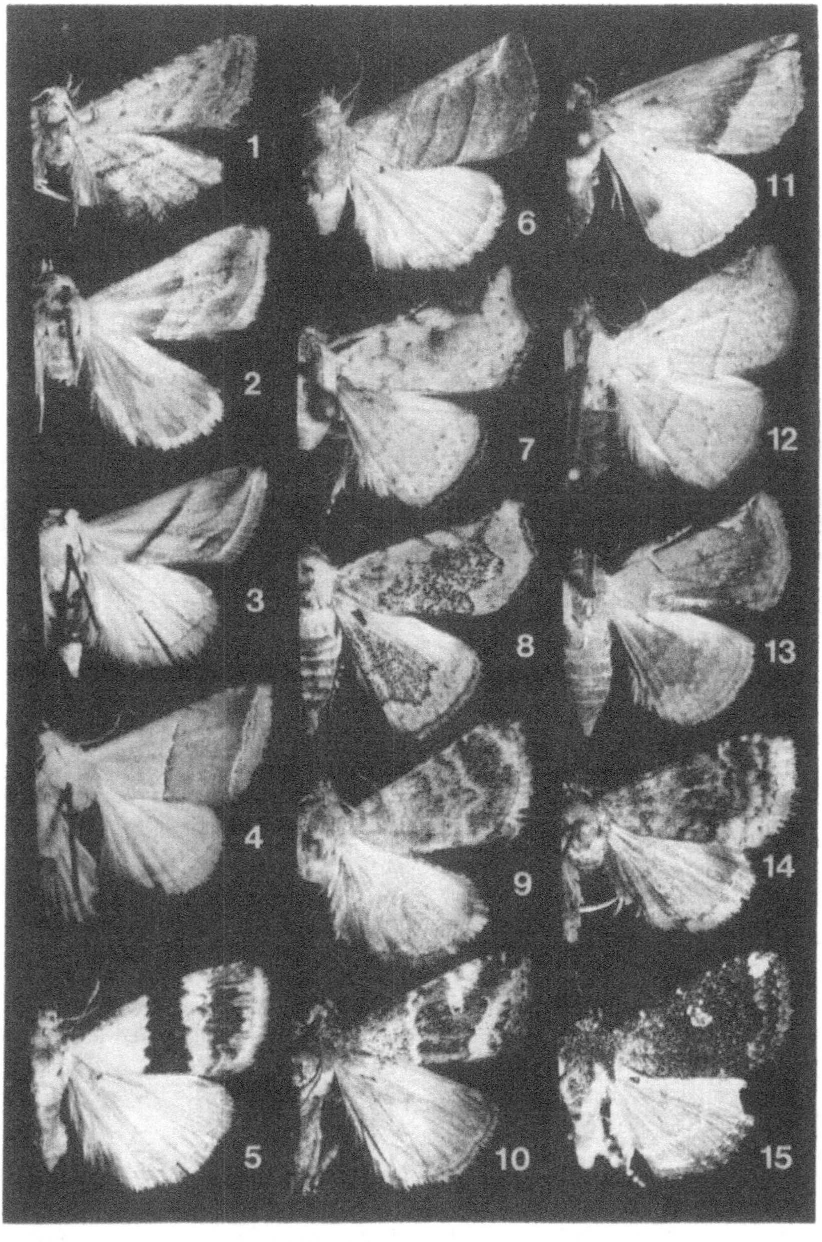


*Plate 77.* 1. *Araeopteron canescens* (5 mm); 2. *Eublemma rivula* (5 mm); 3. *Eublemma cochylioides* (9 mm); 4. *Eublemma anachoresis* (8 mm); 5. *Maliattha ritsemae* (10 mm); 6. *Eublemma baccalix* (9 mm); 7, 8. *Eublemma rufogalactea;* 9. *Eublemma glaucochroa* (7 mm); 10. *Maliattha inconcisoides;* 11. *Eublemma pudica* (7 mm); 12, 13. *Eublemma variochrea;* 14. *Eublemma miasma* (7 mm); 15. *Amyna flavirena.*

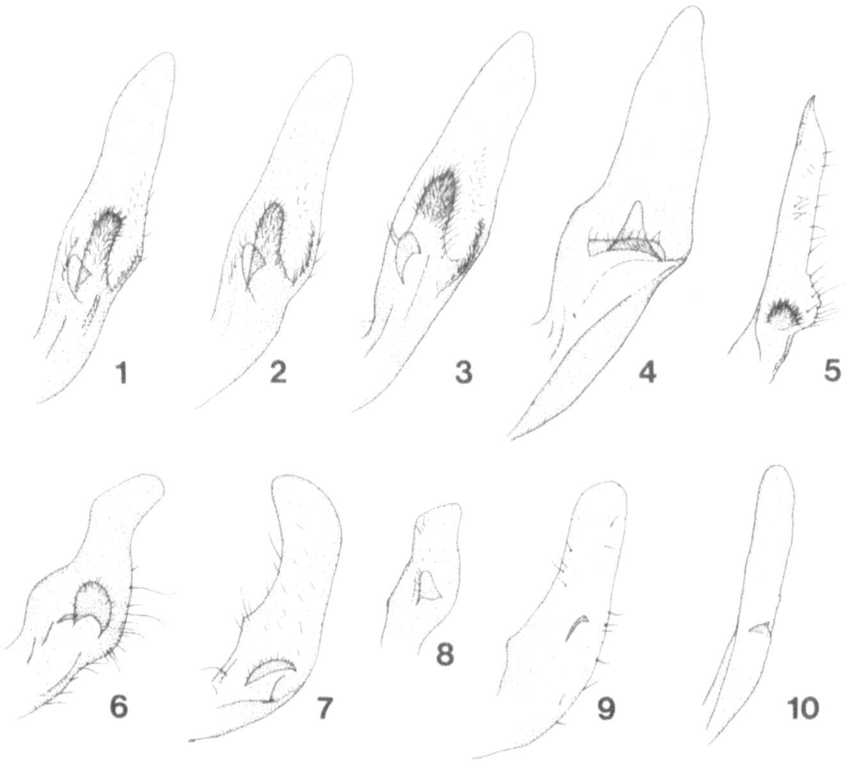


*Fig. 105.* Male valves of *Eublemma* species: *rufogalactea* (1, 2), *variochrea* (3), *baccalix* (4), *anachoresis* (5), *cochylioides* (6), *miasma* (7), *rivula* (8), *glaucochroa* (9), *pudica* (10).

### *Eublemma rivula* (Moore) (Plate 77: 2, Figure 105: 8)

*Eublemma rivula* (Moore); Robinson (1975: 141).

A single male was taken at Sarraméa. *E. rivula* is widespread but poorly recorded in the Old World tropics, Australia and the Pacific, extending to Samoa and the Society Is.

### **Eublemma rufogalactea sp. n.** (Plate 77: 7, 8, Figure 105: 1, 2)

♂♀ 7–8 mm. The ground colour is fawn to pale warm grey, variable; the fasciae of both wings are as illustrated, rufous in tint; the medials are not obvious but irregular and more ochreous (they are straight in the next species discussed); there is a thin rufous line with a thicker band of ground colour distad at the margin of both wings, then the fringes are basally dark brown, distally lilac grey; the submarginals are irregular, wavy, pale fasciae with black flecks on the veins. The forewing has a blue-grey subapical triangle on the costa, bounded with rich brown; the discal spot is diffusely greyish brown. The species is variable, one specimen having a pale rufous tinge broadly over the forewing disc and three having bright rufous and dark brown scales thickly scattered over the wings basally to the postmedial and brown scales banding the abdomen, which is usually uniform.

The species resembles *galactea* Hampson (type examined, Sangir; also in Sulawesi and New Guinea) in the general markings and dark wing fringes, particularly the fawn form, but the subapical triangle on the forewing costa is much larger; in the male genitalia the valve of *galactea* is much longer relatively, distal to the harpe. In *rufogalactea* the wings below grade darker grey over the marginal third; this is more clearly defined, the grey darker in *galactea*. The rufous suffused form is reminiscent of *pectorora* Lucas (the type of the synonymous *novaeguineae* Strand was examined) from Queensland, New Guinea and islands to the south-east, but the rufous suffusion extends to the base, is not bounded by a rufous antemedial as in *pectorora*, and the postmedial boundary to the rufous area has dentations more rounded, not angular, and is more sharply and darkly defined. *E. semirufa* Hampson (type examined, Sikkim; also Java, Bali, Sulawesi) is similar but closer to *pectorora* than *rufogalactea*. The hindwing underside of *pectorora* has a dark margin but not the forewing; the valve of the male genitalia is narrower, longer, more tapering, apically sinuous and with the harpe shorter and broader.

HOLOTYPE ♂. New Caledonia: Poupa, 200 m Site 20 06120/76150 (J. D. HOLLOWAY) 10.6.1971, BM 1971-507. BM noctuid slide 7545.

PARATYPES: 8♂♂, 2♀♀ New Caledonia, Grand Lac site 2 (2), site 3 (1) Mt. des Sources site 11 (1), Anse Longue site 16 (1), Poupa (1), Bambou R. (3), BM noctuid slide 7544 (♂); 1♀ Mt. Koghis, Forêt de Thi 530 m 8.iii.61 (J. SEDLACEK) BPBM.

**Eublemma variochrea sp. n.** (Plate 77: 12, 13, Figure 105: 3)

♂♀ 8–10 mm. The ground colour varies from beige to ochreous yellow-fawn; the fringes are tipped lilac grey. The forewing costal subapical mark is more elongate, less triangular than in the previous species, bounded dark brown towards the dorsum and broadly rufous at the costa; the costa basad is more brightly ochreous. The fasciae follow a generally similar course to those of the previous species but the submarginals and postmedials are less distinct and the medial is well-defined, straight on both wings, that of the forewing oblique, flexed acutely basad one quarter from the costa. The medials are usually pale, edged darker basad, but they are broadly darker brown in four specimens, including the holotype. The genitalia are very similar to those of the previous species. In facies the species is closer to *abrupta* Walker, a widespread Indo-Australian tropical species, but it lacks the dark blotch on the hindwing tornus of *abrupta* and has the medial closer to the forewing subapical costal marking. The genitalia of *abrupta* show greater dissimilarity, having a broad and curved valve.

HOLOTYPE ♀. New Caledonia: Ouatou R., 320 m Site 23 05863/76151 (J. D. HOLLOWAY) 13.6.1971, BM 1971-507. BM noctuid slide 8831.

PARATYPES: 14 ex. New Caledonia, Grand Lac site 1 (1), site 2 (3), Plaine des Lacs site 5 (1), Pic du Pin site 7 (2), R. Bleue site 9 (1), Mt. des Sources site 14 (1), Poupa (1), Bambou R. (1), Kavatch (1), Mt. Krapé site 55 (2), BM noctuid slides 7543, 8827(♂); 1♂ Le Crouen 31.i.1963 (N. L. H. KRAUSS) BPBM.

**Maliattha inconcisoides sp. n.** (Plate 77: 10, Figure 106: 3)

♂♀ 8–9 mm. The ground is fawn-white, marked with brown as illustrated. The species is intermediate in appearance between *inconcisa* Butler (1882, *Ann, Mag. nat. Hist.* (5), 10: 229, type examined; *M. furcata* Warren, 1913, Gross-Schmett. Erde, 11: 282, **syn. n.,** type examined) from the Bismarcks, Solomons and New Hebrides, and *comes* Warren (type examined) from New Guinea (and one similar specimen from Queensland). On the forewing the basal brown suffusion and the narrow band of pale ground separating it from the medial brown area are as in *comes*, but the medial band is more broadly based on the dorsum than in either of the other species; the postmedial runs more or less obliquely to the dorsum (curving in and ultimately perpendicular to it in the other species), this obliqueness being continued anteriorly into the apical diffuse brown patch as in *inconcisa* (this brown patch extends much further towards the tornus in *comes*); the area of ground colour based on the dorsum distal to the postmedial is more extensive than in *comes* but narrower than in *inconcisa*.

The male genitalia have the valves broader, more apically broadened and splayed than in either of the other species; both harpes are stouter than in *inconcisa*, resembling *comes*, the left straight (curved in *inconcisa*, straight, but longer and more distal in *comes*), the right shorter than the valve, curved dorsad (straight or slightly sinuous in the other species, more distal, extending to the valve apex in *comes* and beyond it in *inconcisa*).

HOLOTYPE ♂. New Caledonia: Petit Gouli, 30 m Site 28 05852/76047 (J. D. HOLLOWAY) 20.6.1971, BM 1971-507.

PARATYPES: 5♂♂, 6♀♀ New Caledonia, various localities, BM noctuid slide 7548; 3 ex. Tao 8–10.ii.1963 Malaise trap (C. YOSHIMOTO & N. L. H. KRAUSS) BPBM; 2 ex. Bourail 4–6.ii.1963 (J. L. GRESSITT, C. YOSHIMOTO & N. KRAUSS) light trap, BPBM; 2 ex. Le Crouen 31.i.1963 (C. YOSHIMOTO & N. KRAUSS) light trap, BPBM; 1♂ Tipindjé 10.i.1963 (C.YOSHIMOTO & N. KRAUSS) light trap, BPBM; 1♀ Mt. Koghis, Forêt de Thi 539 m 8.iii.61 (J. SEDLACEK) BPBM.

The species was recorded as follows during the survey: R. Bleue site 9 (2); Anse Longue site 16 (2), site 32 (4); Petit Gouli (2); Sarraméa (2); Kopéto (1); R. Amoa (1). It may be commoner on the west side of the island.

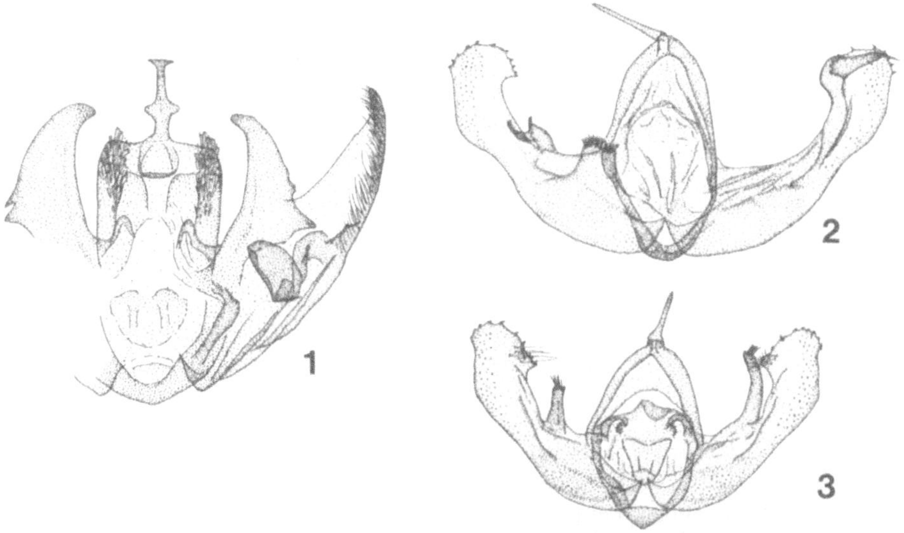


*Fig. 106.* Male genitalia of *Amyna flavirena* (1), *Maliattha ritsemae* (2), and *M. inconcisoides* (3).

*Maliattha ritsemae* (Snellen) (Plate 77: 5, Figure 106: 2)

*Maliattha ritsemae* (Snellen); ROBINSON (1975: 142); HOLLOWAY (1977: 73).

*M. ritsemae* extends from Sundaland, the Lesser Sundas, the Philippines and Sulawesi through New Guinea and northern Australia to Samoa, Tonga, and the Society and Austral Is. It is vagrant to Norfolk I. It occurs in the Loyalty Is. and was recorded as follows in New Caledonia: Pic du Pin site 7 (1); Anse Longue site 16 (2); Dothio site 19 (1); Ouambéo R. (1). P COCHEREAU took it frequently in Nouméa. BPBM: Bourail, Mt. Koghis, Le Crouen.

*Oruza cariosa* (Lucas) (Plate 78: 4)

*Oruza cariosa* (Lucas); ROBINSON (1975: 142).

*O. cariosa* was described from Queensland but has also been recorded from the Northern Territory, the Lesser Sundas (Alor), New Guinea, Sudest I. and Fiji. A single female of *cariosa* was taken at Sarraméa during the survey and it has also been recorded from Le Crouen, Bourail and Forêt de Thi (BPBM).

**Oruza kavatcha sp. n.** (Plate 78: 1, 2, Figure 104: 1)

♂♀ 8–9 mm. The general coloration is rufous brown, the forewing costa in some specimens broadly rufous fawn, this extending to the base of the hindwing and the thorax as illustrated; dorsal to the subapical costal triangle there is a diffuse, elongate rufous streak. The species is variable, some specimens being uniform and others having the fawn forewing costa or merely a rufous fawn forewing subapical triangle; greyish scales suffuse the forewing tornal area in some specimens and these also have greyish flecks distal to the hindwing postmedial in the spaces; one specimen has paler fawn patches in the three most dorsal spaces distal to the forewing postmedial, this being repeated to a lesser extent on the hindwing.

The sinuous posterior margin to the forewing costal pale area and the prominent, distally directed tooth to the hindwing postmedial distinguish the species from similar congeners such as *stragulata* Pagenstecher (type examined, widespread in Indo-Australian tropics), which has a distinctive double forewing postmedial, and *bipunctata* Warren (type examined, Queensland), a larger species (12 mm) with the forewing costal band extending more broadly to the apex. Both these species have a square saccus in the male genitalia, that of *kavatcha* being rounded; *kavatcha* has a costal process on the valve adjacent and equal to the harpe but this is twice as long, extending to the valve apex, in *stragulata* and absent in *bipunctata*.

HOLOTYPE ♂. New Caledonia: Kavatch 40 m Site 50 04857/77068 (J. D. HOLLOWAY) 18.7.1971, BM 1971-507. BM noctuid slide 7546.

PARATYPES. 16 ex. New Caledonia, various localities.

The species was recorded as follows: Dothio site 19 (1); Poupa (1); Koh (1); Ouatou R. (2); Bréhoa R. (2); Kopéto (1); Mt. Mou site 33 (1); Bambou R. (1); Cap Colnett (1); Kavatch (2); Col des Roussettes (1); near Bâ (1).

### **Oruza narabuta sp. n.** (Plate 78: 3)

♂♀ 9–10 mm. The male genitalia of this New Hebrides species are as in *kavatcha* but the wing pattern has significant differences. The forewing costal bar is always present, fawn rather than rufous fawn, the dorsal margin to the band being straight rather than sinuous; there is no rufous patch dorsal to the subapical triangle; the dorsal two thirds of the forewing postmedial and all that of the hindwing are roughly straight with a distinctive rufous band distad and often fawn patches in the spaces towards the dorsum. The hindwing margins are more crenulate and irregular. The submarginal area of the forewing is often slightly lilac in tint, especially near the tornus.

HOLOTYPE ♀, New Hebrides: Efate I., Narabut Camp 30.vi–7.vii.1971, (G. S. ROBINSON) Roy. Soc. Expedition.

PARATYPES (all New Hebrides): 2♂♂, 2♀♀ as holotype, BM noctuid slide 8825 (♂); 1♀ Santo, Mt. Tabwemasana, Nokowula-3700 ft 1.ix.1971; 2♂♂ Santo, Apouna R. Camp 3, 1000 ft 9–11.ix.1971.

### *Trissernis greeni* Holloway

*Trissernis greeni* Holloway, 1977: 73.

A single specimen of this Norfolk I. species was taken at Port Boisé site 74. Further subspecies occur in the New Hebrides.

*Plate 78.* 1, 2. *Oruza kavatcha;* 3. *Oruza narabuta;* 4. *Oruza cariosa* (8 mm); 5. *Corgatha concavata* (8 mm); 6. *Corgatha figuralis paucifasciata;* 7. *Codalithia subtilis;* 8. *Codalithia quincuncialis;* 9. *Penicillaria maculata richardsoni*; 10. *Lophoptera hemithyris* (12 mm); 11. *Gyrtona divitalis* (10 mm); 12. *Nigramma polionota* (13–14 mm).

## Noctuidae: Euteliinae

### *Paectes cristatrix* (Guenée) (Figure 107: 2)

*Paectes cristatrix* (Guenée); Robinson (1975: 146).

The species occurs from Sri Lanka to New Guinea and Queensland with subspecies *fijiensis* Robinson in Fiji and Tonga. The typical race occurs in the

Solomons and Lifu. It was recorded as follows in New Caledonia: Amoa R. (1); Probokonoro (3).

### *Pataeta carbo* (Guenée) (Figure 107: 1)

*Pataeta carbo* (Guenée); ROBINSON (1975: 146).

The species is recorded from Australia (Northern Territory to New South Wales), Tenimber, Saleyer, Amboina, New Guinea, the Loyalty Is. and Fiji. New Caledonian records are as follows: Grand Lac site 2 (1), site 3 (1); Mt. des Sources site 14 (1); R. Kuenthio (2); Koh (3); Sarraméa (1); Bambou R. (1); Kavatch (1); R. Amoa (1); Probokonoro (2); near Bâ (1). BMNH: Tinchialit. MNHN: Nouméa.

### *Penicillaria jocosatrix* Guenée (Figure 107: 4)

*Penicillaria jocosatrix* Guenée; ROBINSON (1975: 145).

The type of ***Penicillaria*** Guenée is ***nugatrix*** Guenée, obviously congeneric with *jocosatrix*, type species for the younger generic name, ***Bombotelia*** Hampson (NYE, 1975). The older name is therefore used here.

A single male of *jocosatrix* was taken at Probokonoro. The species was recorded from Nouméa by VIETTE (1949) and has been reared from mango by ORSTOM entomologists. The larva has been recorded from other Anacardiaceae in India (SINGH *et al.*, 1954–1960) and from mango in Fiji (ROBINSON). The species is widespread in the Indo-Australian tropics, extending to northern Australia, Micronesia, the New Hebrides, Fiji and Tonga.

### **Penicillaria maculata richardsoni ssp. n.** (Plate 78: 9, Figure 107: 3)

*Penicillaria maculata* Butler, 1889, *Illustr. typical Specimens Lepid. Heterocera Colln Br. Mus.*, 7: 12, 71, type examined.

♂♀ 15–17 mm. This subspecies is larger than typical *maculata* which ranges from 13–15 mm and is recorded from Sri Lanka, India, the Himalaya, Borneo, New Guinea, St. Aignan and New Ireland. The forewing is a more uniform purplish red, the subbasal more oblique with the basal area less markedly darker than the ground colour distad; the area between the postmedial and the posterior section of the submarginal lacks the pale violet suffusion of the typical race. The hindwing discal spot is more prominent. The male genitalia do not differ from those of the typical race.

HOLOTYPE ♂. New Caledonia: Grand Lac, 260 m Site 3 06965/75355 (J. D. HOLLOWAY) BM 1971-507. BM noctuid slide 7496.

PARATYPES: 8 ex. New Caledonia: various localities; 1♂ Mt. Koghis BPBM.

The species was recorded as follows in New Caledonia: Grand Lac site 3 (1); Mt. des Sources site 11 (1); Probokonoro (5); Mt. Krapé site 55 (3); Mt. Aoupinié (1).

The subspecies has also been recorded in the New Hebrides (Santo, Efate) by G. S. ROBINSON.

*Fig. 107.* Male genitalia of *Pataeta carbo* (1) and male genitalia and aedeagi of *Paectes cristatrix* (2), *Penicillaria maculata richardsoni* (3), and *P. jocosatrix* (4).

### *Phlegetonia delatrix* (Guenée)

*Phlegetonia delatrix* (Guenée); ROBINSON (1975: 148).

There is a single specimen of this species from Nouméa in the MNHN. It is widespread in the Indo-Australian and Pacific tropics, extending from Northern Territory to New South Wales in Australia and being recorded as far east as the Marquesas and Rapa I.

## NOCTUIDAE: STICTOPTERINAE

### *Gyrtona divitalis* Walker (Plate 78: 11, Figure 109: 3)

*Gyrtona divitalis* Walker; ROBINSON (1975:150).

A single specimen was taken at Grand Lac site 2 and Miss CHEESMAN took two at Tinchialit. The species ranges from Sundaland through the Philippines, Sulawesi, New Guinea and the Bismarcks to the Northern Territory, Fiji, Samoa and the Society Is. The male holotype (Sarawak) has a small spine subapically on the valve costa in the male genitalia. the New Caledonian male dissected lacked this spine and had a relatively much shorter saccus and fused gnathus and a longer, narrower uncus. A specimen from Sulawesi lacked the costal spine but was intermediate in the other characteristics.

### *Lophoptera hemithyris* (Hampson) (Plate 78: 10, Figure 108: 4)

*Lophoptera hemithyris* (Hampson); ROBINSON (1975: 151).

The species is widespread in the Oriental tropics and has been recorded from Sulawesi, Kei, New Guinea, the New Hebrides and Fiji. It was recorded as follows in New Caledonia: Grand Lac site 2 (1); Dothio site 18 (1); Mt. Krapé site 56 (2).

### *Lophoptera squammigera* Guenée (Plate 79: 1, Figure 108: 2)

*Lophoptera squammigera* Guenée, 1852, *Hist. nat. Insectes* (*Lépid.*), 7: 55, the type species of the genus.

The species is recorded from the Indian region, the Philippines, Sulawesi, the Moluccas, Australia (Queensland, New South Wales), New Guinea, the Bismarcks, Solomons, New Hebrides, Loyalties and New Caledonia. It was recorded as follows: Mt. des Sources site 14 (1); Anse Longue site 16 (1); R. Kuenthio (2).

### *Nigramma polionota* (Hampson) **comb. n.** (Plate 78: 12, Figure 108: 5)

*Gyrtona polionota* Hampson, 1905, *Ann. Mag. nat. Hist.* (7), 16: 539, type examined.

*Nigramma perstrialis* Hampson, 1918, *Novit. zool.*, 25: 192, **syn. n.,** type examined; ROBINSON (1975: 152).

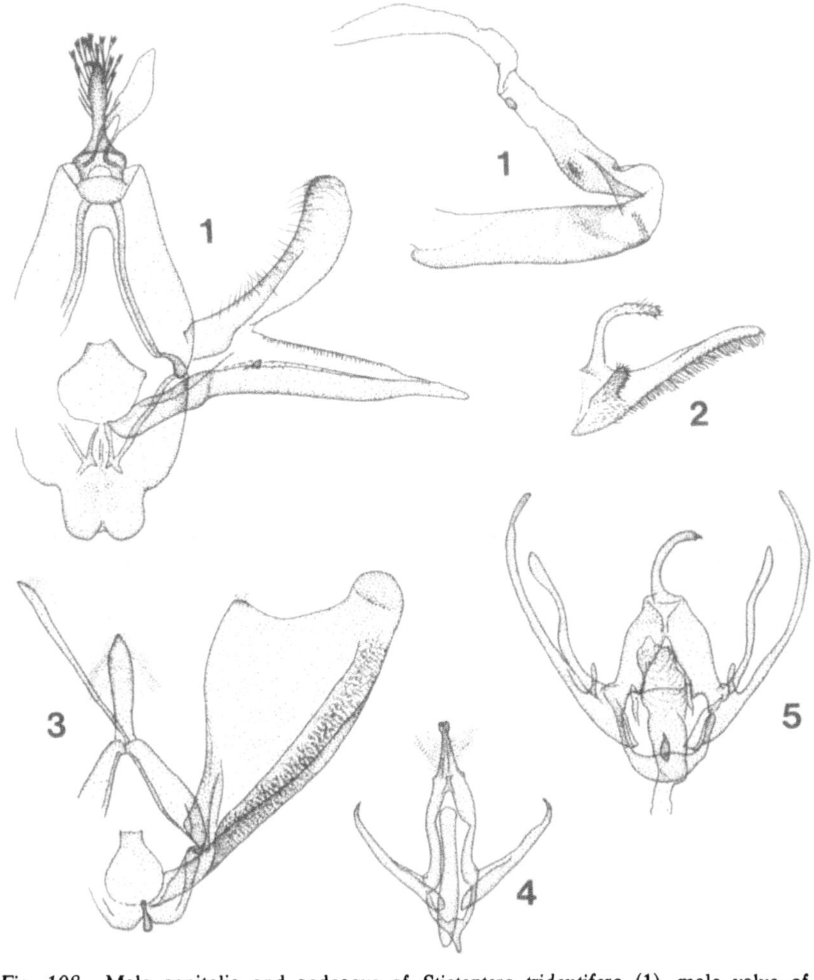


*Fig. 108.* Male genitalia and aedeagus of *Stictoptera tridentifera* (1), male valve of *Lophoptera squammigera* (2), and male genitalia of *Gyrtona divitalis* (3), *Lophoptera hemithyris* (4) and *Nigramma polionota* (5).

Dissection of male *polionota* and *perstrialis* from their shared type locality (Woodlark I.) indicated that they are conspecific with each other and with variable forms from Manus I., islands south-east of New Guinea (including Woodlark), the Solomons, New Hebrides (Santo), Fiji and Samoa. The typical form has a uniform pinkish brown forewing with darker fasciation (the lower specimen illustrated). Other specimens are darker brown, marked with cream as in the upper specimen illustrated (similar specimens fly on Woodlark I.), whereas *perstrialis* forms have the forewing broadly striped longitudinally with dark brown and fawn (illustrated by Robinson). The sinuous dorsal arm of the valve in the male genitalia is diagnostic, the single large cornutus of the aedeagus being exhibited also by forms from Sula Mangoli, Amboina and Buru that are similar in facies but have the dorsal arm of the valve straight.

The species was recorded as follows: Grand Lac site 2 (1); Probokonoro (2). BPBM: Mt. Koghis (3). MNHN: La Coulée (1).

### *Stictoptera describens* (Walker)

*Stictoptera describens* (Walker); ROBINSON (1975: 154).

*S. describens* has been recorded widely in the Oriental tropics and from Sulawesi, the S. Moluccas, New Guinea, N. Queensland, the New Hebrides, New Caledonia and Fiji. Singletons were recorded at Grand Lac sites 2 and 3.

### **Stictoptera tridentifera sp. n.** (Plate 79: 2, Figure 108: 1)

♂♀ 21–24 mm. This species is highly variable in the facies of the forewing and thorax, the commonest form being the lowest illustrated in the Plate, with fawn, grey and rufous markings on a medium brown ground. Most *Stictoptera* species exhibit such variability with distinct forms recurring from species to species that reliable discriminatory characters of facies are often hard to find. The New Caledonian species has, in most specimens, three or four black submarginal triangles subdorsally that are distinctive (the New Guinea species *bisexualis* Hampson has similar black triangles but a more produced forewing apex); the exterior margin of the orbicular and the adjacent marking posteriorly consist of uplifted pale scales, thinly edged black distad, another fairly distinctive feature.

The typical form is often paler, suffused with creamy-fawn; other forms are more distinctive and are described below with some indication of their parallels in other, mainly Oriental species.

(a) The forewing is suffused with uniform, warm, medium grey with prominent black basal, submarginal and reniform marks (fifth forewing illustrated). This form has parallels in *ferrifera* Walker, *atrifera* Hampson and *terribilis* Holloway. It appears to be restricted to the male, perhaps the female equivalent being (e) below.

(b) The forewing is dark vinous grey with fawn patches basally, apically and in the anal angle (second forewing). This form occurs in both sexes and has parallels in *cucullioides* Guenée.

(c) The forewing has broad cream blotches on a uniform brown ground (first illustration) or greyish blotches on a typical ground. No other species has cream blotches of this form but greyish patches are found in some forms of *signifera* Walker, *stygia* Hampson and *stygia sensu* ROBINSON (1975; the Fijian and New Hebrides forms attributed to *stygia* have slightly different genitalia and variants not found in typical *stygia*), and *cucullioides*. Both sexes exhibit this form.

(d) The forewing is markedly two-toned with the boundary at the straight medial (third forewing illustrated). There are parallels in *signifera*, *stygia* and *cucullioides*; the variation appears to be restricted to the female.

(e) This is possibly the female equivalent of (a). The forewing is generally typical but the basal, medial and submarginal fasciae are picked out in black (fourth illustration). A vaguely similar form occurs in *ferrifera*.

(f) The forewing is broadly streaked creamish longitudinally with the lower part of the reniform prominently paler (sixth illustration). A parallel form occurs in *atrifera*. The form appears to be restricted to the female.

(g) This is another female form slightly modified from the typical one with the medial accentuated and the area basal to it suffused with creamish fawn. This is paralleled in *signifera* and *cucullioides*.

The generation time for Fijian *stygia* at least is only of the order of a month (ROBINSON, 1975), so this highly variable section of *Stictoptera* might provide ideal material for genetic studies as there is evidently some sex linkage involved; also it would be feasible to compare the factors controlling the various morphs from species to species.

*Stictoptera* species are best characterised by the male genitalia where the ornamentation of the vesica, the shape and setation of the uncus, and the shape of the dorsal arm of the valve and of the saccus usually provide diagnostic characters. *S. tridentifera* is unique amongst the species examined in having a crest of tridentate setae to the uncus whereas *vitiensis* Hampson, *stygia* and Fijian *stygia* resemble *signifera* in having two lateral rows of setae ventrally and subbasally. The four species in the *ferrifera* complex discussed by HOLLOWAY (1976) have a straight uncus with a loose scattering of setae; the uncus in *stygia* and *signifera* is somewhat sigmoid and in *cucullioides* it is distinctly broadened towards the base. The dorsal arm of the valve is broader apically in *tridentifera*, parallel-sided in *vitiensis* and more or less so in *stygia*. The vesica has one group of larger spines, compared with three in *vitiensis* and none in *stygia*. The saccus is deep with the terminal projection prominent and bilobed; in *stygia* and *vitiensis* the saccus is shallow with a small monolobed projection, square in the latter.

HOLOTYPE ♂. New Caledonia: Grand Lac 250 m Site 4 06980/75367 (J. D. HOLLOWAY) 19.5.1971, BM 1971-507. BM noctuid slide 7502.

PARATYPES: 115 ex. New Caledonia, various survey localities, BM noctuid slide 7503 (♂); 1♂, 1♀ Mt. Koghis 500 m 26.ii.1963 (C. M. YOSHIMOTO) BPBM; 2♀♀ Mt. Koghis, Forêt de Thi 530 m 8.iii.61 (J. SEDLACEK) BPBM; 1♂ Mt. Koghis 600 m 20.v.1971 (P. COCHEREAU) MNHN; 1♂, 3♀♀ La Coulée 20.1.1977 (A. DELOBEL) MNHN.

The species was recorded as follows during the survey: Grand Lac site 1 (1), site 2 (14), site 3 (10), site 4 (6); Plaine des Lacs site 5 (3); Pic du Pin site 7 (15); R. Bleue site 8 (1), site 10 (4); Mt. des Sources site 11 (2), site 12 (3), site 14 (1); R. Kuenthio (1); Dothio site 18 (1), site 19 (5); Poupa (6); Bréhoa R. (11); Col d'Amieu (10); Col Toma (19); Sarraméa (4); Kopéto (2); Mt. Dogny (2); Mt. Mou site 33 (17), site 34 (2); Col d'Amoss site 47 (2); Cap

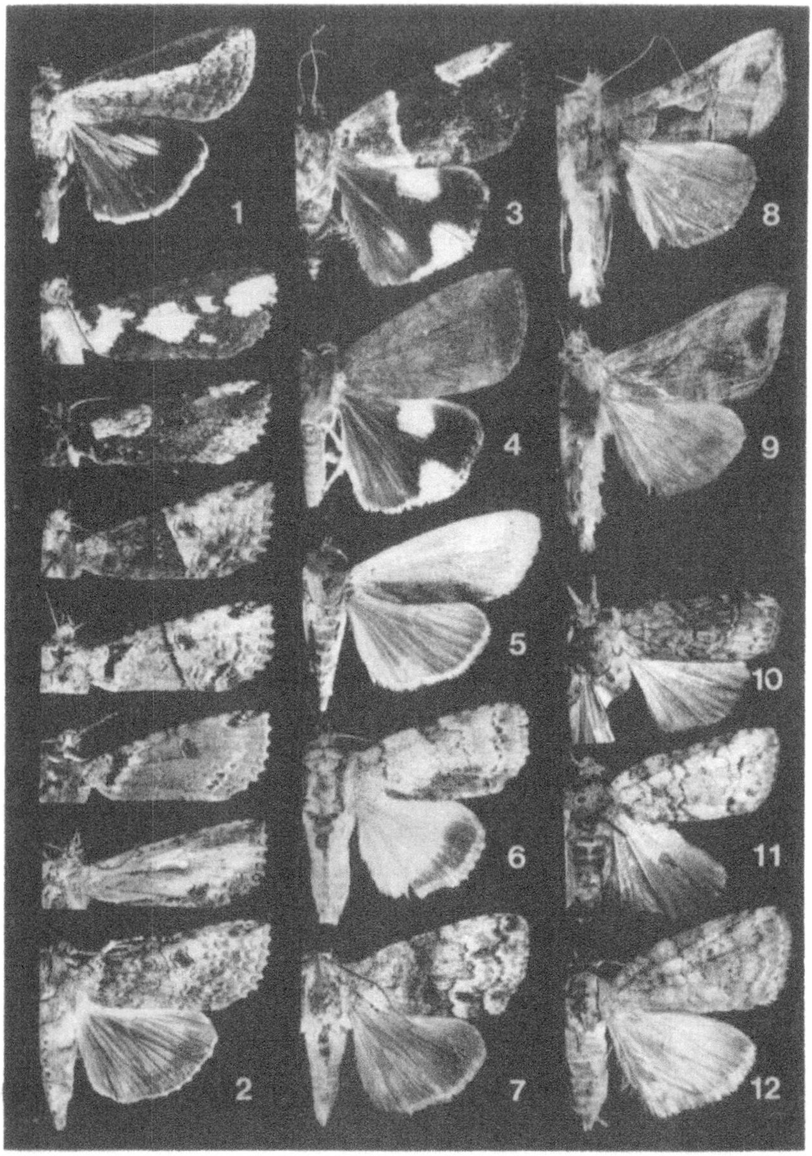

*Plate 79.* 1. *Lophoptera squammigera* (16 mm); 2. *Stictoptera tridentifera*, showing forewing variability; 3, 4. *Maceda savura paniensis;* 5. *Westermannia poupa* ♂; 6. *Blenina donans* (16–18 mm); 7. *Blenina lichenopa;* 8. *Ctenoplusia placida caledonica;* 9. *Anadevidia peponis* (18 mm); 10. *Nycteola pseudonilotica;* 11. *Nycteola canoides;* 12. *Characoma vallata* (9 mm).

Colnett (1); R. Amoa (4) Ste. Therèse (2); Probokonoro (2); Col des Roussettes (2); Mt. Krapé site 55 (32), site 56 (3); Col de Boa (1); near Bâ (5); Mt. Panié site 60 (4); Mt. Aoupinié (10); Col Mouirange (3); Mt. Pouédihi (2); Yaté Gorge (3); Grand Lac site 72 (1); Port Boisé site 73 (2), site 74 (1); R. Koué (5); Anna-Madeleine (1); Les Dalmates site 77 (1), site 78 (3); Mt. Koghis site 80 (3).

It is common on both main rock substrates, perhaps associated to a large degree with rainforest.

## NOCTUIDAE: SARROTHRIPINAE

### *Blenina donans* Walker (Plate 79: 6, Figure 109: 1)

*Blenina donans* Walker, 1857, *List Specimens lepid Insects Colln Br. Mus.*, 13: 1215, type examined.
*Blenina metachrysa* Turner, 1902, *Proc. Linn. Soc. N. S. Wales*, 27: 89, **syn. n.**
*Blenina triphaenopsis* Hampson, 1905, *Ann. Mag. nat. Hist.* (7), 16: 545, **syn. n.**, type examined.

These three names apply to insects with grey forewings and yellow hindwings banded brown marginally, from Sri Lanka, Queensland and Singapore respectively. No difference could be detected in the genitalia of males from these localities and externally the specimens are very similar or within the range of variation for any of the localities concerned. They are therefore regarded here as conspecific.

The species occurs in the Indian region and from Sundaland through the Philippines, Sulawesi, Lesser Sundas and Moluccas to New Guinea and northern Australia. VIETTE (1949) recorded the species from the Loyalty Is. and it was taken during the survey as follows: Grand Lac site 2 (3), site 72 (2); Pic du Pin site 7 (3); Mt. des Sources site 11 (4), site 12 (1); Dothio site 19 (1); Poupa (1); Bréhoa R. (5); Col Toma (7); Mt. Dogny (1); Mt. Mou site 33 (4), site 36 (1); Probokonoro (10); Mt. Krapé site 55 (6), site 56 (2); Mt. Panié site 60 (6), site 61 (1); Mt. Aoupinié (2); R. Koué (1); Mt. Koghis site 80 (1). BPBM: Nouméa, Mt. Koghis. MNHN: La Coulée.

### *Blenina lichenopa* (Meyrick) (Playe 79: 7, Figure 109: 2)

*Ceparcha lichenopa* Meyrick, 1897, *Trans. ent. Soc. Lond.*, 1897: 374, type examined.

The following named forms are considered here to be subspecies of *lichenopa* (Queensland), having only slightly different male genitalia and similar facies, though each race is variable.

*Blenina lichenopa nigrans* Bethune-Baker, **stat. n.** *Blenina nigrans* Bethune-Baker, 1906, *Novit. zool.*, 13: 222, type examined. *Blenina varians* Bethune-Baker, 1906, *Ibid*: 223, type examined, **syn. n.** This subspecies has a generally

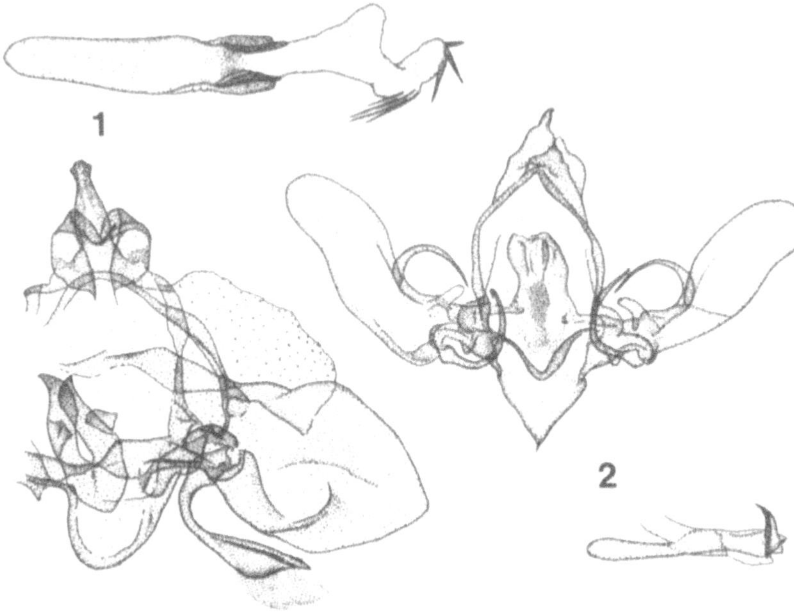


*Fig. 109.* Male genitalia of *Blenina donans* (1) and *B. lichenopa* (2).

darker green forewing ground than the typical race and is often suffused blackish. New Guinea, Bismarck Is.

*Blenina lichenopa solomonis* Swinhoe, **stat. n.** *Blenina solomonis* Swinhoe, 1905, *Ann. Mag. nat. Hist.* (7), 16: 151. The forewings have a more transversely banded appearance in many specimens, with alternation of greenish brown and grey. Solomons; similar specimens have been taken in the New Hebrides.

*Blenina lichenopa vatu* Robinson, **stat. n.** *Blenina vatu* Robinson, 1975: 157, type examined. This race is similar externally to typical *lichenopa* but there are minor genitalic differences, notably an atypically narrow basal sclerotised band to the juxta. Fiji, Tonga.

New Caledonian specimens were most similar to typical *lichenopa* and were taken as follows: Dothio site 19 (2); Koh (4); Ouatou R. (4); Bréhoa R. (14); Ouambéo R. (1); Anse Longue site 32 (1); Bambou R. (3); Kavatch (3); Probokonoro (10); Mt. Krapé site 55 (3). BPBM: La Foa. MNHN: Col d'Amieu.

## *CALATHUSA* Walker

*Calathusa* Walker, 1858, type species *basicunea* Walker (Nye, 1975);
*Steneugoa* Hampson, 1914, *Cat. Lep. Phalaenae Colln Br. Mus.*, (Suppl.), 1: 369, **syn. n.**

*Steneugoa*, a monotypic genus originally placed in the Lithosiinae, has the first abdominal spiracle within the tympanal organ and is therefore a noctuid (A. WATSON, *pers. comm.*). The facies and the structure of the male genitalia of the type species, *nubilosa* Hampson, indicate that the genus should be sunk to *Calathusa*.

Though centred in Australia, this genus extends strongly to New Caledonia where there are five species. There are two in each of the New Hebrides and New Guinea and one in each of Fiji (not described by ROBINSON), the Solomons, Bismarcks and Sulawesi. All these species are described or discussed below. The forewings are elongate, usually greyish with prominent stigmata and dark fasciae, the hindwings usually pale with a broad, diffuse dark band marginally. The male antennae are strongly ciliated, those of the female very much less so. The male retinaculum is not significantly bar-shaped but the forewing stigmata contain raised scales. The genitalia of the males are simple relative to those of other sarrothripines, the valves short, basally rather rounded, varying mainly in the apical processes from species to species.

*Calathusa stenophylla* (Turner) (Plate 80: 1, Figure 110: 5)

*Corula stenophylla* Turner, 1902, *Proc. Linn. Soc. N. S. Wales*, 27: 100, cross-checked photographically with the type, courtesy of Dr. I. F. B. COMMON.

The species is known from southern Queensland and New South Wales. Three specimens were taken in New Caledonia, one at Anse Longue site 16 and two at Ouen Toro site 64.

**Calathusa brehoa sp. n.** (Plate 80: 11, Figure 110: 1)

♂♂ 11 mm. The forewing ground is white, shaded grey and fasciated black as illustrated. The general forewing fasciation and shading and the enlarged stigmata are as in *octogesima* Turner (Queensland) but *octogesima* is larger (14 mm), with forewings a generally darker grey, less mottled, and the hindwings and abdomen fawnish white instead of white. The hindwing grey marginal band is wider and more definite in *octogesima*. The genitalia resemble those of *octogesima* but the valve apex is double instead of single, less elongate and less incurved; the aedeagus vesica lacks the two small cornuti of the Australian species.

HOLOTYPE ♂. New Caledonia: Bréhoa R., 320 m Site 24 05853/76139 (J. D. HOLLOWAY) 14.6.1971, BM 1971-507. BM noctuid slide 7537.

PARATYPE ♂. New Caledonia: Poupa, 200 m Site 20 06120/76150 (J. D. HOLLOWAY) 10.6.1971, BM 1971-507.

**Calathusa uniformis sp. n.** (Plate 80: 6, Figure 110: 2)

♂♀ 14 mm. The thorax and forewings are grey, the hindwings and abdomen pale fawn, the hindwings with a brownish grey margin. The black-flecked form

illustrated is less frequent, only two out of six specimens taken exhibiting them. The aedeagus contains a single cornutus which, with the generally similar facies, places the species close to *basicunea* Walker (type examined; Queensland, New South Wales). But the forewing fasciation, obvious in *basicunea,* is almost absent in *uniformis* and the postmedial is more angled opposite the cell. There are similarly dark-flecked forms amongst the BMNH material of *basicunea.* The apex of the valve in *uniformis* lacks the prominent inturned and angled projection of *basicunea.*
HOLOTYPE ♂. New Caledonia: Kopéto, 160 m Site 30 05102/76542 (J. D. HOLLOWAY) 24.6.1971, BM 1971-507. BM noctuid slide 7537.
PARATYPES: 4♂♂, 1♀ New Caledonia, R. Bleue site 10 (2), R. Kuenthio (1), Kopéto (1), Tiebaghi site 41 (1); 1♂ La Coulée 28.10.76 (A. DELOBEL) MNHN.

**Calathusa basicunoides sp. n.** (Plate 80: 7, Figure 111: 1)

♂♀ 13 mm. The male is very similar externally to *basicunea,* the fasciation resembling most closely that of the type of *acronyctoides* Walker, a synonym of *basicunea.* The postmedials and antemedials of *basicunoides* tend to be angled rather than sinuous. The hindwing is paler and less suffused greyish from the margin, often with a prominent discal spot. The female forewing is suffused dark grey and the hindwing also is darker. The male genitalia are similar to those of *uniformis* but the valve apex is more bulbous, less evidently bilobed.
HOLOTYPE ♂. New Hebrides: Aneityum I., Agathis Camp-1150 ft, 19–21.vii.1971 (G. S. ROBINSON) Roy. Soc. Expedition. BM noctuid slide 8841.
PARATYPES. New Hebrides: 1♂, as holotype; 1♂ Efate I., Vila 23.vi.1971 (G. S. ROBINSON); 1♂, 1♀ Erromango I., Ipota 9–13.viii.1971 (G. S. ROBINSON); 1♂ Santo, Luganville 1.ix.1953 (J. D. BRADLEY); 2♂♂ Red Crest, Aneityum, 3 miles N.E. of Anelgauhat, 1200 ft June 1955 (E. CHEESMAN).

**Calathusa humboldti sp. n.** (Plate 80: 8, Figure 111: 3)

♂♀ 12 mm. This species is very similar to *basicunoides* in genitalia structure but the valves are more incurved apically and the aedeagus cornutus is shorter. Externally it is generally smaller, the forewing more fawn in ground colour. The forewing fasciation is relatively reduced or absent, but similar to that of *basicunoides* when present. The costal section of the antemedial and the irregular submarginal are generally evident, prominent in some specimens, the former occasionally with dark brown diffusing broadly along the costa distad as in the holotype (illustrated; the submarginal is unusually faint).
HOLOTYPE ♂. (New Guinea) Humboldt Bay, Sept–Oct 1892 (W. DOHERTY). BM noctuid slide 8864.
PARATYPES. ♀♀ as holotype.

The species also occurs on islands off the coast of New Guinea, namely Rook, Vulcan and St. Aignan.

**Calathusa argentea sp. n.** (Plate 80: 9, Figure 111: 2)

♂♀ 14–16 mm. The forewing has black-brown markings on a silvery grey ground. The hindwing is pale brown with a broad whitish area over the dorsum diffusing into the brown. The forewing fasciation is similar to that of *basicunoides* but the antemedial is straighter, more oblique and the postmedial is less angled, more smoothly curved and without the veins being darkened basad as occurs in *basicunoides*; the medial area is generally darker than the ground. The hindwing has no discal spot.

The male genitalia are typical of the *basicunoides* group with simple, inturned valves and a large cornutus in the vesica, but there is also a second cornutus and the valve is apically broader than in the other species.

HOLOTYPE ♂. Sidaonta, Paloe, W. Celebes (Sulawesi) 4500 ft June 1937 J. P. A. KALIS). BM 1939-1, Rothschild Bequest. BM noctuid slide 8873.

PARATYPES. Sulawesi, same general data as holotype: 6♂♂, 3♀♀ G. Tompoe (2700 ft); 1♂, 1♀ Lindoe Paloe (3700 ft); 1♂ G. Rangkoenan (1800 ft); 5♂♂, 3♀♀ Koelawi Paloe (3100 ft).

**Calathusa basirufa sp. n.** (Plate 80: 12, Figure 110: 3)

♂♀ 9–12 mm. The ground colour of the forewings is fawn and is marked and shaded with dull brown as illustrated; there is a rufous tinge to the forewing antemedial, postmedial and the area between the reniform and postmedial in many specimens. The female specimen illustrated has a reddish tinge to the basal area of the hindwing, which may be a characteristic of fresh specimens. The hindwing has a broad grey border. The male genitalia differ from those of all congeners examined in the relatively small and simple valves.

HOLOTYPE ♂. New Caledonia: R. Amoa, 10 m Site 51 05300/76859 (J. D. HOLLOWAY) 19.7.1971, BM 1971-507. BM noctuid slide 7539.

PARATYPES. 6♀♀ New Caledonia, R. Bleue site 8, Ouatou R., Petit Gouli, Cap Colnett, Ouen Toro site 64.

There are distinct subspecies in the Loyalty Is. and New Hebrides as follows.

**Calathusa basirufa suzukii ssp. n.** (Plate 80: 13, 14)

♂ 12 mm. The forewings are fawn, only lightly marked as illustrated. The subspecies is distinguished by the two dark brown bars at the costa on the discal side of the postmedial and antemedial and by the lack of a basal indentation of the postmedial just exterior to the reniform. A female specimen also lacks this indentation but the forewing is darkish grey, ashy in places and the antemedial is almost straight, though irregular.

HOLOTYPE ♂. Loyalty Is.: Ouvéa, May 1974 (H. SUZUKI).

PARATYPE ♀. Loyalty Is.: Fayaoué, Ouvéa, ii.1963 (N. L. H. KRAUSS) BPBM.

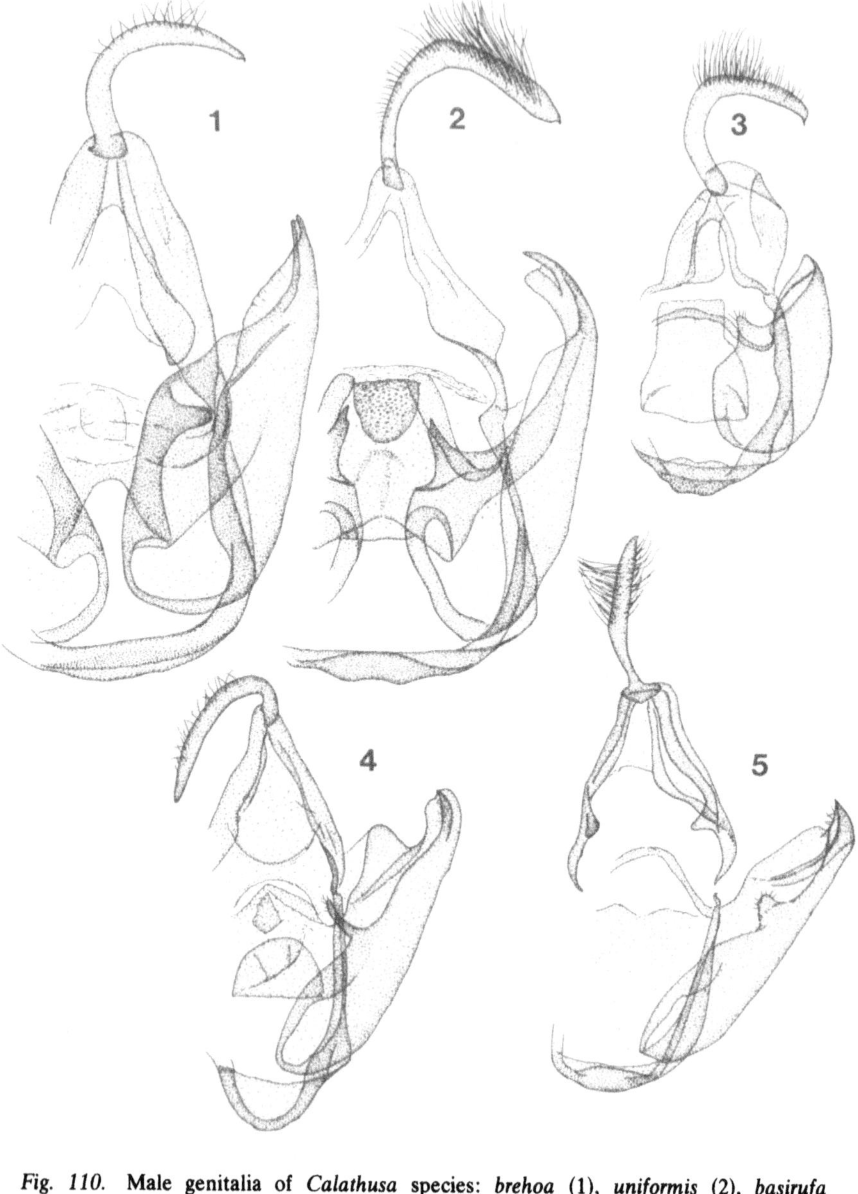


*Fig. 110.* Male genitalia of *Calathusa* species: *brehoa* (1), *uniformis* (2), *basirufa basirufa* (3), *mesospila* (4) and *stenophylla* (5).

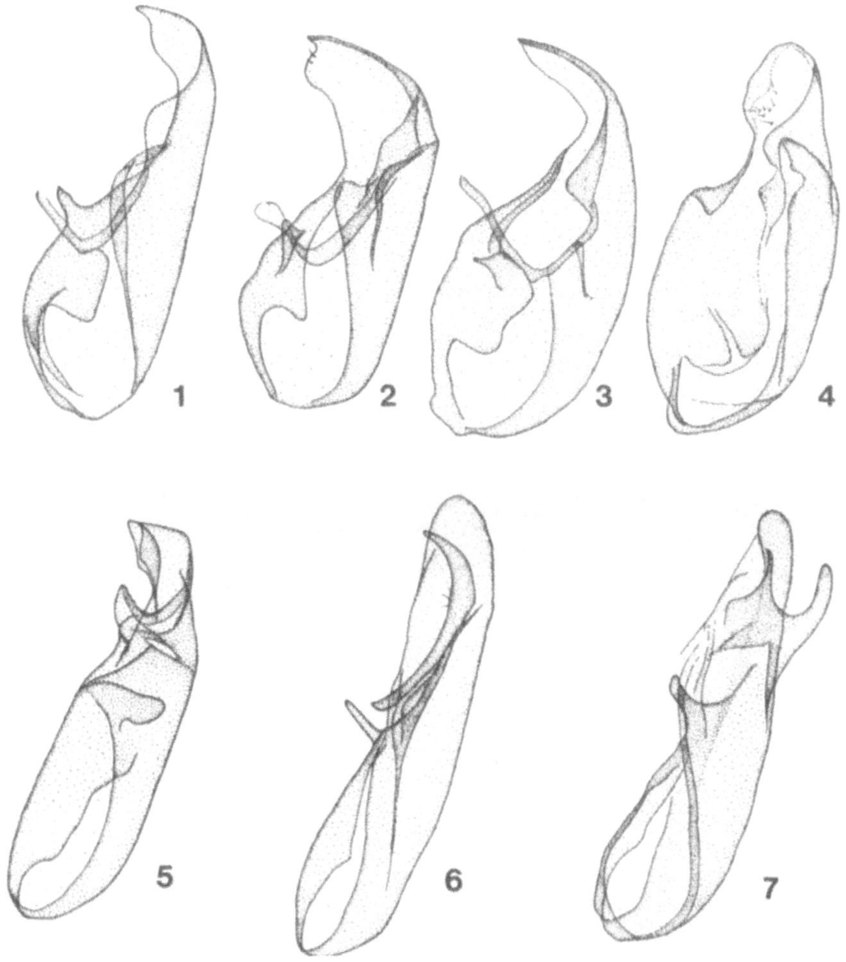


*Fig. 111.* Valves of the male genitalia of *Calathusa* species: ***basicunoides*** (1), ***argentea*** (2), ***humboldti*** (3), ***talesea*** (4), ***nubilosa*** (5), ***striata*** (6), ***cinerea*** (7).

**Calathusa basirufa ipota ssp. n.** (Plate 80: 15)

♂♀ 11–13 mm. This subspecies is very similar to the typical race but the hindwings are almost uniform grey, slightly darker marginally. The forewings are grey, suffused with blackish brown and much less rufous than in *basirufa*.
HOLOTYPE ♂. New Hebrides: Erromango I., Ipota 9–13.viii.1971 (G. S. ROBINSON) Roy. Soc. Expedition. BM noctuid slide 8840.
PARATYPES, New Hebrides: 8♂♂, 5♀♀ as holotype; 1♂ Santo, Apouna R.

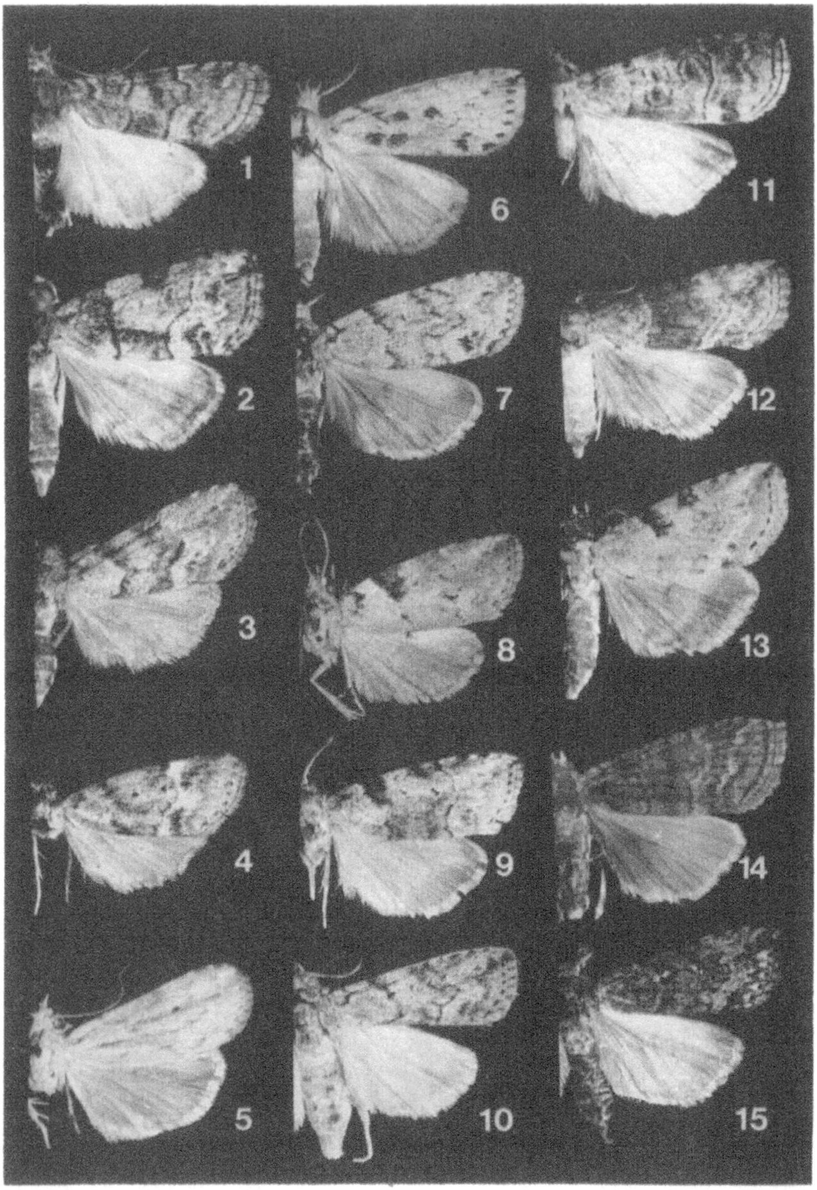


*Plate 80.* *Calathusa* species. 1. *stenophylla* (9 mm); 2. *mesospila* (8 mm); 3. *talesea;* 4. *nubilosa* (13 mm); 5. *striata;* 6. *uniformis;* 7. *basicunoides;* 8. *humboldti;* 9. *argentea;* 10. *cinerea;* 11. *brehoa;* 12. *basirufa basirufa;* 13 (♂), 14 (♀). *basirufa suzukii;* 15. *basirufa ipota.*

Camp 2 450 ft 26–27.viii.1971; 1♂, 1♀ Efate I., Vila, 1971. All Roy. Soc. Expedition (G. S. ROBINSON).

*Calathusa mesospila* (Turner) (Plate 80: 2, Figure 110: 4)

*Corula mesospila* Turner, 1902, *Proc. Linn. Soc. N. S. Wales*, 27: 99, BMNH specimen from Turner collection examined.

This Australian species (S. Queensland, New South Wales) was recorded in New Caledonia as follows: Dothio site 18 (1); Kopeto (2); Kavatch (1).

**Calathusa cinerea sp. n.** (Plate 80: 10, Figure 111: 7)

*Calathusa sp.;* Robinson (1975: 158).

♂♀ 11–13 mm. The male forewing is shades of ashy grey with darker, almost black markings and fasciation; the hindwing is fawnish white, greyish at the margin. The female has the forewing darker, more uniform, the fasciae slightly broader, less sharp, the postmedial and antemedial dull rufous rather than black; the hindwing is pale brown, grading darker over the marginal third. The forewing postmedial is shaped as in *basicunea* but the antemedial is curved concave based rather than oblique. The male genitalia have a diagnostic bifid projection subapically on the valve; the aedeagus vesica has no cornuti.

HOLOTYPE ♂. Fiji: Lautoka (H. PHILLIPS) 8.9.1928, pres. by Comm. Inst. Ent., BM 1950–525. BM noctuid slide 6950.

PARATYPES. Fiji: 1♀ as holotype, 28.5.1930; 1♀ no locality; 1♀ Suva 23.12.27 (H. W. SIMMONDS); 1♂ Viti Levu, Suva, December 1974; 3♂♂ Wakaya I., 13–15.viii.1974 (G. S. ROBINSON) BM 1974-462.

The species has also been taken in Tonga.

**Calathusa talesea sp. n.** (Plate 80: 3, Figure 111: 4)

♂♀ 12–14 mm. The facies and colouring are similar to that of female *cinerea* but the species is larger, somewhat less ashy grey, more fawn, and with a diagnostic row of dark spots submarginally on the forewing (not obvious in the illustration). The hindwings are shades of pale, dull brown. The forewing medial area is variable, either uniform with the rest of the wing or darker; one specimen has a paler central area. The male genitalia have the valves very broad apically compared with other species.

HOLOTYPE ♂. New Britain: Talesea, January 1925 (A. F. EICHHORN). BM noctuid slide 8868.

PARATYPES. 6♀♀ as holotype.

*Calathusa nubilosa* (Hampson) **comb. n.** (Plate 80: 4, Figure 111: 5)

*Steneugoa nubilosa* Hampson, 1914, *Cat. Lep. Phalaenae Colln Br. Mus.* (Suppl.) 1: 639, type examined.

♂♀ 11–13 mm. The forewing facies resemble *talesea* but the moth is smaller; there are the same forewing submarginal dots but the prominent dark spots at the reniform and orbicular are distinctive. The male genitalia have the valves broad and complex apically, folded and lobed as illustrated.

The species was described from New Guinea. There is material in the BMNH from Sudest I. and Rossell I. that is generally smaller (10 mm) but otherwise identical in facies and genitalia.

**Calathusa striata sp. n.** (Plate 80: 5, Figure 111: 6)

♂ 10 mm. The forewing is bone white, marked in rufous fawn with diffuse dark brown longitudinal streaks in the centre of the cell and just posterior to it. The hindwings are dull, pale brown, diffusing paler basad. The genitalia have the valves much more elongate than in other species, with a distinctive harpe as illustrated.

HOLOTYPE ♂. (Solomons) Guadalcanar, v.01 (A. S. MEEK). BM noctuid slide 8867.

*Characoma vallata* (Meyrick) (Plate 79: 12, Figure 112: 3)

*Hypothripa vallata* Meyrick, 1902, *Trans. ent. Soc. Lond.*, 1902: 47, type examined.

*C. vallata* is an Australian species (Northern Territory to New South Wales) and was recorded as follows in New Caledonia: Grand Lac site 2 (7), site 4 (1); Mt. des Sources site 11 (2), site 12 (2); R. Kuenthio (7); Poupa (2); Koh (3); Ouatou R. (2); Bréhoa R. (2); Col d'Amieu (2); Col Toma (2); Petit Gouli (1); Sarraméa (2); Anse Longue site 32 (3); Mt. Mou site 33 (6); Bambou R. (3); Col d'Amoss site 47 (1); Cap Colnett (3); Kavatch (6); R. Amoa (2); Probokonoro (8); Mt. Krapé site 55 (2); Ouen Toro site 64 (1), site 81 (1), site 82 (1); Col Mouirange (1); Port Boisé site 74 (1); R. Koué (1); Anna-Madeleine (1); Les Dalmates site 77 (5). BPBM: Bourail, Nouméa. It is a widespread species, perhaps most characteristic of dry, open habitats off the ultramafics.

**Nycteola canoides sp. n.** (Plate 79: 11, Figure 112: 1)

♂♀ 10 mm. The species is close to *cana* Hampson (New Guinea, type examined, 12 mm) but smaller. The forewing has similar dark brownish grey fasciation and grey shading on a white or slightly fawn ground. The species differs in the more zig-zag double postmedial, especially near the costa, and in the greater distance between the double antemedial and double subbasal fasciae; the fasciation is generally more definite. The hindwing is diffusely grey over the apical two-thirds in *canoides* but only over the apical third in *cana*. The genitalia are as illustrated.

HOLOTYPE ♂. New Caledonia: Mt. Krapé, 460 m Site 55 05276/76375 (J. D. HOLLOWAY) 22.7.1971, BM 1971-507. BM noctuid slide 7534.

PARATYPES. New Caledonia: 1♂ Sarraméa. 160 m Site 29 05888/76065 (J. D. HOLLOWAY) 22.6.1971; 1♂ St. Louis? ex white cocoon on *Eugenia jambosa* leaflet, Nov. 9, 1940 (FXW) BPBM; 1♀ Mt. Koghis 10.1.77 (A. DELOBEL) MNHN.

**Nycteola pseudonilotica sp. n.** (Plate 79: 10, Figure 112: 2)

♂ 9 mm. The insect is generally brownish grey, fasciated darker and paler on the forewing as illustrated. In facies the species bears some resemblance to the pantropical *Characoma nilotica* Rogenhofer (photo of type examined) but the forewing postmedial is more evenly curved, the antemedial more oblique, thus closer to the postmedial at the dorsum; the narrow, dark submarginal of *nilotica* is lacking. The hindwing is a uniform grey, but off-white with a grey-brown margin diffusing basad along the veins in *nilotica*. The male genitalia of *nilotica* are similar to those of *vallata* (q.v.), but those of *pseudonilotica* are typical of *Nycteola*.

The closest relative within *Nycteola* is probably *exophila* (Meyrick) from Queensland (type examined). Compared with *exophila*, *pseudonilotica* is browner, less grey, with the forewing antemedial more curved, concave basad, and the postmedial not markedly sigmoid. The male genitalia have a more prominent peniculus than *exophila*, a short, broad, square uncus (cf. narrower, longer), a broader saccus and larger, more complex valves with a diagnostic, single, strap-like seta at the apex.

HOLOTYPE ♂. New Caledonia: Sarraméa, 160 m Site 29 05888/76065 (J. D. HOLLOWAY) 22.6.1971, BM 1971-507. BM noctuid slide 7533.

There is also a specimen with hindwings and abdomen missing from Col Toma.

## NOCTUIDAE: CHLOEPHORINAE

### *Earias flavida* Felder

*Earias flavida* Felder; ROBINSON (1975: 165).

This species is widespread in the Indo-Australian tropics, extending to Queensland, Samoa and Tonga. It was recorded in New Caledonia as follows: Poupa (1); Bréhoa R. (1); Mt. Mou site 33 (1).

### *Earias luteolaria* Hampson

*Earias luteolaria* Hampson; ROBINSON (1975: 166).

The distribution is similar to that of the previous species but the species is recorded more widely through northern Australia and occurs in the Loyalty Is. In New Caledonia the larva is a pest of *Gossypium* (COHIC, 1956) and was recorded during the survey as an adult as follow: Mt. des Sources site 14 (1); Cap Colnett (1); Amoa R. (1); Mt. Krapé site 55 (1).

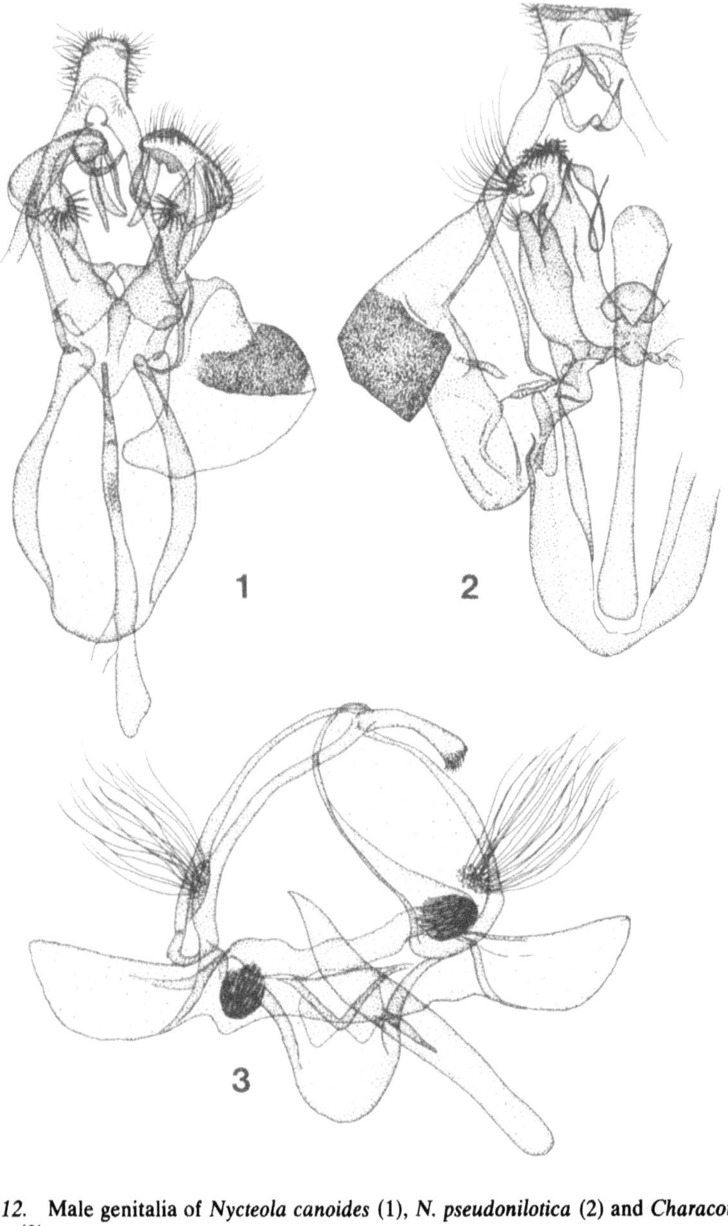


*Fig. 112.* Male genitalia of *Nycteola canoides* (1), *N. pseudonilotica* (2) and *Characoma vallata* (3).

*Earias perhuegeli* Holloway (Figure 113: 2)

*Earias perhuegeli* Holloway, 1977: 78.

This species was referred in the past to *huegeli* Rogenhofer (ROBINSON, 1975: 166) but the type of *huegeli* resembles *vittella* (Fabricius). *E. perhuegeli* is widespread in Australia and extends eastwards into the Pacific as far as the Austral Is., the Marquesas and the Gilbert Is. It is vagrant to Norfolk I. The larva has been recorded as feeding on *Gossypium* and *Hibiscus rosa-sinensis* in New Caledonia (COHIC, 1956) and VIETTE (1949) recorded the species from Nouméa. Two adults were taken at Anse Longue site 16 and one at Kaala-Gomen during the survey and there is one from Bourail in the BPBM. It is probably a west coast species.

*Earias vittella* (Fabricius)

*Earias vittella* (Fabricius); ROBINSON (1975: 167).

The distribution is similar to that of *luteolaria* but the species has not been recorded from New Caledonia. There is a single specimen from Lifu in the BMNH. This species was misspelt *vitella* by ROBINSON (1975) and HOLLOWAY (1976), following an unjustified emendation by SHERBORN in *Index Animalium*.

**Maceda savura paniensis ssp. n.** (Plate 79: 3, 4, Figure 113: 1)

*Maceda savura* Robinson (1968: 249); ROBINSON (1975: 169), type examined.

♂♀ 11–17 mm. Compared with the typical race (Fiji, New Hebrides), *paniensis* has the antemedial of the forewing irregularly rather than smoothly curved; the postmedial generally meets the dorsum obtusely rather than acutely distad; the hindwing white areas are more extensive, the anterior almost twice the area of that of *savura*; the marginal white area of the hindwing extends narrowly along the margin on each side of the patch more than in *savura*, with the fringes flecked dark brown. The male genitalia are as in *savura*.

HOLOTYPE ♂. New Caledonia: Mt. Krapé site 55 460 m 05276/76375 (J. D. HOLLOWAY) 22.7.1971, BM 1971-507. BM noctuid slide 7526.

PARATYPES: 25 ex., New Caledonia, various survey localities; 29 ex. N. New Caledonia, Tinchialit, 2020 ft 22.viii–10.ix.1949 (L. E. CHEESMAN) BM 1950-1.

The species was recorded as follows: Mt. Mou site 33 (1); Probokonoro (2); Mt. Krapé site 55 (4); Mt. Panié site 60 (16), site 61 (3); Mt. Aoupinié (2). It is probably a rainforest species and is particularly abundant in localities on the Panié range. In Fiji it is also a rainforest species (ROBINSON).

*Maurilia iconica* (Walker)

*Maurilia iconica* (Walker); ROBINSON (1975: 169).

This is a widespread Indo-Australian tropical species that extends to northern Australia and east to Samoa, Tonga and the Cook Is. It has been recorded as follows in New Caledonia: Pic du Pin site 7 (1); Port Boisé site 74 (1); Nouméa (P. COCHEREAU).

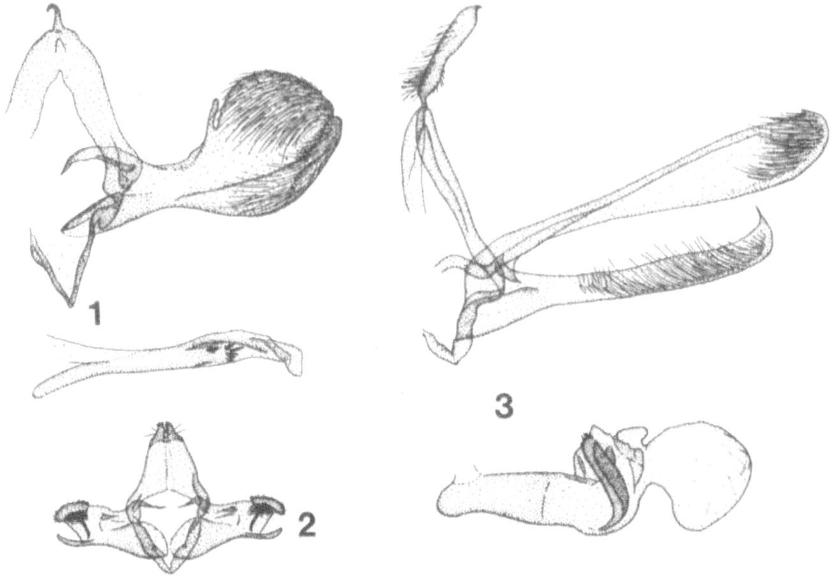


*Fig. 113.* Male genitalia and aedeagi of *Maceda savura paniensis* (1), *Westermannia poupa* (3), and male genitalia of *Earias perhuegeli* (2).

**Westermannia poupa sp. n.** (Plate 79: 5, Figure 113: 3)

♂ 12 mm, ♀♀ 13–14 mm. The head and palps are white, the thorax and abdomen dark grey; these are all creamy white in the closest relative, *argentata* Butler, from northern Australia. The forewings are glossy white as in *argentata* but with a submarginal row of black dots and a broad band of dark grey along the dorsum, banded obliquely paler especially in the male. The hindwing is off-white, marginally (except for white fringes) dark grey-brown which grades into the ground basad, especially along the veins, but less so than in *argentata*. In the male genitalia the valves are almost completely divided as illustrated; in *argentata* the valve is rounded, entire, though much more heavily sclerotised on the dorsal and ventral margins. The two species are similar in the structure of the uncus, tegumen, transtillae and saccus. The aedeagi are also similar though the apex of the vesica in *argentata* is not so produced or bulbous.
HOLOTYPE ♂. New Caledonia: Mt. Krapé, 460 m Site 55 05276/76375 (J. D. HOLLOWAY) 22.7.1971, BM 1971-507. BM noctuid slide 7526.
PARATYPES. New Caledonia: 1♀ as holotype; 3♀♀ Poupa, site 20.

Two further specimens from Poupa were not retained. The species may prove to be associated with *Acacia spirorbis*, common at both localities where it was taken.

### *Xanthodes congenita* (Hampson)

*Xanthodes congenita* (Hampson); ROBINSON (1975: 170).

A single specimen from Lifu is amongst the BMNH series of this species. It is widespread in Australia except the extreme south and has been recorded from Fiji. The facies and male genitalia indicate *congenita* to be close to, possibly conspecific with, *intersepta* Guenée from the Oriental Region (India, Burma, Andamans, Java). *X. graellsii* Feisthamel, from Africa and southern Europe, has similar facies but distinctive male genitalia.

## NOCTUIDAE: CATOCALINAE

### *Achaea janata* (*Linnaeus*)

*Achaea janata* (Linnaeus); VIETTE (1950a: 147); ROBINSON (1975: 171); HOLLOWAY (1977: 80).

Widespread in the Indo-Australian and Pacific tropics, *janata* was recorded as follows in New Caledonia: Grand Lac site 1 (1), site 2 (5), site 3 (2); Plaine des Lacs site 5 (1); Pic du Pin site 7 (10); Mt. des Sources site 11 (6), site 12 (8), site 14 (53); Anse Longue site 15 (7), site 16 (18), site 32 (4); R. Kuenthio (4); Dothio site 18 (1), site 19 (3); Poupa (2); Koh (9); Bréhoa R (1); Col d'Amieu (5); Col Toma (1); Mt. Mou site 33 (7), site 34 (1); Bambou R. (1); near Ba (1); Col Nassiboué (1); Les Dalmates site 77 (1). BPBM: Nouméa, Col des Roussettes. MNHN: Mt. Koghis. The species declined in abundance markedly during the survey period. It is widespread and may prove to be commonest in more open habitats. It was noted frequently by day amid the leguminous thicket at Anse Longue.

New Caledonian host plants are *Musa sapientum*, *Ricinus communis* (COHIC, 1956) and *Excoecaria agallocha* (P. COCHEREAU, ORSTOM records). The last two are recorded host plants on Norfolk I. (HOLLOWAY, 1977); numerous host plants have been noted in Fiji (ROBINSON).

### *Achaea serva* (Fabricius)

*Achaea serva* (Fabricius); VIETTE (1950a: 146); ROBINSON (1975: 171).
*Achaea pentasema* Prout, 1919, *Ann. Mag. nat. Hist.* (9), 3: 181.

This species extends from the Oriental tropics to northern Australia, Tonga and Samoa. In New Caledonia it was recorded more from rainforest localities than its congener as follows: Grand Lac site 3 (1); Mt. des Sources site 14 (2); Poupa (1); Koh (6); Ouatou R. (1); Bréhoa R. (11); Ouambéo R. (1); Col d'Amieu (17); Col Toma (3); Sarraméa (2); Kopéto (3); Mt. Mou site 33 (10); Cap Colnett (1); Ste. Therèse (5); Probokonoro (11); Mt. Krapé site 55 (9), site 56 (2); Houailou (1); near Bâ (1); Yaté Gorge (2); Port Boisé site 73 (1), site 74 (3); Les Dalmates site 77 (1); Mt. Koghis site 80 (1). BMNH: Tinchialit. BPBM: Isle of Pines.

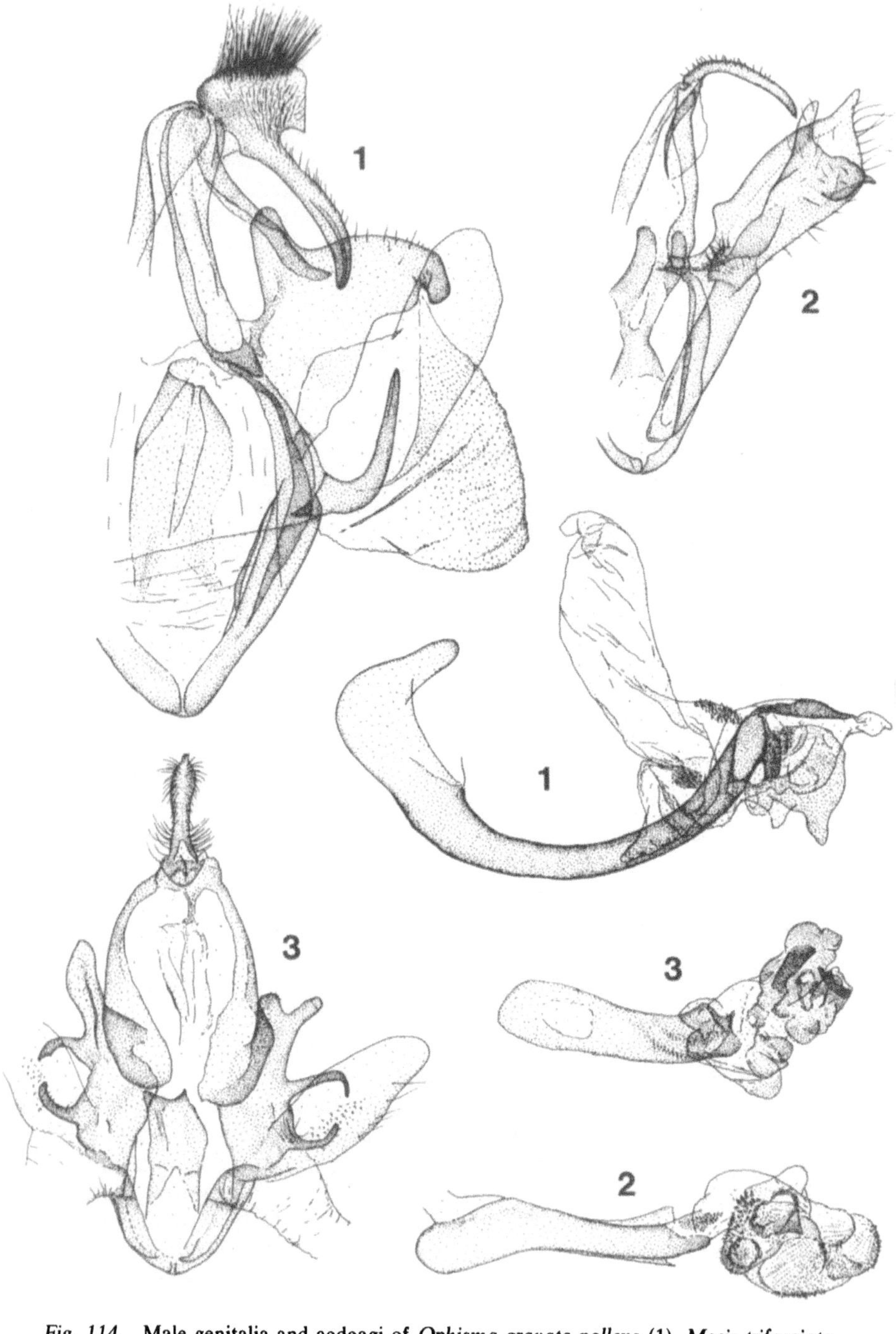


*Fig. 114.* Male genitalia and aedeagi of *Ophisma gravata pallens* (1), *Mocis trifasciata* (2) and *Chalciope alcyona* (3).

*Chalciope alcyona* (Druce) (Figure 114: 3)

*Chalciope alcyona* (Druce); ROBINSON (1975: 174).

The male genitalia of this species resemble those of the widespread Oriental *mygdon* (Cramer), but the chocolate area of the forewing is more elongate with a straighter margin anteriorly; the cream bar bisecting the chocolate area is at a much more acute angle to the dorsum. *C. alcyona* is known from New Guinea, northern Australia, Fiji and Samoa, the Bismarcks, Solomons, Moluccas, Lesser Sundas and Sulawesi. It was recorded as follows in New Caledonia: Grand Lac site 1 (1), site 3 (1); Pic du Pin site 7 (2); Dothio site 18 (1); Koh (1); Col d'Amieu (1); Petit Gouli (1); Kopéto (2); Col d'Amoss site 46 (1); R. Amoa (7); Mt. Krapé site 55 (5); Houailou (1); Port Boisé site 73 (1). BPBM: Nouméa. MNHN: Mt. Koghis.

*Grammodes oculicola* Walker

*Grammodes oculicola* Walker; VIETTE (1950a: 152); ROBINSON (1975: 175).

VIETTE recorded *oculicola* from Nouméa and the Loyalty Is., and there is a specimen from Mt. Koghis in the BPBM. It is found throughout most of Australia except the south and occurs in the New Hebrides, Fiji, Tonga, and the Cook and Austral Is.

*Mocis frugalis* (Fabricius)

*Mocis frugalis* (Fabricius); VIETTE (1950a: 155); ROBINSON (1975: 176); HOLLOWAY (1977: 80).

*M. frugalis* is widespread in the Old World and Pacific tropics and was recorded as follows in New Caledonia: Grand Lac site 3 (2), site 72 (1); Pic du Pin site 7 (4); Mt. des Sources site 14 (6); Anse Longue site 15 (1); R. Kuenthio (11); Dothio site 19 (3); Poupa (4); Ciu (1); Koh (11); Bréhoa R. (2); Col d'Amieu (2); Sarraméa (3); Mt. Mou site 33 (3); Grottes de Koum (2); Bambou R. (5); Col d'Amoss site 47 (1); Cap Colnett (1); Kavatch (4); R. Amoa (23); Ste. Therèse (2); Probokonoro (7); Col des Roussettes (5); Mt. Krapé site 55 (4); Port Boisé site 73 (1), site 74 (2); Anna-Madeleine (1); Ouen Toro site 81 (1). VIETTE: Nouméa, Canala, Voh. MNHN: La Tamoa. BPBM: Nouméa, La Foa.

*Mocis trifasciata* (Stephens) (Figure 114: 2)

*Mocis trifasciata* (Stephens); VIETTE (1950a: 156); ROBINSON (1975: 176); HOLLOWAY (1977: 81).

The species has been recorded from the Moluccas, New Guinea, Queensland, the Bismarcks, Solomons, New Hebrides, Loyalties, Fiji, Rotuma, Samoa, Tonga, the Society Is., Austral Is. and the Marquesas. Like its congener this species shows a preference for open secondary habitats and grassland on sedimentary rock in New Caledonia; both are grass feeders as larvae. It was recorded as follows: Grand Lac site 1 (2), site 2 (2), site 3 (2); Pic du Pin site 7 (4); Mt. des Sources site 14 (13); Anse Longue site 15 (1), site 16 (3); R. Kuenthio (3); Dothio site 19 (4); Poupa (6); Koh (54); Ouatou R. (3); Bréhoa R. (3); Col d'Amieu (14); Petit Gouli (2); Sarraméa (4); Mt. Mou site 33 (4); Bambou R. (3); Cap Colnett (4); Kavatch (3); R. Amoa (20); Ste. Therèse (1); Probokonoro (34); Col des Roussettes (1); Mt. Krapé site 55 (1); near Bâ (3); Col Nassiboué (1); Port Boisé site 73 (2); Ouen Toro site 81 (1). BPBM: Le Crouen, Tipindjé, Nouméa.

*Ophisma gravata pallens* (Lucas) (Plate 83: 12, Figure 114: 1)

*Ophisma gravata* Guenée, 1852, *Hist. nat. Insectes* (*Lépid.*), 7: 237, type examined.

*Grammodes pallens* Lucas, 1892, *Proc. Linn. Soc. N. S. Wales* (2), 7: 260.

The typical race has been recorded from India, the Himalaya, China, S.E. Asia, the Philippines and New Guinea. Subspecies *pallens* from Queensland and New South Wales has similar male genitalia but is paler, the forewing lacking the thin, straight medial line and the hindwing having a less developed brown band. A single specimen was taken in New Caledonia at the R. Bambou site.

*OPHIUSA* Ochsenheimer

This genus, with its type species *tirhaca* (Cramer) (Nye, 1975), is most properly applied to those Pacific species usually referred to *Anua* Walker (e.g. by Viette (1950a) and Robinson (1975)); this treatment follows that of Berio (1965) who studied the genitalia of numerous African Catocalinae. The type species of *Anua* is *amplior* Walker, a subjective synonym of *finifascia* Walker (Nye, 1975), a species with an unmodified uncus and simple valves. *O. disjungens*, *O. discriminans* and *O. parcemacula*, discussed below, are closely related to *tirhaca*; *O. coronata* and *O. hituensis* have somewhat different genitalia and are related to each other; *O. trapezium* has genitalia somewhat similar to *parcemacula* and *tirhaca* but is rather different in facies.

*O. tirhaca* has been recorded as feeding on *Myrtus communis* and *Pistacia spp.* as a larva in the Middle East (E. P. Wiltshire, *pers. comm.*) and on a variety of shrubs (*Combretum*, *Erica*, *Rhus Leptospermum*, *Eucalyptus*, *Daphne*) in South Africa (Pinhey, 1975). Sevastopulo (1941) recorded the species from *Pistacia*, *Cistus* and *Rhus* in India and Singh *et al.* (1954–1960) added *Terminalia* for that locality. Sugi (1976) gave *Myrtus* and *Psidium* as Japanese host plants.

*O. disjungens* or its subspecies *tongaensis* has been reared from *Eucalyptus*, *Carea* and other Myrtaceae in India (Sevastopulo; as its subspecies *indiscriminata*), from *Styphelia* in Guam (Swezey, 1946) and from *Eucalyptus* and *Psidium* (both Myrtaceae) in Fiji (Robinson).

Both *O. discriminans* and *O. parcemacula* have been reared from *Melaleuca quinquenervia* (niaouli; Myrtaceae) in New Caledonia, so it would appear that this group of species has a definite association with the Myrtaceae. The *O. kenricki* Bethune-Baker group of species discussed by ROBINSON (1975) is also related, possibly originally derived from *discriminans*, but there is no information on their biology.

The other *Ophiusa* species probably have a wider range of hosts; ROBINSON listed *Storckiella, Quisqualis* and *Terminalia* for *O. coronata* in Fiji.

*Ophiusa coronata* (Fabricius)

*Ophiusa coronata* (Fabricius); VIETTE (1950a: 145); ROBINSON (1975: 172).

This species was not taken during the survey but VIETTE recorded it from Nouméa and it is known as a citrus pest, the adult piercing ripe fruit in a manner similar to *Othreis* species (P. COCHEREAU, *pers. comm.*). COHIC (1956) gave *Terminalia catappa* as a New Caledonian host plant. The species is widespread in the Indo-Australian tropics, Micronesia, northern Australia and eastwards into the Pacific as far as the Society Is.

*Ophiusa disjungens disjungens* (Walker) (Figures 115: 1, 117: 2)

*Ophiodes disjungens* Walker, 1858, *List Specimens lepid. Insects Colln Br. Mus.*, 14: 1360, type examined.

The species has a wide Indo-Australian distribution, represented in the Oriental Region by ssp. *indiscriminata* (Hampson) **stat. n.** (*Minucia indiscriminata* Hampson, 1893, *Illust. typical Specimens Lepid. Heterocera Colln Br. Mus.*, 9: 111, type examined) which has an ochreous forewing not suffused greyish as in the typical form and with a prominent apical grey area that is indistinct in the typical race. The typical race occurs in the Lesser Sundas, New Guinea, Australia (Northern Territory to New South Wales, Western Australia), New Caledonia and the Loyalty Is. New Caledonian specimens differ slightly from typical Australian *disjungens* in the reduction of the hindwing black area; in a small proportion of specimens the forewing is suffused blackish, one specimen taken having the forewing almost entirely black. In the New Hebrides, Fiji, Samoa and Tonga the large subspecies *tongaensis* (Hampson) **stat. n.** (*Anua tongaensis* Hampson, 1913, *Cat. Lepid. Phalaenae Colln Br. Mus.*, 12: 434, type examined) replaces the typical race; it also occurs in Guam according to SWEZEY (1946). The forewing of *tongaensis* is relatively broader and Fijian specimens generally have the reniform slightly darker than the ground. The male genitalia of the three races exhibit no significant differences.

The species was recorded as follows in New Caledonia: Grand Lac site 2 (1); Pic du Pin site 7 (1); Mt. des Sources site 11 (1); Dothio site 19 (1); Poupa (3); Koh (18); Ouatou R. (3); Bréhoa R. (3); Col d'Amieu (4); Sarraméa (1); Mt. Mou site 33 (22), site 34 (3), site 36 (1); Tiebaghi site 40 (1), site 41 (1); Bambou R. (5); Col d'Amoss site 47 (2); Mt. Ignambi (1); Cap Colnett (2); R. Amoa (1); Probokonoro (37); Mt. Krapé site 55 (8), site 56 (1); near Bâ (16); Mt. Aoupinié (4); Les Dalmates site 77 (2). BPBM: Mt. Koghis, Col Mouirange.

*Fig. 115.* Male genitalia of *Ophiusa disjungens disjungens* (parts, 1), *O. discriminans intacta* (2) and *O. tirhaca* (3).

*Ophiusa discriminans intacta* (Hampson) **stat. n.** (Plate 81: 1, Figures 115: 2, 117: 4)

*Ophiodes discriminans* Walker, 1858, *List Specimens lepid. Insects Colln Br. Mus.*, 14: 1358, type examined.
*Anua intacta* Hampson, 1913, *Cat. Lepid. Phalaenae Colln Br. Mus.*, 12: 437, type examined.

Typical *discriminans* is recorded from India, S.E. Asia, Sundaland and the Philippines. Subspecies *intacta*, lacking the broad black submarginal of the

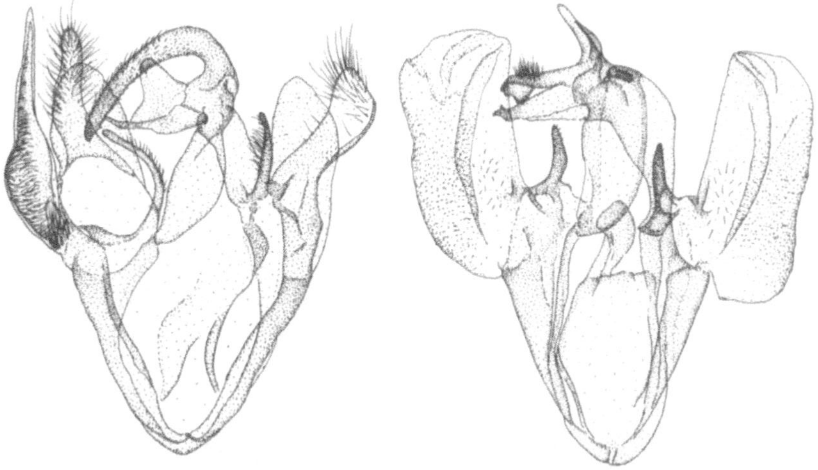

*Fig. 116.* Male genitalia of *Ophiusa hituensis* (left) and *O. trapezium* (right).

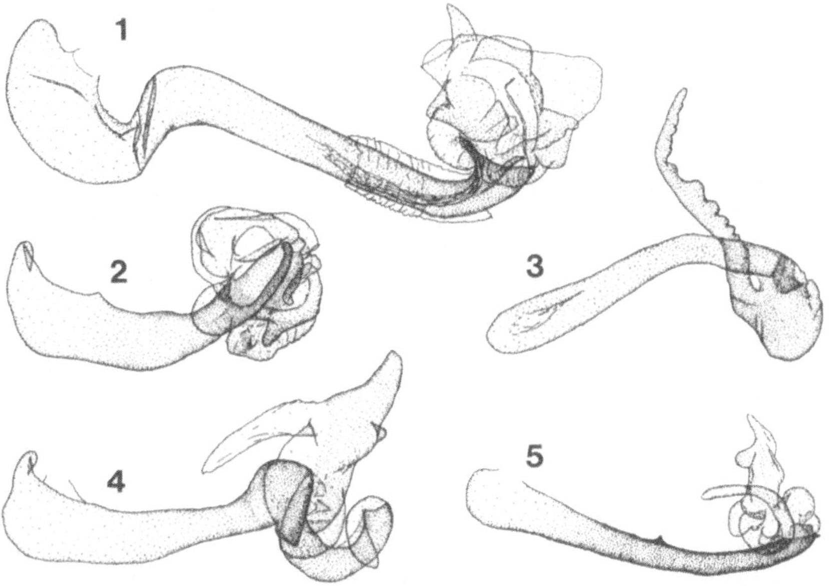


*Fig. 117.* Aedeagi of *Ophiusa* species: *tirhaca* (1), *disjungens* (2), *hituensis* (3), *discriminans* (4), *trapezium* (5).

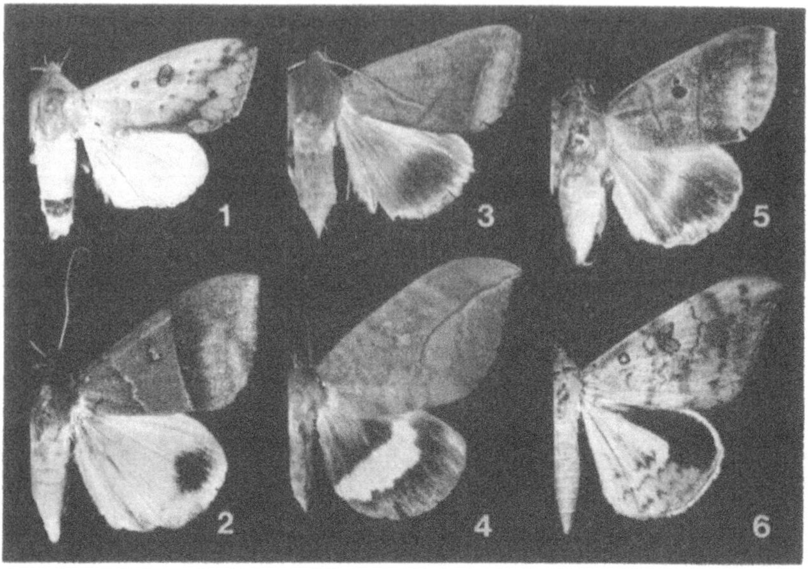

*Plate 81.* 1. *Ophiusa discriminans intacta* (28 mm); 2. *Thyas miniacea rubrior;* 3. *Ophiusa trapezium* ♀ (22 mm); 4. *Phyllodes imperialis dealbata;* 5. *Ophiusa hituensis* (32 mm); 6. *Oxyodes scrobiculata obscurior.*

hindwing, having a narrower left process of the tegumen in the male genitalia, a narrower left harpe that is not spatulate and a right valve broader apically, is recorded from northern Australia, Kei, the Moluccas, New Guinea (*sublutea* Warren is synonymous), the Bismarcks, the New Hebrides and New Caledonia.

New Caledonian records were as follows: Pic du Pin site 7 (1); Dothio site 19 (1); Koh (2); Ouatou R. (2); Bréhoa R. (1); Col d'Amieu (4); Sarraméa (1); Mt. Mou site 33 (3); Mt. Ignambi (1); Probokonoro (45); Col des Roussettes (1); Mt. Krapé site 55 (2); near Bâ (2); Mt. Aoupinié (1). BMNH: Tinchialit. It is less frequent than *disjungens*, possibly more closely associated with rainforest.

A small larva was swept from niaouli (*Melaleuca quinquenervia*) at Col Toma on 19 June. It was perhaps 1.5 cm long and dark grey-brown as in the left hand illustration on Plate 55 (taken two weeks later when 2 cm). It reared to maturity on niaouli, becoming paler with successive instars. The right hand illustration of Plate 55 shows the penultimate instar, marked with brown on fawn, tinged with ochre. It is similar in appearance to the larvae of *disjungens* and *tirhaca*. On 6 August it was 5 cm long, usually lying pressed to the stem of the foodplant but moving in geometrid fashion when disturbed. It underwent four moults from discovery. On 18 August it was 7 cm long, and soon after it pupated, spinning together niaouli leaves. Pupation was on 26 August and the adult emerged late in September. The precise date of emergence is unknown, the pupa having been freighted back to Britain whilst the author visited Norfolk I.

*Ophiusa hituensis* (Pagenstecher) (Plate 81: 5, Figures 116: left, 117: 3)

*Ophisma hituensis* Pagenstecher, 1884, *Jb. nassau. Ver. Naturk.*, 37: 235.

This species is probably a Papuan Region derivative of *coronata*, being recorded from Sulawesi, Kei, the Moluccas, New Guinea, Northern Territory and N. Queensland, the Bismarcks, Solomons and New Hebrides. It was recorded as follows in New Caledonia: Koh (1); Col d'Amieu (2); Col Toma (3); Bambou R. (1); R. Amoa (1); Ste. Therèse (1); Mt. Aoupinié (1).

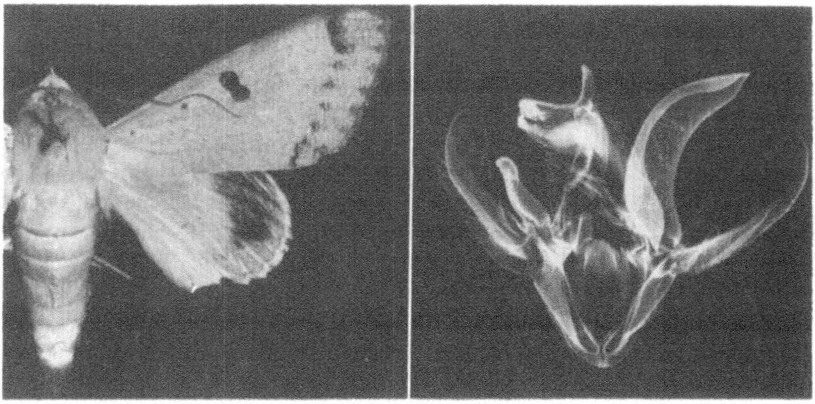

*Plate 82.* Facies (30 mm) and male genitalia of *Ophiusa parcemacula*. The aedeagus is not shown but is similar to that of *O. trapezium*.

*Ophiusa parcemacula* (Lucas) (Plate 82)

*Ophiodes parcemacula* Lucas, 1891, *Proc. Linn. Soc. N. S. Wales* (2), 5: 304.

A single male of this species, reared from *Melaleuca quinquenervia* at La Coulée, was sent to the author by A. DELOBEL and will be deposited in the MNHN. The species is widespread in Australia and has been recorded from New Guinea.

*Ophiusa tirhaca* (Cramer) (Figures 115: 3, 117: 1)

*Noctua tirhaca* Cramer, 1780, *Uitlandsche Kapellen*, 2: 116.

The species occurs in Africa, the Mediterranean area and most of the Oriental tropics and subtropics. It has been recorded from Sulawesi, the Moluccas, most of Australia except the south, and Vulcan I. The distribution is somewhat similar to that of *Hypocala deflorata* (q.v.). One specimen was taken at Probokonoro during the survey and it has been recorded infrequently in Nouméa by ORSTOM entomologists.

*Ophiusa trapezium* (Guenée) (Plate 81: 3, Figures 116: right, 117: 5)

*Ophiodes trapezium* Guenée, 1852, *Hist. nat. Insectes* (*Lépid.*), 7: 231, type examined.
*Ophisma circumferens* Walker, 1865, *List Specimens lepid. Insects Colln Br. Mus.*, 33: 956, **syn. n.,** type examined.

The species is sexually dimorphic; *trapezium* was described from a male and *circumferens* from a female. A female is illustrated. The male is larger, the forewing more ochreous with a prominent, broad, marginal band.

The species has been recorded from India, the Himalaya, Sundaland, New Guinea, Queensland and the Bismarcks. It was recorded in New Caledonia as follows: Col Toma (1); Cap Colnett (2); Ste. Therèse (3).

**Parallelia lateritica sp. n.** (Plate 83; 10, 11, Figures 118: 1, 119: 4)

♂♀ 22–24 mm. The male has the forewing and thorax shades of rufous and the hindwing and abdomen rufous fawn as illustrated; the females are usually a more leaden, purplish red. The paler area of the forewing from the base to the straight medial is suffused with pale bluish grey, especially at the costa. The head and patagia are rich rufous in both sexes. The species is somewhat similar in facies to the much larger (30 mm) Fijian *koroensis* Robinson, but the forewing margin is not obtusely angled centrally and the forewing antemedial is concave distad instead of straight or slightly convex.

In the male genitalia the Fijian species has a prominently trifid dorsobasal process to the valve, compared to a simple, rounded one, and about ten moderate cornuti in the aedeagus vesica instead of one large one; the valve in *koroensis* is much rounder, the rod-like harpe much longer; the uncus lacks the dorsal 'peak' of *lateritica*. No other *Parallelia* examined (e.g. the species listed by HOLLOWAY (1976)) have such large cornuti in the vesica. The two species also differ from the rest of the '*joviana*' type of species (a broader aggregation than the complex discussed below), characterised by symmetrical genitalia, in lacking the chocolate area on the costa subapically. They are probably most closely related to each other but are nevertheless significantly different.
HOLOTYPE ♂. New Caledonia: Bréhoa R., 320 m Site 24 05853/76139 (J. D. HOLLOWAY) 15.6.1971, BM 1971-507. BM noctuid slide 7427.
PARATYPES. 20 ex. New Caledonia various localities.

The species is a member of the ultramafic association and was recorded as follows: Grand Lac site 2 (7), site 3 (3), site 72 (1); Pic du Pin site 7 (7); Bréhoa R. (1); Mt. Mou site 33 (2); Ste. Therèse (1); Mt. Krapé site 55 (2); R. Koué (1).

The Fijian *koroensis* is associated with montane *Agathis* and *Podocarpus* forest (ROBINSON, 1975).

The *joviana* complex of species

Three New Caledonian *Parallelia*, *cuneilineata*, *vitiensis* and *solomonensis*, are very closely related to each other and to *P. joviana* (Stoll). *P. joviana* (type locality the Coromandel coast of India) occurs widely in the Oriental tropics with only minor variation in facies and genitalia. It was illustrated by HOLLOWAY (1976). The dorsobasal process of the valve is diagnostic, elongate, tapering, flexed at right-angles centrally, bearing only small lateral projections. A distinct race (undescribed) occurs in New Guinea and the Goodenough Is., having a paler ground colour and a smoothly concave posterior two-thirds to the exterior margin of the forewing postmedial chocolate band.

*P. cuneilineata* (Warren) is restricted to New Caledonia, the Loyalties and, with the related *duplicata* Robinson from Fiji, is distinguished by the notching of the lobes of the forewing chocolate band and chocolate subapical mark. These two species have somewhat similar dorsobasal processes to the valves of the male genitalia.

The other two species in the group overlap in distribution through virtually all the Papuan Region. *P. vitiensis* (Butler) resembles *joviana* in external appearance but has distinctive male genitalia. *P. solomonensis* is also distinctive in its male genitalia and has different facies, except in the Solomons and Bismarcks where it converges in appearance with *vitiensis*. Externally there are no really reliable distinguishing characters for the two species in these localities but in *vitiensis* the hindwing postmedial below tends to be angled two-thirds anteriorly instead of curved; the corresponding fascia on the forewing is also slightly more angled. In the female genitalia the ductus bursae in *solomonensis* is more scobinate in the distal half and is angled in the basal half; it is straight in *vitiensis*.

### *Parallelia cuneilineata* (Warren) (Plate 83: 6, 7, Figures 118: 4, 119: 3)

*Ophiusa cuneilineata* Warren, *Novit. zool.*, 22: 153, type examined.

The species is variable, the forewing being relatively uniform or with the postmedial band increasingly chocolate brown distad. It occurs in Lifu and the New Hebrides (Santo, Malekula, Erromango, Aneityum). In the latter locality the dorsobasal process of the valve is slightly broader and the apical spine of the aedeagus single instead of bifid.

It was recorded in New Caledonia as follows: Grand Lac site 2 (1); Mt. des Sources site 14 (1); Koh (3); Col d'Amieu (3); Col Toma (1); Mt. Mou site 33 (1); R. Amoa (1); Probokonoro (16); near Bâ (1); Col Nassiboué (1); Port Boisé site 74 (1). BMNH: Tinchialit. MNHN: Mt. Koghis.

### *Parallelia solomonensis solomonensis* Hampson

*Parallelia solomonensis* Hampson, 1913, *Cat. Lepid. Phalaenae Colln Br. Mus.*, 12: 572, type examined.

The race from the Solomons and Bismarcks resembles *vitiensis* from these localities externally, being distinguished by the genitalia. Compared with the other subspecies the posterior projection of the forewing postmedial is much less prominent; the aedeagus vesica is ornamented as in ssp. *bicacuminata*. A large form of this subspecies with the forewing postmedial edged more broadly white distad occurs on Rennell I. (FLETCHER, 1957: 49).

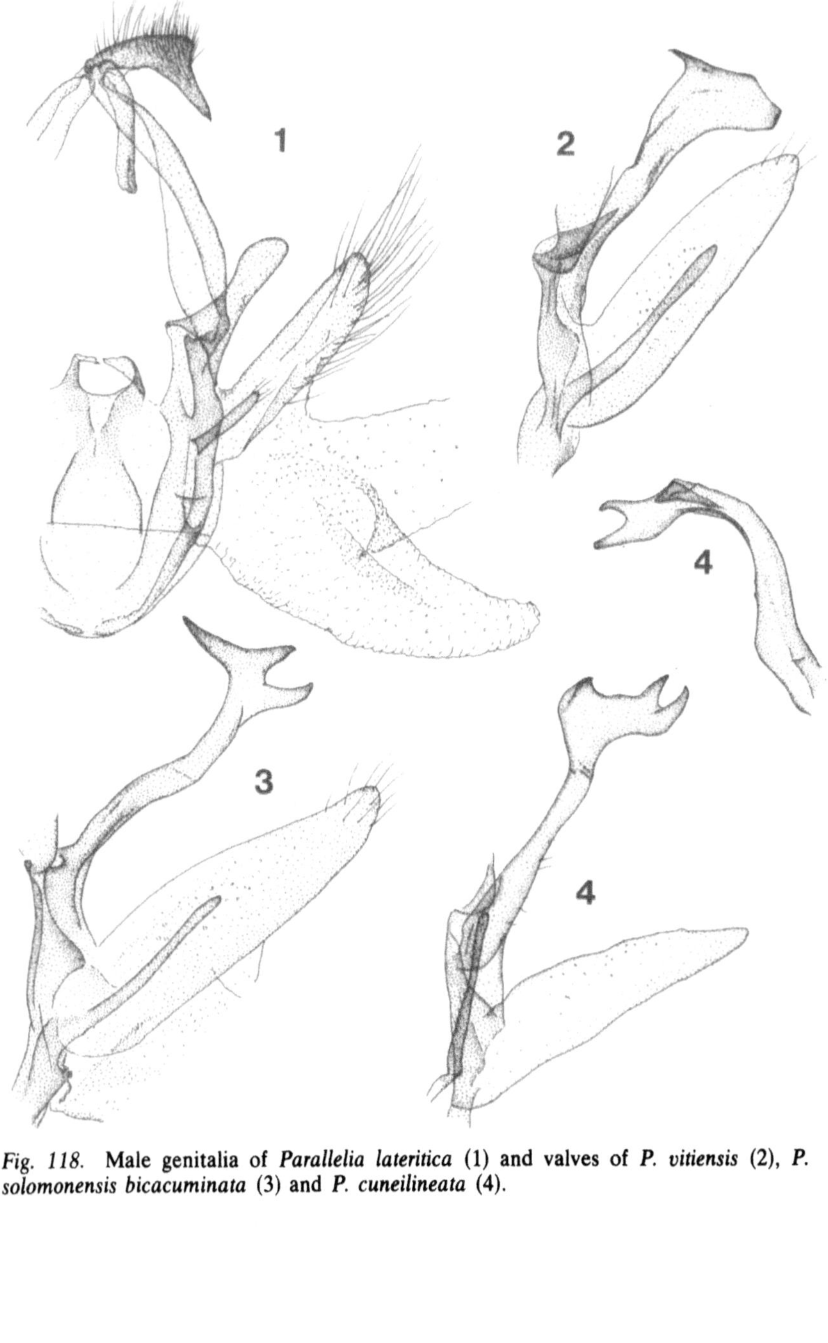


*Fig. 118.* Male genitalia of *Parallelia lateritica* (1) and valves of *P. vitiensis* (2), *P. solomonensis bicacuminata* (3) and *P. cuneilineata* (4).

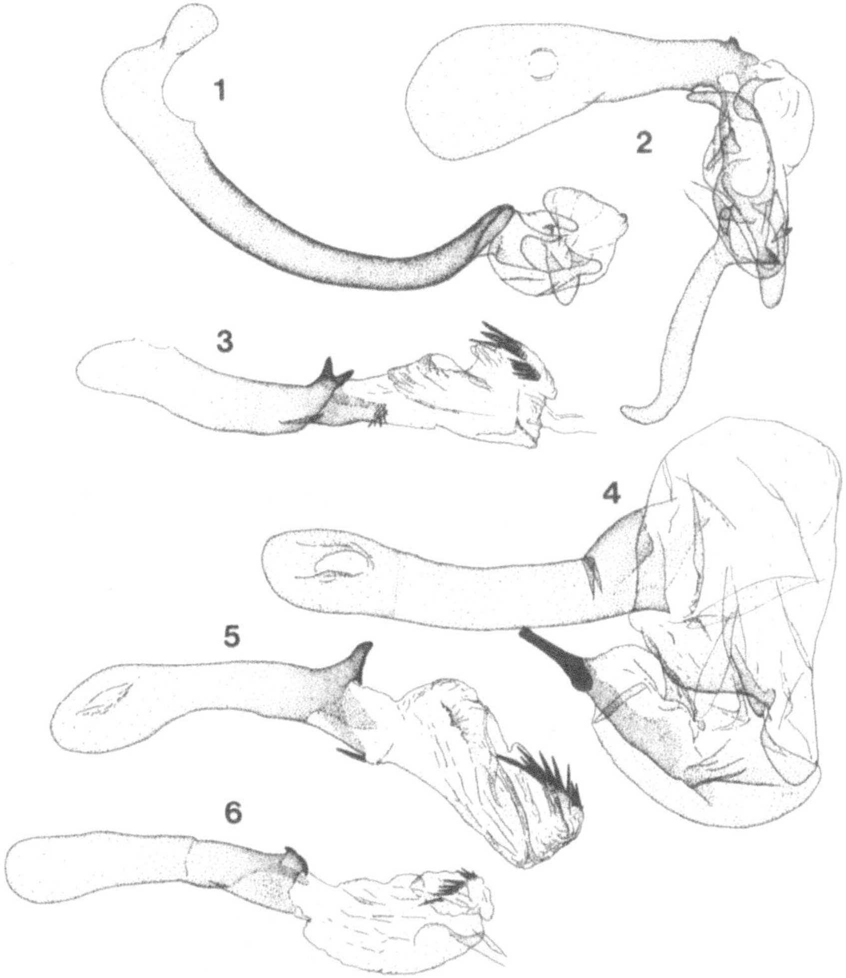


*Fig. 119.* Aedeagi of *Parallelia* species: *redunca* (1), *propyrrha* (2), *cuneilineata* (3), *lateritica* (4), *solomonensis bicacuminata* (5) and *vitiensis* (6).

**Parallelia solomonensis bicacuminata ssp. n.** (Plate 83: 2, Figures 118: 3, 119: 5)

*Parallelia joviana* (Stoll) *sensu* VIETTE (1950a: 151).

♂♀ 22–24 mm. The thorax and forewing have the pale areas pinkish fawn, variably suffused with medium grey-brown, and with chocolate brown areas as illustrated. The ground is paler than in *solomonensis* and the posterior projection of the postmedial band is almost as prominent as the anterior one. The genitalia are as illustrated.

HOLOTYPE ♂. New Caledonia: Mt. des Sources, 500 m Site 14 06636/75488 (J. D. HOLLOWAY) 31.5.1971, BM 1971-507. BM noctuid slide 7433.
PARATYPES: 7♂♂ New Caledonia, various localities; 1♂, 1♀ N. Caledon. Numéa (♀ slide 8737); 6♂♂, 3♀♀ Lifu; 1♀ Col d'Amieu (A. DELOBEL) 1977 MNHN.

The species was recorded as follows in New Caledonia: Mt. des Sources site 14 (1); Koh (1); Col d'Amieu (1); Sarraméa (1); Kopéto (1); Mt. Mou site 33 (2); Probokonoro (2).

An aberrant female was taken at Sarraméa (Plate 83: 4), perhaps a hybrid with *vitiensis*.

### **Parallelia solomonensis hebridesia ssp. n.** (Plate 83: 3)

♂♀ 22–25 mm. The forewing is heavily suffused with grey-brown, perhaps more than in typical *solomonensis*, but with a narrow pale band basad from the postmedial bar. In a few specimens the forewing postmedial band is tinged rufous. In the male genitalia the dorsobasal projection of the valve is broader than in the New Caledonian race and has the margin between the two most apical spurs convex instead of concave (the specimen illustrated by VIETTE (1950a) has this feature).
HOLOTYPE ♂. New Hebrides: Red Crest, Aneityum, 3 m N.E. of Anelgauhat, 1200 ft, May 1955 (E. CHEESMAN) BM 1955-217. BM noctuid slide 6282.
PARATYPES: 2♂♂, 3♀♀ as holotype; 1♀ Santo Is., Hog Harbour, 17.vi.1925 (P. A. BUXTON; 1♂ Efate I., Narabut Camp, 30.vi.-7.vii.1971 (G. S. ROBINSON) Roy. Soc. Expedition; 2♂♂, 1♀ Aneityum I., Agathis Camp 1150 ft 19–21.vii.1971 (G. S. ROBINSON) Roy. Soc. Expedition; 1♀ Tanna x.1930 (L. E. CHEESMAN) BM 1931-30; 1♀ Erromango viii.1930 (L. E. CHEESMAN) BM 1930-477.

### **Parallelia solomonensis papuana ssp. n.** (Plate 83: 1)

♂♂ 25 mm. Externally this subspecies is similar to *bicacuminata* except that the posterior projection of the postmedial band is less developed, though not so reduced as in the typical race. The subspecies is distinguished primarily by four cornuti central to the aedeagus vesica that do not occur in other races.
HOLOTYPE ♂ (New Guinea) Milne Bay (A. S. MEEK) 1900–45. BM noctuid slide 8718.
PARATYPE ♂. Papua: Kokoda 1200 ft viii–ix.1933 (L. E. CHEESMAN) BM 1933–427.

The race also occurs in Queensland.

### *Parallelia vitiensis* (Butler) (Plate 83: 8, Figures 118: 2, 119: 6)

*Parallelia vitiensis* (Butler); ROBINSON (1975: 180), type examined.

Dissection of material in the BMNH has indicated this species to be much more widely distributed than indicated by ROBINSON. It occurs in Sulawesi, the

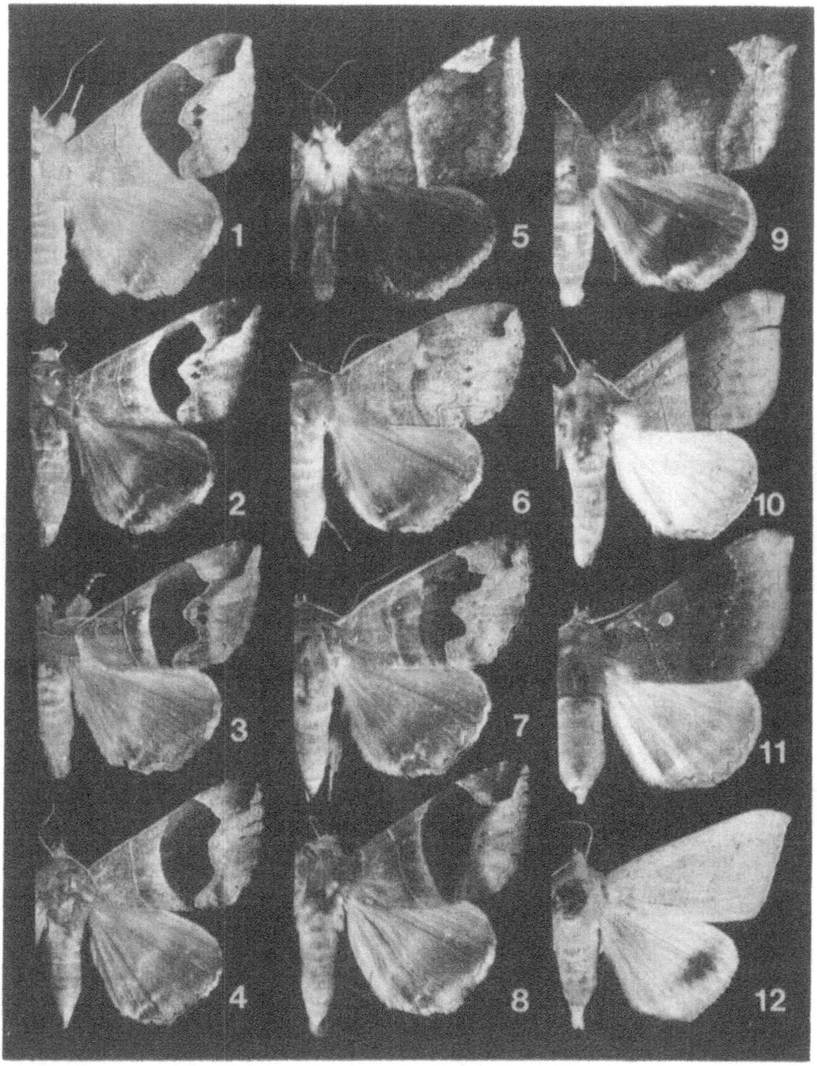


*Plate 83. Parallelia* species (1–11). 1. *solomonensis papuana;* 2. *solomonensis bicacuminata;* 3. *solomonensis hebridesia;* 4. Sarraméa ♀; 5. *propyrrha* (20 mm); 6, 7. *cuneilineata* (19–21 mm); 8. *vitiensis* (19–21 mm); 9. *redunca* (28 mm); 10 (♂), 11 (♀). *lateritica;* 12. *Ophisma gravata pallens* (25 mm).

N. Moluccas, Kei, New Guinea, Witu I., the Bismarcks and Solomons as well as the New Hebrides, New Caledonia, Fiji, Rotuma, Samoa and Tonga. From the Solomons westwards the posterior projection of the forewing postmedial dark band is more prominent generally than in more easterly forms.

The species was recorded as follows in New Caledonia: Grand Lac site 3 (1); Ouatou R. (1); Sarraméa (1); Probokonoro (3); Ouen Toro site 64 (1). MNHN: Nouméa.

*Parallelia (Pindara) prisca* (Walker)

*Parallelia prisca* (Walker); VIETTE (1950a: 150) ROBINSON (1975: 179).

The posteriorly angled (cf. curved) apical chocolate brown mark on the forewing costa distinguishes the species from very closely related taxa such as *serratilinea* Bethune-Baker from New Guinea and *illibata* (Fabricius) from the Oriental Region, the latter the type species of *Pindara* Moore. The species is found in the New Hebrides, the Loyalty Is., Fiji, Samoa and Tonga. It was recorded as follows in New Caledonia: R. Bleue site 9 (1); Koh (4); Bréhoa R. (3); Col d'Amieu (15); Col Toma (1); Sarraméa (4); Mt. Mou site 33 (1); R. Amoa (1); Probokonoro (11); Mt. Aoupinié (1). It is probably associated with rainforest.

*Parallelia propyrrha* (Walker) (Plate 83: 5, Figures 119: 2, 120: 1)

*Naxia propyrrha* Walker, 1858, *List Specimens lepid. Insects Colln Br. Mus.*, 14: 1407, type examined.

The male genitalia show this species to be close to *feneratrix* (Guenée) from Sundaland, illustrated by HOLLOWAY (1976: Figure 186). The species was described from Queensland but it also occurs in New South Wales, Fiji, Lifu, Ouvéa, the Isle of Pines and was recorded in New Caledonia as follows: Poupa (2); Sarraméa (1); Kopeto (1); Ouen Toro site 82 (1). VIETTE (1951b): Ponérihouen. MNHN: Nouméa, La Coulée. Many of the survey localities were in areas of abundant *Acacia spirorbis*.

*Parallelia (Bastilla) redunca* (Swinhoe) (Plate 83: 9, Figures 119: 1, 120: 2)

*Ophiusa redunca* Swinhoe, 1900, *Cat. east and Aust. Lepid. Heterocera*, 2: 141. VIETTE (1950a: 148).

This taxonomically rather isolated species, the type of *Bastilla* Swinhoe, is restricted to the Papuan Region. It is known from Kei, the Moluccas, New Guinea, Queensland, New South Wales, the Bismarcks, Solomons, Rennell, the New Hebrides and New Caledonia. It occurs on Lifu and was recorded in New Caledonia as follows: Grand Lac site 3 (1); Dothio site 19 (1); Koh (5); Col d'Amieu (6); Bambou R. (1); Ste. Therèse (1); Probokonoro (7); Mt. Aoupinié (2).

**Phyllodes imperialis dealbata ssp. n.** (Plate 81: 4, Figure 121: 1)

*Phyllodes imperialis* Druce, 1888, *Ann. Mag. nat. Hist.* (6), 2: 241. VIETTE (1950a: 141).

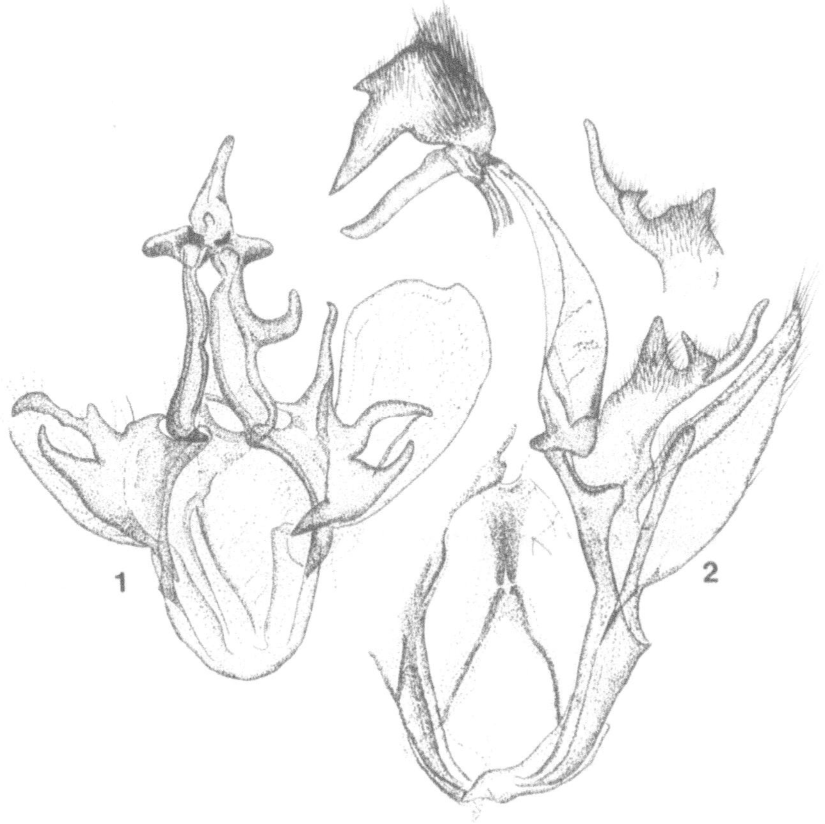


*Fig. 120.* Male genitalia of *Parallelia propyrrha* (1) and *P. redunca* (2).

♂♀ 77 mm. This subspecies differs from typical *imperialis* from New Guinea, Queensland, the Bismarcks, Solomons, New Hebrides and Lifu in the absence of diffuse white marginal triangles in the spaces of the hindwing (except very faintly in the two most anterior ones). The hindwing pink band extends five-sixths of the way to the costa from the anal angle instead of two-thirds. There is no white at the costa postmedially on the forewing. On the underside the ground colour is paler, greyer, and the hindwing dorsal dark area surrounding the pink bar is narrower distally but more extensive basally than in the typical race; on the forewing the dark, basal two-thirds containing white patches is black instead of brown and sharply defined at a more crenulate margin by white edging distad. The genitalia of the male are as in *imperialis*.
HOLOTYPE ♀. New Caledonia: Bréhoa R., 320 m Site 24 05853/76139 (J. D. HOLLOWAY) 14.6.1971, BM 1971-507.

PARATYPE ♂. New Caledonia: Poupa, 200 m Site 20 06120/76150 (J. D. HOLLOWAY) 10.6.1971, BM 1971-507. BM noctuid slide 7423.

A further very worn specimen was taken at Col d'Amieu.

*Thyas miniacea* (Felder)

*Thyas miniacea* (Felder); VIETTE (1950a: 143); ROBINSON (1975: 181); type examined.

Dissection has indicated that *regia* Lucas (1894, *Proc. Linn. Soc. N. S. Wales* (2), 8: 151) from Queensland should be placed as a subspecies (**stat. n.**) of the Fijian *miniacea*. The latter is slightly smaller than *regia*, the forewing fasciation finer and more sharply defined. Forms best referred to *regia* occur in the Lesser Sundas, Moluccas, New Guinea, Australia (Northern Territory, Queensland), the Bismarcks and Solomons. The typical race flies in Fiji and Samoa. A distinct subspecies, described below, occurs in New Caledonia and two specimens from the New Hebrides appear intermediate between the New Caledonian race and *miniacea*, somewhat closer to the former.

**Thyas miniacea rubrior ssp. n.** (Plate 81: 2, Figure 121: 2)

♂♀ 46–49 mm. This race differs from others in the diffuse and indistinct forewing submarginal, extensive pale greyish suffusion of the forewing, and the redder (cf. orange) hindwing which also has a much reduced black submarginal area. The genitalia are as in the typical race.

HOLOTYPE ♂. New Caledonia: Mt. Aoupinié, 520 m Site 62 05298/76594 (J. D. HOLLOWAY) 30.7.1971, BM 1971-507. BM noctuid slide 7424.

PARATYPES: 10 ex. New Caledonia, various localities; 1♀ Hienghène, Oct 3–6, 1940 (F.X.W.) BPBM.

The species was recorded as follows during the survey: Mt. des Sources site 14 (1); Ciu (1); Koh (2); Bréhoa R. (1); Col d'Amieu (5); Sarraméa (1); Mt. Mou site 33 (3); R. Amoa (1); Ste. Therèse (1); Probokonoro (7); Mt. Krapé site 55 (1), site 56 (1); Mt. Aoupinié (2). VIETTE: Nouméa.

*Trigonodes cephise* (Cramer)

*Trigonodes cephise* (Cramer); VIETTE (1950a: 153); ROBINSON (1975: 181).

This is a widespread Indo-Australian tropical species that extends to northern Australia and Samoa. A single female was taken at the R. Amoa site and there is a specimen from Ponérihouen in the BPBM.

*Trigonodes hyppasia* (Cramer)

*Trigonodes hyppasia* (Cramer); ROBINSON (1975: 182); HOLLOWAY (1977: 81).

This species extends throughout the Old World tropics, reaching its eastern limits in Tonga, New South Wales, Norfolk and Lord Howe Is. It may be vagrant to Norfolk I. (HOLLOWAY, 1977). It was recorded as follows in New

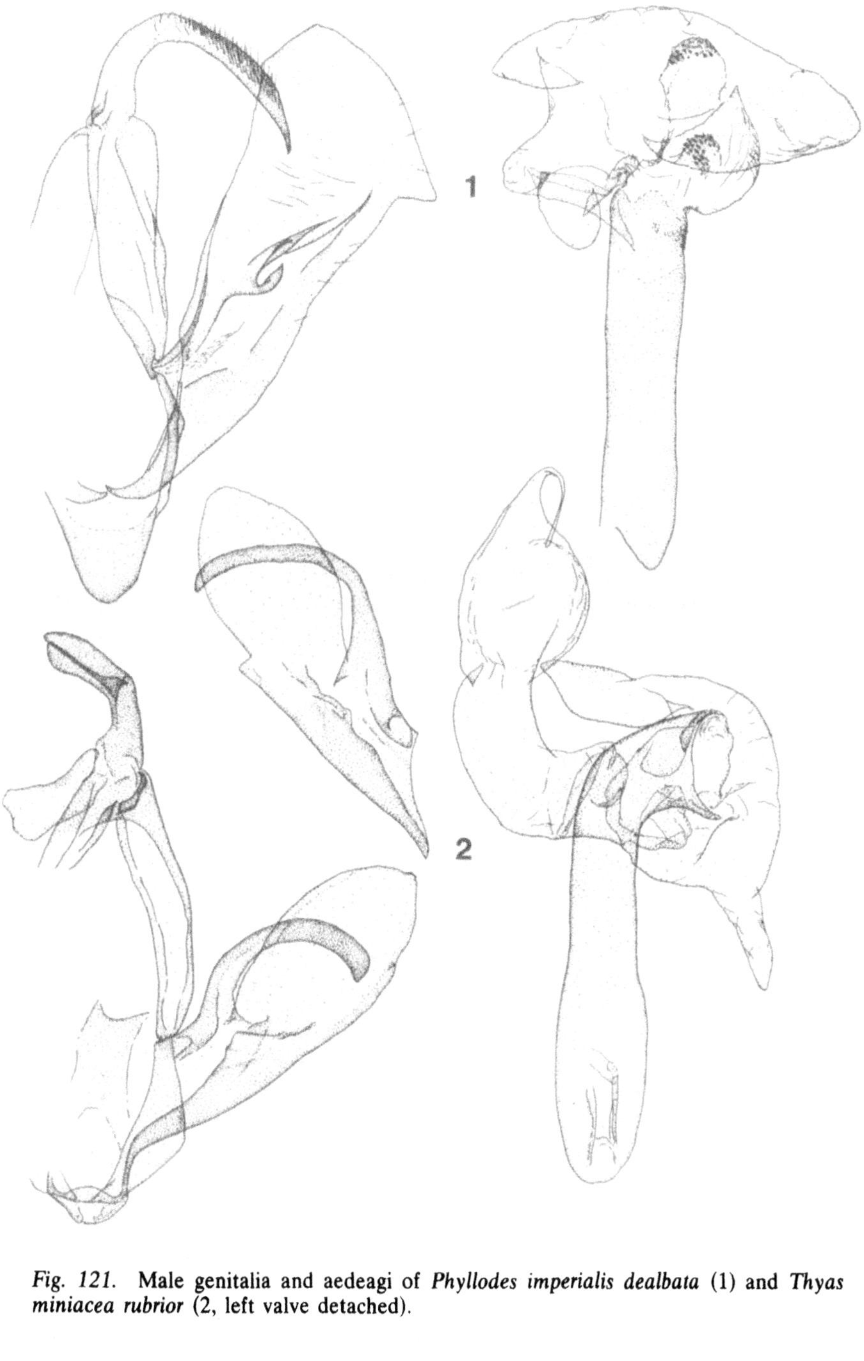


*Fig. 121.* Male genitalia and aedeagi of *Phyllodes imperialis dealbata* (1) and *Thyas miniacea rubrior* (2, left valve detached).

Caledonia: R. Kuenthio (2); Poupa (1); Koh (1); Col d'Amieu (3); Petit Gouli (1); Kopéto (1); Cap Colnett (2); R. Amoa (6); near Bâ (1); Anna-Madeleine (1). This is probably a species characteristic of open vegetation and pasture like the *Mocis* species.

## NOCTUIDAE: PLUSIINAE

### *Anadevidia peponis* (Fabricius) (Plate 79: 9, Figure 122: 2)

*Noctua peponis* Fabricius, 1775, *Syst. Ent.*: 608.

A single specimen of *peponis* was taken on Mt. Aoupinié. The species is widespread in the Indian region, S.E. Asia and Sundaland, but further east has only been recorded from New Guinea, Queensland and New South Wales. It is the type species of *Anadevidia* Kostrowicki.

### *Argyrogramma signata* (Fabricius)

*Noctua signata* Fabricius, 1794, *Ent. Syst.* 3(2): 81.

This species extends throughout the Indo-Australian tropics to Queensland and Melanesia, having been recorded recently from Tonga and the Cook Is. (it was not noted for Fiji by ROBINSON (1975)). It was recorded as follows in New Caledonia: Mt. des Sources site 14 (1); Dothio site 19 (1); Poupa (1); Koh (1); Ouatou R. (2); Bréhoa R. (1); Col d'Amieu (1); Probokonoro (1).

### *Chrysodeixis eriosoma* (Doubleday)

*Chrysodeixis eriosoma* (Doubleday); ROBINSON (1975: 185, as *chalcites* (Esper)); HOLLOWAY (1977: 81).

*C. eriosoma* extends throughout the Indo-Australian tropics and subtropics, extending as far east as Easter I. It occurs in the Loyalty Is. and was recorded as follows in New Caledonia: Grand Lac site 2 (18), site 3 (6), site 4 (3); Plaine des Lacs site 5 (2); Pic du Pin site 7 (1); R. Bleue site 9 (2), site 10 (2); Mt. des Sources site 11 (8), site 12 (4), site 13 (2), site 14 (38); Dothio site 19 (1); Poupa (5); Koh (11); Ouatou R. (2); Bréhoa R. (6); Col d'Amieu (28); Sarraméa (1); Mt. Dogny (3); Mt. Mou site 33 (84), site 34 (5), site 36 (3); Bambou R. (1); Col d'Amoss site 45 (1); Mt. Ignambi (1); Cap Colnett (1); R. Amoa (9); Probokonoro (10); Mt. Krapé site 55 (2); Mt. Panié site 60 (1); Mt. Aoupinié (1); Les Dalmates site 77 (2); Mt. Koghis site 80 (1). VIETTE (1949): Nouméa. BPBM: St. Louis Valley, Forêt de Thi.

The general population level declined somewhat during the survey but the species would appear to be widespread, perhaps more abundant away from the ultramafics. It can be a pest of vegetables and has been recorded from tomato in New Caledonia (ORSTOM records). ROBINSON (1975) listed numerous foodplants for Fiji.

*Chrysodeixis illuminata* (Robinson) **comb. n.**

*Plusia illuminata* Robinson (1968: 250); ROBINSON (1975: 186).

This species is here transferred to *Chrysodeixis* Hübner to be consistent with the gradual revision of the Plusiinae being undertaken by DUFAY.

The species was described from Fiji and has been recorded throughout the Papuan Region (Timor, Batjan, New Guinea, the Bismarcks, Solomons, New Hebrides, Fiji, Rotuma and Samoa). It has recently been taken in Tonga and the Cook Is. In New Caledonia it was recorded as follows: Grand Lac site 2 (1); Pic du Pin site 7 (1); Dothio site 19 (2); Ouatou R. (3); Bréhoa R. (7); Col d'Amieu (4); Petit Gouli (1); Sarraméa (4); Cap Colnett (1); Probokonoro (26); Mt. Aoupinié (1); Port Boisé site 74 (1). It is perhaps more characteristic of rainforest than *eriosoma*.

*Ctenoplusia albostriata* (Bremer & Grey)

*Ctenoplusia albostriata* (Bremer & Grey); ROBINSON (1975: 184); HOLLOWAY (1977: 83).

This species is widespread in the Oriental Region and as far east as the Moluccas but is only of sporadic occurrence in the records from further east. It occurs in Queensland and New South Wales, is vagrant to New Zealand, partially or wholly vagrant to Norfolk I., and is resident in Fiji, the New Hebrides and on Rapa I. In New Caledonia it was taken as follows: Dothio site 19 (3); Ouatou R. (1); Bréhoa R. (1); Col d'Amieu (1); Mt. Mou site 34 (1). MNHN: Nouméa.

**Ctenoplusia placida caledonica ssp. n.** (Plate 79: 8, Figure 122: 1)

*Plusia placida* Moore, 1884, *Lep. Ceylon*, 3: 73.

♂♂ 16–17 mm. The wings are patterned with brown on fawnish grey, the brown areas of the forewing with a dull greenish iridescence. The subspecies differs from the typical form (Indian region, Java, and a large, greenish, possibly distinct form from New Guinea) in that the forewing antemedial border is convex distad instead of straight, the postmedial is more sinuous, flexed strongly at spaces 3–5 instead of being almost straight; the straight discal band of the hindwing above is more prominent. In the male genitalia the cucullus is less produced ventrally, the central pair of cornuti in the aedeagus vesica is smaller, and there is an additional long cornutus posterior to the basal row.

HOLOTYPE ♂. New Caledonia, Ouatou R., 320 m site 23 05863/76151 (J. D. HOLLOWAY) 13.6.1971, BM 1971-507.

PARATYPES. 14♂♂ New Caledonia: Ouatou R. (6); Bréhoa R. (3); Ouambéo R. (1); Col d'Amieu (1); Sarraméa (1); Grottes de Koum (2). BM noctuid slide 7436.

The species would appear to be associated with rainforest on sedimentary rock.

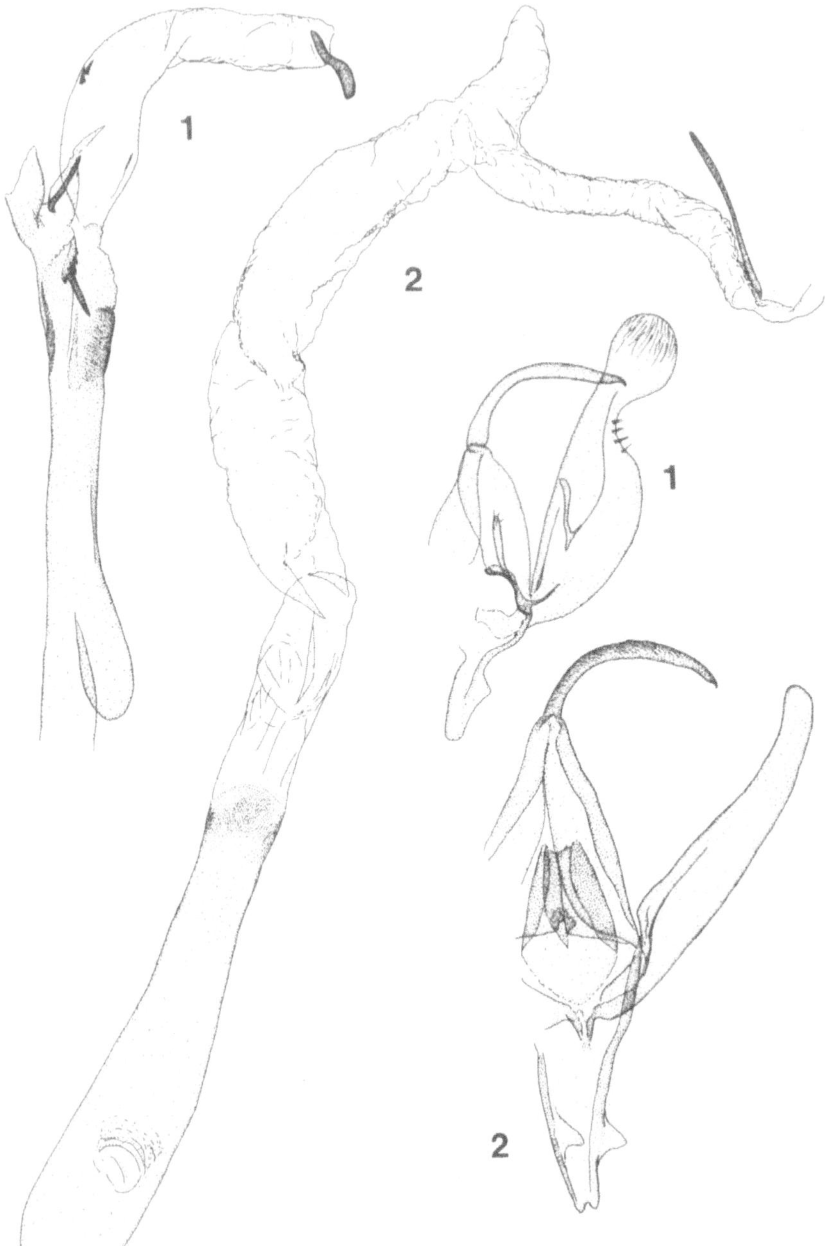


*Fig. 122.* Aedeagi and male genitalia of *Ctenoplusia placida caledonica* (1) and *Anadevidia peponis* (2).

In Queensland there occurs *Plusia chillagoes* Lucas (1900, *Proc. Roy. Soc. Qd*, 15: 150) which is almost certainly referable to *placida*. A related species, *fracta* Walker, occurs in Africa.

## NOCTUIDAE: OPHIDERINAE

### *Aedia leucomelas* (Linnaeus) (Figure 123: 1)

*Aedia leucomelas* (Linnaeus); ROBINSON (1975: 187).

This species is recorded from S. Europe to Australia, Fiji and Samoa. HOLLOWAY (1976) suggested there might be a complex of species in the Indo-Australian area but examination of specimens from Europe, Borneo, Australia (ssp. *acronyctoides* (Guenée)), New Caledonia, the New Hebrides and Fiji suggests that there is probably only one species, variable in both facies and characters of the male genitalia, particularly the number and dispersion of setae in the corona and the lobing and sclerotisation of the aedeagus vesica.

Fijian specimens are slightly larger and paler grey in the forewing than Australian specimens. Those from the New Hebrides are as large as Fijian ones, dark on the forewing with the white area reduced to a spot on the reniform and one posteriorly, one-third from the dorsum. New Caledonian specimens resemble the Australian race in size and appearance but in the two specimens taken the white areas of the forewing were more extensive.

Single specimens were taken at Bréhoa R. and Col d'Amieu.

### *Anomis flava* (Fabricius)

*Anomis flava* (Fabricius); ROBINSON (1975: 189); HOLLOWAY (1977: 84).

This Old World tropical species extends into the Pacific as far east as Rapa I. It was confused with *A. lyona* (q.v.) in field counts in New Caledonia but would appear to be less common, single specimens being retained from Pic du Pin site 7, Mt. des Sources site 14, Bréhoa R. and Sarraméa.

### *Anomis involuta vitiensis* (Butler) (Figure 124: 2)

*Anomis involuta vitiensis* (Butler); ROBINSON (1975: 192); HOLLOWAY (1977: 85).

Subspecies *vitiensis* is restricted to eastern Melanesia but typical *involuta* ranges widely in the Indo-Australian tropics, to Australia and Norfolk I., recurring further east in Polynesia, including Rapa I. The Melanesian subspecies is represented in the Loyalty Is. and was recorded as follows in New Caledonia: Mt. des Sources site 11 (1), site 14 (8); R. Kuenthio (2); Dothio site 19 (2); Poupa (1); Koh (2); Ouatou R. (4); Bréhoa R. (1); Col d'Amieu (2); Col Toma (1); Anse Longue site 32 (1); Mt. Mou site 33 (9); Kaala-Gomèn (1); Tiebaghi site 41 (1); Bambou R. (2); Cap Colnett (2); Probokonoro (4); Col Nassiboué (1); Port Boisé site 74 (5). BPBM: Tontouta R.

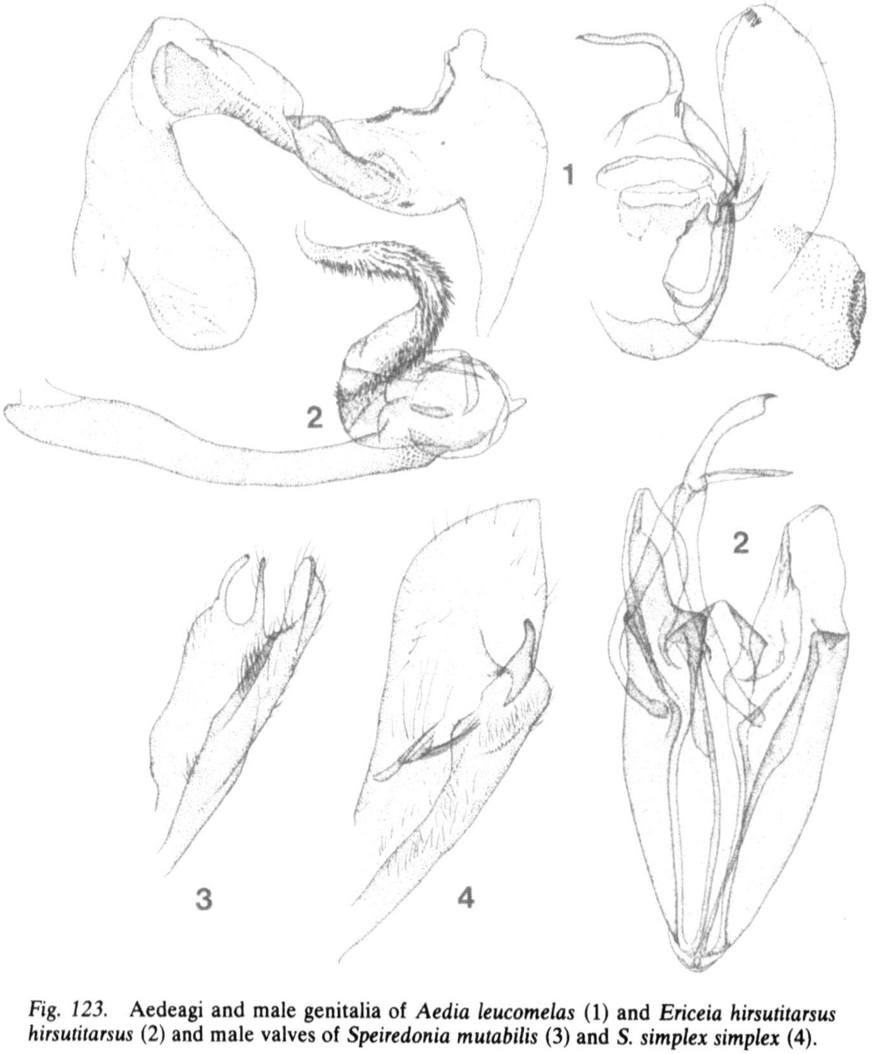


*Fig. 123.* Aedeagi and male genitalia of *Aedia leucomelas* (1) and *Ericeia hirsutitarsus hirsutitarsus* (2) and male valves of *Speiredonia mutabilis* (3) and *S. simplex simplex* (4).

COHIC (1956) recorded the larva as feeding on *Hibiscus* and *Thespesia* (both Malvaceae) in New Caledonia.

### *Anomis lyona* (Swinhoe)

*Anomis lyona* (Swinhoe); HOLLOWAY (1977: 86).

*A. lyona* is widespread in the Indo-Australian tropics but Pacific records are limited to Norfolk I. (as a vagrant), New Caledonia and Samoa. Specimens

*Fig. 124.* Male genitalia and aedeagi of *Anomis nigritarsis xanthochroa* (1, vesica separate, left of aedeagus), *A. involuta vitiensis* (2) and *Hyperlopha cristifera* (3).

were retained as follows: Mt. des Sources site 14 (1); Poupa (1); Ouatou R. (1); Bréhoa R. (4); Mt. Mou site 33 (2); Bambou R. (1); Mt. Aoupinié (1). BPBM: Nouméa.

### *Anomis nigritarsis xanthochroa* (Butler) (Figure 124: 1)

*Anomis nigritarsis xanthochroa* (Butler); Robinson (1975: 190).

The species is widespread in the Indo-Australian tropics but the name *xanthochroa* refers to forms from the Solomons, New Hebrides, Fiji, Rotuma, Samoa and Tonga. In New Caledonia it was recorded as follows: Pic du Pin site 7 (1); R. Kuenthio (1); Koh (4); Col d'Amieu (1); Petit Gouli (1); Sarraméa (1); Mt. Mou site 33 (1); Bambou R. (1); Cap Colnett (1); R. Amoa (7); Ste. Therèse (2); Probokonoro (2); Houailou (1). MNHN: La Coulée. BPBM: specimen reared from *Hibiscus tiliaceus*. VIETTE (1951a) referred to this or the next species as *Cosmophila fulvida* (Guenée). *Hibiscus tiliaceus* is also a Fijian host plant (ROBINSON).

### *Anomis revocans* (Walker)

*Anomis revocans* (Walker); ROBINSON (1975: 191); HOLLOWAY (1976: Fig. 234).

This species was less frequent than the previous during the survey, single specimens being taken at Kopéto and Probokonoro. P. COCHEREAU sent the author numerous specimens of each that had been taken feeding from fruit in Nouméa. The species ranges from Sundaland to northern Australia, Fiji and Rotuma, though no specimens from the Bismarcks and Solomons were observed in the BMNH.

A fully grown larva was found on *Hibiscus tiliaceus* at Kavatch. It was maroon with thin, pale grey dorsal and lateral lines and broad yellow dorsolateral bands, edged and broken with darker brown; there are pale grey spots just dorsally to the yellow band. The larva spun up a leaf to pupate and emerged four weeks later.

### *Anticarsia irrorata* (Fabricius) (Figure 132: 5)

*Anticarsia irrorata* (Fabricius); ROBINSON (1975: 193); HOLLOWAY (1977: 86).

The species is widespread in the Indo-Australian and Pacific tropics to Queensland and Rapa I. It is vagrant to Norfolk I. and the Kermadecs. It occurs in Lifu and was recorded as follows in New Caledonia: Sarraméa (1); Probokonoro (2). VIETTE (1951b): Nouméa.

### *Arcte coerula* (Guenée)

*Arcte coerula* (Guenée); VIETTE (1950a: 140, under *Cocytodes*); ROBINSON (1975: 194).

*A. coerula* is widespread through the Indo-Australian tropics, extending to Queensland, Lifu, Fiji and Samoa. It was recorded as follows in New Caledonia: Dothio site 19 (1); Koh (1); Bréhoa R. (1); Col d'Amieu (1); Mt. Dogny (1); Mt. Mou site 34 (1), site 36 (1); Mt. Aoupinié (1). VIETTE: Ponérihouen, Baraoua (near Bourail), Canala.

The larva has been recorded from several hosts, mainly of the family Urticaceae such as *Pipturus* (New Caledonia; P. COCHEREAU), *Cypholophus* (New Hebrides; ROBINSON, 1975), *Debregeasia* (India), and *Boehmeria* (India; SEVASTOPULO, 1941).

*Avatha discolor* (Fabricius) (Figure 145: 3)

*Avatha discolor* (Fabricius); ROBINSON (1975: 173).

The facies and male genitalia of *discolor* indicate it to be allied to '*Athyrma*' species from the Oriental Region (as discussed and illustrated by HOLLOWAY (1976)). As *discolor* is the type species of *Avatha* Guenée, probably most Oriental '*Athyrma*' should be combined with *Avatha*. The type species of *Athyrma* is the South American *adjutrix* (Cramer), with very different male genitalia.

The species ranges through the Indo-Australian tropics to northern Australia, Fiji, the Cook Is., Society Is., and the Marquesas. It was recorded as follows in New Caledonia: R. Bleue site 9 (1); Koh (5); Ouatou R. (3); Bréhoa R. (5); Col d'Amieu (2); Col Toma (6); Sarraméa (1); R. Amoa (1); Probokonoro (3). MNHN: La Coulée. It appears to be a rainforest species.

**Avitta australis sp. n.** (Plate 86: 9, Figure 125: 3)

♂♀ 20–22 mm. The insect is a dark-grey brown, the forewings tinged purplish in fresh specimens; the forewing pattern is typical of the *subsignans* Walker group.

In species from this group there is, in the male, a prominent tuft of scales from the base of the fore tibia (covering the rather basal epiphysis), usually retained within the general scaling of the femur but occasionally released, when it expands into a brush. It is especially prominent in the unnamed New Hebrides species. The male antennae are slightly flattened in the horizontal plane. The structure of the male genitalia is also distinctive within the genus: the sacculus of the valve is reduced, and the valve has a distinctive basal process to the costa.

*A. subsignans* Walker from India (type examined) has the valve costal process shaped like an inverted, out-turned, shod foot; the valve sacculus and the uncus are reduced relative to those of more eastern forms. Papuan Region species differ from Oriental ones in having the forewing fasciae somewhat straighter and, on the underside, the discal spots of the wings are relatively larger and the postmedial fasciae fainter. The only named Papuan Region species is *longicorpus* Prout from Ceram (type examined, a male, 27 mm).

The male genitalia of *australis* have the uncus markedly sigmoid and the costal process of the valve cornute as illustrated. In *longicorpus* the uncus is not sigmoid and the costal process is triangular (the triangle inverted) apically with an external spur at one angle of the triangle and a rounded lobe near the junction of the valve. In Australia and New Guinea occurs a species with the uncus not sigmoid but the costal process is intermediate in form between those of *australis* and *longicorpus*, the triangular part reduced somewhat, the external spur more prominent, upcurved. The New Caledonian *australis* is probably a derivative of this, as the Solomons and New Hebrides species have characteristic sclerotised processes from the valve sacculus absent from *australis* and *longicorpus* but present in a different (broadly bifid) form in *subsignans*. Males examined from New Ireland and Nissan I. (north-west of Bougainville in the Solomons) have a triangular costal process but this supports three spurs.

*A. subsignans* is the generic type species (NYE, 1975).

HOLOTYPE ♂. New Caledonia: Port Boisé, 20 m Site 74 07039/75275 (J. D. HOLLOWAY) 13.8.1971, BM 1971-507. BM noctuid slide 8755.

PARATYPES: 3♂♂. New Caledonia, Pic du Pin site 7(1), Col d'Amieu (2), slides 7516, 8755; 2♂♂ (slides 7695, 7696), 2♀♀ LIFU, Loyalty Is.

*Avitta ophiusalis lunifera* (Druce) **stat. n.** (Figure 125: 2)

*Bocana ophiusalis* Walker, 1858, *List Specimens lepid. Insects Colln Br. Mus.*, 16: 173, syntypes examined.

*Toxocampa lunifera* Druce, 1888, *Proc. zool. Soc. Lond.*, 1888: 575, type examined; ROBINSON (1975: 195).

Examination of the male genitalia indicates that *ophiusalis* and *lunifera* are conspecific. The dark brown lunulate reniform is variable in size but generally much more pronounced in E. Melanesian specimens to which the name *lunifera* refers (t. loc. Solomon Is.).

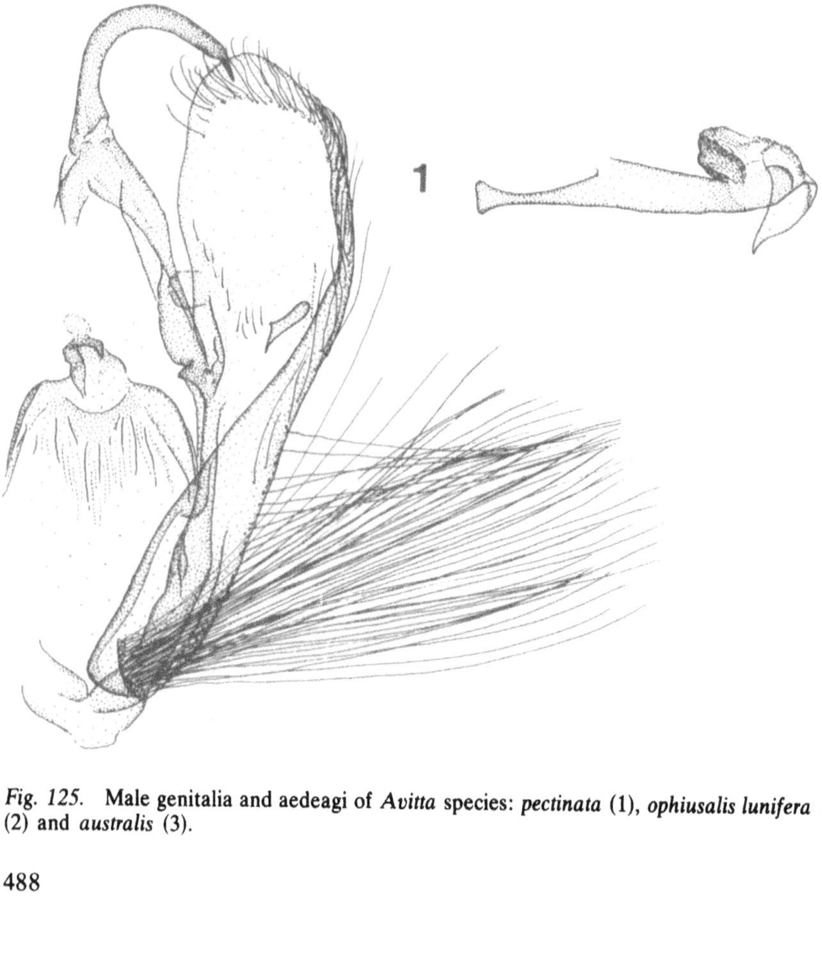


*Fig. 125.* Male genitalia and aedeagi of *Avitta* species: *pectinata* (1), *ophiusalis lunifera* (2) and *australis* (3).

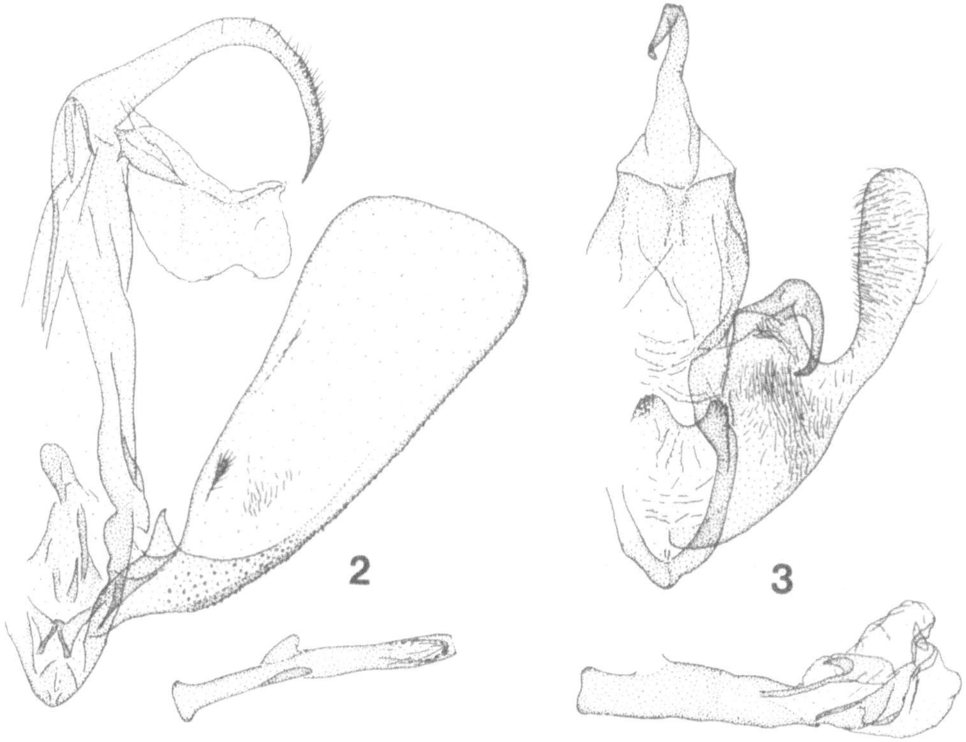

*Fig. 125 (cont.)*

The combination of these names leads to synonymy in generic nomenclature for *ophiusalis* is the type species of *Pantura* Moore and *lunifera*, as its subjective synonym *fluviatilis* Lucas, is the type species of *Imleanga* Lucas (NYE, 1975). *Imleanga* must therefore sink to *Pantura* **syn. n.** *Pantura* might be taken to refer at a subgeneric level to those *Avitta* species with large valves without prominent costal processes, but more work needs to be done on the genus to confirm this, for the name *Asta* Walker, type species *quadrilinea* Walker (placed in *Avitta* in the BMNH arrangement), is also involved and has priority over *Pantura*.

The typical race, *ophiusalis*, has been recorded from the Indian region, the Himalaya, China, Japan and Sundaland; *lunifera* is known from Queensland, the Solomons, New Hebrides and Fiji. It was taken as follows in New Caledonia: Pic du Pin site 7 (1); Col Mouirange (1).

**Avitta pectinata sp. n.** (Plate 86: 10, Figure 125: 1)

♂♀ 20 mm. The head, all of the thorax and the forewings are dark purplish brown, the hindwings and abdomen a dark grey-brown. The antennae of the male are bipectinate (filiform in *rufifrons* Moore, its closest relative, extending from India to New Guinea), but the pectinations are very short. The male fore tibial tuft is one-third the length of that of species of the *subsignans* group (virtually absent in *rufifrons*). As in *rufifrons*, the discal spots and fasciae of the underside are barely evident. The fasciation of the forewing upperside is similar to that of *rufifrons* but the medial is absent, the antemedial straighter and the submarginal reduced to sharp markings apically and at the anal angle. In *rufifrons* the head and patagia are rufous, the latter edged finely with white.

The male genitalia differ in the presence of a harpe to the valve and in the broader base to the valve. The ventral fringe of setae to the valve is also diagnostic. The genitalia of *rufifrons* were illustrated by HOLLOWAY (1976: Figure 251).

HOLOTYPE ♂. New Caledonia: Col Toma, 460 m Site 27 05813/76123 (J. D. HOLLOWAY) 19.6.1971, BM 1971-507. BM noctuid slide 8754.

PARATYPES: 1♀ New Caledonia, Col de Boa, 210 m Site 57; 1♀ Népoui, Aug. BPBM; 1♂ Lifu, Loyalty Is., BMNH.

*Calyptra argyrosigna* (Moore) (Plate 85: 4)

*Oraesia argyrosigna* Moore, 1884, *Lepid. Ceylon*, 3: 77.

The species is recorded from the Indian region, Sundaland and New Guinea. Two specimens were taken at Probokonoro and there is material from Lifu in the BMNH. The related *provocans* (Walker) flies in Africa.

**COLNETTIA Gen. n.**

The male antennae are finely bipectinate and the female antennae filiform. The palps are twice as long as the head, the third segment one-third the length of the second. The venation is as in Figure 126. The legs have no distinctive features.

In facies the forewing is somewhat reminiscent of that of *Oglasa obliquilinea* Hampson (New Guinea) but the barring of the hindwing tornus is diagnostic and the arrangement of the fasciation generally is unusual within the Noctuidae.

The male and female genitalia are as illustrated (Figures 127, 131: 3); the heavily sclerotised, incurved narrow apices of the valves of the former and the corresponding reinforcement, two pouches almost, of the eighth sternite of the latter are diagnostic.

The type species is hereby designated as *brinoni*, described below, The genus is named for JAMES COLNETT, the first European to sight New Caledonia (on COOK's voyage), and the species is named for M. MARCEL BRINON, the forester who guided the author and Dr. J. L. GRESSITT up Mt. Panié.

**Colnettia brinoni sp. n.** (Plate 85: 11, 12, Figures 126, 127, 131: 3)

♂♀ 15–18 mm. The forewings are dull reddish brown, scattered with pale grey and dark brown scales. The forewing reniform is a transverse dark brown dash,

*Fig. 126.* Wing venation of *Colnettia brinoni.*

often obscure and broken; the postmedial is oblique, curved parallel to the margin; the antemedial is more curved, irregular, concave basad; both fasciae are variable, usually ochreous with an associated row of dark brown dots, but often only the latter is present (as in the holotype); the medial area varies in tone from darker to lighter than the apical and basal areas. The hindwing is

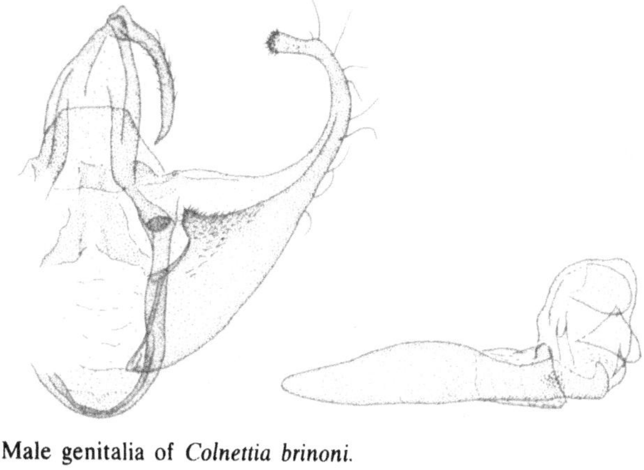

*Fig. 127.* Male genitalia of *Colnettia brinoni.*

straw colour, grading reddish distally; the postmedial is evident at the tornus only, consisting of the same elements as that of the forewing.

HOLOTYPE ♂. New Caledonia: Plaine des Lacs, 240 m Site 5 07020/75362 (J. D. HOLLOWAY) 21.5.1971, BM 1971-507. BM noctuid slide 8805.

PARATYPES. 5♂♂, 1♀ New Caledonia: Grand Lac site 4, Mt. Mou site 33, Cap Colnett, Mt. Krapé site 56, Col Mouirange, Anna-Madeleine. Slides 7415, 8806 (♂♂), 8810 (♀), 7621 (venation).

Apart from Cap Colnett the species was taken only in ultramafic localities.

*Fig. 128.* Male genitalia of *Crioa acronyctoides.*

*Crioa acronyctoides* Walker (Plate 86: 8, Figure 128)

*Crioa acronyctoides* Walker, 1857, *List Specimens lepid, Insects Colln Br. Mus.*, 13: 1111, type examined.

*Gadirtha aroa* Bethune-Baker, 1908, 1908, *Novit. zool.*, 15: 199, **syn. n.**, type examined.

Specimens of *acronyctoides* from Queensland and *aroa* from New Guinea are identical. *Crioa* contains several other Australian species, *acronyctoides* being the type species (NYE, 1975). The form of the uncus of the male genitalia indicates affinities with *Felinia* Guenée (e.g. *precedens* Walker). A single male was taken at Ouen Toro site 64. There is a further specimen from New Caledonia in the BMNH without a precise locality, one in the BPBM from the Col des Roussettes and one from Erromango in the New Hebrides in the BMNH.

*Egone bipunctalis* Walker (Plate 85: 10, Figures 129, 130, 131: 2)

*Egone bipunctalis* Walker, 1863, *List Specimens lepid. Insects Colln Br. Mus.*, 27: 4, unique female type examined.
*Dnopheropis scotaea* Turner, 1902, *Proc. Linn. Soc. N. S. Wales*, 27: 104, colour photo of type (ANIC) and photo of male genitalia examined, **syn. n.**

Both *bipunctalis* and *scotaea* were described from Australia and were originally both described in new genera (NYE, 1975). Therefore *Dnopheropis* Turner must sink subjectively to *Egone* Walker **syn. n.**

In Australia the species occurs in Queensland and New South Wales. In New Caledonia it was taken only in the Plaine des Lacs area, most commonly near the herbaceous marshy association. Records were as follows: Grand Lac site 2 (3), site 3 (1), site 4 (6), site 71 (8); Pic du Pin site 7 (11); R. Koué (1); Anna-Madeleine (3).

*Fig. 129.* Venation of *Egone bipunctalis.*

*Ericeia hirsutitarsus hirsutitarsus* Holloway (Plate 86: 5, Figure 123:)

*Ericeia hirsutitarsus* Holloway, 1975: 88.

This, by far the commoner New Caledonian *Ericeia*, occurs also in the Loyalty Is. and Norfolk I. A distinct subspecies, *robinsoni* Holloway, flies in the New Hebrides. The males are distinguished from those of the next species

*Fig. 130.* Male genitalia of *Egone bipunctalis.*

by the thick mass of long scales on the hind tarsi. Both sexes are generally greyer, less ochreous orange, the postmedial below on both wings being generally more distinct, less crenulate. Where the forewing submarginal is invested with white (in a minority of specimens), it extends over the whole length rather than just over the apical section as in *inangulata.* It is very variable, some specimens being suffused strikingly with pale, ashy grey as in the female illustrated.

The species were not separated in field counts, but orange-ochre specimens were noted to be in a minority, though attributed to general variation within one species. It was therefore felt justifiable to treat the specimens counted as being *hirsutitarsus* and to include that species in the numerical analyses.

Field records and specimens retained totalled as follows: Grand Lac site 1 (1), site 2 (51), site 3 (17), site 4 (1); Plaine des Lacs site 5 (5); Pic du Pin site 7 (42); R. Bleue site 8 (2), site 10 (4); Mt. des Sources site 11 (12), site 12 (6), site 14 (46); Anse Longue site 15 (2), site 16 (5), site 32 (3); R. Kuenthio (1); Dothio site 18 (5), site 19 (1); Poupa (9); Koh (53); Ouatou R. (6); Bréhoa R. (15); Ouambéo R. (2); Col d'Amieu (126); Col Toma (21); Petit Gouli (27); Sarraméa (40); Kopéto (58); Mt. Dogny (11); Mt. Mou site 33 (99), site 34 (8), site 36 (3); Kaala-Gomèn (2); Tiebaghi site 40 (10), site 41 (14), site 42 (8); Grottes de Koum (11); Bambou R. (88); Col d'Amoss site 45 (2), site 46 (8), site 47 (2); Cap Colnett (18); Kavatch (3); Amoa R. (21); Ste. Therèse (2); Probokonoro (58); Col des Roussettes (10); Mt. Krapé site 55 (90), site 56 (15); Col de Boa (19); Houailou (15); near Bâ (26); Mt. Aoupinié (11); Col Nassiboué (4); Ouen Toro site 64 (6), site 82 (2); Col Mouirange (4); Mt. Pouédihi (1); Port Boisé site 73 (2), site 74 (20); R. Koué (1); Anna-Madeleine (1); Les Dalmates site 77 (16), site 78 (6); Mt. Koghis site 79 (2), site 80 (5). The species is one of the most abundant and widespread in New Caledonia, being recorded from virtually all habitat types. It is similarly widely distributed on Norfolk I.

### *Ericeia inangulata levuensis* Prout **stat. n.**

*Hulodes inangulata* Guenée, 1852, *Hist. nat. Insectes* (*Lepid.*), 7: 210, type examined; VIETTE (1951a, b) recorded *inangulata* from New Caledonia.

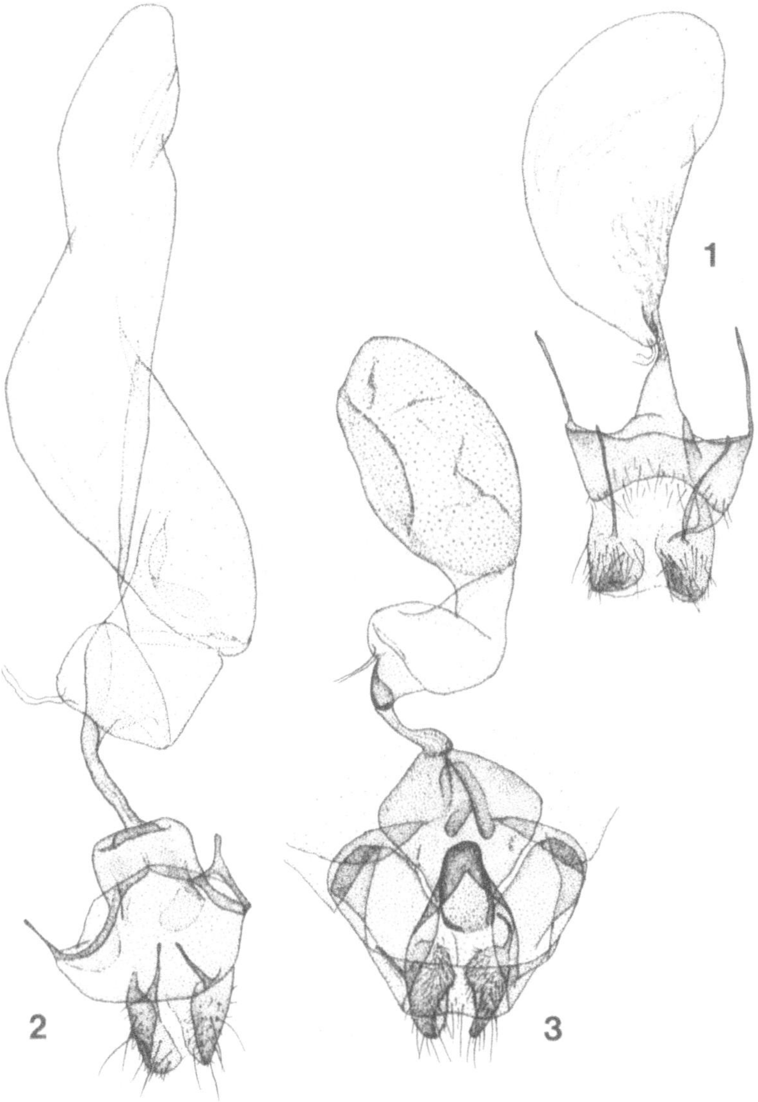


*Fig. 131.* Female genitalia of *Pseudocoarica caledonica* (1), *Egone bipunctalis* (2) and *Colnettia brinoni* (3).

***Ericeia levuensis*** Prout, 1929, *Ann. Mag. nat. Hist.* (10) 3: 597, type examined; ROBINSON (1975: 198).

No differences could be detected in the male genitalia of *inangulata* and *levuensis*. The latter is therefore placed as a subspecies of the former; the facies are also very similar and even this separation is thus open to criticism.

The typical race extends from the Indian region through S.E. Asia to Sundaland, the Philippines, Sulawesi, the Moluccas and New Guinea, whilst *levuensis* is recorded from Fiji, the New Hebrides, New Caledonia and Samoa. Male specimens from the Philippines in the BMNH that had been matched with the original illustration of *intextilia* Schultze had genitalia as in *inangulata*, as did *certilinea* Prout from the Solomons. It would thus appear that *inangulata* is widespread in the Indo-Australian tropics.

*E. dysmorpha* Prout (Solomons: Choiseul) and *E. epitheca* Swinhoe (Rossel I.) have identical male genitalia. They are asymmetrical as in *inangulata* but differ in that the ventral projections of the valves are narrow, rounded instead of square; the apex of the right valve is narrower and that of the left valve has a triangular (instead of rounded) lateral scobinate lobe. These species are larger than *inangulata*, similar in facies though the straight medial fascia of the hindwing is half way between the base of the wing and the darker postmedial-submarginal fasciae instead of being nearer the latter; the white dots on the forewing reniform are more prominent, though neither of these distinguishing characteristics are fully reliable. These two species may represent an earlier invasion of the Papuan Region by *inangulata* stock.

*E. inangulata* was recorded as follows in New Caledonia: Grand Lac site 3 (1); R. Bleue site 10 (1); Col Toma (2); Mt. Mou site 33 (1), site 36 (1); Tiebaghi site 41 (1), site 42 (2); R. Amoa (5); Ste. Therèse (1); Port Boisé site 74 (1).

The larva has been recorded from a variety of Leguminosae in India (SINGH *et al.*, 1954–1960) and from *Cassia* of that family in Fiji (ROBINSON).

### *Eudocima salaminia* (Cramer)

*Eudocima salaminia* (Cramer); VIETTE (1948: 210); ROBINSON (1975: 198).

This species was not taken during the survey, but there are several specimens in the ORSTOM collection, Nouméa, and specimens from Mt. Dogny and Houailou in the BPBM. The species is found throughout the Indo-Australian tropics to North-eastern Australia, Fiji, Samoa and Tonga. It has not been recorded from the Loyalty Is.

### *Hyperlopha cristifera* (Walker) (Plate 85: 1, Figure 124: 3)

*Hyperlopha cristifera* (Walker); ROBINSON (1975: 200).

This species is widespread in the Indo-Australian tropics, extending to Queensland, the Solomons, Lifu, Fiji and Samoa. It was recorded in New Caledonia as follows: R. Bleue site 8 (2), site 10 (1); Sarraméa (1).

### *Hypocala deflorata australiae* Butler (Figure 132: 1)

*Hypocala deflorata australiae* Butler; ROBINSON (1975: 200); HOLLOWAY (1977: 90).

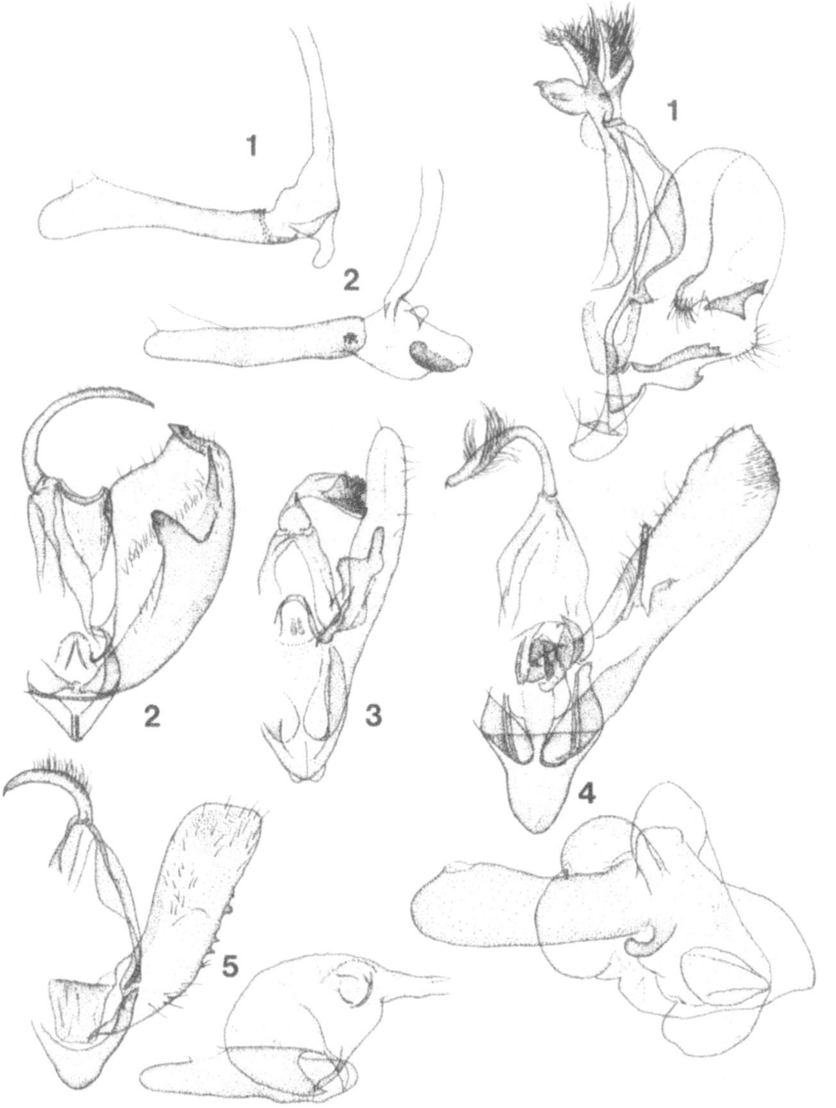


*Fig. 132.* Male genitalia and aedeagi (except 3) of *Hypocala deflorata australiae* (1), *Plusiodonta theresae* (2), *Hypospila bolinoides* (3), *Plecoptera violacea* (4), *Anticarsia irrorata* (5).

Other subspecies of *deflorata* occur in Africa, the Indian region to China, and there is a representative in Hawaii; *australiae* is recorded from Queensland, the New Hebrides, Rotuma, Fiji and Samoa. It is vagrant to New Zealand and Norfolk I. In New Caledonia it was recorded as follows: Grand Lac site 2 (2); Pic du Pin site 7 (3); Mt. des Sources site 12 (1), site 14 (3); Koh (3); Bréhoa R. (1); Col d'Amieu (1); Mt. Mou site 33 (1); Grottes de Koum (1). BPBM: Isle of Pines. The overall population of *australiae* may have declined during the course of the survey.

*Hypospila bolinoides* Guenée (Plate 84: 1, Figure 132: 3)

*Hypospila bolinoides* Guenée, 1852, *Hist. nat. Insectes* (*Lepid.*), 7: 358, types examined.

*H. bolinoides* occurs throughout the Indo-Australian tropics to New Guinea, the Bismarcks and Queensland; it probably spread to New Caledonia from the last locality. It is distinguished from other Melanesian species by the three subequal lobes on the harpe of the male genitalia, the ventral lobe being somewhat reduced in the New Caledonian male dissected. In the New Hebrides, Fiji, Rotuma and Samoa occur races of *H. similis* Tams (discussed by ROBINSON, 1975: 201), where the dorsal harpe lobe is reduced, the harpe being flexed ventrad at that position, terminating in a bilobed apex. Three distinct species occur in the Solomons and are described below for their biogeographical interest.

New Caledonian records of *bolinoides* were as follows: Pic du Pin site 7 (3); Poupa (1); Koh (1); Kavatch (2); R. Amoa (2); Ste. Therèse (1); Probokonoro (7); Mt. Krapé site 55 (3); near Bâ (1); Port Boisé site 74 (4); Ouen Toro site 81 (1).

**Hypospila pseudobolinoides sp. n.** (Plate 84: 2, Figure 133: 3)

♂♂ 20–22 mm, ♀♀ 19–22 mm. The general coloration is dull fawn and brown. The species resembles *bolinoides* closely in both sexes but the white forewing reniform is more prominent and the thin, wavy postmedials and antemedials are more clearly defined. The male genitalia are distinctive: the harpe is elongate with a dorsolateral projection instead of being trilobed; the uncus apex is much less narrow.

The species resembles *similis* in its degree of distinction from *bolinoides*. The other two species are much more distinctive.

HOLOTYPE ♂. (Solomons) Isabel I. 4.vi-9.vii.01 (A. S. MEEK). BM noctuid slide 8769.

PARATYPES: 3♂♂, 5♀♀, data as holotype.

Other material has been examined and found to be of this species from other islands in the Solomons group as follows: Feni, Nissan, Bougainville, Treasury, Choiseul, Vella Lavella, Kulambangra, Rendova, Guizo, New Georgia, Guadalcanal, Florida, San Christobal and Rennell (noted by FLETCHER (1957: 53) as *bolinoides*).

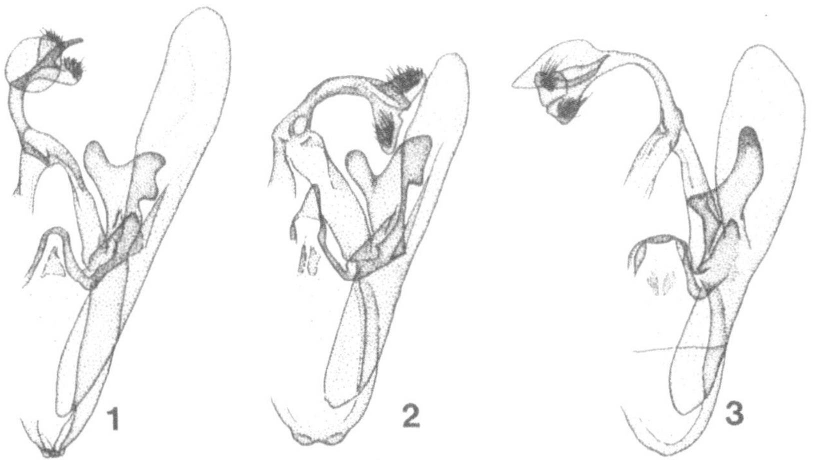


*Fig. 133.* Male genitalia of *Hypospila* species; *ochracea* (1), *elongata* (2) and *pseudobolinoides* (3).

**Hypospila elongata sp. n.** (Plate 84: 3, Figure 133: 2)

♂♂ 20–21 mm. The antennae are almost as long as the forewing (usually four-fifths in other species). The forewing is narrow, apically produced, rendering the postmedial more angular; the area basal to the postmedial on both wings is suffused purplish and distally there is a yellowish band bordered distad by the straight submarginals; otherwise the coloration is as in *bolinoides*. Below, the submarginals are more continuous, less punctate than in other Solomons species.

The male genitalia have the valves narrower than in the other Solomons species; the apical and ventral lobes of the harpe are united to give a square end directed ventrad, the dorsal lobe being very prominent. The saccus is bilobed. The uncus is without the broad dorsal ellipsoid plate of other species. The transtillae are well developed, fused centrally into a triangular structure.

♀ 21 mm. The female is similar to that of *pseudobolinoides* but with an even more prominent white reniform and an intense purple suffusion to both wings except at the submarginals. Below, the submarginals are continuous as in the male.

HOLOTYPE ♂. Solomon Is.: Rendova, Febr. 1904 (A. A. MEEK). BM noctuid slide 8763.

PARATYPES. Solomon Is.: 1♂, 1♀ as holotype; 1♂ Guadalcanar iv.01 (A. S. MEEK), slide 8778; 2♂♂ Florida I. Jan 01 (MEEK), slide 8777; 1♂ Treasury I. 5–10.viii.1901 (MEEK): 1♂ Guadalcanal, Honiara, 20–25.vi.1971 (H. S. ROBINSON).

**Hyposplia ochracea sp. n.** (Plate 84: 4, Figure 133: 1)

♂♀ 19–20 mm. The male forewing is broader than in other species. Both sexes are a richer brown than other species, invested with yellow in patches subapically and tornally in the marginal area of the forewing and in a band basad from the submarginals (broad in the male); the basal part of the wings (to the postmedials) is suffused dark brown with a purple iridescence; the forewing antemedial is wavy as in other species but lacks a prominent lobe extending out over the posterior of the cell.

The male genitalia have a harpe somewhat as in *elongata* but the ventral portion is less pronounced, bilobed; the transtillae are less developed and the uncus has an ellipsoid dorsal plate.

HOLOTYPE ♂. Guizo I., Solomons, Nov. 1903 (A. S. MEEK). BM noctuid slide 8764.

PARATYPES. . Solomon Is.: 1♂ as holotype; 3♂♂, 1♀ north side of Choiseul I., Dec 1903 (A. S. MEEK), slide 6147 (♂); 1♂ Vella Lavella, March 1908 (A. S. MEEK); 1♂ Feni I., E. of New Ireland, July 1924 (A. F. EICHHORN), slide 8765: 1♀ Isabel I., 4.vi–9.vii.01 (A. S. MEEK).

## *LACERA* Guenée

*Lacera* Guenée, 1852, *Hist. nat. Insectes* (*Lépid.*), 7: 337, type species *alope* Cramer.

Previous authors (e.g. VIETTE (1951a) and ROBINSON (1975)) have referred all Indo-Australian *Lacera* to *alope* (Cramer) (*Phalaena alope* Cramer, 1780, *Uitlandsche Kapellen*, 3: 168), of which the original type locality was given as Surinam. The genus does not occur in South America and hence this locality must be erroneous. The original illustration is too poor to enable it to be matched with certainty with any of the species discussed below, but it is probably one of the *noctilio* (Fabricius) group, judging from the prominent tails on the hindwing. The types have not been located in the BMNH or the RNH, Leiden, and may therefore be lost. It is proposed here that *alope* should refer to the common African species discussed briefly below until such a time as a more thorough search can be made for the type than was possible here; the name *noctilio* is available for the common Indo-Australian long-tailed species and this would seem the most convenient arrangement.

Examination of material in the BMNH has indicated that there are seven *Lacera* species in the Indo-Australian area, three of which are endemic to island groups in the south-west Pacific, two more extending into the area. A review of the genus is therefore presented.

The genus divides into two main species groups on the basis of several well-defined characters of facies and genitalia: (1) the *procellosa* Butler group with narrow forewings, a largely dark reniform with a paler dot posteriorly within it, elongate male genitalia, a vesica very much shorter than the aedeagus, and a ductus seminalis arising ventrally from the bursa; (2) the *noctilio* (Fabricius) group with broad forewings, a large, completely pale

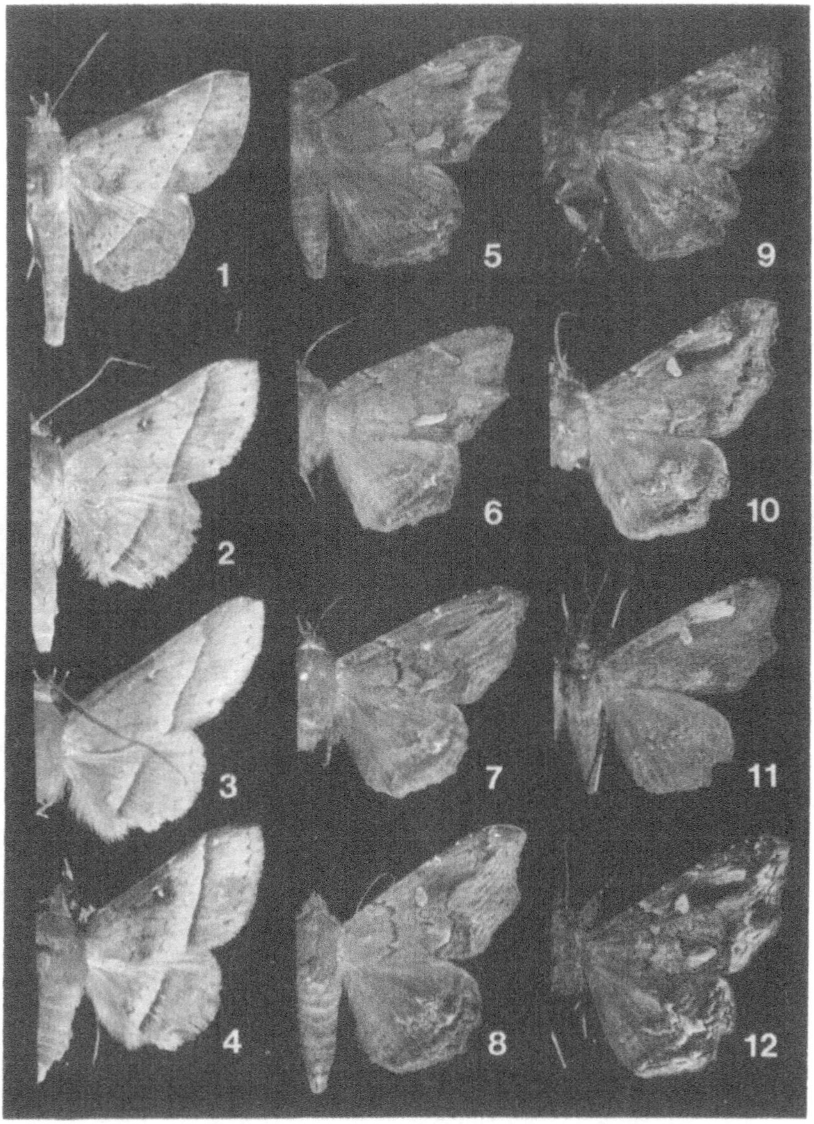


*Plate 84. Hypospila* (1–4) and *Lacera* (5–12) species. 1. *bolinoides* (18 mm); 2. *pseudobolinoides*; 3. *elongata*; 4. *ochracea*; 5. *procellosa* (17 mm); 6. *asinuosa*; 7. *uniformis*; 8. *noctilio* (24 mm); 9. *violacea*; 10 (♂), 11 (♀). *vinacea*; 12. *contrasta*.

reniform, short male genitalia, the vesica almost equal in length to the aedeagus, and the ductus seminalis arising dorsally. The first group contains three species in the Indo-Australian area and the second contains four. The African species referred to *alope* resembles *noctilio* closely, but may be distinguished by the forewing antemedial which is centrally oblique to the dorsum instead of perpendicular to it.

In India 'L. alope' was recorded as feeding on *Acacia* and *Caesalpinia* as a larva by SINGH *et al.* (1954–1960).

(1) The *procellosa* Butler group

*Lacera procellosa* Butler (Plate 84: 5, Figures 134: 2, 135: 5, 136: 6)

*Lacera procellosa* Butler, 1879, *Ann. Mag. nat. Hist.* (5), 4: 368, type examined.

A diagnosis of *procellosa* relative to the similarly sized *uniformis* is presented in the description of the latter. The type locality of *procellosa* is Japan but a New Guinea specimen of similar facies proved to have identical male genitalia. Other material of comparable facies was examined from Sri Lanka, India, China, Taiwan, the Himalaya, Borneo (HOLLOWAY, 1976), Sumatra, Java, the Philippines and Sulawesi.

**Lacera uniformis sp. n.** (Plate 84: 7, Figures 134: 3, 135: 6, 136: 4)

♂♀ 23–27 mm. The upperside is dark brown, much more uniform in shade than in *procellosa*, the pale dot on the reniform being more prominent. The forewing lacks the prominent, pale subdorsal lobe of the postmedial of *procellosa*, though this area is often slightly paler; the submarginal area by the central angle of the forewing margin is only slightly darker than the ground, if at all, whereas in *procellosa* this is generally distinctly darker, with the submarginal pale markings paler than anteriorly; conversely, on the forewing below, the pale violet submarginal markings are broken by dark brown opposite this angle, but in *procellosa* they are usually entire. The female forewing postmedial below is distinctively broadly edged cream basad as far as the reniform, the male having this fascia finely tinged with pale violet. The hindwing below is more strongly suffused with violet than in *procellosa*, especially towards the anal angle, with the medial and postmedial dark brown bands more prominent and distinctly sinuous posteriorly.

The male genitalia are distinguished from those of *procellosa* by the large double corema to the valve, the double harpe at the valve costa, the peak to the uncus, the larger rounder sacculus and the lesser scobination to the aedeagus vesica. The female bursa is rectangular.

HOLOTYPE ♂. S. India: Larwar 8.11.16 (T. R. BELL), BM 1934-394, BM noctuid slide 8769.

PARATYPES. 5♂♂, 8♀♀ S. India, various localities (T. R. BELL) BM 1934-394.

There is further material of this species in the BMNH from India, Burma, Malaya, Singapore, Borneo (Mt. Kinabalu, though not recorded by HOLLOWAY (1976) and possibly only at lower elevations), Balabac, the Philippines, Sulawesi, Java, Bali, Obi, New Guinea, Queensland, Bismarcks, Solomons, New Hebrides (Santo). Apart from variability in the male vesica, the genitalia are constant throughout this range (Queensland and New Hebrides specimens dissected).

**Lacera asinuosa sp. n.** (Plate 84: 6, Figures 134: 6, 135: 4, 136: 5)

♂♀ 21–24 mm. This species is uniform medium brown above, paler than in other species; the forewing postmedial central portion is diagnostically at right angles to the dorsum, lacking the sinuous character of that of other species; costally and basally the postmedial is angled basad and edged pale fawn basally. The hindwing below has distinct bands: a double dark brown postmedial, a broad cream submarginal and a narrower dull violet marginal. Basally the hindwing is suffused violet on fawn and brown. The genitalia are as in *procellosa* but the uncus of the male is somewhat broader and more hirsute.
HOLOTYPE ♂. (Java): Singolangoe, Tengger, 5000 ft April 1934 (J. P. A. KALIS) Rothschild Bequest BM 1939-1. BM noctuid slide 8791.
PARATYPES: 51 ex. Java (as holotype and Djoenggo Ardjeono (4500 ft)) (J. P. A. KALIS): 1♀ Nogkodjadjar, E. Java, 4000 ft. January 1934 (A. M. R. WEGNER), slide 8792; 1♂ Git-Git, E. Bali, 5000 ft, May, 1936 (J. P. A. KALIS).
The species would appear to be a montane duplex of *procellosa*.

(2) The *noctilio* (Fabricius) group

*Lacera noctilio* (Fabricius) (Plate 84: 8, Figures 134: 4, 135: 1, 136: 3)

*Noctua noctilio* Fabricius, 1794, *Ent. Syst.* 3(2): 12, photographs of ♂ type (India orientalis) and its genitalia examined.
*Lacera capella* Guenée, 1852, *Hist. nat. Insectes* (*Lépid.*), 7: 337, type examined (Java), **syn. n.**
Specimens of this species have been examined from Sri Lanka, India, Taiwan, Indochina, Burma, the Philippines, Borneo, Java, the Lesser Sundas, Sulawesi, the N. Moluccas, Timor, Tenimber, New Guinea, Queensland, the Bismarcks, Solomons, New Hebrides (Santo, Malekula, Aneityum), Fiji (Ovalau, Taveuni, Nandarivatu, Nananu I Ra, and probably many of the other localities given for *alope* by ROBINSON (1975)), Samoa (Upolu) and Tonga (Vavau). Three vagrant specimens were taken on Norfolk I. in 1977.
The male genitalia do not vary significantly through this range but several east Melanesian specimens, especially the two Tongan examples, have a much less sinuous forewing postmedial.
The species occurs in New Caledonia but was confused in the field with *vinacea*, described below. Specimens were retained as follows: Plaine des Lacs

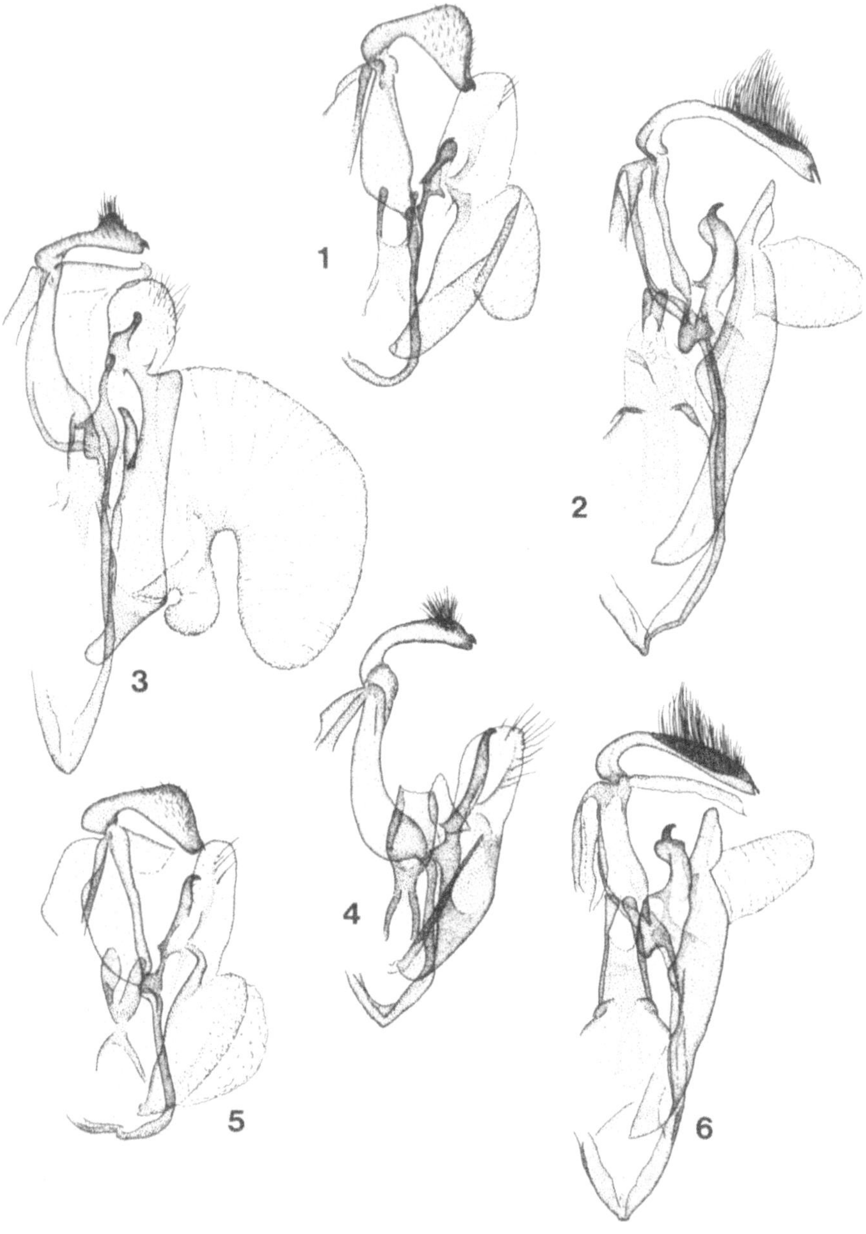


*Fig. 134.* Male genitalia of *Lacera* species: *contrasta* (1), *procellosa* (2), *uniformis* (3), *noctilio* (4), *vinacea* (5), *asinuosa* (6).

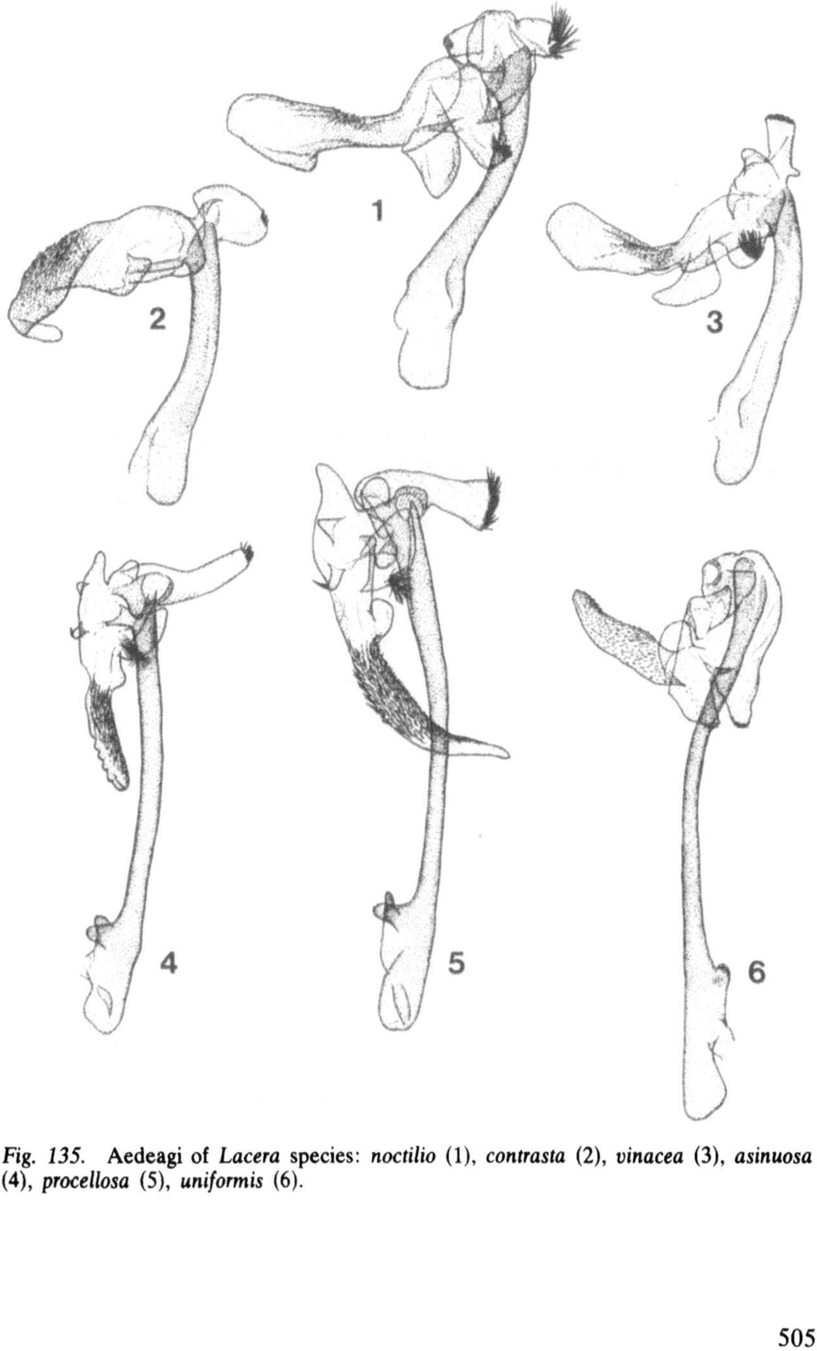


*Fig. 135.* Aedeagi of *Lacera* species: *noctilio* (1), *contrasta* (2), *vinacea* (3), *asinuosa* (4), *procellosa* (5), *uniformis* (6).

site 5 (1); R. Bleue site 9 (1); R. Kuenthio (1); Poupa (2); Ciu (1); Petit Gouli (1).

The three species described below are probably derivatives of a previous invasion of the south-west Pacific by *noctilio.*

**Lacera vinacea sp. n.** (Plate 84: 10, 11, Figures 134: 5, 135: 3, 136: 2)

♂♀ 22–25 mm. The hindwing is more deeply two-tailed than in *noctilio.* The ground colour is vinous brown, marked with creamy fawn. It resembles *noctilio* generally but the wings, especially in the almost uniform female, are much less contrastedly marked, though several specimens of both sexes have the anterior lobe of the forewing postmedial filled with cream colour. The submarginals of both wings are crenulate or punctate, edged with black dots distad in the male (these consist of oblique pale flecks on the forewing of *noctilio* with no black distad, and the hindwing has a continuous pale line with continuous black distad). Below, the submarginals are crenulate rather than dashed or entire, more uniform than in *noctilio* and lacking the submarginal white fleck on the costa of the forewing of that species.

The male genitalia have a very deep uncus, markedly corematous valves with the costal process much shorter than the corona (equal in *noctilio*); the aedeagus vesica and the female genitalia are generally similar to those of *noctilio.*

HOLOTYPE ♂. New Caledonia: Dothio, 400 m Site 19 06133/76135 (J. D. HOLLOWAY) 8.6.1971, BM 1971-507. BM noctuid slide 8776.

PARATYPES. 6♂♂, 9♀♀ New Caledonia: Mt. des Sources site 11 (1); Ciu (1); Bréhoa R. (1); Col d'Amieu (5); Kopéto (1); Mt. Mou site 33 (3); Mt. Krapé site 55 (2); near Bâ (1).

The species possibly favours rainforest. VIETTE (1951a) illustrated this species as *alope.*

**Lacera contrasta sp. n.** (Plate 84: 12, Figures 134: 1, 135: 2)

♂♂ 25 mm. The wings are deep, rich brown, marked with cream colour, tinged with mauvish grey basally. The cream markings distinguish the species from *noctilio* which flies with it; the basal lobe of the forewing postmedial is more produced, closer to the margin. Below, the forewing has narrow marginal black lines in spaces, continuous with each other (they are broader and set obliquely, not continuous, in *noctilio*); both wings are heavily marked with cream colour, especially the hindwing where it forms an extensive basal suffusion. The genitalia have a deep uncus and the valves corematous as in *vinacea,* but the costal process of the valve is shorter and stouter; the aedeagus vesica is much more simple than in either of the previous two species.

HOLOTYPE ♂. Fiji: 11.4.1966 (H. S. ROBINSON). BM noctuid slide 8761.

PARATYPE ♂. Fiji: Nandarivatu 27–30.ix.1968 (H. S. & G. S. ROBINSON).

*Fig. 136.* Female genitalia of *Lacera* species: *violacea* (1), *vinacea* (2), *noctilio* (3), *uniformis* (4), *asinuosa* (5), *procellosa* (6).

**Lacera violacea sp. n.** (Plate 84: 9, Figure 136: 1)

♀ 22 mm. The wings are dark fawn, brown and blackish, sparsely speckled with pale violet scales; the submarginals are punctate rather than oblique streaks or continuous, thus resembling the previous two species and differing from *noctilio*; the forewing antemedial is oblique to the dorsum and perpendicular to the costa, less curved anteriorly than in *noctilio*; the angle of the forewing and the posterior tail of the hindwing have a darker area submarginally, absent in *noctilio*; the forewing medial dark bar is more prominent posteriorly. Below, the species is distinguished from *noctilio* by the marked light violet suffusion of the hindwing exterior to the postmedial.

The female genitalia have the ductus seminalis dorsal as in the rest of the group but this is more central, less spiralled than in *noctilio* or *vinacea*; the ductus bursae is shorter and the lamella postvaginalis is more broadly bilobed anteriorly, the lobes not turned inwards.

HOLOTYPE ♀. New Hebrides: Santo, Mt. Tabwemasana Ridge 4500 ft 2.ix.1971 (G. S. ROBINSON) Roy. Soc. Expedition. BM noctuid slide 8774.

*Leptotroga costalis armstrongi* Tams **stat. n.** (Plate 85: 9, Figure 137)

*Acharya costalis* Moore, 1883, *Proc. zool. Soc. Lond.*, 1883: 28, type examined.

*Leptotroga armstrongi* Tams, 1935, *Ins. Samoa*, 3(4): 228, type examined. ROBINSON (1975: 202).

*L. armstrongi* Tams is here associated with *costalis* (Moore) through its similarity of facies and male genitalia. Minor differences occur in the genitalia as can be seen in the paratype male of the former and the Java male of the latter illustrated; these may prove to be clinal when further material is available. Subspecies *armstrongi* is taken here to refer to forms from Samoa, Fiji and Rotuma (ROBINSON); a single male was taken in New Caledonia at Ouen Toro site 81. The typical race is known from S.E. Asia, the Andamans (t. loc.), Java, the Moluccas, Queensland, the Bismarcks and Solomons.

**Ophyx maculosus sp. n.** (Plate 86: 4, Figure 138)

♂♀ 22–25 mm. The ground colour is ochreous fawn, marked with rich brown as illustrated. The species is very similar to its only congener, *ochroptera* Guenée (Queensland), but the wings are irregularly speckled instead of transversely banded in different shades; as in *ochroptera*, only the male has dark brown patagia. The male genitalia are similar to those of *ochroptera*, but the valve is more rectangular, less triangular, and the saccus is more rounded.

HOLOTYPE ♂. New Caledonia: Bréhoa R., 320 m Site 24 05853/76139 (J. D. HOLLOWAY) 15.6.1971, BM 1971-507. BM noctuid slide 7509.

PARATYPES: 14♂♂, 4♀♀ New Caledonia, various survey localities; 1♂ Mt. Koghis 600 m 20.v.1971 (P. COCHEREAU) MNHN.

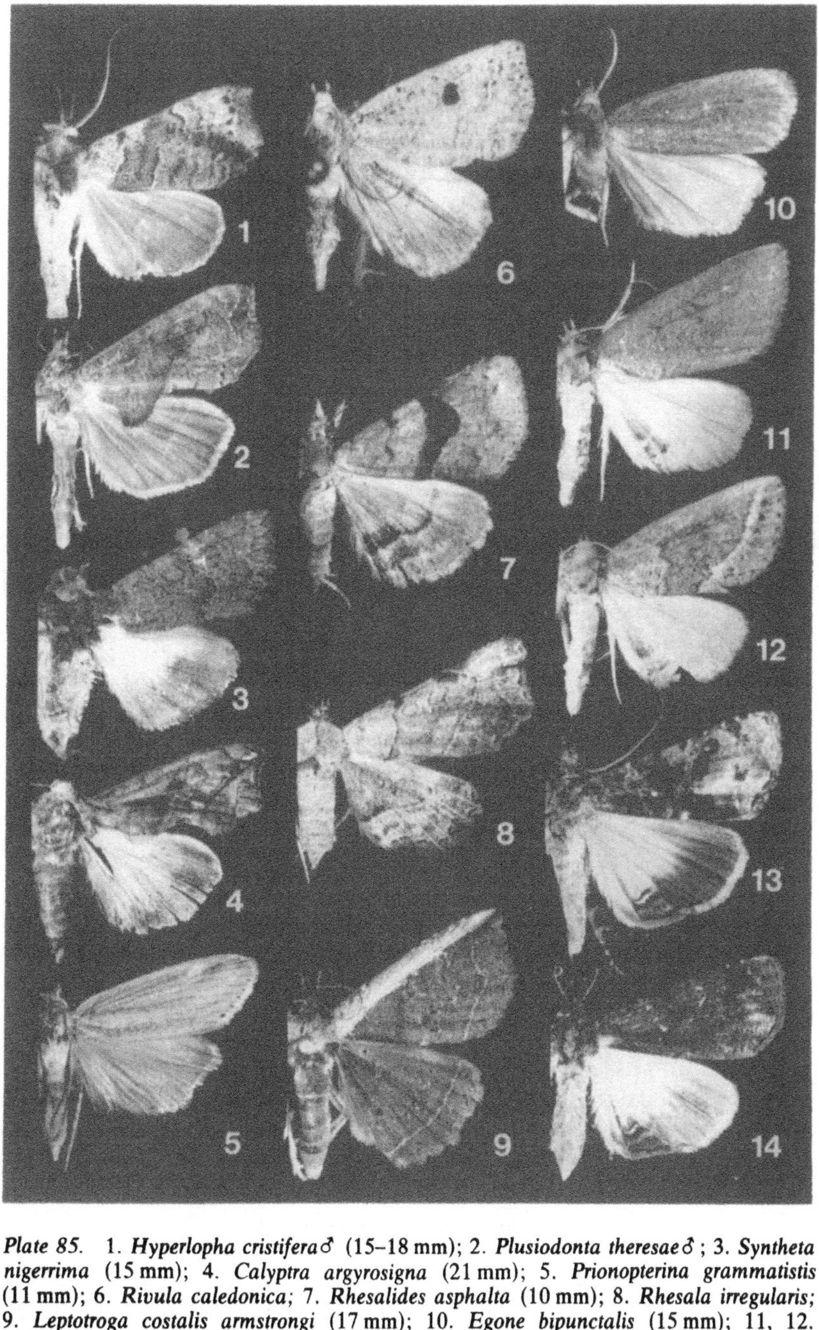

*Plate 85.* 1. ***Hyperlopha cristifera*** ♂ (15–18 mm); 2. ***Plusiodonta theresae*** ♂; 3. ***Syntheta nigerrima*** (15 mm); 4. ***Calyptra argyrosigna*** (21 mm); 5. ***Prionopterina grammatistis*** (11 mm); 6. ***Rivula caledonica;*** 7. ***Rhesalides asphalta*** (10 mm); 8. ***Rhesala irregularis;*** 9. ***Leptotroga costalis armstrongi*** (17 mm); 10. ***Egone bipunctalis*** (15 mm); 11, 12. ***Colnettia brinoni;*** 13 (♂), 14(♀). *Pseudocoarica caledonica.*

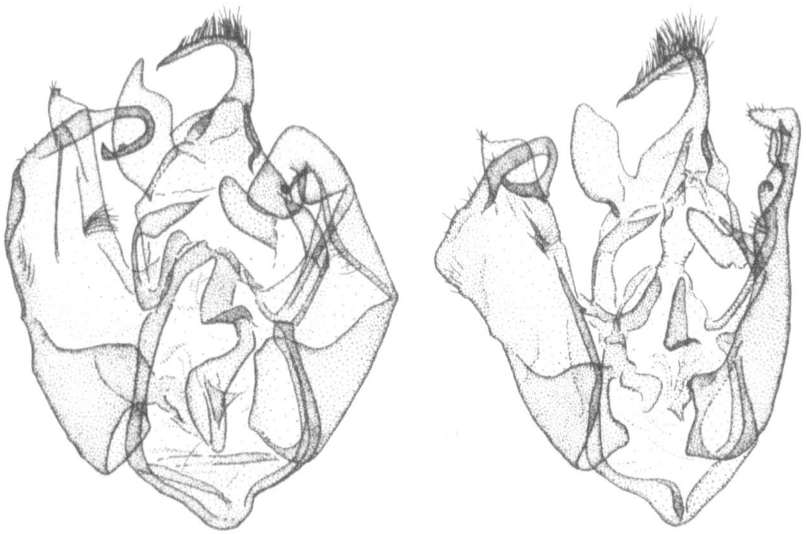

*Fig. 137.* Male genitalia of *Leptotroga costalis costalis* (right, Java) and *L. c. armstrongi* (left, Samoan paratype).

The species was recorded as follows: Dothio site 19 (1); Ciu (1); Koh (2); Bréhoa R. (7); Col d'Amieu (1); Col Toma (5); Sarraméa (1); Grottes de Koum (1); Probokonoro (2). These are mainly rainforest localities on sedimentary rock.

### *Othreis fullonia* (Clerck)

*Othreis fullonia* (Clerck); VIETTE (1951a: 213); ROBINSON (1975: 205).

The species is widespread in the Old World tropics to northern Australia, Samoa and the Society Is. The larva feeds on *Erythrina* species and the adult is a pest of citrus, piercing the fruit to suck juice. An account of its biology is given by ROBINSON (1975). During the survey the adult population was low (P. COCHEREAU, *pers. comm.*) and records were mainly from rainforest localities as follows: Grand Lac site 2 (2); Bréhoa R. (6); Col d'Amieu (4); Mt. Mou site 33 (2); Amoa R. (2); Probokonoro (4); Mt. Krapé site 55 (1); Mt. Aoupinié (1). VIETTE (1949): Nouméa, Canala.

### *Othreis materna* (Linnaeus)

*Othreis materna* (Linnaeus); VIETTE (1950a: 216); ROBINSON (1975: 206).

This species has as wide a distribution as *fullonia* in the Old World but has been more patchily recorded. It extends to northern Australia and the Society Is. but no specimens from Samoa, Tonga or the New Hebrides have been seen.

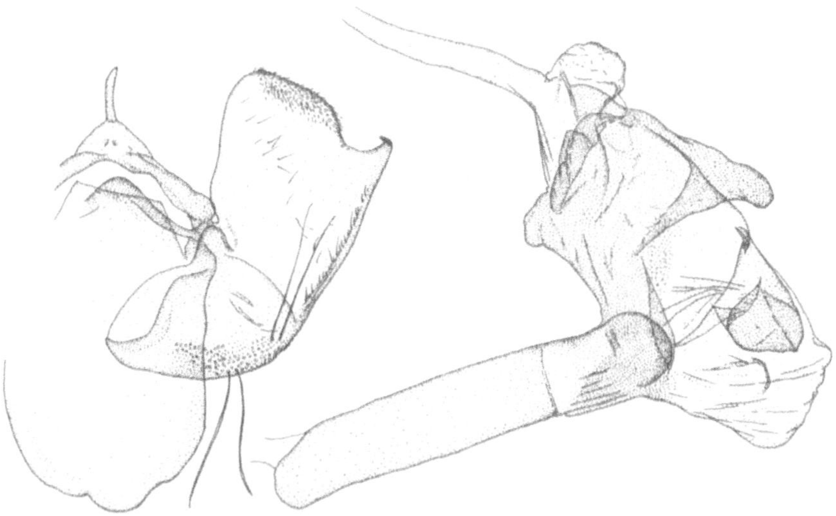

*Fig. 138.* Male genitalia of *Ophyx maculosus.*

It is vagrant to Norfolk I. and New Zealand. One specimen was taken at Probokonoro during the survey and there is another from Houailou in the BPBM, but it is also known as a citrus pest in New Caledonia and is probably much more widespread.

**Oxyodes scrobiculata obscurior ssp. n.** (Plate 81: 6, Figure 139)

*Oxyodes scrobiculata* (Fabricius); ROBINSON (1975: 207).

♂♀ 22–29 mm. The New Caledonian race differs from others of *scrobiculata* in the intense grey-brown suffusion to the forewings and the dorsal half of the hindwings. Other subspecies are more uniformly yellowish (Moluccas, New Guinea, Queensland, Bismarcks, Solomons, Rotuma, Samoa, Tonga), though *tanymekes* Tams from Fiji and *novaehebridensis* Viette from the New Hebrides show some darker suffusion, but not to the same degree. The male genitalia are typical of these yellower Melanesian races, typified by *ochreata* Rothschild, lacking the dorsal spine to the uncus of typical *scrobiculata*, a greyer insect from the Oriental tropics. The genitalia of the species show close similarity in many respects, such as the coremata, juxta structure and costal process to the valve, to several catocaline genera discussed earlier.

HOLOTYPE ♂. New Caledonia: Pic du Pin, 240 m Site 7 06879/75383 (J. D. HOLLOWAY) 23.5.1971, BM 1971-507. BM noctuid slide 7573.

PARATYPES. 4♂♂, 6♀♀ New Caledonia: various localities.

The species was recorded as follows: Grand Lac site 2 (4); Pic du Pin site 7 (4); Dothio site 18 (1), site 19 (2); Poupa (17); Koh (29); Ouatou R. (3); Bréhoa R (17); Col d'Amieu (11); Petit Gouli (1); Sarraméa (1); Mt. Dogny

*Fig. 139.* Male genitalia of *Oxyodes scrobiculata obscurior.*

(1); Mt. Mou site 33 (2); R. Amoa (1); Probokonoro (11); Col des Roussettes (2); Mt. Krapé site 55 (12), site 56 (1). The population probably declined during the course of the survey.

A single specimen of this subspecies was taken on Uvéa, Loyalty Is., by Mr. H. SUZUKI in May, 1974.

*Pantydia capistrata* Lucas (Plate 86: 1, Figure 140: 1)

*Pantydia capistrata* Lucas, 1894, *Proc. Linn. Soc. N. S. Wales,* (2), 8: 143.

*P. capistrata* is an Australian species (excluding Tasmania) found also in New Guinea. It was recorded from the following localities in New Caledonia, most having *Acacia spirorbis* present to a greater or lesser extent: Dothio site 18 (1); Poupa (3); Petit Gouli (2); Mt. Mou site 33 (1); Kaala-Gomen (1); Tiebaghi site 41 (2), site 42 (1); Bambou R. (4); Mt. Krapé site 56 (1); Col de Boa (1); Houailou (6); Ouen Toro site 64 (2), site 81 (5), site 82 (8). BPBM: (Nouméa) Anse Vata.

*Pantydia metaspila* (Walker) (Plate 86: 2, Figure 140: 2)

*Pantydia metaspila* (Walker): ROBINSON (1975: 208).

This species has a rather scattered distribution in the Indo-Australian

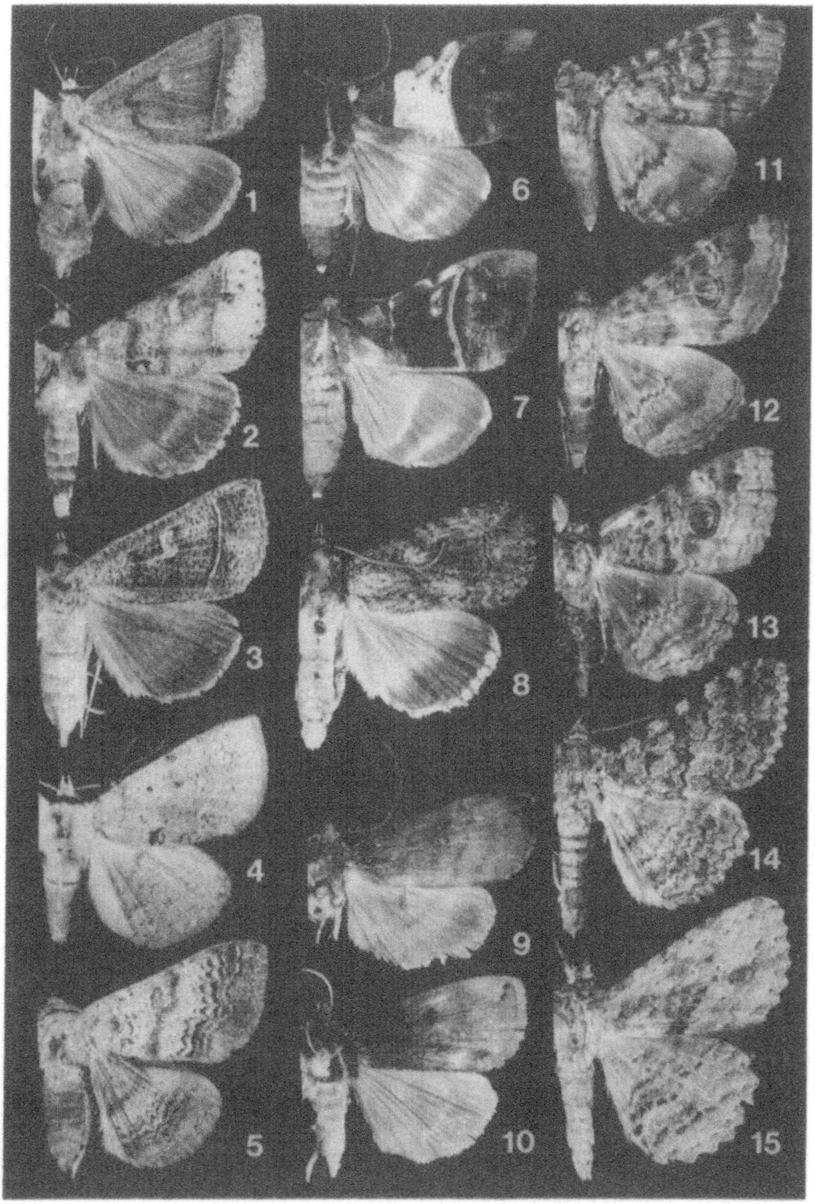


*Plate 86.* 1. *Pantydia capistrata* (20 mm); 2. *Pantydia metaspila* (18 mm); *Pantydia sparsa* (16 mm); 4. *Ophyx maculosus;* 5. *Ericeia hirsutitarsus hirsutitarsus* ♀, ashy grey form (25 mm); 6, 7. *Serrodes mediopallens* (32 mm); 8. *Crioa acronyctoides* (16 mm); 9. *Avitta australis;* 10. *Avitta pectinata;* 11. *Speiredonia mutabilis* (33 mm); 12. *Speiredonia simplex simplex* (30–21 mm); 13. *Speiredonia zamis layardi* (33–35 mm); 14. *Polydesma boarmioides* (25 mm); 15. *Veia pectinata.*

tropics, recorded from Sri Lanka, China, Taiwan, Sundaland, the Philippines, New Guinea, Queensland, the Solomons, New Hebrides and Fiji. ROBINSON collected a larva from *Vigna* (Leguminosae) on Rotuma; the genus *Pantydia* may prove to have general associations with this family, for *capistrata* is generally found where *Acacia spirorbis* occurs in New Caledonia and the larva of *sparsa* has been recorded on *Medicago sativa* in Australia (COMMON, *in litt.*), though it also feeds on *Exocarpos* (Santalaceae).

*P. metaspila* was taken as follows: Anse Longue site 32 (1); Grottes de Koum (2); R. Amoa (1); Probokonoro (1); Houailou (2); near Bâ (1). BMNH: Tinchialit. BPBM: Isle of Pines.

### *Pantydia sparsa* Guenée (Plate 86: 3, Figure 140: 3)

*Pantydia sparsa* Guenée; HOLLOWAY (1977: 91).

This is another species found throughout Australia. It is recorded from Norfolk I., possibly only vagrant, and was taken as follows in New Caledonia: Koh (1); Bambou R. (1); Probokonoro (1). BMNH: Tinchialit. VIETTE (1951a): Nouméa. MNHN: Port Laguerre. BPBM: Forêt de Thi.

### *Plecoptera violacea* (Pagenstecher) (Figure 132: 4)

*Heliothis violacea* Pagenstecher, 1884, *Jb. nassau. Ver. Naturk.*, 37: 223.

*P. violacea* is known from Sundaland, the Moluccas, New Guinea, the Bismarcks, Solomons and New Hebrides. It was taken as follows in New Caledonia: Sarraméa (1); R. Amoa (5); Probokonoro (6).

### **Plusiodonta theresae sp. n.** (Plate 85: 2, Figure 132: 2)

♂♂ 17 mm. The forewings are warm grey, almost uniform, flecked with greenish gold scales basally and with a fine, greenish gold submarginal parallel to the angled margin except at the tornus. The hindwings are uniform grey. The structure of the male genitalia indicates affinities with the Fijian *P. dimorpha* Robinson where the male forewing, though similarly coloured, is more markedly fasciated except the submarginal gold fascia is reduced or absent. The hindwing of *dimorpha* is basally straw, grading grey distally, with a faint medial band of grey also. The female of *dimorpha* has large areas of gold on the forewing and the female of *theresae* may prove to be similar.

In the male genitalia, compared with *dimorpha*, the harpe is more distal, narrower, the intermediate projection from the valve is spine-like instead of spatulate, and the apical projections are less well-developed; the aedeagus vesica has a diagnostic heavy sclerotisation.

HOLOTYPE ♂. New Caledonia: Ste. Therèse, 100 m Site 52 05238/76815 (J. D. HOLLOWAY) 19.7.1971, BM 1971-507.

PARATYPE ♂. New Caledonia: Mt. Krapé, 460 m Site 55 05276/76375 (J. D. HOLLOWAY) 22.7.1971, BM 1971-507. BM noctuid slide 7574.

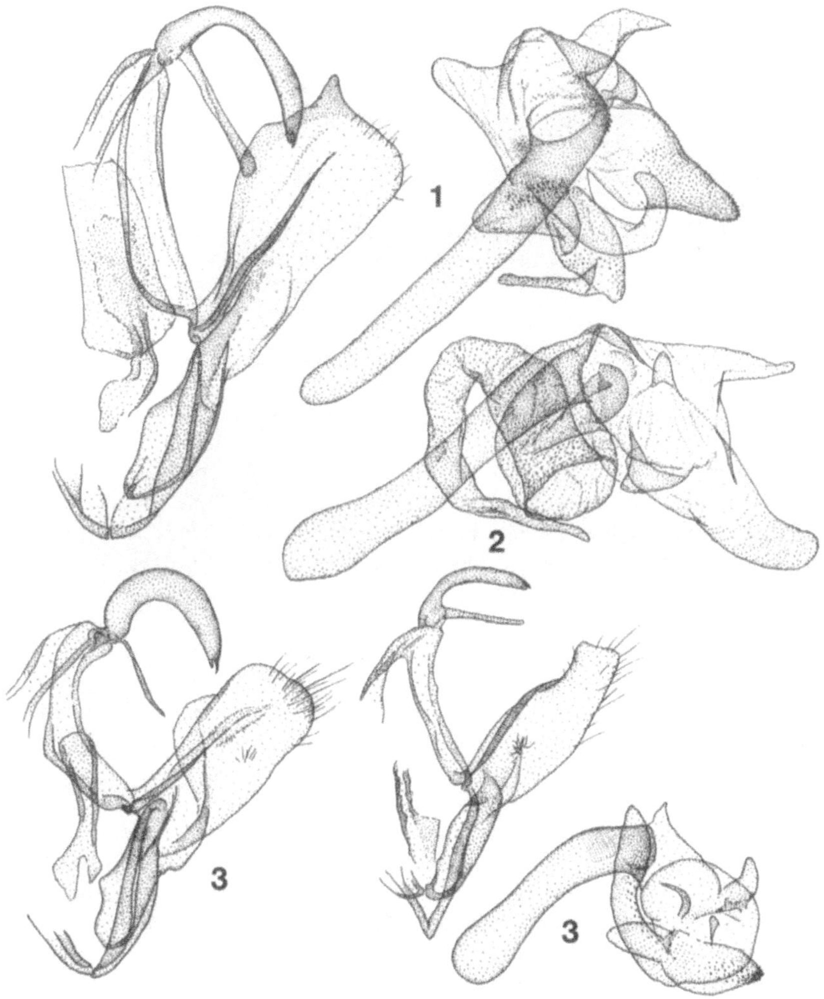


*Fig. 140.* Male genitalia and aedeagi of *Pantydia* species: *capistrata* (1), *metaspila* (2), *sparsa* (3).

### *Polydesma boarmioides* Guenée (Plate 86: 14)

*Polydesma boarmioides* Guenée; ROBINSON (1975: 209).

The distribution of *boarmioides* extends throughout the Indo-Australian tropics to Queensland, Samoa, the Society Is. and Hawaii. In New Caledonia records were as follows: Grand Lac site 2 (2); Koh (2); Ouambéo R. (1); Col d'Amieu (2); Mt. Mou site 34 (1); Mt. Krapé site 56 (1). VIETTE (1949, 1951a, as *umbricola* Boisduval): Nouméa, Baraoua (near Bourail). BPBM: La Foa, Le Crouen, Nouméa.

*Prionopterina grammatistis* (Meyrick) (Plate 85: 5, Figure 149: 1)

*Prionophora grammatistis* Meyrick, 1897, *Trans. ent. Soc. Lond.*, 1897: 374, type examined.

A single male of this Queensland species was taken at R. Kuenthio. It also occurs in New Guinea. It resembles species of *Acrapex* (q.v.) in general appearance but is readily distinguished by the more elongate, rounded forewings.

## PSEUDOCOARICA Gen. n.

The antennae of the male are markedly bipectinate, those of the female slightly so. The palps are stout, curved upwards in front of the face. The tegulae of both sexes are made prominent by broadly spatulate, somewhat curved scales; the base of the abdomen dorsally bears one or two tufts of similar scales. The legs are dark brown with narrow paler bands distally on the tibiae and each tarsal segment. The wing pattern is as illustrated (Plate 85) and described in the specific description. There is sexual dimorphism, the female forewing being very dark brown, the markings of the lighter brown, more variegated male being almost obscured. The hindwings have a broad marginal band and distinctive multiple banding at the tornus, setting the genus apart from most other ophiderines. The bluish patch at the centre of the male forewing and the sparse bluish suffusion elsewhere are produced by four-pronged scales that are dark brown with a distinctive pale blue edging. The genitalia are as illustrated (Figures 131, 142).

In its general shape, the tornal markings to the hindwing and the tufts at the base of the abdomen, the genus bears some resemblance to the monotypic Himalayan genus *Coarica* Moore (females only are known). It is distinguished from *Coarica* by its forewing venation (Figure 141), where veins $R_{1+2}$ and $R_3$ are fused a short distance from their origin on the cell to give a distinctive second areole.

The type species is hereby designated as *caledonica*, described below.

**Pseudocoarica caledonica sp. n.** (Plate 85: 13, 14, Figures 131: 1, 141, 142)

♂♂ 15–17 mm, ♀♀ 18–21 mm. The hindwings of both sexes are basally pale straw colour, distally dark brown. The male forewings have rufous, brown and blackish-brown markings of a straw ground, the medial area, especially at the costa, irrorated with blue-edged scales. The head, thorax and abdominal tufts are rufous in the male, more blackish brown in the female, matching the general tone of the forewings.

HOLOTYPE ♂. New Caledonia Mt. Aoupinié, 520 m Site 62 05298/76594 (J. D. HOLLOWAY) 29.7.1971, BM 1971-507. BM noctuid slide 7411.

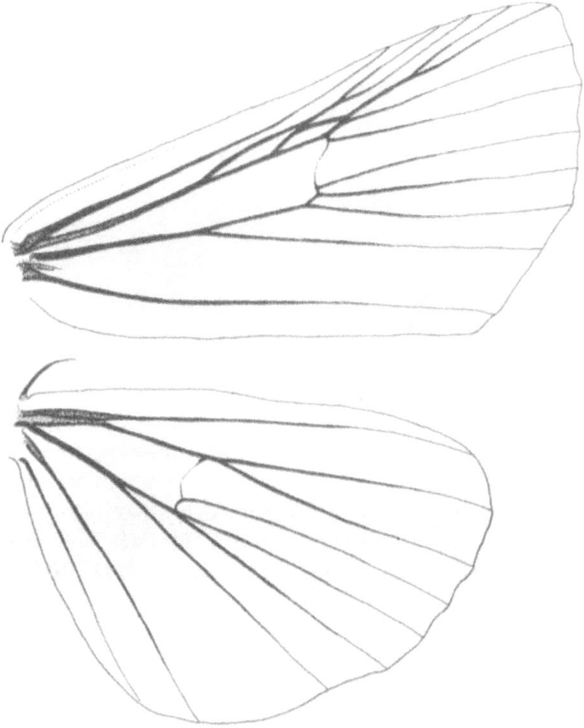

*Fig. 141.* Wing venation of *Pseudocoarica caledonica.*

PARATYPES. 7♂♂, 3♀♀ New Caledonia, various localities. BM noctuid slides 8808 (♀), 7620 (venation).

The species is generally rare and appears restricted to rainforest. It was recorded as follows: Pic du Pin site 7 (1); Dothio site 19 (1); Ouambéo R. (1); Col Toma (1); Mt. Aoupinié (5); Port Boisé site 74 (1); Mt. Koghis (1).

**Rhesala irregularis sp. n.** (Plate 85: 8, Figure 143)

♂♀ 12–13 mm. The ground colour is beige in the two males, greyer in the slightly larger female. The fasciation is darker brown, the stigmata edged white in a manner characteristic of the genus. The males have the forewing medial area slightly darker brown than the ground. The margins of both wings are markedly irregular, distinguishing the species from the type of the genus, *imparata* Walker (type examined, a female from Sri Lanka). The fasciation of the wings is smoothly curved and angled rather than wavy as in *imparata.* The genus contains a complex of species, mostly undescribed, extending throughout the Indo-Australian tropics; none have irregular margins to the forewings except this, the most easterly species.

*Fig. 142.* Male genitalia of *Pseudocoarica caledonica.*

The male genitalia of *irregularis* are typical, with elongate valves, a simple but small harpe basally on the ventral margin and a basal corema. The uncus supports an array of round-ended setae dorsally, these also occurring along slender, lateral, backwardly directed processes, one each side, arising from the uncus at the distal termination of the setose area. A male from Assam in the BMNH series of *imparata* (slide 8862) lacked the basal corema to the valve and the prominent setae to the uncus, but had the very characteristic harpe. A male from New Britain (slide 8863) had the corema, uncus setation and uncus lateral processes well developed.
HOLOTYPE ♂. New Caledonia: Port Boisé, 15 m Site 73 07039/75275 (J. D. HOLLOWAY) 13.8.1971, BM 1971-507. BM noctuid slide 7577.
PARATYPES. New Caledonia: 1♂ Mt. Krapé site 55; 1♀ Grand Lac site 2; 1♀ Népoui Valley, July 1940, BPBM.

**Rhesala irregularis circuluncus ssp. n.**

Indeterminate ophiderine B, ROBINSON, 1975: 216.
♂♀ 12–14 mm. This subspecies differs from the typical race in the less sharply angled, more rounded postmedials to both wings. In the male genitalia the lateral processes to the uncus are more prominent, curving outwards and upwards to meet above, almost forming a circle; the section of the uncus distal to the setose area flexes at right angles upwards and then curves down in a semicircle. The facies is very variable (described by ROBINSON).
HOLOTYPE ♂. Fiji: Nandarivatu, 16–20.xii.1968 (H. S. & G. S. ROBINSON). BM noctuid slide 7229.
PARATYPES. 4♂♂, 9♀♀ Fiji: Nandarivatu (6); Vanua Levu, Delaikoro VHF Sta. (2); Nausori Highlands (1); Dobuilevu (1); Waimanu R., Vakuvula (2); Colo-i-Suva (1).
The subspecies also occurs in the New Hebrides (Santo, Efate, Aneityum).

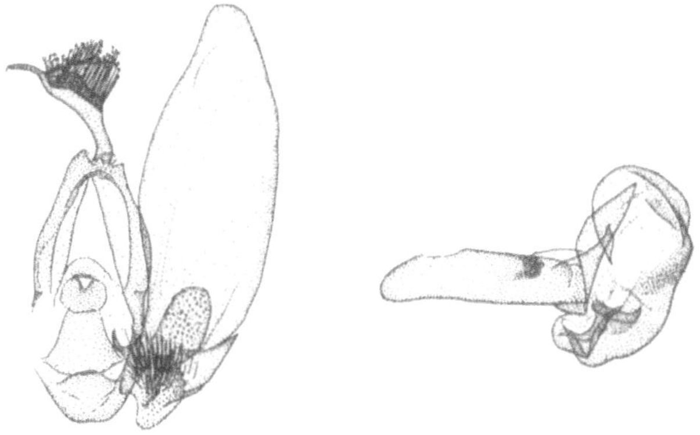

*Fig. 143.* Male genitalia of *Rhesala irregularis irregularis.*

*Rhesalides asphalta* (Swinhoe) **comb. n.** (Plate 85: 7, Figure 144: 1)

*Rhesala asphalta* Swinhoe, 1901, *Ann. Mag. nat. Hist.* (7), 7: 500, type examined.
*Rhesala albizziae* Prout, 1929, *Ann. Mag. nat. Hist.* (10), 3: 598, **syn. n.,** type examined. ROBINSON (1975: 210).

As *asphalta* is very dissimilar to typical *Rhesala* Walker and generally resembles the type species of *Rhesalides* Prout, *keiensis* Prout, in facies and genitalia (Java, Kei, Little Kei, Timor, Tioor, type examined), it is transferred here from *Rhesala* to *Rhesalides*. *R. keiensis* has much rounder valves with prominent basal coremata but is similar in other respects; the uncus, v-shaped in cross section, is not as broad as in *asphalta*. The forewing postmedial in *keiensis* is more sharply angled, with the portion posterior to the angle markedly concave distad.

The type locality of *asphalta* is Queensland and that of *albizziae* is Fiji, the species also occurring in the Tonga group. The two forms are inseparable significantly on the basis of facies or male genitalia and are therefore treated as conspecific. There is much variability, specimens with uniform, dull brown forewings flying with others strongly banded with dark brown in the medial area (illustrated). ROBINSON (1975) recorded the species from the New Hebrides but specimens determined by him as such in the BMNH have a different wing shape and require further examination.

The species was recorded in New Caledonia as follows: Koh (4); Col d'Amieu (1); Sarraméa (21); Anse Longue site 32 (1); Ouen Toro site 64 (1), site 81 (1). The trap at Sarraméa was set near a clump of large *Albizzia* (Leguminosae) trees, recorded as a host plant in Fiji (ROBINSON), which may explain the large catch from that locality. BPBM: Le Crouen, La Foa, Tao.

Two other named forms must be regarded as subspecies of *asphalta* as discussed below.

*Rhesalides asphalta nigra* (Bethune-Baker) **comb. & stat. n.**

*Magulaba nigra* Bethune-Baker, 1906, *Novit. zool.*, 13: 212, type examined.
This insect is purplish grey but with markings similar to those of *asphalta*. The male genitalia are very similar, the uncus slightly smaller, the dorsal margin of the valve not so curved and the dorsobasal process of the valve more broadly rounded apically. The subspecies occurs in New Guinea.

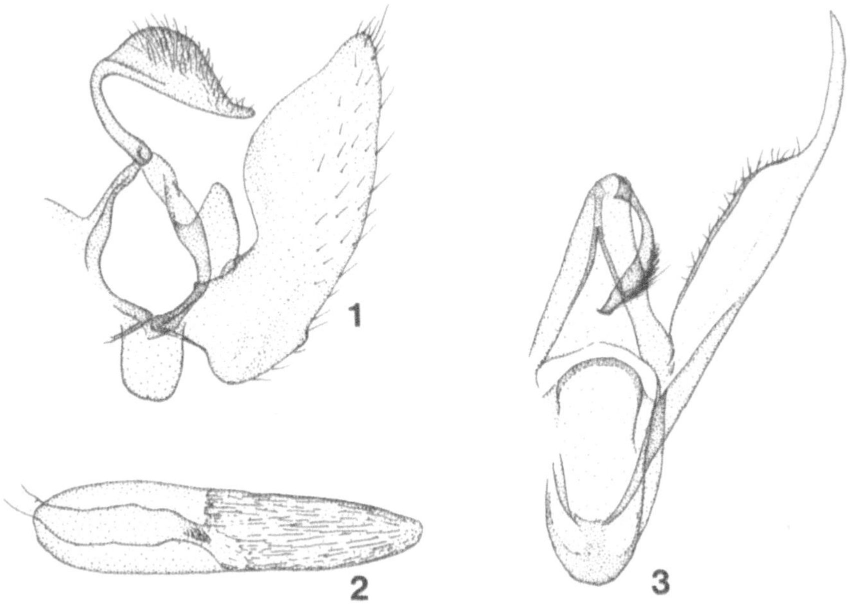


*Fig. 144.* Male genitalia of *Rhesalides asphalta* (1) and aedeagus (2) and male genitalia (3) of *Rivula caledonica*.

*Rhesalides asphalta admiraltensis* Hampson **stat. n.**

*Rhesalides admiraltensis* Hampson, 1926, *Descr. new Gen. Sp. Lepid. Phalaenae Noctuinae Br. Mus.*, p. 274, type examined.
This subspecies, from the Admiralty Is. and others of the Bismarck group, has a more fawnish ochreous ground colour to the wings than *asphalta*. The male genitalia resemble those of *nigra*.

**Rivula caledonica sp. n.** (Plate 85: 6, Figure 144: 2, 3)

♂♀ 9–11 mm. The ground colour of the wings is straw. The forewing is speckled and marked brown, with a prominent black reniform (faint in a few specimens). The hindwing is suffused greyish. The mid tibiae bear tufts of long ochre and black scales. The species is larger than *aroa* Bethune-Baker (8–9 mm) from New Guinea, *aroa* having a less prominent reniform, the forewing postmedial is more distal, less angled, more smoothly curved, and the hindwings lack the discal spot and basal grey suffusion of *caledonica*. On the underside, *caledonica* is more heavily suffused with dark brown on the forewing, and on the hindwing the discal dot and postmedial are much more prominent than in *aroa*. The narrow valve of the male genitalia, especially slender over the apical third, is also distinctive.

HOLOTYPE ♂. New Caledonia: Ouatou R., 320 m Site 23 05863/76151 (J. D. HOLLOWAY) 13.6.1971, BM 1971-507.

PARATYPES. New Caledonia: 7♂♂ (Ouatou R. (1), Bréhoa R. (1), Col Toma (1), Mt. Krapé site 55 (1), Mt. Aoupinié (3)), BM noctuid slide 7582; 1♀ Mt. Koghis 27.i.1963 (C. YOSHIMOTO & N. L. H. KRAUSS) BPBM.

The species has only been taken in areas with extensive rainforest

### *Serrodes campana callipepla* Prout (Figure 145: 2)

*Serrodes campana callipepla* Prout; ROBINSON (1975: 214).

This is the eastern Melanesian subspecies, with a generally greyer forewing medial area, of a widespread species recorded throughout the Oriental tropics and from New Guinea, Queensland, New South Wales, the New Hebrides, the Loyalty Is., Fiji and Samoa. In New Caledonia it was taken as follows: Grand Lac site 3 (1); Mt. des Sources site 12 (1); R. Kuenthio (1); Koh (10); Bréhoa R. (4); Col d'Amieu (8); Col Toma (1); Sarraméa (3); Mt. Mou site 33 (9); Tiebaghi site 41 (1); Mt. Ignambi (1); R. Amoa (3); Probokonoro (11); Col des Roussettes (2); Mt. Panié site 60 (5), site 61 (1). BPBM: Le Crouen. It is probably a rainforest species. P. COCHEREAU has recorded the adult sucking from citrus fruit in a similar manner to *Othreis* species.

### *Serrodes mediopallens* Prout (Plate 86: 6, 7, Figure 145: 1)

*Serrodes mediopallens* Prout; ROBINSON (1975: 215).

The type locality of *mediopallens* is New Guinea and there are specimens in the BMNH from Luzon, Kei, Queensland, the New Hebrides, Lifu, Fiji and Samoa. Compared with *campana*, the forewing postmedial is markedly concave distad, the medial area uniform instead of mottled or paler in the basal half, the reniform is 8-shaped, the centre of each circle pale (in *campana* the circles are larger, more fused, never pale-centred). The medial area is of various shades from black through grey and fawn to cream. The male genitalia are distinctive as illustrated.

In New Caledonia the species was much commoner than *campana*, again found primarily in rainforest localities as follows: Mt. des Sources site 14 (1); R. Kuenthio (3); Poupa (1); Koh (17); Bréhoa R. (18); Ouambéo R. (3); Col d'Amieu (72); Col Toma (8); Sarraméa (1); Mt. Mou site 33 (2); Bambou R.

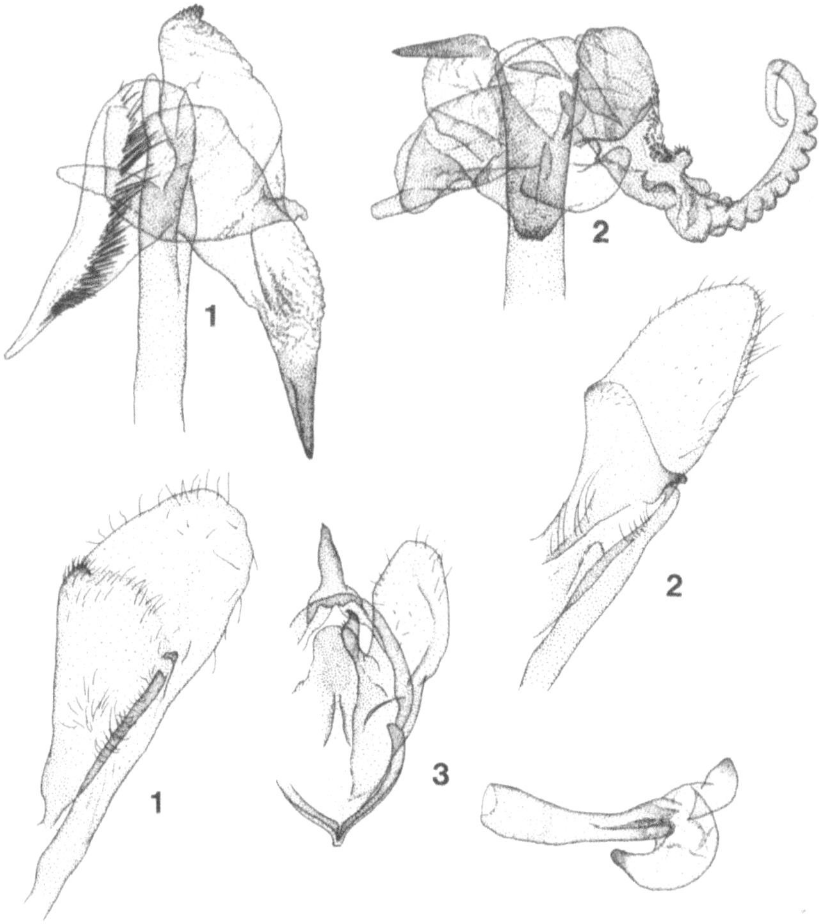


*Fig. 145.* Aedeagus vesica and valve of *Serrodes mediopallens* (1) and *S. campana callipepla* (2), and male genitalia and aedeagus of *Avatha discolor* (3).

(1); Col d'Amoss site 47 (1); Mt. Ignambi (1); R. Amoa (3); Ste. Therèse (3); Probokonoro (60); Col des Roussettes (4); Mt. Krapé site 55 (13); near Bâ (4); Mt. Panié site 61 (1). BPBM: Mt. Koghis, Népoui, Thi Valley.

The adults also pierce and suck from citrus fruit.

## *SPEIREDONIA* Hübner

It is apparent from NYE (1975) that the name *Speiredonia* Hübner, type species *feducia* (Stoll), must refer to those ophiderines usually referred to

*Sericia* Guenée (e.g. ROBINSON, 1975; HOLLOWAY, 1976), type species *spectans* Guenée. *Speiredonia* is usually applied to catocalines of similar appearance such as *retorta* (Clerck), for which the name *Spirama* Guenée is available. The two ophiderine type species have very similar facies, indicating that *Sericia* should sink to *Speiredonia* at the generic level. The male genitalia of *feducia* have valves with a lateral harpe as in *zamis* (Stoll) below, but those of *spectans* have a valve divided apically into several sclerotised projections, resembling *mutabilis* (Fabricius). Thus the name *Sericia* may be applicable at a subgeneric level.

*Speiredonia zamis layardi* Hampson **comb. & stat. n.** (Plate 86: 13)

*Phalaena zamis* Stoll, 1790, *Uitlandsche Kapellen* (Suppl.): 162.
*Sericia zamis* (Stoll) and synonymy; FLETCHER (1957: 50).
*Sericia layardi* Hampson, 1926, *Descr. new Gen. Sp. Lepid. Phalaenae Noctuinae Br. Mus.*: 121, type examined.

In facies and male genitalia *layardi* generally resembles typical *zamis* from the Indian region, though lacking the paler banding to the forewing of the type. In this last it is generally closer to material from Sundaland and Melanesia. The general purplish tinge to the wings distinguishes New Caledonian specimens and justifies the retention of the name *layardi* at the subspecific level. VIETTE (1951a) illustrated the male genitalia of this species as *feducia*, but *feducia* is not related.

The species occurs throughout the Indo-Australian tropics to northern Australia, the Bismarcks and Solomons. Singletons were taken at Mt. Mou site 33 and Mt. Krapé site 55 during the survey and others have been noted from Plum Farm (BMNH) and Bourail (BPBM).

The next species may prove to be merely another, albeit distinctive, subspecific grouping of *zamis*, but further collecting needs to be done in New Caledonia and the Loyalty Is. to discover if the two forms are sympatric.

*Speiredonia simplex simplex* Butler **comb. rev.** (Plate 86: 12, Figure 123: 4)

*Speiredonia simplex* Butler, 1877, *Ann. Mag. nat. Hist.* (8), 16: 198, type examined.
*Speiredonia simplex obalauae* (Bethune-Baker); ROBINSON, 1975: 213, **stat. & comb. n.**

This species, though similar to the previous, differs in several respects. In the male genitalia there is a slight angle on the sacculus basal to the harpe and the valve apex is somewhat more acute. The fascia immediately distal to the ocellus on the forewing and the corresponding one on the hindwing is pale, not highly crenulate, separating dark ground basad from paler distad; in *zamis* these fasciae are more crenulate and the thin pale fascia has an adjacent fine darker one distad before the paler ground. Subspecies *obalauae* (New Hebrides, Fiji, Tonga) is strongly tinged purple instead of brownish, slightly larger, and has a pronounced lobe distad on the postmedial at the dorsum posterior to the reniform 'ocellus'. Typical *simplex* has been taken only in the Loyalty Is. Several specimens from Lifu in the BPBM had been taken in a cave.

**Veia pectinata sp. n.** (Plate 86: 15, Figure 146)

♂♂ 20 mm, ♀♀ 17 mm. The male ground colour is reddish brown, that of the female paler, more ochreous. The markings are very similar to those of *umbrosa* Walker (New Guinea, Bismarcks) but the dark postmedials are relatively more basal, slightly more sinuous. The male antennae are markedly bipectinate (filiform in *umbrosa*). The underside is darker and more heavily fasciated than in *umbrosa*.

The male genitalia have the valves large, broad, membranous ventrally instead of relatively small and narrow (*umbrosa*); the costal processes to the valves are asymmetrical, whereas in *umbrosa* they are more distal, symmetrical, narrow, outcurved spines; the uncus is relatively half the length of that of *umbrosa*; the aedeagus vesica has a cornutus, lacking in *umbrosa*.

HOLOTYPE ♂. New Caledonia: Port Boisé, 20 m Site 74 07039/75276 (J. D. HOLLOWAY) 13.8.1971, BM 1971-507. BM noctuid slide 8191.

PARATYPES. New Caledonia: 1♂ Sarraméa; 1♀ Mt. Aoupinié; 1♀ Poindimié 7.ii.63 (N. L. H. KRAUSS) BPBM; 1 ex. + pupa, Yahoué Valley Aug 29, 1940 (F.X.W.) BPBM.

*Fig. 146.* Male genitalia of *Veia pectinata*.

Ophiderinae *sp. indet.* A (Plate 87: 5, Figure 147)

A single male (BM noctuid slide 7581) was taken at Anse Longue site 16, on 5.6.1971. It measures 7 mm, is of robust build and has filiform antennae. The palps are small, upturned. The forewings are straw colour with a diffuse, straight, dark brown postmedial arising from two-thirds along the dorsum, running obliquely to bisect the apex; a paler brown bar extends basally posteriorly to the cell as far as a more broken, less oblique antemedial. The hindwings are very pale yellow. In colour, size and shape it is reminiscent of the genus *Rivula* Guenée but the mid-leg is not larger than the others and the genitalia are extremely distinctive as illustrated.

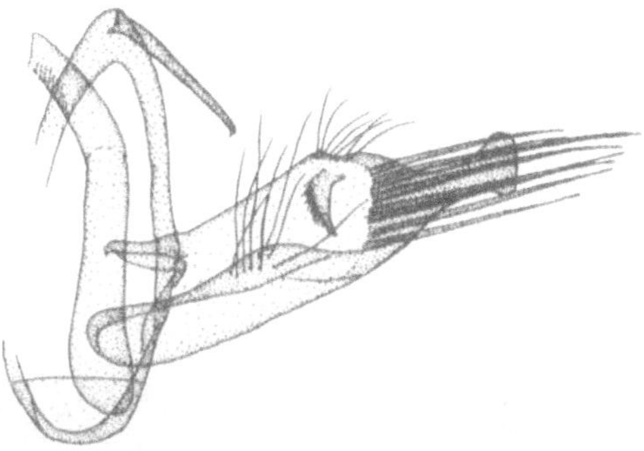

*Fig. 147.* Male genitalia of indeterminate ophiderine A.

### Ophiderinae *sp. indet.* B (Plate 87: 4)

A single female was taken at Ouen Toro site 64. It is similar in appearance to the previous insect but is a little larger (8 mm) with the forewing irrorated with rufous scales. both postmedial and antemedial are oblique, in each case consisting of a row of dark brown dots on the veins edged by an entire yellow fascia away from the medial area; both fasciae are angled sharply basad two-thirds anteriorly.

## Noctuidae: Hypeninae

### *Adrapsa ablualis* Walker (Plate 87: 12, Figure 148: 3)

*Adrapsa ablualis* Walker, 1859, *List Specimens lepid. Insects Colln Br. Mus.*, 16: 170, type examined.

The type species of its genus, *ablualis* has been recorded from Sri Lanka, Sundaland, the Philippines, Sulawesi, New Guinea, Queensland, the Bismarcks, New Hebrides and New Caledonia. It was recorded as follows: Grand Lac site 3 (1); Dothio site 19 (5); Koh (3); Ouatou R. (3); Bréhoa R. (5); Col d'Amieu (2); Kopéto (1); Probokonoro (1). BPBM: Tao.

*Plate 87.* 1. ***Harita nebulosa*** (12 mm); 2. ***Harita nodyna*** (13 mm); 3. ***Harita irregularis;*** 4. indeterminate ophiderine A; 5. indeterminate ophiderine B; 6. ***Simplicia toma*** ♂; 7. ***Hypena fijiensis*** (12 mm); 8. ***Hypena probokonora;*** 9. ***Lysimelia litoralis;*** 10. indeterminate hypenine; 11. ***Anoratha paritalis*** ♀ (20 mm); 12. ***Adrapsa ablualis*** (13 mm); 13. ***Cofimpacia chusaroides;*** 14. ***Cofimpacia luteata;*** 15. ***Cofimpacia violacea.***

*Anoratha paritalis* (Walker) (Plate 87: 11)

*Hypena paritalis* Walker, 1859, *List Specimens lepid. Insects Colln Br. Mus.*, 16: 65.

Singleton females were recorded from Col d'Amieu and Mt. Koghis site 79 during the survey. Elsewhere the species has a patchy but wide distribution, being recorded from Japan, the Indian region, Sundaland and New Guinea.

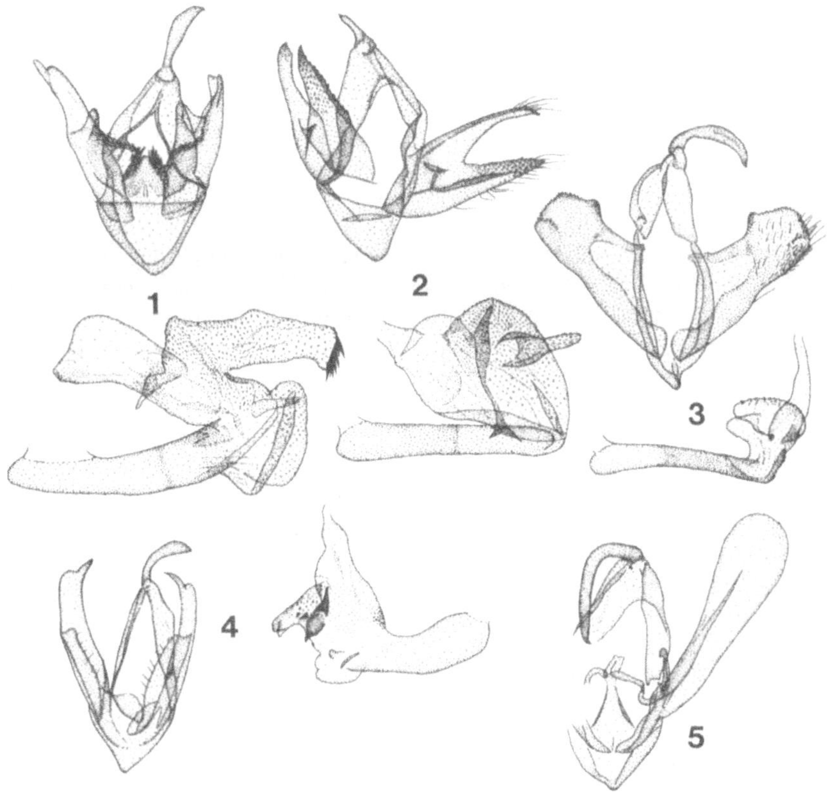


*Fig. 148.* Male genitalia and aedeagi (except 5) of *Nodaria cornicalis* (1), *Hydrillodes norfolki obscurior* (2), *Adrapsa ablualis* (3), *Progonia micrastis* (4) and *Bocana manifestalis* (5).

*Bocana manifestalis* Walker (Figure 148: 5)

*Bocana manifestalis* Walker; ROBINSON (1975: 217).

This species is widespread from the Indian region and China to New Guinea, Queensland, the Bismarcks, Solomons, New Hebrides, Lifu, Fiji, Samoa and Tonga. It was recorded in New Caledonia as follows: Mt. des Sources site 14

(1); Bréhoa R. (1); Ouambéo R. (1); Col d'Amieu (1); Col Toma (3); Sarraméa (2); Mt. Mou site 33 (3); Probokonoro (11); Mt. Krapé site 55 (1); Houailou (1); Mt. Aoupinié (1); Ouen Toro site 64 (1), site 82 (1); Mt. Koghis site 79 (3). MNHN: La Coulée. BPBM: Le Crouen, Mt. Koghis, Tao.

## COFIMPACIA Gen. n.

The antennae are filiform, more markedly ciliate in the male. The palps are directed anteriorly, twice as long as the head. The mid- and hind-legs are much longer than the darker fore-legs.

The forewing lacks an areole; $R_2$ arises from the centre of $R_{3+4}$ which divides subapically; $R_5$ arises separately from the cell, as do the median and cubital veins. In the hindwing Cula and $M_3$ are fused for a quarter of their length but the other veins arise separately from the cell.

The hindwings are unfasciated but the forewings of all species have a common pattern of fasciation reminiscent of a number of species of the genus *Chusaris* Walker. There are subbasal and antemedial zig-zag fasciae darker than the ground, arising at right-angles to the dorsum to traverse the forewing. There is a fainter, more diffuse medial fascia that may be partially fused with the postmedial. The latter is dark, constituted as the antemedial, and would be similarly transverse but for a section one-third of its length that is displaced distad opposite the reniform, which is also diffusely dark. There is an irregular row of pale spots in the spaces submarginally and broad dark triangles at the margin.

The male and female genitalia are as illustrated (Figures 149, 151) and the sinuous valve of the male with its characteristic, central subventral fold is probably diagnostic. In typical *Chusaris* (*retatalis* Walker) the male valves are simply bifurcate and the aedeagus vesica contains a single cornutus.

The genus is named in appreciation of the help given the author by the Cofimpac mining consortium and its principal participant, International Nickel. The type species is hereby designated as *luteata*, described below. The genus contains three species endemic to New Caledonia, recorded from rainforest or ultramafic localities. Its affinities are obscure.

### **Cofimpacia chusaroides sp. n.** (Plate 87: 13, Figure 151: 1)

♂♂ 10–13 mm. The forewings are brown, marked with darker brown, the submarginal spots being white. The postmedial is separate from the medial, is not in line with the reniform anteriorly and posteriorly, and is divided down its length by a narrow grey band. The hindwings are medium greyish brown. The valves of the male genitalia are distinctively angled dorsally and the aedeagus vesica has prominent cornuti and scobination basally; in *luteata* (q.v.) the structures are much less modified.

HOLOTYPE ♂. New Caledonia: Mt. Panié, 800 m Site 60 04764/77257 (J. D. HOLLOWAY) 27.7.1971, BM 1971-507. BM noctuid slide 7584.

PARATYPES. New Caledonia: 1♂ Ouambéo R., 1♂ Sarraméa.

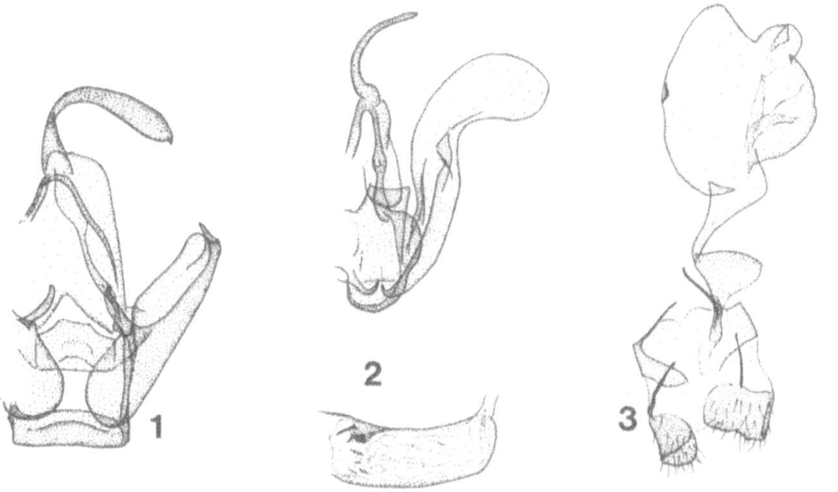


*Fig. 149.* Male genitalia of *Prionopterina grammatistis* (1), male genitalia and aedeagus of *Cofimpacia luteata* (2) and female genitalia of *Cofimpacia violacea* (3).

**Cofimpacia luteata sp. n.** (Plate 87: 14, Figure 149: 2)

♂♀ 8–9 mm, smaller than *chusaroides*. The forewings are mottled with shades of brownish grey and marked with brownish black. Anteriorly and posteriorly the postmedial forms a line with the reniform and is partially fused with the medial. The submarginal spots are creamish white. The veins and postmedial are picked out with yellow, though this is reduced in the holotype. The hindwings are medium pale grey.
HOLOTYPE ♂. New Caledonia: Grand Lac, 250 m Site 2 06975/75356 (J. D. HOLLOWAY) 14.5.1971, BM 1971-507. BM noctuid slide 9520.
PARATYPES. New Caledonia: 1♂ Col. Mouirange; 1♀ Pic du Pin site 7, slide 9521.

**Cofimpacia violacea sp. n.** (Plate 87: 15, Figure 149: 3)

♀ 8 mm. The forewings are grey, the fasciae narrower, less broken than in the other two species, the medial not evident. The forewings are irregularly suffused with violet scales which are concentrated in the submarginal spots and in a narrow band distad from the postmedial.
HOLOTYPE ♀. New Caledonia: Les Dalmates, 330 m Site 78 06714/75420 (J. D. HOLLOWAY) 18.8.1971, BM 1971-507. BM noctuid slide 9522.

*Dichromia quinqualis* Walker (Figure 150: 3)

*Dichromia quinqualis* Walker; ROBINSON (1975: 219).

The species occurs in Sundaland, the Lesser Sundas, Moluccas, New Guinea, Australia (south-west, Northern Territory to New South Wales), Lord Howe I., the New Hebrides and Fiji. It was recorded as follows in New Caledonia: Pic du Pin site 7 (1); Mt. des Sources site 14 (2); Dothio site 19 (1); Poupa (3); Kopéto (3); Mt. Mou site 33 (5); R. Amoa (2); Probokonoro (2); Port Boisé site 73 (1), site 74 (3). BMNH, BPBM: Isle of Pines. MNHN: Maré, Loyalty Is.

*Harita nodyna* (Bethune-Baker) (Plate 87: 2, Figure 150: 4)

*Harita nodyna* (Bethune-Baker); ROBINSON (1975: 220).

The species was described from New Guinea and occurs also in the Bismarcks, Solomons, New Hebrides, Fiji and New Caledonia, where it was recorded as follows: Dothio site 18 (2); Poupa (2); R. Amoa (1); near Bâ (1).

A specimen placed over the name *brachyphylla* Turner (*Placerobela brachyphylla* Turner, 1903, *Trans. Roy. Soc. S. Aust.*, 14) in the BMNH has genitalia identical to those of *nodyna*, and has the long, straight palps, expanding gently to a square end over the long terminal segment, that characterise the genus, of which the type species is the Indian *rectilinea* Moore. The specimen concerned is from the TURNER collection, determined by TURNER as *brachyphylla*, a Queensland species. The specimen is a more uniform brown on the forewing than specimens of *nodyna*, and has the forewing subapical white fleck relatively reduced. Further material of *brachyphylla* should be examined, including the type. Should the similarities and differences outlined above prove consistent, then *brachyphylla* would take priority for the specific name, *nodyna* referring to the Melanesian subspecies.

**Harita irregularis sp. n.** (Plate 87: 3, Figure 150: 1)

♂♀ 14 mm. The forewing is brown of varying shades as illustrated, the hindwing greyer. The species differs from the closely allied *nodyna* in the irregularly zig-zag forewing postmedial and the more prominent dark orbicular and antemedial. There is generally a paler area distad from the postmedial, suffused vinous in some specimens and white in the Lifu female listed; this area is suffused violet grey and more uniform with the basal area in *nodyna*. The submarginal row of dark spots is more prominent, flecked white distad. The dark transverse striations of the forewing are sparser, more irregular than in *nodyna*. The male genitalia are very similar but the valves are less rounded apically, more produced dorsally.

HOLOTYPE ♂. New Caledonia: Koh, 140 m Site 22 05866/76177 (J. D. HOLLOWAY) 12.6.1971, BM 1971-507. BM noctuid slide 7565.

PARATYPES: 4♂♂, 2♀♀ New Caledonia, various localities; 1♀ Tipindjé 10.ii.1963, Malaise trap (C. YOSHIMOTO & N. KRAUSS) BPBM; 1♀ Lifu, Loyalty Is., Rothschild Bequest, BM 1939-1.

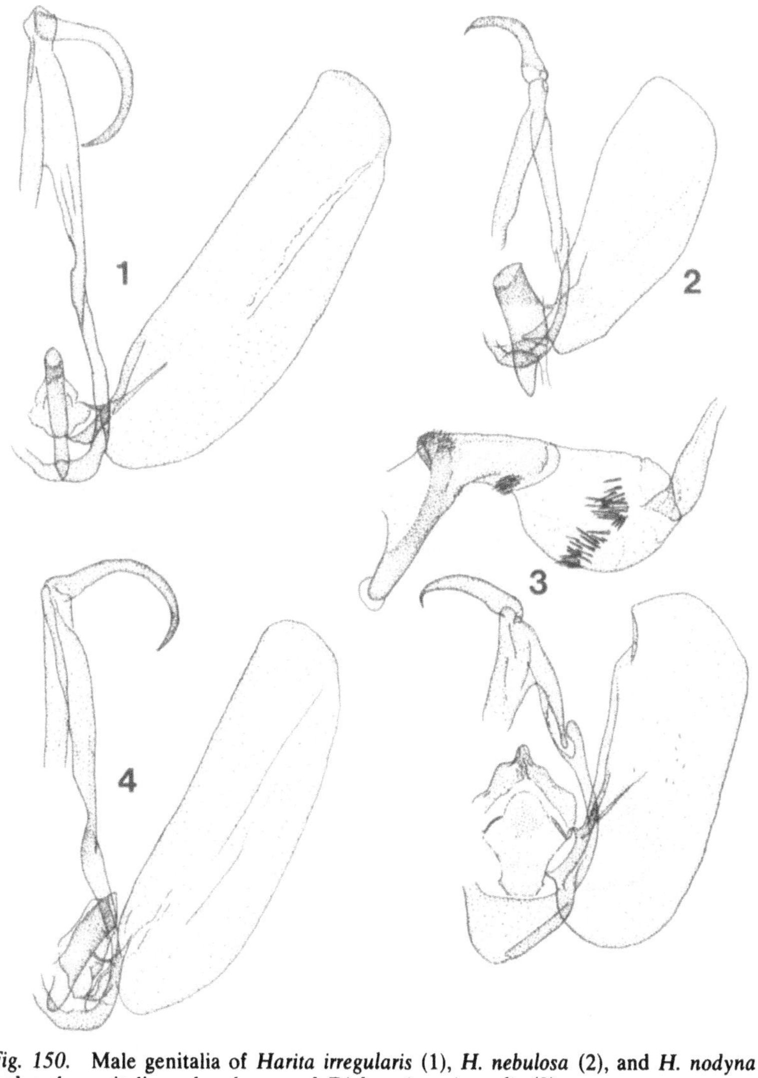


*Fig. 150.* Male genitalia of *Harita irregularis* (1), *H. nebulosa* (2), and *H. nodyna* (4), and male genitalia and aedeagus of *Dichromia quinqualis* (3).

The species was recorded as follows in New Caledonia: Pic du Pin site 7 (1); Koh (3); Ouambéo R. (2); Sarraméa (2); Ouen Toro site 81 (1).

A male and female were taken by G. S. ROBINSON in the New Hebrides (Santo: Mt. Tabwemasana).

*Harita nebulosa* (Moore) (Plate 87: 1, Figure 150: 2)

*Acosmetia nebulosa* Moore, 1881, *Proc. zool. Soc. Lond.*, 1881: 350, type examined.

This species lacks the characteristic *Harita* palps, these being shorter, porrect, the terminal segment relatively short and tapering. The genitalia have elongate valves, longer than in most *Hypena* but shorter than in *nodyna*.

The species occurs in Sri Lanka, India, the Himalaya and Taiwan, recurring in Queensland and New Caledonia, where it was recorded as follows: Grand Lac site 2 (2), site 3 (1); Pic du Pin site 7 (1); Mt. des Sources site 11 (2), site 14 (7); R. Kuenthio (1); Koh (1); Col d'Amieu (1); Sarraméa (1); R. Amoa (1); Col de Boa (1). BPBM: Anse Vata (Nouméa).

*Hydrillodes norfolki obscurior* Holloway (Figure 148: 2)

*Hydrillodes norfolki obscurior* Holloway, 1977: 96.

*H. norfolki* is part of the *surata* Meyrick complex discussed by HOLLOWAY (1977). The typical race flies on Norfolk I. In both localities it is the most abundant species in light-trap samples, especially on warm, humid, drizzly nights (e.g. Probokonoro). The larvae feed on vegetable detritus and large numbers were found by the author in bracken used to make a mattress during the survey, presumably from eggs laid by adults which had escaped from the trap.

The species was recorded as follows: Grand Lac site 1 (5), site 2 (22), site 3 (5), site 4 (1); Pic du Pin site 7 (7); R. Bleue site 8 (3), site 9 (4), site 10 (5); Mt. des Sources site 11 (22), site 12 (4), site 14 (158); Anse Longue site 16 (7), site 32 (17); Dothio site 18 (2); Poupa (2); Koh (70); Ouatou R. (6); Bréhoa R. (4); Ouambéo R. (13); Col d'Amieu (33); Col Toma (24); Petit Gouli (1); Sarraméa (9); Kopéto (1); Mt. Dogny (2); Mt. Mou site 33 (647), site 34 (161); Col de Boguen (7); Kaala-Gomen (2); Tiebaghi site 40 (1), site 41 (20), site 42 (1); Grottes de Koum (13); Bambou R. (7); Col d'Amoss site 45 (22), site 46 (3), site 47 (23); Mt. Ignambi (1); Cap Colnett (9); Kavatch (9); R. Amoa (33); Ste. Therèse (32); Probokonoro (690); Col des Roussettes (22); Mt. Krapé site 55 (89), site 56 (18); Col de Boa (23); Houailou (8); near Bâ (31); Mt. Panié site 60 (2), site 61 (11); Mt. Aoupinié (15); Col Nassiboué (54); Ouen Toro site 64 (1); Mt. Pouédihi (3); R. Fausse Yaté (3); Yaté Gorge (4); Port Boisé site 73 (8), site 74 (6); Anna-Madeleine (2); Les Dalmates site 77 (11), site 78 (21); Mt. Koghis site 79 (2), site 80 (1). BPBM: Tipindjé, Le Crouen, La Foa, Tao, Bourail, Mt. Koghis, Col des Roussettes, Népoui, Nouméa. MNHN: La Coulée (reared from carrot).

*Hypena fijiensis* Robinson (Plate 87: 7)

*Hypena fijiensis* Robinson, 1875: 223.

One female of *fijiensis* was taken in a Malaise trap on Mt. Panié at 300 m, and there is a female from Ouvéa, Loyalty Is. in the BPBM. The species was described from Fiji and occurs in the New Hebrides. It is very close to *sublividalis* Snellen (Oriental Region, New Guinea) and *caerulealis* Walker (Queensland) but is distinguished by facies and characteristics of the male genitalia. Confirmation of the status of the New Caledonian insect must wait on the capture of a male.

*Hypena laceratalis* Walker

*Hypena laceratalis* Walker; ROBINSON (1975: 224); HOLLOWAY (1977: 98).

This lantana-feeding species is known throughout the Oriental tropics, occurs in Australia (Northern Territory to New South Wales), Norfolk I., the New Hebrides, Fiji, Rotuma, Samoa, Tonga and the Cook Is., and was recorded in New Caledonia as follows: Grand Lac site 4 (1); Mt. des Sources site 14 (8); Anse Longue site 16 (2); Dothio site 19 (1); Grottes de Koum (5); Kavatch (2); Probokonoro (7); Mt. Krapé site 56 (1): Ouen Toro site 81 (2). BPBM: Nouméa, Tipindjé.

*Hypena masurialis ferriscitalis* Walker

*Hypena masurialis ferriscitalis* Walker; ROBINSON (1975: 225); HOLLOWAY (1977: 99).

A single female of *masurialis* was taken in Nouméa near Ouen Toro by T. W. DAVIES, the specimen being in the CAS collection. The species is widespread in the Indo-Australian tropics, extending to New South Wales, Lord Howe I., Norfolk I., Samoa and Tonga.

**Hypena probokonora sp. n.** (Plate 87: 8, Figure 151: 4)

♂♀ 12 mm. The forewings are shades of brown as illustrated, the hindwings greyer brown. The species is close to *walkeri* Collenette (Tahiti, unique female type examined) in general size and appearance, but the palps are relatively shorter, the forewing postmedial is better defined, nearer the margin, the forewing subapical markings are more prominent, a dark antemedial is present (almost lacking in *walkeri*), enclosing the slightly darker brown medial area in conjunction with the postmedial, and the submarginal is evenly curved concave distad (irregular in *walkeri*). The genitalia resemble those of *longfieldae* Collenette (Marquesas, Rapa I.) in that the valve has a ventral, basally directed thorn-like projection from the margin; *longfieldae* has very different facies, notably a zig-zag forewing postmedial.

HOLOTYPE ♂. New Caledonia: Probokonoro, 70 m Site 53 05447/76357 (J. D. HOLLOWAY) 20.7.1971, BM 1971-507. BM noctuid slide 7585.

PARATYPE ♀. Data as holotype.

Two further worn specimens were taken at Probokonoro but were not retained.

*Luceria cooki* Holloway (Figure 151: 3)

*Luceria cooki* Holloway, 1977: 100.

This species occurs on Norfolk I. and was recorded in New Caledonia as follows: Sarraméa (2); Col des Roussettes (1); Ouen Toro site 64 (1), site 81 (4), site 82 (1). BMNH: Pouébo. BPBM: Col des Roussettes, Tao, Le Crouen. The species is one of the rarer members of the widespread ecological association in Norfolk I., showing some preference for coastal localities. It seems to show a similar preference in New Caledonia.

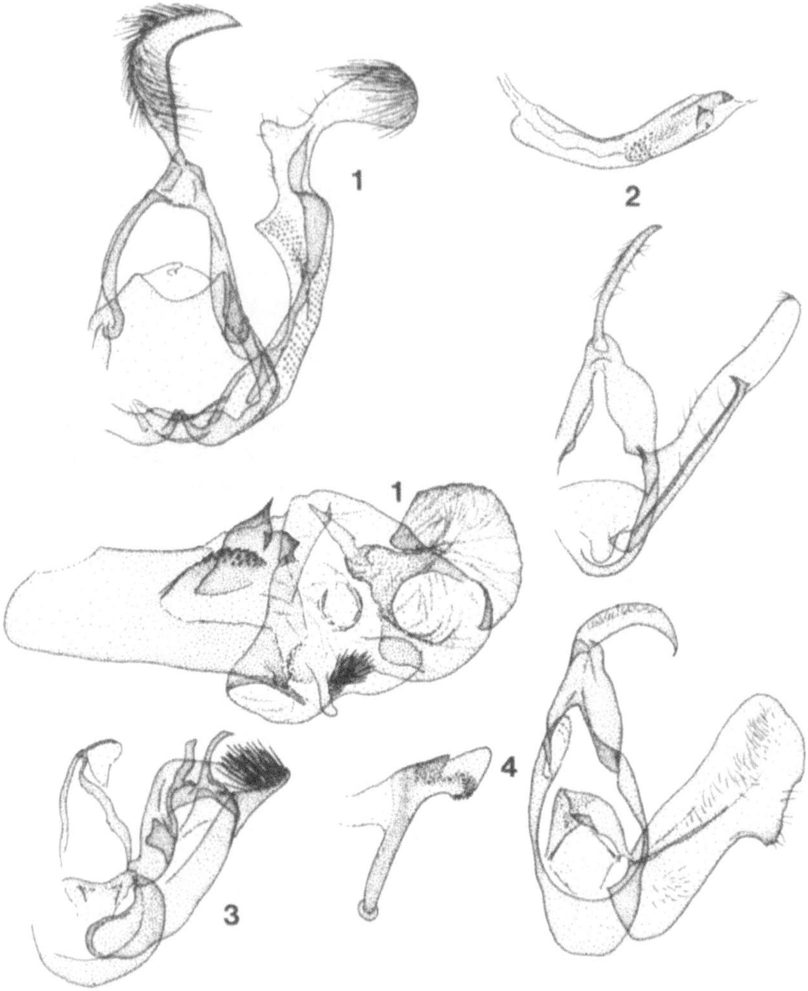


*Fig. 151.* Male genitalia and aedeagi (except 3) of *Cofimpacia chusaroides* (1), *Lysimelia litoralis* (2), *Luceria cooki* (3) and *Hypena probokonora* (4).

**Lysimelia litoralis sp. n.** (Plate 87: 9, Figure 151: 2)

♂♀ 7–9 mm. The wings are marked and shaded brown on a pale grey ground. The closest named species is *neleusalis* Walker (Sarawak, type examined, the type species of the genus) which is widespread in the Oriental Region and includes *maculipex* Hampson (Khasia Hills, type examined) and *lysimelioides* Hampson (Sri Lanka, type examined) **stat. n.** as subspecies (the genitalia of these subspecies differ in the male only in that there is a small, broad-based triangular projection from the dorsal margin of the valve). *L. litoralis* is generally greyer, with the forewing reniform only a dark dot anteriorly in most specimens instead of being c-shaped (though a posterior dot occurs in the specimen illustrated); the white submarginal is less regular and less well defined, flexing basad through the adjacent dark area to separate a small dark triangle apically; the dark postmedial is angled opposite the cell in *litoralis* but rounded in *neleusalis*. The hindwing submarginal is almost entire in *neleusalis* but only really prominent towards the dorsum in *litoralis*.

In the male genitalia the valve apex is rounded in *litoralis* but produced into an incurved spine in *neleusalis*; the latter lacks the central, ventral harpe of the former and has the aedeagus vesica adorned with a mass of even-sized, moderate, slender cornuti. There are closer relatives to *litoralis* in the Oriental Region, described and discussed below but previously unnamed except in manuscript (*alstoni* is an A. E. PROUT manuscript name).

HOLOTYPE ♂. New Caledonia: Dothio, 400 m Site 19 06133/76135 (J. D. HOLLOWAY) 8.6.1971, BM 1971-507. BM noctuid slide 7586.

PARATYPES. 7♂♂, 2♀♀ New Caledonia, various localities.

The species was taken in New Caledonia as follows: Anse Longue site 16 (2); Dothio site 19 (2); Poupa (1); Bréhoa R. (1); Port Boisé site 73 (1); Ouen Toro site 81 (6), site 82 (4).

Two BMNH specimens from Queensland (one labelled Coomooboolaroo, Duaringa) were determined as this species.

### Lysimelia alstoni sp. n.

♂♀ 8–10 mm. This species is similar to *litoralis* in appearance but the forewing reniform has the posterior dot emphasised rather than the anterior one; the submarginal cuts off a larger dark triangle at the apex; the postmedial is more acutely angled; the basal edge of the dark marginal line is straight rather than wavy; both forewing and hindwing postmedials are more distant from the submarginal. The male genitalia have a broader aedeagus, not curved, with a greater number of large, broad-based cornuti; the valve has a constriction above a larger harpe.

HOLOTYPE ♂. (Sri Lanka): Galboda, 8.02, Mackwood Colln BM 1927-341. BM noctuid slide 8851.

PARATYPES. (Sri Lanka): 1♀ as holotype; 1♀ Haldamulla; 1♂, 1♀ Haputale; 1♀ Nilambe. All Mackwood Colln.

Other material of this species in the BMNH is from Sri Lanka and India (Travancore). Singleton specimens from Singapore, Pulo Laut, Luzon, Borneo and Sulawesi suggest a complex of related forms may occur throughout the Indo-Australian area. The Philippines male had genitalia intermediate in character between those of *alstoni* and *litoralis* but further material is needed to clarify the picture.

*Nodaria cornicalis* (Fabricius) (Figure 148: 1)

*Phalaena cornicalis* Fabricius, 1794, *Ent. Syst.* 3(2): 229.

This species is found throughout the Oriental tropics and has been recorded from New Guinea, Queensland, the Bismarcks, Solomons and New Caledonia. It is a species of the west coast and *Acacia spirorbis* association and was recorded as follows: Grand Lac site 4 (1); Anse Longue site 16 (1), site 32 (2); Poupa (2); Ouatou R. (1); Col d'Amieu (1); Col Toma (2); Sarraméa (3); Kopéto (1); Mt. Dogny (1); Kaala-Gomén (3); Amoa R. (1); Ste. Therése (1); Mt Krapé site 55 (1), site 56 (2); Col Nassiboué (6); Ouen Toro site 64 (1), site 81 (3), site 82 (7); Port Boisé site 73 (5), site 74 (1); Les Dalmates site 55 (1), BPBM: Népoui Valley.

*Progonia micrastis* (Meyrick) (Figure 148: 4)

*Progonia micrastis* (Meyrick); ROBINSON (1975: 236).

*P. micrastis* occurs in the Indian region, Sundaland, Queensland, the Solomons, New Hebrides, Fiji, Rotuma and Samoa. It was recorded in New Caledonia as follows: Mt. des Sources site 14 (1); Ouatou R. (1); Bréhoa R. (3); Ouambéo R. (1); Col d'Amieu (1); Sarraméa (7); Kaala-Gomèn (3); Grottes de Koum (11); Bambou R. (5); Cap Colnett (1); R. Amoa (3); Ste. Therèse (1); Col des Roussettes (1); Mt Krapé site 55 (1); Col Nassiboué (5); Ouen Toro site 64 (2), site 81 (9), site 82 (12); Pic du Pin site 69 (1); Port Boisé site 73 (1); Les Dalmates site 77 (1), site 78 (1). BMNH: Pouébo. BPBM: Le Crouen, Tao, Népoui valley. The species is a member of the west coast and *Acacia spirorbis* association.

*Schrankia boisea* Holloway

*Schrankia boisea* HOLLOWAY, 1977: 109.

The unique male type of this species was taken at Port Boisé site 74. It stands apart from other Pacific *Schrankia* taxonomically, notably in the coarse scaling of the wings and the unusually small genitalia.

*Simplicia caeneusalis* (Walker) (Figure 152: 2)

*Simplicia caeneusalis* (Walker); ROBINSON (1975: 237); HOLLOWAY (1977: 112).

The species occurs throughout the Indo-Australian tropics to Queensland and Samoa, with records from Lifu, the Society and Austral Is., the Marquesas and Rapa. A distinct subspecies flies on Norfolk I. In New Caledonia *caeneusalis* was recorded as follows: Grand Lac site 2 (3); Mt. des Sources site 14 (3); Ouatou R. (2); Bréhoa R. (1); Col d'Amieu (1); Kopéto (2); Anse Longue site 32 (1); Mt. Mou site 33 (5); Bambou R. (1); R. Amoa (5); Probokonoro (5); Ouen Toro site 64 (1), site 81 (3), site 82 (1). BPBM: Nouméa, Mt. Koghis, Népoui, Tipindjé.

**Simplicia toma sp. n.** (Plate 87: 6, Figure 152: 1)

♂♀ 17 mm. The male is shades of pale, rufous brown, the forewing more rufous ochre basally; the hindwing is a straw colour broadly along the costa. The female is a more even, darker, greyer brown. Similar dimorphism occurs in *turpatalis* Walker (Indian region; Sri Lanka type examined) but in *turpatalis*

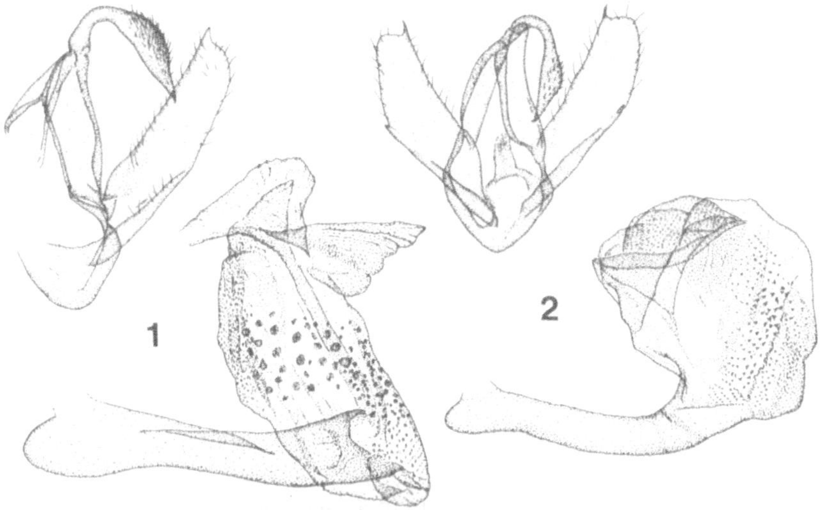


*Fig. 152.* **Male genitalia and aedeagi of *Simplicia toma* (1) and *S. caeneusalis* (2).**

the male is generally greyer in tint, the forewing postmedial more definite, the dark brown marginal markings on the forewing veins are faint only, more transversely elongate, less triangular than in *toma*; the forewing antemedial in *toma* is more oblique, being acute basad at the dorsum.

Compared with those of *turpatalis*, the male genitalia of *toma* have the uncus deeper, the apical spine of the valve shorter, the valve narrowing to the apex slightly instead of being broadly rounded; the aedeagus vesica has a central ring of sclerotised patches as compared with a lateral, triangular patch of evenly spaced, stout, squarish spines. The eighth sternite is unmodified but in *turpatalis* it has a narrowly defined marginal band of sclerotisation and a small, setose corema.

HOLOTYPE ♂. New Caledonia: Col Toma, 460 m Site 27 05813/76123 (J. D. HOLLOWAY) 19.6.1971, BM 1971-507. BM noctuid slide 7588.
PARATYPE ♀. Data as holotype.

### *Trigonistis toroensis* Holloway

*Trigonistis toroensis* Holloway, 1977: 115.

The unique male type of *toroensis* was taken at Ouen Toro site 82. The nearest relative is *T. andersoni* Holloway, an infrequent species of the Norfolk I. forest association. More distant relatives occur in Australia and possibly Fiji.

### Hypeninae *sp. indet.* (Plate 87: 10)

The single, rather worn female was taken at Ouen Toro site 81. It measures 13 mm. The forewing ground is straw colour, suffused and fasciated with shades of dark brown. The hindwing is dull medium brown. The palps are long, directed anteriorly, the third segment two-thirds the length of the second, both bearing a crest of scales dorsally. The species bears some resemblance to the genus *Mecistoptera* Hampson, especially in forewing shape.

## AGARISTIDAE

### **Eutrichopidia macchia sp. n.** (Plate 68: 12, Figure 153)

♂♀ 23–27 mm. The species is very similar to its only congener, *E. latinus* (Donovan) from south-east Australia and Tasmania, having similar wing pattern, hairy eyes, and orange hairs on the mid tibia, the last two being generic characteristics. The wings and body are predominantly black and the abdomen is tipped with orange. The submarginal and marginal markings of the hindwing are white in *macchia* but cream in *latinus*. The forewings are apically and tornally edged with white; a thin, irregular orange band extends from subtornally to just distad of the centre of the costa, a band that is paler, twice as broad and slightly more distad at the costa in *latinus*; the submarginal row of pale blue flecks on the veins is more prominent than in *latinus* and there is pale blue present subbasally on the costa in some specimens. The genitalia are similar to those of *latinus*.

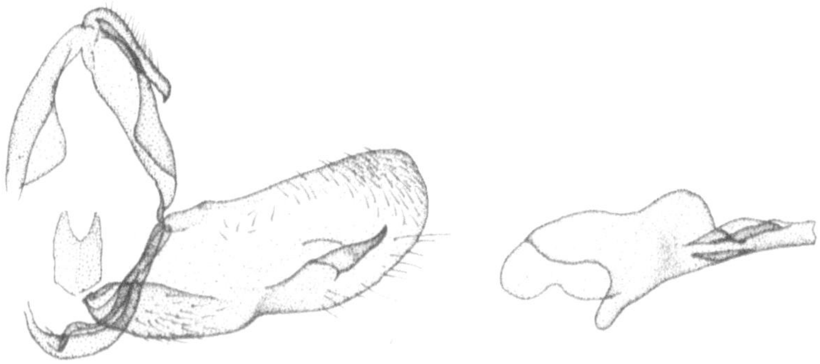

*Fig. 153.* **Male genitalia of *Eutrichopidia macchia*.**

HOLOTYPE ♂. New Caledonia: Grand Lac, 250 m 06978/75357, to lantana (J. D. HOLLOWAY) 12.5.1971, BM 1971-507. BM noctuid slide 8589.
PARATYPES: 3♀♀ as holotype, slide 8570; 1♀ Mt. Koghis 26.1.63 (YOSHIMOTO) BPBM.

The species was observed flying by day at Pic du Pin and the Yaté Gorge. It would appear to be restricted to the southern ultramafics.

*E. latinus* has as larval host plants *Haloragis teucrioides* of the family Haloragaceae (KIRIAKOFF, 1976, *Das Tierreich*, 97: 128) and *Hibbertia* (COMMON, 1970). Both genera occur in New Caledonia and *Hibbertia* has a secondary centre of radiation there, its species being important components of various ultramafic vegetation types.

Since going to press, the acronictine noctuid *Spodoptera exempta* Walker has been recorded in New Caledonia (DELOBEL, in press and *in litt.*). Though not recorded previously, the species appeared in numbers at various points on the west coast in 1978, the larvae causing significant damage to pasture. Since the initial outbreaks, four or five subsequent generations have been noted. DELOBEL suggested that a cyclone in January could have brought the species from Australia. It occurs widely in open habitats in the Old World tropics; its biology and distribution was reviewed by BROWN & DEWHURST (1975).

Two omissions from the systematic section were noted in proof. They are as follows.

*Asota caricae melanesiensis* Viette (1951b). This subspecies of a widespread Indo-Australian tropical species was described from the New Hebrides and New Caledonia (Nouméa). It was not recorded during the survey.

*Speiredonia mutabilis* (Fabricius); ROBINSON (1975: 212). This species extends throughout the Indo-Australian tropics as far east as Fiji and Tonga. A single male was taken at the Col de Boa locality. It is illustrated in Plate 86: 11, and its genitalia are shown in Figure 123.

In October, 1978 I received New Caledonian Lepidoptera from M. J. GUTIERREZ of ORSTOM, Nouméa, which included the following species new to New Caledonia.

*Hypena iridis* Butler (1887, *Ann. Mag. nat. Hist.*, (5) 19: 435, Solomons type examined). This species extends from Sundaland and the Moluccas to Queensland, and the Solomons, with a record also from Samoa. A single specimen from La Coulée in New Caledonia was received. The name *iridis* is probably a junior synonym of *H. albifascialis* Pagenstecher (1884, *Jb. nassau. ver. Naturk.*, 37: 244), but the identity of the latter is confusing. Colour slides in the BMNH of two syntypes from the Wiesbaden Museum (from 'many' in the original description) suggest the original series is mixed, including: a specimen from the Kei Is. labelled 'Originalexemplar' that is probably conspecific with *Harita nodyna* Bethune-Baker; a specimen from Amboina (1910, Dr. Pgstchr) (20) that correlates most closely with *iridis* in appearance and matches the figure in the original description of *albifascialis* in possessing a pale stigma in the forewing discal area. It would probably be less confusing to designate as lectotype an Amboina specimen such as this relating to *iridis* because: it compares with the original figure; the original paper was devoted to Amboina; and the name *Harita nodyna* has been used currently by ROBINSON (1975) and in this book.

# X. LIST OF NEW TAXA, NEW SYNONYMS AND NEW COMBINATIONS

## NEW GENERA

| | |
|---|---|
| GEOMETRIDAE: | *Caledophia* |
| | *Caledasthena* |
| | *Papuanticlea* |
| SPHINGIDAE: | *Compsulyx* |
| ARCTIIDAE: | *Fabresema* |
| | *Thumathoides* |
| NOCTUIDAE: | *Codalithia* |
| | *Colnettia* |
| | *Pseudocoarica* |
| | *Cofimpacia* |

## NEW SPECIES

| | |
|---|---|
| LIMACODIDAE: | *Miresa obscurimargo* |
| GEOMETRIDAE: | *Adeixis baeckeae* |
| | *Adeixis major* |
| | *Adeixis montana* |
| | *Caledophia pauli* |
| | *Derambila pellucida* |
| | *Oenochroma unifasciata* |
| | *Chloeres boisensis* |
| | *Chlorocoma octoplagiata* |
| | *Comostola thalassias* |
| | *Hemithea acaudata* |
| | *Thalassodes microchloropis* |
| | *Thalassodes pseudochloropis* |
| | *Thalassodes viridimargo* |
| | *Anisodes dalmatensis* |
| | *Anisodes kohensis* |
| | *Anisodes lateritica* |
| | *Anisodes paragloria* |
| | *Gnamptoloma rubra* |
| | *Idaea fruticeti* |
| | *Idaea lateritica* |
| | *Idaea paniensis* |
| | *Scopula caledonica* |
| | *Scopula quinquefasciata* |
| | *Caledasthena montana* |
| | *Chloroclystis callinda* |
| | *Chloroclystis lunifera* |
| | *Chloroclystis macroaedeagus* |
| | *Eoasthena sarramea* |

| | |
|---|---|
| | *Eois tiebaghi* |
| | *Euphyia perseverata* |
| | *Gonanticlea pulcherrima* |
| | *Gymnoscelis pseudofluctuosa* |
| | *Horisme grisearia* |
| | *Micrulia pacifica* |
| | *Papuanticlea ochroria* |
| | *Poecilasthena dimorpha* |
| | *Casbia serpentina* |
| | *Cleora acacicola* |
| | *Cleora dothionis* |
| | *Ctimene nocturnignis* |
| | *Nadagara serpentina* |
| | *Parosteodes grisea* |
| | *Petelia trifascia* |
| | *Symmetroctena variegata* |
| | *Syneora strixoides* |
| URANIIDAE: | *Micronia cynthiae* |
| | *Dirades robinsoni* |
| | *Epiplema yoshimotoi* |
| | *Epiplema lateritica* |
| | *Epiplema spatulata* |
| NOTODONTIDAE: | *Cerura amoa* |
| | *Neola octofera* |
| LYMANTRIIDAE: | *Euproctis dimorphissima* |
| | *Orgyia sarramea* |
| ARCTIIDAE: | *Eilema diliensis* |
| | *Eilema caledonica* |
| | *Fabresema gerardi* |
| | *Fabresema elisabethae* |
| | *Fabresema catherinae* |
| | *Fabresema sylviae* |
| | *Fabresema valeriae* |
| | *Fabresema flavibasalis* |
| | *Fabresema costiplaga* |
| | *Fabresema cheesmanae* |
| | *Fabresema grisea* |
| | *Fabresema sarramea* |
| | *Fabresema bicornuta* |
| | *Fabresema suffusa* |
| | *Fabresema toma* |
| | *Fabresema obliqua* |
| | *Macaduma biangulata* |
| | *Macaduma cristata* |
| | *Schistophleps subtilis* |
| | *Thallarcha multifasciata* |
| | *Thumathoides fumosa* |
| | *Spilosoma lateritica* |
| HYPSIDAE: | *Asota caledonica* |
| NOLIDAE: | *Nola bitransversata* |

NOCTUIDAE:

*Leucania latericia*
*Tiracola circularis*
*Tiracola tabwemasana*
*Acrapex pacifica*
*Athetis postpuncta*
*Callopistria dimorpha*
*Callopistria falcata*
*Chasmina basiflava*
*Eutamsia quadrispina*
*Hypoperigea lunulata*
*Leucogonia amarginata*
*Molvena caledonica*
*Amyna flavirena*
*Codalithia quincuncialis*
*Codalithia subtilis*
*Eublemma rufogalactea*
*Eublemma variochrea*
*Maliattha inconcisoides*
*Oruza kavatcha*
*Oruza narabuta*
*Stictoptera tridentifera*
*Calathusa brehoa*
*Calathusa uniformis*
*Calathusa basicunoides*
*Calathusa humboldti*
*Calathusa argentea*
*Calathusa basirufa*
*Calathusa cinerea*
*Calathusa talesea*
*Calathusa striata*
*Nycteola canoides*
*Nycteola pseudonilotica*
*Westermannia poupa*
*Parallelia lateritica*
*Avitta australis*
*Avitta pectinata*
*Hypospila pseudobolinoides*
*Hypospila elongata*
*Hypospila ochracea*
*Lacera uniformis*
*Lacera asinuosa*
*Lacera vinacea*
*Lacera contrasta*
*Lacera violacea*
*Ophyx maculosus*
*Plusiodonta theresae*
*Pseudocoarica caledonica*
*Rhesala irregularis*
*Rivula caledonica*
*Veia pectinata*

| | |
|---|---|
| | *Cofimpacia chusaroides* |
| | *Cofimpacia luteata* |
| | *Cofimpacia violacea* |
| | *Harita irregularis* |
| | *Hypena probokonora* |
| | *Lysimelia litoralis* |
| | *Lysimelia alstoni* |
| | *Simplicia toma* |
| AGARISTIDAE: | *Eutrichopidia macchia* |

NEW SUBSPECIES

| | |
|---|---|
| GEOMETRIDAE: | *Encryphia frontisignata acornuta* |
| | *Aeolochroma hypochromaria caledonica* |
| | *Comostola leucomerata caledonica* |
| | *Uliocnemis biplagiata reducta* |
| | *Idaea scintillans flavior* |
| | *Gymnoscelis sara hamata* |
| | *Horisme xylinata orientalis* |
| | *Casbia rectaria lilacina* |
| | *Casbia rectaria rosacea* |
| | *Cleora illustraria caledonica* |
| | *Elphos subrubida aoupiniensis* |
| HYPSIDAE: | *Agape lifuensis hebridesia* |
| | *Digama marmorea clinchorum* |
| NOCTUIDAE: | *Acrapex pacifica santonis* |
| | *Corgatha figuralis paucifasciata* |
| | *Penicillaria maculata richardsoni* |
| | *Calathusa basirufa suzukii* |
| | *Calathusa basirufa ipota* |
| | *Maceda savura paniensis* |
| | *Parallelia solomonensis bicacuminata* |
| | *Parallelia solomonensis hebridesia* |
| | *Parallelia solomonensis papuana* |
| | *Phyllodes imperialis dealbata* |
| | *Thyas miniacea rubrior* |
| | *Ctenoplusia placida caledonica* |
| | *Oxyodes scrobiculata obscurior* |
| | *Rhesala irregularis circuluncus* |

NEW SYNONYMS: GENERA

*Steneugoa* Hampson, of *Calathusa* Walker (Noctuidae)
*Imleanga* Lucas, of *Pantura* Moore (Noctuidae)
*Dnopheropis* Turner, of *Egone* Walker (Noctuidae)

NEW SYNONYMS: SPECIES
(GENUS OF ORIGINAL DESCRIPTION QUOTED)

*Brachycola decolorata* Warren, of *Anisodes niveopuncta* Warren (Geometridae)

*Miresa sobrina* Druce and *Miresa alma* Druce, of *Ochrogaster circumfumata* Felder (Thaumetopoeidae)
*Agape novapommeranicola* Strand, of *Agape stapialis* Gaede (Hypsidae)
*Callopistria meridionalis nauticorum* Tams, of *Eriopus maillardi* Guenée (Noctuidae)
*Trilophota tegulata* Wileman & West, of *Erastria miasma* Hampson (Noctuidae)
*Maliattha furcata* Warren, of *Maliattha inconcisa* Butler (Noctuidae)
*Nigramma perstrialis* Hampson, of *Gyrtona polionota* Hampson (Noctuidae)
*Blenina metachrysa* Turner and *Blenina triphaenopsis* Hampson, of *Blenina donans* Walker (Noctuidae)
*Blenina varians* Bethune-Baker, of *Blenina nigrans* Bethune-Baker (Noctuidae)
*Ophisma circumferens* Walker, of *Ophiodes trapezium* Guenée (Noctuidae).
*Gadirtha aroa* Bethune-Baker, of *Crioa acronyctoides* Walker (Noctuidae)
*Dnopheropis scotaea* Turner, of *Egone bipunctalis* Walker (Noctuidae)
*Lacera capella* Guenée, of *Noctua noctilio* Fabricius (Noctuidae)
*Rhesala albizziae* Prout, of *Rhesala asphalta* Swinhoe

## STATUS CHANGES

| | |
|---|---|
| GEOMETRIDAE: | *Eumelea rosalia degener* Warren |
| | *Comostola leucomerata flavifimbria* Warren |
| | *Comostola leucomerata rubeibasis* Prout |
| | *Anisodes lautokensis* Prout |
| | *Collix ghosha subligata* Warren |
| | *Cleora lacteata ecdees* Prout |
| URANIIDAE: | *Epiplema desistaria instabilata* (Walker) |
| ARCTIIDAE: | *Eilema nitens minima* (Hampson) |
| HYPSIDAE: | *Agape lifuensis* (Rothschild) |
| NOLIDAE: | *Nola major formosana* Wileman & West |
| NOCTUIDAE: | *Tiracola rufimargo concolor* Prout |
| | *Tiracola versicolor* Prout |
| | *Blenina lichenopa nigrans* Bethune-Baker |
| | *Blenina lichenopa solomonis* Swinhoe |
| | *Blenina lichenopa vatu* Robinson |
| | *Ophiusa disjungens indiscriminata* (Hampson) |
| | *Ophiusa disjungens tongaensis* (Hampson) |
| | *Ophiusa discriminans intacta* (Hampson) |
| | *Thyas miniacea regia* Lucas |
| | *Avitta ophiusalis lunifera* (Druce) |
| | *Ericeia inangulata levuensis* Prout |
| | *Leptotroga costalis armstrongi* Tams |
| | *Rhesalides asphalta nigra* (Bethune-Baker) |
| | *Rhesalides asphalta admiraltensis* Hampson |
| | *Speiredonia zamis layardi* (Hampson) |
| | *Speiredonia simplex obalauae* (Bethune-Baker) |
| | *Lysimelia neleusalis lysimelioides* Hampson |

## NEW COMBINATIONS
## (ORIGINAL GENUS BRACKETED WITH AUTHOR)

| | |
|---|---|
| GEOMETRIDAE: | *Papuanticlea horia* (Prout; *Gonanticlea*) |
| | *Papuanticlea subcaesia* (Warren; *Gonanticlea*) |
| | *Papuanticlea onaea* (Prout; *Gonanticlea*) |
| | *Papuanticlea semiflava* (Warren; *Anticlea*) |
| SPHINGIDAE: | *Compsulyx cochereaui* (Viette; *Compsogene*) |
| NOTODONTIDAE: | *Lasioceros euteles* (West; *Gargetta*) |
| ARCTIIDAE: | *Eilema nitens* (Walker; *Lithosia*) |
| | *Eilema nitens minima* (Hampson; *Lexis*) |
| | *Macaduma atriplaga* (Hampson; *Heliosia*) |
| NOCTUIDAE: | *Agrotis fumipennis* (Warren; *Euxoa*) |
| | *Molvena melaleuca* (Hampson; *Thalatha*) |
| | *Nigramma polionota* (Hampson; *Gyrtona*) |
| | *Calathusa nubilosa* (Hampson; *Steneugoa*) |
| | *Chrysodeixis illuminata* (Robinson; *Plusia*) |
| | *Rhesalides asphalta* (Swinhoe; *Rhesala*) |
| | *Rhesalides asphalta nigra* (Bethune-Baker; *Magulaba*) |
| | *Speiredonia zamis layardi* (Hampson; *Sericia*) |
| | *Speiredonia simplex* Butler (revived) |
| | *Speiredonia simplex obalauae* (Bethune-Baker; *Sericia*) |

# REFERENCES


Abele, L. G., & Patton, W. K., 1976. The size of coral heads and the community biology of associated decapod crustaceans. *J. Biogeog.*, **3:** 35–47.

Alexander, C. P., 1929. New or little known Tipulidae from New Caledonia and Samoa. *Encycl. Ent., Diptera*, **5**: 83–92.

Alexander, C. P., 1934. New or little known Tipulidae from eastern Asia (Diptera), XIX, XX. *Philipp. J. Sci.*, **54:** 309–342, 433–471.

Alexander, C. P., 1945. New or little known Tipulidae from New Caledonia (Diptera). *Proc. Hawaii ent. Soc.*, **12:** 235–244.

Alexander, C. P., 1948a. The craneflies of New Caledonia (Diptera, Tipulidae). *Trans. R. ent. Soc. Lond.*, **99:** 361–393.

Alexander, C. P., 1948b. New or insufficiently known crane-flies from New Caledonia (Diptera: Tipulidae), Part I. *Ann. ent. Soc. Amer.*, **41:** 137–148.

Ashton, P. S., 1971. The plants and vegetation of Bako National Park. *Malay. Nat. J.*, **24:** 151–162.

Aubréville, A., 1967. Sapotacées. In *Flore de La Nouvelle-Calédonie et Dépendances.* Paris: Muséum National d'Histoire Naturelle.

Audley-Charles, M. G., 1974. Sulawesi. In Spencer (1974), pp. 365–378.

Audley-Charles, M. G., 1975. The Sumba fracture: a major discontinuity between eastern and western Indonesia. *Tectonophysics*, **26:** 213–288.

Audley-Charles, M. G., Carter, D. J., & Milsom, J., 1972. Tectonic development of eastern Indonesia in relation to Gondwanaland dispersal. *Nature Phys. Sci.*, **239:** 35–39.

Avery, D. F., & Tanner, W. W. 1970. Speciation in the Fijian and Tongan iguana *Brachylophus* (Sauria, Iguanidae) with the description of a new species. *Gt Basin Nat.*, **30:** 166–172.

Avias, J., 1967. Overthrust structure of the main ultrabasic New Caledonian massives. *Tectonophysics*, **4:** 531–541.

Avias, J., 1973. Major features of the New Guinea-Louisiade-New Caledonia-Norfolk arc system. In Coleman (1973), pp. 113–126.

Axelrod, D. I., 1974. Plate tectonics in relation to the history of the angiosperm flora of India. *Birbal Sahni Inst., Palaeobot. spec. Publ.*, **1:** 5–18.

Bain, J. H. C., 1973. A summary of the main structural elements of Papua-New Guinea. In Coleman (1973), pp. 147–161.

Balgooy, M. M. J. Van, 1969. A study on the diversity of island floras. *Blumea*, **17:** 140–178.

Balgooy, M. M. J. Van, 1971. Plant-geography of the Pacific, as based on a census of Phanerogam genera. *Blumea*, Suppl. **6:** 1–222.

Barnard, P. D. W., 1973. Mesozoic floras. In *Organisms and Continents through Time. Spec. Pap. Palaeont.*, **12:** 175–187.

Beaglehole, J. C. (ed.), 1955–1969. *The Journals of Captain James Cook on his Voyages of Discovery.* Cambridge: Hakluyt Society.

Beard, J. S., 1977. Tertiary evolution of the Australian flora in the light of latitudinal movements of the continent. *J. Biogeog.*, **4:** 111–118.

Béland, P., 1975. Enquête sur les oiseaux de Calédonie. *Nature calédonienne*, **8:** 31–34, **9:** 33–35.

Belkin, J. N., 1962. *Mosquitoes of the South Pacific*. 2 vols. Berkeley & Los Angeles: University of California Press.

Berio, E., 1965. Le Catocaline africane a tibie spinose del Museo di Tervuren. *Annali Mus. civ. Stor. nat. Giacomo Doria*, **75:** 181–332.

Berland, L., 1934. Les araigneés du Pacifique. *Mém. Soc. Biogéogr.*, **4:** 155–180.

Berlioz, J., 1962. Les caractères de la faune avienne en Nouvelle-Calédonie. *C. R. Somm. Séanc. Soc. Biogéogr.*, **39:** 65–69.

Beveridge, A. E., 1975. Kauri forests in the New Hebrides. *Phil. Trans. R. Soc. Lond.*, B., **272:** 369–383.

Bianchi, F. A., 1945. Introduction to the Thysanoptera of New Caledonia. *Proc. Hawaii ent. Soc.*, **12:** 249–278.

Bianchi, F. A., 1952. Additions to the Thysanoptera of New Caledonia. *Proc. Hawaii ent. Soc.*, **14:** 385–390.

Braithwaite, A. F., 1975. The phytogeographical relationships and origin of the New Hebrides fern flora. *Phil. Trans. R. Soc. Lond.* B., **272:** 293–313.

Breuning, S., 1951, 1953. Notes systématiques sur les longicornes de Nouvelle Calédonie. Two parts. *Bull. Inst. r. Sci. nat. Belg.*, **27**(32), **29**(16).

Brindle, A., 1976a. The Dermaptera of New Caledonia. *Pacif. Ins.*, **17:** 61–86.

Brindle, A., 1976b. The Dermaptera of the New Hebrides. *Rec. S. Aust. Mus.*, **17:** 221–238.

Brookfield, H. C., & Hart, D., 1971. *Melanesia, a Geographical Interpretation of an Island World*. London: Methuen.

Brothers, R. N., 1954. The relative Pleistocene chronology of the south Kaipara district, New Zealand. *Trans. R. Soc. N. Z.*, **82:** 677–694.

Brothers, R. N., 1973. Tertiary orogenesis and high-pressure metamorphism in New Caledonia. In Coleman (1973), p. 469.

Brown, D. A., Campbell, K. S. W., & Crook, K. A. W., 1968. *The Geological Evolution of Australia and New Zealand*. London: Pergamon.

Brown, E. S., & Dewhurst, C., 1975. The genus *Spodoptera* in Africa and the Near East. *Bull. ent. Res.*, **65**: 221–262.

Brown, W. C., 1965. New frogs of the genus *Cornufer* (Ranidae) from the Solomon Islands. *Breviora*, **218:** 1–16.

Brown, W. L. Jr., & Taylor, R. W., 1970. Superfamily Formicoidea. In *The Insects of Australia*, pp. 951–959. Melbourne: CSIRO.

Brownlie, G., 1965. The geographical affinities of the south Pacific island fern floras. *Pacif. Sci.*, **19:** 219–223.

Bull, P. C., & Whitaker, A. H., 1975. The amphibians, reptiles, birds and mammals. In G. Kuschel (ed.), *Biogeography and Ecology in New Zealand*, pp. 231–276. Monographiae Biologicae 27. The Hague: W. Junk.

Burke, K. C., & Wilson, J. T., 1976. Hot spots on the Earth's surface. *Sci. Amer.*, **235**(2): 46–57.

Burtt, B. L., 1962. Studies on the Gesneriaceae of the Old World. XXIV: Tentative keys to the tribes and genera. *Notes R. bot. Gdn Edinb.*, **24:** 205–220.

Campion, H., 1921. Odonata collected in New Caledonia by the late Mr Paul D. Montague. *Ann. Mag. nat. Hist.* (9), **8:** 33–67.

Carey, S. W., 1970. Australia, New Guinea and Melanesia in the current revolution in concepts of the evolution of the Earth. *Search*, **1:** 178–189.

Carlquist, S., 1967. The biota of long distance dispersal V. Plant dispersal to Pacific islands. *Bull. Torrey bot. Club.*, **94:** 129–162.

Carlquist, S. 1974. *Island Biology*. New York and London: Columbia University Press.

CARTER, D. J., AUDLEY-CHARLES, M. G., & BARBER, A. J., 1976. Stratigraphical analysis of island arc-continental margin collision in eastern Indonesia. *Jl geol. Soc. Lond.*, **132:** 179–198.

CHAPPELL, J., 1974. Relationships between sealevels, $^{18}O$ variations and orbital perturbations during the past 250,000 years. *Nature, Lond.*, **252:** 199–202.

CHASE, C. G., 1971. Tectonic history of the Fiji Plateau. *Bull. geol. Soc. Am.*, **82:** 3087–3110.

CHEW W.-L., 1975. The phanerogamic flora of the New Hebrides and its relationships. *Phil. Trans. R. Soc. Lond.* B., **272:** 315–328.

CHRISTOFFEL, D. A. & VAN DER LINDEN, W. J. M., 1972. Macquarie Ridge-New Zealand Alpine Fault transition. In HAYES (1972), pp. 235–242.

CIESIELSKI, P. F., & WEAVER, F. M., 1974. Early Pliocene temperature changes in the Antarctic seas. *Geology, Boulder, Colo.*, **2:** 511–515.

CLARKE, J. F. GATES, 1971. *The Lepidoptera of Rapa Island.* Smithsonian Contributions to Zoology, 56.

CLIFFORD, H. T., & STEPHENSON, W., 1975. *An Introduction to Numerical Classification.* London: Academic Press.

CLIMO, F. M., 1975. The land snail fauna. In G. KUSCHEL (ed.) *Biogeography and Ecology in New Zealand*, pp. 459–492. Monographiae Biologicae **27.** The Hague: W. Junk.

COHIC, F., 1956. *Parasites animaux des plantes cultivées en Nouvelle-Calédonie et Dépendances.* Nouméa: Institut français d'Océanie.

COHIC, F., 1959. Report of a visit to the Chesterfield Islands. *Atoll Res. Bull.*, **63:** 1–11.

COLEMAN, F., 1972. *Frequencies, tracks and intensities of tropical cyclones in the Australian Region, 1909–1969.* Commonwealth of Australia, Department of the Interior, Bureau of Meteorology.

COLEMAN, P. J., 1966. The Solomon Islands as an island arc. *Nature Lond.*, **211:** 1249–1251.

COLEMAN, P. J. (ed.), 1973. *The Western Pacific: Island Arcs, Marginal Seas, Geochemistry.* Perth: University of Western Australia Press.

COLEMAN, P. J., & HACKMAN, B. D., 1974. Solomon Islands. In SPENCER (1974), pp. 453–461.

COLEMAN, R. G., 1967. Glaucophane schists from California and New Caledonia. *Tectonophysics*, **4:** 479–498.

COMMON, I. F. B., 1958. The Australian cutworms of the genus *Agrotis* (Lepidoptera: Noctuidae). *Aust. J. Zool.*, **6:** 69–88.

COMMON, I. F. B., 1963. *Australian Moths.* Brisbane: Jacaranda Press.

COMMON, I. F. B., 1970. Lepidoptera. In *The Insects of Australia*, pp. 765–866. Melbourne: CSIRO.

COMMON, I. F. B., & WATERHOUSE, D. F., 1972. *Butterflies of Australia.* Sydney: Angus & Robertson.

COOKSON, I. C., 1964. Some early angiosperms from Australia: the pollen record. In CRANWELL (1964), pp. 81–84.

CORNER, E. J. H., 1965. Checklist of *Ficus* in Asia and Australasia. *Gdns Bull., Singapore*, **21:** 1–186.

CORNER, E. J. H., 1967. *Ficus* in the Solomon Islands and its bearing on the post-Jurassic history of Melanesia. *Phil. Trans. R. Soc. Lond.* B, **253:** 23–159.

CORNER, E. J. H., 1970. *Ficus* subgen. *Pharmacosycea* with reference to the species of New Caledonia. *Phil. Trans. R. Soc. Lond.* B, **259:** 383–433.

CORNER, E. J. H., 1975. *Ficus* in the New Hebrides. *Phil. Trans. R. Soc. Lond.* B, **272:** 343–367.

COUPER, R. A., 1960a. New Zealand Mesozoic and Cainozoic plant microfossils. *N. Z. Geol. Surv. Palaeont. Bull.*, **32:** 1–88.

COUPER, R. A., 1960b. Southern hemisphere Mesozoic and Tertiary Podocarpaceae and Fagaceae and their palaeogeographic significance. *Proc. R. Soc. Lond.* B, **152:** 491–500.

CRANWELL, L. M., 1963. *Nothofagus:* living and fossil. In J. L. GRESSITT (ed.), *Pacific Basin Biogeography*, pp. 387–400. Honolulu: Bishop Museum Press.
CRANWELL, L. M. (ed.), 1964. *Ancient Pacific Floras.* Honolulu: University of Hawaii Press.
CRANWELL, L. M., 1964a. Rapa Island coal and its microfossils: a preliminary report. In CRANWELL (1964), pp. 43–47.
CRANWELL, L. M., 1964b. Antarctica: cradle or grave for its *Nothofagus.* In CRANWELL (1964), pp. 87–93.
D'ABRERA, B., 1971. *Butterflies of the Australian Region.* Melbourne: Lansdowne Press.
DARLINGTON, P. J. JR., 1965 *Beiogeography of the Southern End of the World.* Cambridge, Massachusetts: Harvard University Press.
DAVID, T. W. E., 1950. *The Geology of the Commonwealth of Australia.* 2 vols. London: Arnold.
DAWSON, J. W., 1963. New Caledonia and New Zealand—a botanical comparison. *Tuatara,* **11:** 178–193.
DELOBEL, A., in press. Note sur une pullulation de *Spodoptera exempta* Walker sur la côte occidentale de la Nouvelle-Calédonie (Lépidoptères Noctuidae). *Cahiers ORSTOM, Sér. Biologie.*
DETTMANN, M. E., 1973. Angiospermous pollen from Albian to Turonian sediments of eastern Australia. *Spec. Publs geol. Soc. Aust.,* **4:** 3–34.
DIAMOND, J. M., 1970. Ecological consequences of island colonisation by southwest Pacific birds, II. The effect of species diversity on total population density. *Proc. natn. Acad. Sci. U.S.A.,* **67:** 1715–1721.
DIAMOND, J. M., 1975. Assembly of species communities. In CODY, M. L., & DIAMOND, J. M. (eds.), *Ecology and Evolution of Communities,* pp. 342–444. Cambridge, Mass., and London: Belknap Press, Harvard University.
DIAMOND, J. M., in press. Continental and insular speciation in Pacific land birds. *Syst. Zool.*
DIAMOND, J. M., & MARSHALL, A. G., 1976. Origin of the New Hebridean avifauna. *Emu,* **76:** 186–200.
DIAMOND, J. M., & MAYR, E., 1976. Species-area relation for birds of the Solomon Archipelago. *Proc. natn. Acad. Sci. U.S.A.,* **73:** 262–266.
DICKINSON, W. R., 1967. Tectonic development of Fiji. *Tectonophysics,* **4:** 543–553.
DILLON, L. S., & DILLON, E. S., 1943. Cerambycidae of Fiji. *Bull. Bernice P. Bishop Mus.,* **206:** 1–114.
DUBOIS, J., GUILLON, J. H., LAUNAY, J., RECY, J., & TRESCASES, J. J., 1973. Structural and other aspects of the New Caledonia-Norfolk area. In COLEMAN (1973).
DUBOIS, J., LAUNAY, J., & RECY, J., 1975. Some new evidence on lithospheric bulges close to island arcs. *Tectonophysics,* **26:** 189–196.
DUFFELS, J. P., 1977. A revision of the genus *Diceropyga* Stål, 1870 (Homoptera, Cicadidae). *Mon. Ned. ent. Ver.,* **8.**
DUGDALE, J. S., 1961. Larval characters of taxonomic significance of New Zealand Ennomines (Lepidoptera: Geometridae). *Trans. Roy. Soc. N. Z., Zool.,* **1:** 215–233.
DUGDALE, J. S., 1964. Insects of Campbell Island. Appendix: Lepidoptera: Geometridae. *Pacif. Ins. Monogr.,* **7:** 607–623.
DUGDALE, J. S., 1971a. Entomology of the Aucklands and other islands south of New Zealand: Lepidoptera, excluding non-crambine Pyralidae. *Pacif. Ins. Monogr.,* **27:** 55–172.
DUGDALE, J. S., 1971b. Genera of New Zealand Cicadidae (Homoptera). *N. Z. Jl Sci.,* **14:** 856–882.
DUGDALE, J. S., 1975. The insects in relation to plants. In G. KUSCHEL (ed.), *Biogeography and Ecology in New Zealand,* pp. 561–589. Monographiae Biologicae, **27.** The Hague: W. Junk.
DUMBLETON, L. J., 1961. The Aleyrodidae (Hemiptera-Homoptera) of New Caledonia. *Pacif. Sci., Honolulu,* **15:** 114–136.

DUMBLETON, L. J., 1966. Genitalia, classification and zoogeography of the New Zealand Hepialidae (Lepidoptera). *N. Z. Jl Sci.*, **9**: 920–981.

EDMUNDS, G. F., Jr., 1972. Biogeography and evolution of Ephemeroptera. *Ann. Rev. Ent., Palo Alto,* **17:** 21–42.

ELIOT, J. N., 1973. The higher classification of the Lycaenidae (Lepidoptera): a tentative arrangement. *Bull. Br. Mus. nat. Hist.* (Ent.), **28:** 371–505.

EVANS, J. W., 1974. New Caledonian leafhoppers and the systematic position of *Kosmiopelix* Kirkaldy and *Euacanthella* Evans (Homoptera: Cicadelloidea). *Pacif. Ins.,* **16:** 165–175.

FAUVEL, A., 1903–1907. Faune analytique des coléoptères de la Nouvelle-Calédonie. *Rev. Ent., Caen,* **22:** 203–379, **23:** 113–208, **24:** 209–244, **25:** 29–100, **26:** 149–152.

FELTEN, H., 1964. Flughunde der Gattung *Pteropus* von Neukaledonien und den Loyalty-Inseln (Mammalia, Chiroptera). *Senckenberg biol.,* **45:** 671–683.

FENNAH, R. G., 1969. Fulgoroidea (Homoptera) from New Caledonia and the Loyalty Islands. *Pacif. Ins. Monogr.,* **21:** 1–116.

FERRAR, P., STANDFAST, H. A., & DYCE, A. L., 1975. A survey of blood-sucking and synanthropic Diptera and dung insects on Norfolk Island, South Pacific. *J. Aust. ent. Soc.,* **14:** 7–13.

FIELD, J. G., & MCFARLANE, G., 1968. Numerical methods in marine ecology. I. A quantitative 'similarity' analysis of rocky shore samples in False Bay, South Africa. *Zoologica Africana,* **3:** 119–137.

FLEMING, C. A., 1975. The geological history of New Zealand and its biota. In G. KUSCHEL (ed.), *Biogeography and Ecology in New Zealand,* pp. 1–86. Monographiae Biologicae **27**. The Hague: W. Junk.

FLEMING, C. A., 1976. New Zealand as a minor source of terrestrial plants and animals in the Pacific. *Tuatara,* **22:** 30–37.

FLETCHER, D. S., 1957. Macroheterocera of Rennell Island. *Nat. Hist. Rennell I., Br. Solomon Is.,* **2:** 31–66.

FORSTER, R. R., 1975. The spiders and harvestmen. In G. KUSCHEL (ed.) *Biogeography and Ecology in New Zealand,* pp. 493–506. Monographiae Biologicae **27**. The Hague: W. Junk.

FOSBERG, F. R., 1967. A classification of vegetation for general purposes. In G. PETERKEN, *Guide to the Check Sheet for IBP Areas,* pp. 73–120. IBP Handbook **4**. Oxford and Edinburgh: Blackwell Scientific Publications.

GASKIN, D. E., 1970. The origins of the New Zealand fauna and flora: a review. *Geogrl Rev.,* **40:** 414–434.

GASKIN, D. E., 1974a. The species of *Pareromene* Osthelder (Pyralidae: Crambinae: Diptychophorini) from the western South Pacific, with further notes on the New Zealand species. *J. Ent.* (*B*), **43:** 159–184.

GASKIN, D. E., 1974b. The species of *Pareromene* Osthelder (Pyralidae: Crambinae: Diptychophorini) from Malaysia, Indonesia and New Guinea. *J. Ent.* (*B*), **43:** 185–208.

GASKIN, D. E., 1975. Reconsideration of New Zealand biogeographical problems, with special reference to the Mesozoic. *Mém. Mus. nat. Hist. nat.,* Nlle Serie (A), **88:** 87–97.

GERVASIO, F. C., 1973. Geotectonic development of the Philippines. In COLEMAN (1973), pp. 307–324.

GILL, E. D., 1961. The climates of Gondwanaland in Kainozoic times. In A. E. M. NAIRN (ed.), *Descriptive Palaeoclimatology,* pp. 332–353. New York: Interscience.

GILL, J. B., 1970. Geochemistry of Viti Levu, Fiji and its evolution as an island arc. *Contr. Miner. Petrol.,* **27:** 179–203.

GILL, J. & GORTON, M., 1973. A proposed geological and geochemical history of eastern Melanesia. In COLEMAN (1973), pp. 543–566.

GILL, J. B., & MCDOUGALL, I., 1973. Biostratigraphic and geological significance of Miocene-Pliocene volcanism in Fiji. *Nature, Lond.,* **241:** 176–180.

GILLISON, A. N., 1975. Phytogeographical relationships of the northern islands of the New Hebrides. *Phil. Trans. R. Soc. Lond.* B, **272:** 385–390.

GILLISON, A. N., & BEVERIDGE, A. E., 1974. Vegetation types of the New Hebrides—the northern islands and some phytogeographic relationships. Paper given at the Informal Meeting to discuss the results of the Royal Society/Percy Sladen New Hebrides Expedition, 1971.

GLENIE, R. C., SCHOFIELD, J. C., & WARD, W. T., 1968. Tertiary sea levels in Australia and New Zealand. *Palaeogeogr., Palaeoclimatol., Palaeoecol.*, **5:** 141–163.

GODLEY, E. J., 1975. Flora and vegetation. In G. KUSCHEL (ed.) *Biogeography and Ecology in New Zealand*, pp. 177–230. Monographiae Biologicae **27**. The Hague: W. Junk.

GOOD, R., 1950. Madagascar and New Caledonia; a problem in plant geography. *Blumea*, **6:** 470–479.

GOOD, R., 1958. The biogeography of Australia. *Nature, Lond.*, **181:** 1763–1765.

GOOD, R., 1960. On the geographical relationships of the angiosperm flora of New Guinea. *Bull. Br. Mus. nat. Hist.* (Bot.), **2,** 205–226.

GOOD, R., 1963. On the biological and physical relationships between New Guinea and Australia. In J. L. GRESSITT (ed.) *Pacific Basin Biogeography*, pp. 301–309. Honolulu: Bernice P. Bishop Museum

GOOD, R., 1974. *The Geography of the Flowering Plants* (4th edn.). London: Longmans.

GORHAM, S. W., 1965. Fiji frogs. *Zool. Beitr.*, **11:** 381–435.

GORHAM, S. W., 1968. Fiji frogs—life history data from field work. *Zool. Beitr.*, **14:** 427–446.

GREEN, D., & CULLEN, D. J., 1973. The tectonic evolution of the Fiji region. In COLEMAN (1973), pp. 127–145.

GREENSLADE, P. J. M., 1969. Insect distribution patterns in the Solomon Islands. *Phil. Trans. R. Soc. Lond.* B, **225:** 271–284.

GREER, A. E., 1974. The generic relationships of the sciurid lizard genus *Leiolopisma* and its relatives. *Aust. J. Zool., Suppl. Ser.*, **31:** 1–67.

GRESSITT, J. L., 1956. Some distribution patterns of Pacific island faunae. *Syst. Zool.*, **5:** 11–32, 47.

GRESSITT, J. L., 1958. Zoogeography of insects. *Ann. Rev. Ent., Palo Alto*, **3:** 207–230.

GRESSITT, J. L., 1960a. Zoogeography of Pacific Coleoptera. *Proc. 11th Int. Congr. Ent. Wien*, **1:** 463–465.

GRESSITT, J. L., 1960b. Hispine beetles from New Caledonia (Chrysomelidae). *Pacif. Ins.*, **2:** 101–121.

GRESSITT, J. L., 1961. Problems in the zoogeography of Pacific and Antarctic insects. *Pacif. Ins. Monogr.*, **2:** 1–94.

GRESSITT, J. L., 1974. Insect biogeography. *Ann. Rev. Ent., Palo Alto*, **19:** 293–321.

GRIFFITHS, J. R., & VARNE, R., 1972. Evolution of the Tasman Sea, Macquarie Ridge and Alpine Fault. *Nature, Phys. Sci.*, **235:** 83–86.

GUILLAUMIN, A., 1952. Les caractères de la végétation néo-calédonienne. *C. R. somm. Séanc. Soc. Biogéogr.*, **29:** 82–86.

GUILLAUMIN, A., 1964. L'endemisme en Nouvelle-Calédonie. *C. R. somm. Séanc. Soc. Biogéogr.*, **41:** 67–74.

GUILLAUMIN, A., & VIROT, R., 1962. Taxonomie et distribution des *Araucaria* de Nouvelle-Calédonie. *C. R. somm. Séanc. Soc. Biogéogr.*, **39:** 71–75.

GUILLON, J. H., 1974. New Caledonia. In SPENCER (1974), pp. 447–452.

HACKMAN, B. D., 1973. The Solomon Islands fractured arc. In COLEMAN (1973), pp. 179–191.

HAIR, J. B., 1963. Cytogeographical relationships of the southern podocarps. In J. L. GRESSITT (ed.), *Pacific Basin Biogeography*, pp. 401–414. Honolulu: Bernice P. Bishop Museum.

HAMILTON, W. 1971. Plate tectonic evolution of Indonesia. *Geol. Soc. Amer., abstracts with programmes*, **3:** 589.

HARRIS, W. K., 1974. Palynology of Palaeocene sediments at site 214, Ninetyeast Ridge. In C. C. VON DER BORCH *et al.* (eds.), *Initial Reports of the Deep Sea Drilling Project*, **22:** 503–519. Washington: Government Printing Office.

HAWKINS, J. W. JR., 1976. Petrology and geochemistry of basaltic rocks of the Lau Basin. *Earth planet, Sci. Lett.*, **28:** 283–297.

HAYASHI, M., 1961. Cerambycidae from New Caledonia (Coleoptera), Part 1. *Bull. Osaka Mus. nat. Hist.*, **13:** 7–65.

HAYES, D. E., (ed.), 1972. *Antarctic oceanology II. The Australian-New Zealand Sector.* Antarctic Res. Ser. Washington, **19**.

HAYES, D. E., *et al.*, 1973. Leg 28 deep-sea drilling in the southern ocean. *Geotimes*, **18**(6): 19–24.

HAYES, D. E., & RINGIS, J., 1973. Sea floor spreading in the Tasman Sea. *Nature, Phys. Sci.*, **235:** 454–458.

HAYES, D. E., & TALWANI, M., 1972. Geophysical investigation of the Macquarie Ridge complex. In HAYES (1972), pp. 211–234.

HEATH, J., 1965. A genuinely portable M. V. light trap. *Entomologist's Rec. J. Var.*, **77:** 236–238.

HERMES, J. J., 1974. West Irian. In SPENCER (1974), pp. 476–490.

HOCHSTEIN, M. P., SCHOFIELD, J. C., & SHOR, G. G. JR., 1974. Tonga-Kermadec-Lau. In SPENCER (1974), pp. 417–423.

HOLLOWAY, J. D., 1969. A numerical investigation of the biogeography of the butterfly fauna of India and its relation to continental drift. *Biol. J. Linn. Soc.*, **1:** 373–385.

HOLLOWAY, J. D. 1970. The biogeographical analysis of a transect sample of the moth fauna of Mt. Kinabalu, Sabah, using numerical methods. *Biol. J. Linn. Soc.*, **2:** 259–286.

HOLLOWAY, J. D., 1973a. Problems with small islands. In *Organisms and Continents Through Time. Spec. Pap. Palaeont.*, **12:** 107–112.

HOLLOWAY, J. D., 1973b. The taxonomy of four groups of butterflies (Lepidoptera) in relation to general patterns of butterfly distribution in the Indo-Australian area. *Trans. R. ent. Soc. Lond.*, **125:** 125–176.

HOLLOWAY, J. D., 1974a. The biogeography of Indian butterflies. In M. S. MANI (ed.), *Ecology and Biogeography in India, pp.* 473–499. Monographiae Biologicae **23**. The Hague: W. Junk.

HOLLOWAY, J. D., 1974b. The endemic Satyridae (Lepidoptera: Rhopalocera) of New Caledonia. *J. Ent. (B)*, **43:** 89–101.

HOLLOWAY, J. D., 1976. *Moths of Borneo, with special reference to Mt. Kinabalu.* Kuala Lumpur: Malayan Nature Society.

HOLLOWAY, J. D., 1977. *The Lepidoptera of Norfolk Island, their biogeography and ecology.* Series Entomologica **13**. The Hague: W. Junk.

HOLLOWAY, J. D. & JARDINE, N., 1968. Two approaches to zoogeography: a study based on the distributions of butterflies, birds and bats in the Indo-Australian area. *Proc. Linn. Soc. Lond.*, **179:** 153–188.

HOLLOWAY, J. D., & PETERS, J. V., 1976. The butterflies of New Caledonia and the Loyalty Islands. *J. nat. Hist.*, **10:** 273–318.

HUDSON, G. V., 1928, 1939. *Butterflies and Moths of New Zealand*, and Supplement. Wellington: Ferguson & Osborn.

HUGHES, N. F., 1973. Mesozoic and Tertiary distributions, and problems of land-plant evolution. In *Organisms and Continents Through Time. Spec. Pap. Palaeont.*, **12:** 188–198.

HUTCHINSON, J., 1973. *The Families of Flowering Plants* (3rd edn). Oxford: Clarendon Press.

ILLIES, J., 1965. Phylogeny and zoogeography of the Plecoptera. *Ann. Rev. Ent., Palo Alto*, **10:** 117–140.

JACKSON, E. D., SILVER, E. A., & DALRYMPLE, G. B., 1972. Hawaiian-Emperor chain and its relation to Cenozoic circumpacific tectonics. *Bull. geol. Soc. Am.*, **83:** 601–618.

JAFFRÉ, T., 1974. La végétation et la flore d'un massif de roches ultrabasiques de Nouvelle-Calédonie: le Koniambo. *Candollea*, **29:** 427–456.
JAFFRÉ, T., BROOKS, R. R., LEE, J., & REEVES, R. D., 1976. *Sebertia acuminata:* a hyperaccumulator of nickel from New Caledonia. *Science, N.Y.*, **193:** 579–580.
JAFFRÉ, T., & LATHAM, M., 1974. Contribution à l'étude des relations sol-végétation sur un massif des roches ultrabasiques de la côte ouest de la Nouvelle-Calédonie: le Boulinda. *Adansonia*, ser. 2, **14:** 311–336.
JARDINE, N., 1972. Computational methods in the study of plant distributions. In D. H. VALENTINE (ed.), *Taxonomy, Phytogeography and Evolution*, pp. 381–393. London and New York: Academic Press.
JARDINE, N., & SIBSON, R., 1968a. A model for taxonomy. *Math. Biosci.*, **2:** 465–482.
JARDINE, N., & SIBSON, R., 1968b. The construction of hierarchic and non-hierarchic classifications. *Comput. J.*, **11:** 177–184.
JOHNSON, B. D., POWELL, C. McA., & VEEVERS, J. J., 1976. Spreading history of the eastern Indian Ocean and Greater India's northward flight from Antarctica and Australia. *Bull. geol. Soc. Am.* **87:** 1560–1566.
JOHNSON, L. A. S., & BRIGGS, B. G., 1963. Evolution in the Proteaceae. *Aust. J. Bot.*, **11:** 21–61.
JOHNSON, L. A. S., & BRIGGS, B. G., 1975. On the Proteaceae—the evaluation and classification of a southern family. *Bot. J. Linn. Soc.*, **70:** 83–182.
JONES, J. G., 1971. Australia's Caenozoic drift. *Nature, Lond.*, **230:** 237–239.
JONES, J. G., & McDOUGALL, I., 1973. Geological history of Norfolk and Philip Islands, southwest Pacific Ocean. *J. geol. Soc. Aust.*, **20:** 239–254.
KARIG, D. E., 1970. Ridges and basins of the Tonga-Kermadec island arc system. *J. geophys. Res.*, **75:** 239–254.
KARIG, D. E., 1972. Remnant arcs. *Bull. geol. Soc. Am.*, **83:** 1057–1068.
KARIG, D. E., & MAMMERICKX, J., 1972. Tectonic framework of the New Hebrides island arc. *Mar. Geol.*, **12:** 187–205.
KATILI, J. A., 1975. Volcanism and plate tectonics in the Indonesian island arcs. *Tectonophysics*, **26:** 165–188.
KEAST, A., 1971. Continental drift and the evolution of the biota on southern continents. *Q. Rev. Biol.*, **46:** 335–378.
KEAST, A., 1973. Contemporary biotas and the separation sequence of the southern continents. In TARLING & RUNCORN (1973), pp. 309–343.
KEMP, E. M., 1972. Reworked palynomorphs from the West Ice Shelf area, East Antarctica, and their possible geological and palaeoclimatological significance. *Mar. Geol.*, **13:** 145–157.
KENNETT, J. P., *et al.*, 1972. Australian-Antarctic continental drift, palaeocirculation changes and Oligocene deep-sea erosion. *Nature Phys. Sci.*, **239:** 51–55.
KERSHAW, A. P., 1974. A long continuous pollen sequence from north-eastern Australia. *Nature, Lond.*, **251:** 222–223.
KIRIAKOFF, S., 1956. Recherches sur les organes tympaniques des lépidoptères en rapport avec la classification. XII. Groupe d'*Epicoma* Hübner. *Biologisch Jaarboek*, **1956:** 218–227.
KIRIAKOFF, S., 1970. Über die Taxonomie der Thaumetopoeidae (Lep., Notodontoidea). *Ent. Zeitschr.*, **80:** 61–60.
KLUGE, A. G., 1967a. Higher taxonomic categories of gekkonid lizards and their evolution. *Bull. Amer. Mus. nat. Hist.*, **135:** 1–60.
KLUGE, A. G., 1967b. Systematics, phylogeny, and zoogeography of the lizard genus *Diplodactylus* Gray (Gekkonidae). *Aust. J. Zool.*, **15:** 1007–1108.
KORONEVA, E. V., 1964. Distribution of spores and pollen of terrestrial plants in bottom sediments of the Pacific Ocean. In CRANWELL (1964), p. 31.
KOSTERMANS, A. J. G. H., 1974. Lauracées. In *Flore de la Nouvelle-Calédonie et Dépendances*, **5**. Paris: Muséum National d'Histoire Naturelle.
KRUSKAL, J. B., 1964a. Multidimensional scaling by optimising goodness-of-fit to a nonmetric hypothesis. *Psychometrika*, **29**: 1–27.

Kruskal, J. B., 1964b. Nonmetric multidimensional scaling: a numerical method. *Psychometrika*, **29:** 115–129.
Ladd, H. S., 1960. Origin of the Pacific Island molluscan fauna. *Am. J. Sci.*, **258A** (Bradley Vol.): 137–150.
Lakhanpal, R. N., 1970. Tertiary floras of India and their bearing on the historical geology of the region. *Taxon*, **19:** 675–694.
Laubenfels, D. J. de, 1969. A revision of the Malesian and Pacific rainforest conifers, I. Podocarpaceae, in part. *J. Arnold Arbor.*, **50:** 274–369.
Laubenfels, D. J. de, 1972. Gymnospermes. In *Flore de la Nouvelle-Calédonie et Dépendances*, **4**. Paris: Museum National d'Histoire Naturelle.
Lee, K. E., 1975. Introductory remarks. A discussion on the results of the 1971 Royal Society-Percy Sladen Expedition to the New Hebrides. *Phil. Trans. R. Soc. Lond.* B., **272:** 269–276.
Lieftinck, M. A., 1975. The dragonflies (Odonata) of New Caledonia and the Loyalty Islands. Part 1. Imagines. *Cah. O.R.S.T.O.M., sér. Hydrobiol.*, **9:** 127–166.
Lillie, A. R., & Brothers, R. N., 1970. The geology of New Caledonia. *N. Z. Jl Geol. Geophys.*, **13:** 145–183.
Linden, W. J. M. van der, 1969. Extinct mid-oceanic ridges in the Tasman Sea and mid-Pacific. *Earth planet. Sci. Lett.*, **6**: 483–490.
MacArthur, R. H., & Wilson, E. O., 1967. *The Theory of Island Biogeography*. Princeton, New Jersey: Princeton University Press.
Mackerras, I. M., 1961. The zoogeography of western Pacific Tabanidae (Diptera). *Pacif. Ins. Monogr.*, **2:** 101–106.
Maddison, P. A., 1976. Records of Lepidoptera from the Cook Islands. *UNDP/FAO Pest and Disease Survey, Technical Document* **11**.
Mallick, D. I. J., 1975. Development of the New Hebrides archipelago. *Phil. Trans. R. Soc. Lond.* B., **272:** 277–285.
Marshall, A. G., & Medway, Lord, 1974. A study of the mangrove community in the New Hebrides. Unpublished paper given at the Informal Meeting of the Royal Society-Percy Sladen New Hebrides Expedition, 1971.
Marshall, J. E., 1977. *Studies in the Hydroptilidae (Trichoptera): morphology, taxonomy and distribution.* Unpublished Ph.D. Thesis (2 vols), University of London.
Martin, H. A., 1973. Upper Tertiary palynology in southern New South Wales. *Spec. Publs geol. Soc. Aust.*, **4:** 35–54.
May, R. M., 1975. Patterns of species abundance and diversity. In Cody, M. L., & Diamond, J. M. (eds.), *Ecology and Evolution of Communities*. Cambridge, Mass.: Harvard University.
Mayr, E., 1940a. Borders and subdivisions of the Polynesian region based on our knowledge of the distribution of the birds. *Proc. 6th Pacific Sci. Congr., 1939, Berkeley*, **4:** 191–195.
Mayr, E., 1940b. The origin and the history of the bird fauna of Polynesia. *Proc. 6th Pacific Sci. Congr., 1939, Berkeley*, **4:** 197–216.
Mayr, E., 1965. The nature of colonization in birds. In H. G. Baker & G. E. Stebbins (eds.), *The Genetics of Colonising Species*, pp. 29–47. New York: Academic Press.
McDowall, R. M., 1968. The status of *Nesogalaxias neocaledonicus* (Weber and de Beaufort) (Pisces, Galaxiidae). *Breviora*, **286:** 1–7.
McDowall, R. M., 1969. Extinction and endemism in New Zealand land birds. *Tuatara*, **17:** 1–12.
McDowell, S. B., 1967. *Aspidomorphus*, a genus of New Guinea snakes of the family Elapidae, with notes on related genera. *J. Zool., Lond.*, **151:** 497–543.
McDowell, S. B., 1969. *Toxicocalamus*, a New Guinea genus of snakes of the family Elapidae. *J. Zool., Lond.*, **159:** 443–511.
McElhinny, M. W., Haile, N. S., & Crawford, A. R., 1974. Palaeomagnetic evidence shows Malay Peninsula was not a part of Gondwanaland. *Nature, Lond.*, **252,** 641–645.

McIntyre, D. J., & Wilson, G. J., 1966. Preliminary palynology of some Antarctic Tertiary erratics. *N. Z. Jl Bot.*, **4:** 315–321.
McKenzie, D. P., & Sclater, J. G., 1971. The evolution of the Indian Ocean since the Late Cretaceous. *Geophys. J. R. astr. Soc.*, **25:** 437–528.
McLellan, I. D., 1975. The freshwater insects. In G. Kuschel (ed.), *Biogeography and Ecology in New Zealand*, pp. 537–560. Monographiae Biologicae **27**. The Hague: W. Junk.
Medway, Lord, & Marshall, A. G., 1975. Terrestrial vertebrates of the New Hebrides: origin and distribution. *Phil. Trans. R. Soc. Lond.* B., **272:** 423–465.
Metcalf, Z. P., 1963. Cicadoidea. *General Catalogue of the Homoptera.* VIII. University of N. Carolina.
Michener, C. D., 1965. A classification of the bees of the Australian and South Pacific Regions. *Bull. Am. Mus. nat. Hist.*, **130:** 1–362.
Milsom, J., 1974. East New Guinea. In Spencer (1974), pp. 463–475.
Molnar, P., & Tapponier, P., 1977. The collision between India and Eurasia. *Sci. Amer.*, **236**(4): 30–41.
Moore, H. E. Jr., 1973a. Palms in the tropical forest ecosystems of Africa and South America. In B. J. Meggers, E. S. Ayensu & W. D. Duckworth (eds.), *Tropical Forest Ecosystems in Africa and South America: A Comparative Review.* Washington, D.C.: Smithsonian Institution Press.
Moore, H. E. Jr., 1973b. The major groups of palms and their distribution. *Gentes Herbarum*, **11:** 27–141.
Moore, H. E. Jr., 1976. New species of *Brongniartikentia* and *Cyphophoenix* (Palmae). *Gentes Herbarum*, **11:** 151–167.
Moore, H. E. Jr., 1978. New genera and species of Palmae from New Caledonia. *Gentes Herbarum*, **11:** 291–309.
Muller, J., 1970. Palynological evidence on early differentiation of angiosperms. *Biol. Rev.*, **45:** 417–450.
Munroe, E., 1965. Zoogeography of insects and allied groups. *Ann. Rev. Ent., Palo Alto*, **10:** 325–344.
Nakata, S., 1961. Some notes on the occurrence of Phasmatodea in Oceania. *Pacif. Ins. Monogr.*, **2:** 107–121.
Nye, I. W. B., 1975. *Generic Names of Moths of the World.* Volume 1. London: British Museum (Natural History).
Packham, G. H., 1973. A speculative Phanerozoic history of the southwest Pacific. In Coleman (1973), pp. 369–388.
Paramonov, S. J., 1958, 1960, 1963. Lord Howe Island: a riddle of the Pacific. *Pacif. Sci.*, **12:** 82–91, **14:** 75–85, **17:** 361–373.
Peake, J. F., 1969. Patterns in the distribution of Melanesian land Mollusca. *Phil. Trans. R. Soc. Lond.* B, **255:** 285–306.
Pinhey, E. C. G., 1975. *Moths of Southern Africa.* Cape Town: Tafelberg.
Preest, D. S., 1963. A note on the dispersal characteristics of the seed of the New Zealand podocarps and beeches and their biogeographical significance. In J. L. Gressitt (ed.), *Pacific Basin Biogeography*, pp. 415–424. Honolulu: Bernice P. Bishop Museum.
Preston, F. W., 1962. The canonical distribution of commonness and rarity. *Ecology*, **43:** 185–215, 410–432.
Prout, L. B., 1928. Geometridae. Part III, Lepidoptera, of *Insects of Samoa and other Samoan Terrestrial Arthropoda*, pp. 117–168. London: British Museum (Natural History).
Quinn, W. H., 1971. Late Quaternary meteorological and oceanographic developments in the equatorial Pacific. *Nature, Lond.*, **229:** 330–331.
Raven, P. H., 1972. Evolution and endemism in the New Zealand species of *Epilobium*. In D. H. Valentine (ed.), *Taxonomy, Phytogeography and Evolution*, pp. 259–274. London and New York: Academic Press.

RAVEN, P. H., & AXELROD, D. I., 1972. Plate tectonics and Australasian biogeography. *Science, N.Y.*, **176:** 1379–1386.
RAVEN, P. H., & AXELROD, D. I., 1974. Angiosperm biogeography and past continental movements. *Ann. Mo. bot. Gdn*, **61:** 539–673.
RIDD, M. F., 1971. South-East Asia as a part of Gondwanaland. *Nature, Lond.*, **234**: 531–533.
RIDD, M. F., 1972. South-East Asia a part of Gondwanaland: reply. *Nature phys. Sci.*, **240:** 140–141.
ROBBINS, R. G., 1961. The montane vegetation of New Guinea. *Tuatara*, **8:** 121–133.
ROBINSON, G. S., 1968. Some new species of Lepidoptera from the Fiji Islands. *Entomologist's Rec. J. Var.*, **80:** 249–255.
ROBINSON, G. S., 1971a. The genus *Utetheisa* Hübner in Fiji with a description of a new species (Lep., Arctiidae). *Entomologist's Rec. J. Var.*, **83:** 123–130.
ROBINSON, G. S., 1971b. The genus *Cleora* Curtis (Lepidoptera: Geometridae) in Fiji. *J. Ent. (B)*, **40:** 71–82.
ROBINSON, G. S., 1975. *Macrolepidoptera of Fiji and Rotuma, a taxonomic and geographic study*. Faringdon: E. W. Classey Ltd.
ROBINSON, G. S., 1976. Biogeography of the New Hebrides Macrolepidoptera. *J. ent. Soc. Aust. (N.S.W.)*, **9:** 47–53.
RODDA, P., 1974. Fiji. In SPENCER (1974), pp. 425–431.
ROSS, H. H., 1956. *The Evolution and Classification of the Mountain Caddis Flies*. Urbana, Illinois: University of Illinois Press.
ROSS, H. H., 1975. A preliminary report on the Helicopsychidae (Trichoptera) of New Caledonia. *Cah. O.R.S.T.O.M., sér. Hydrobiol.*, **9:** 67–80.
ROSS, H. H., 1976. Observations on the Helicopsychidae (Trichoptera) of New Caledonia. *Proc. 1st. Int. Symp. Trichoptera*, 1974: 1–3.
ROUX, J., 1913. Les reptiles de la Nouvelle-Calédonie et des îles Loyalty. In F. SARASIN & J. ROUX (eds.), *Nova Caledonia (A. Zool.)*, **1:** 80–160.
RUTLAND, R. W. R., & WALTER, M. R., 1974. Philippines. In SPENCER (1974), pp. 476–490.
SAHNI, A., & KUMAR, V., 1974. Palaeogene palaeobiogeography of the Indian subcontinent. *Palaeogeogr., Palaeoclimat., Palaeoecol.*, **15:** 209–226.
SANBORN, C. C., & NICHOLSON, A. J., 1950. Bats from New Caledonia, the Solomon Islands and New Hebrides. *Fieldiana, Zool.*, **31:** 313–338.
SCHLANGER, S. O., & GILLETT, G. W., 1976. A geological perspective of the upland biota of Laysan atoll (Hawaiian Islands). *Biol. J. Linn. Soc.*, **8:** 205–216.
SCHLINGER, E. I., 1974. Continental drift, *Nothofagus*, and some ecologically associated insects. *Ann. Rev. Ent., Palo Alto*, **19:** 323–343.
SCHMID, M., 1975. La flore et la végétation de la partie méridionale de l'Archipel des Nouvelles Hébrides. *Phil. Trans. R. Soc. Lond.* B, **272:** 329–342.
SCLATER, J. G., & FISHER, R. L., 1974. Evolution of the east central Indian Ocean, with emphasis on the tectonic setting of the Ninetyeast Ridge. *Bull. geol. Soc. Am.*, **85:** 683–702.
SEVASTOPULO, D. G., 1941. On the foodplants of Indian Agaristidae and Noctuidae (Heterocera). *J. Bombay nat. Hist. Soc.*, **42**(2): 1–10.
SIMPSON, G. G., 1943. Mammals and the nature of continents. *Am. J. Sci.*, **241:** 1–31.
SINGH, BALWANT, BHASIN, G. D., ROONWAL, M. L., & MATHUR, R. N., 1954–1960. A list of insect pests of forest plants in India and the adjacent countries. *Forest Bull., Dehra Dun*, **171:** 10 parts.
SMITH, A. C., 1945. Geographical distribution of the Winteraceae. *J. Arnold Arbor.*, **26:** 48–59.
SMITH, A. C., 1951. The vegetation and flora of Fiji. *Scientific Monthly*, **73:** 3–15.
SMITH, A. C., 1955. Phanerogam genera with distributions terminating in Fiji. *J. Arnold Arbor.*, **36:** 273–292.
SMITH, A. C., 1975. The genus *Macropiper* (Piperaceae). *Bot. J. Linn. Soc.*, **71**: 1–38.

Smith, A. G., Briden, J. C., & Drewry, G. E., 1973. Phanerozoic world maps. In *Organisms and Continents Through Time. Spec. Pap. Palaeont.*, **12:** 1–42.

Smith, J. M. B., 1974. Southern biogeography on the basis of continental drift: a review. *Australian Mammalology*, **1:** 213–230.

Smithers, C. N., & Thornton, I. W. B., 1974. The Myopsocidae (Psocoptera) of New Guinea and New Caledonia. *Trans. R. ent. Soc. Lond.*, **126:** 91–127.

Solem, A., 1958. Biogeography of the New Hebrides. *Nature, Lond.*, **181:** 1253–1255.

Solem, A., 1959. Systematics and zoogeography of the land and fresh-water Mollusca of the New Hebrides. *Fieldiana, Zool.*, **43:** 1–359.

Solem, A., 1961. New Caledonian land and fresh-water snails, an annotated check list. *Fieldiana, Zool.*, **41:** 417–501.

Solomon, M., & Griffith, J. R., 1972. Tectonic evolution of the Tasman Orogenic Zone, Eastern Australia. *Nature phys. Sci.*, **237:** 3–6.

Spencer, A. M. (ed.), 1974. *Mesozoic-Cenozoic Orogenic Belts. Data for Orogenic Studies.* London: Geological Society.

Stauffer, P. H., 1974. Malaya and Southeast Asia in the pattern of continental drift. *Bull. geol. Soc. Malaysia*, **7:** 89–138.

Stauffer, P. H., & Gobbett, D. J., 1972. South-East Asia a part of Gondwanaland? *Nature phys. Sci.*, **240:** 139–140.

Steenis, C. G. G. J. van, 1971. *Nothofagus*, key genus of plant geography, in time and space, living and fossil, ecology and phylogeny. *Blumea*, **19:** 65–98.

Stevens, G. R., 1973. The palaeogeographic history of New Zealand. *N. Z. Ent.*, **5:** 230–239.

Suggate, R. P., 1971. Mesozoic-Cenozoic development of the New Zealand region. *Pacif. Geol.*, **4:** 113–120.

Sugi, S., 1970. The Noctuidae of the Ryukyu Islands. Part I. Trifidae (Lepidoptera). *Tinea*, **8:** 213–229.

Sugi, S., 1976. *Ophiusa separans* (Walker) and *O. indiscriminata* (Hampson), both inhabiting Japan (Noctuidae, Catocalinae). *Japan. Heteroc. J.*, **87:** 441–445.

Sutherland, L., & Ritchie, A., 1974. Defunct volcanoes and extinct horned turtles. Lord Howe issue. *Aust. nat. Hist.*, **18:** 44–49.

Swezey, O. H., 1946. Geometridae, Arctiidae, Agrotidae, and Pyralidae of Guam. *Bull. Bernice P. Bishop Mus.*, **189:** 163–185.

Talbot, G., 1928–1937. *Monograph of the Pierine genus* Delias. London: John Bale, Danielsson and British Museum (Natural History).

Tams, W. H. T., 1935. Heterocera (exclusive of the Geometridae and the microlepidoptera). Part III, Lepidoptera, *Insects of Samoa and other Samoan Terrestrial Arthropoda*, pp. 169–290. London: British Museum (Natural History).

Tarling, D. H., 1972. Another Gondwanaland. *Nature, Lond.*, **238:** 92–93.

Tarling, D. H., & Runcorn, S. K., 1973. *Continental Drift, Sea Floor Spreading and Plate Tectonics.* 2 Vols. London: Academic Press.

Tercinier, G., 1962. *Les Sols de la Nouvelle-Calédonie.* Paris: O.R.S.T.O.M.

Thompson, J. E., 1972. Continental drift and the geological history of Papua-New Guinea. *Australian Petrol. Expl. Assoc. J.*, **12**(2): 64–69.

Thompson, J. E., & Hackman, B. D., 1969. Some geological notes on areas visited by the Royal Society Expedition to the British Solomon Islands, 1965. *Phil. Trans. R. Soc. Lond.* B, **255:** 189–202.

Thorne, R. F., 1963. Biotic distribution patterns in the tropical Pacific. In J. L. Gressitt (ed.), *Pacific Basin Biogeography*, 311–354. *Honolulu: Bernice P. Bishop Museum.*

Thorne, R. F., 1965. Floristic relationships of New Caledonia. *St. Univ. Stud. nat. Hist. Iowa*, **20**(7): 1–14.

Thorne, R. F., 1969. Floristic relationships between New Caledonia and the Solomon Islands. *Phil. Trans. R. Soc. Lond.* B, **255:** 595–602.

THORNTON, I. W. B., & SMITHERS, C. N., 1974. The Philotarsidae (Psocoptera) of New Caledonia. *Pacif. Ins.*, **16:** 177–243.
UYEDA, S., & BEN-AVRAHAM, Z., 1972. Origin and development of the Philippine Sea. *Nature phys Sci.*, **240:** 176–178.
VANE-WRIGHT, R. I., 1967. A re-assessment of the genera *Holorusia* Loew (= *Ctenacroselis* Enderlein), *Ischnotoma* Skuse and *Zelandotipula* Alexander (Diptera: Tipulidae), with notes on their phylogeny and zoogeography. *J. nat. Hist.*, **1:** 511–547.
VIETTE, P. E. L., 1948. Les Ophideres du Pacifique. *Rev. franç. Ent.*, **15:** 209–220.
VIETTE, P. E. L., 1949. Catalogue of the heterocerous Lepidoptera of French Oceania. *Pacif. Sci.*, **3:** 315–337.
VIETTE, P. E. L., 1950a. The Noctuidae Catocalinae from New Caledonia and the New Hebrides. *Pacif. Sci.*, **4:** 139–157.
VIETTE, P. E. L., 1950b. Les Lithosiidae de Nouvelle-Calédonie et des Nouvelles Hébrides. *Ann. Soc. ent. Fr.*, **119:** 81–96.
VIETTE, P. E. L., 1950c. Une nouvelle Geometridae de Nouvelle-Calédonie. *Bull. Soc. zool. Fr.*, **74:** 241–244.
VIETTE, P. E. L., 1951a. Les Noctuidae Noctuinae de la Nouvelle-Calédonie et des Nouvelles Hébrides. *Ann. Soc. ent. Fr.*, **118:** 29–50.
VIETTE, P. E. L., 1951b. Supplément au catalogue des lépidoptères hétérocères de l'Océanie française. *Bull. Soc. ent. Fr.*, **56:** 14–16.
VIETTE, P. E. L., 1961. Découverte de la famille des Hepialidae en Nouvelle-Calédonie. *Bull. Soc. ent. Fr.*, **66:** 106–108.
VIETTE, P. E. L., 1971. Un nouveau sphingide de Nouvelle-Calédonie. *Entomops*, **23:** 237–240.
VIROT, R., 1956. La végétation canaque. *Mem. Mus. nat. Hist. nat., n.S.*, B, Bot. **7:** 1–398.
VIROT, R., 1968. Protéacées. In *Flore de la Nouvelle-Calédonie et Dépendances*, **2.** Paris: Muséum National d'Histoire Naturelle.
VIROT, R., 1975. Epacridacées. In *Flore de la Nouvelle-Calédonie et Dépendances*, **6.** Paris: Muséum National d'Histoire Naturelle.
VOGT, P. R., & CONNOLLY, J. R., 1971. Tasmantid guyots, the age of the Tasman basin, and motion between the Australian plate and the mantle. *Bull. geol. Soc. Am.*, **82:** 2577–2584.
WALKER, E. P., 1964. *Mammals of the World.* Vol 1. Baltimore: Johns Hopkins.
WARDEN, A. J., & MITCHELL, A. H. G., 1974. New Hebrides. In SPENCER (1974), pp. 433–443.
WEISSEL, J. K., & HAYES, D. E., 1972. Magnetic anomalies in the southeast Indian Ocean. In HAYES (1972), pp. 165–196.
WELLMAN, H. W., 1973. New Zealand fault zones and sea-floor spreading. In COLEMAN (1973), pp. 335–348.
WERMUTH, H., 1965. Liste der rezenten Amphibien und Reptilien, Gekkonidae, Pygopodidae, Xantusidae. *Das Tierreich*, **80:** 1–246.
WHALLEY, P. E. S., 1973. The genus *Etiella* Zeller (Lepidoptera: Pyralidae): a zoogeographic and taxonomic study. *Bull. Br. Mus. nat. Hist.* (Ent.), **28:** 1–21.
WHALLEY, P. E. S., 1976. *Tropical Leaf Moths of the Subfamily Striglininae (Lepidoptera: Thyrididae).* London: British Museum (Natural History).
WHALLEY, P. E. S., 1977. Lower Cretaceous Lepidoptera. *Nature, Lond.*, **266:** 526.
WHITMORE, T. C., 1973. Plate tectonics and some aspects of Pacific plant geography. *New Phytol.*, **72:** 1185–1190.
WHITMORE, T. C., 1977. A first look at *Agathis. Tropical Forestry Papers*, **11.** Oxford: Commonwealth Forestry Institute.
WHITTAKER, R. H., 1965. Dominance and diversity in land plant communities. *Science, N.Y.*, **147:** 250–260.
WILLIAMS, C. B., 1947. The logarithmic series and the comparison of island floras. *Proc. Linn. Soc. Lond.*, **158:** 104–108.

WILLIAMSON, M. H., 1972. *The Analysis of Biological Populations.* Edward Arnold.
WILLIAMSON, M. H., 1973. Species diversity in ecological communities. In M. S. BARTLETT & R. W. HIORNS (eds), *The Mathematical Theory of the Dynamics of Biological Populations.* London and New York: Academic Press.
WYNTER-BLYTH, M. A., 1957. *Butterflies of the Indian Region.* Bombay: Bombay Natural History Society.
ZAKLINSKAYA, E. D., 1964. On the relationship between Upper Cretaceous and Paleogene floras of Australia, New Zealand, and Eurasia, according to data from spore and pollen analyses. In CRANWELL (1964), p. 85.
ZEUNER, F. E., 1943. Studies in the systematics of *Troides* Hbn. and its allies. *Trans. zool. Soc. Lond.*, **25:** 107–184.
ZEUNER, F. E., 1959. *The Pleistocene Period, its Climate, Chronology and Faunal Successions.* London: Hutchinson.

# SUBJECT INDEX

This index is not intended to be fully comprehensive for any one item but merely to indicate the principal references to a number of topics not covered by the major chapter headings or subheadings.

# BOTANICAL INDEX

Supraspecific categories only are listed

# ZOOLOGICAL INDEX, EXCLUDING LEPIDOPTERA

Supraspecific categories only are listed

# LEPIDOPTERA INDEX: SUPRASPECIFIC CATEGORIES

Principal taxonomic references are in bold type

# LEPIDOPTERA INDEX: SPECIES AND SUBSPECIES

Principal taxonomic references are in bold type

GPSR Compliance

*The European Union's (EU) General Product Safety Regulation (GPSR) is a set of rules that requires consumer products to be safe and our obligations to ensure this.*

*If you have any concerns about our products, you can contact us on ProductSafety@springernature.com*

In case Publisher is established outside the EU, the EU authorized representative is:

Springer Nature Customer Service Center GmbH
Europaplatz 3
69115 Heidelberg, Germany

**Batch number: 10761697**

Printed by Printforce, the Netherlands